T0310309

Insect Outbreaks Revisited

Insect Outbreaks Revisited

Edited by

Pedro Barbosa
Department of Entomology
University of Maryland
College Park, Maryland

Deborah K. Letourneau
Department of Environmental Studies
University of California
Santa Cruz, California

Anurag A. Agrawal
Department of Ecology and Evolutionary Biology
and Department of Entomology
Cornell University
Ithaca, New York

A John Wiley & Sons, Ltd., Publication

Blackwell Publishing was acquired by John Wiley & Sons in February 2007.
Blackwell's publishing program has been merged with Wiley's global Scientific,
Technical and Medical business to form Wiley-Blackwell.

Registered Office
John Wiley & Sons, Ltd, The Atrium, Southern Gate, Chichester, West Sussex, PO19 8SQ, UK

Editorial Offices
111 River Street, Hoboken, NJ 07030-5774, USA
9600 Garsington Road, Oxford, OX4 2DQ, UK
The Atrium, Southern Gate, Chichester, West Sussex, PO19 8SQ, UK

For details of our global editorial offices, for customer services and for information about
how to apply for permission to reuse the copyright material in this book please see our
website at www.wiley.com/wiley-blackwell.

Library of Congress Cataloging-in-Publication Data

Insect outbreaks revisited / edited by Pedro Barbosa, Deborah K. Letourneau,
Anurag A. Agrawal.
     p.   cm.
   Updates: Insect outbreaks / edited by Pedro Barbosa and Jack C. Shultz. San Diego :
Academic Press, c1987.
   Includes bibliographical references and index.
   ISBN 978-1-4443-3759-4 (cloth)
1. Insect populations. 2. Insects–Ecology. 3. Insect pests. I. Barbosa, Pedro, 1944–
II. Letourneau, DeborahKay. III. Agrawal, Anurag A. IV. Insect outbreaks.
   QL49.15.I572 2012
   595.7–dc23

                    2011046004

A catalogue record for this book is available from the British Library.

Wiley also publishes its books in a variety of electronic formats. Some content that appears in
print may not be available in electronic books.

Set in 10.5/12pt Classical Garamond by SPi Publisher Services, Pondicherry, India
Printed and bound in Malaysia by Vivar Printing Sdn Bhd

1 2012

Cover image: Caterpillar larvae of Buff-tip moth (Phalera bucephala) on oak in the UK.
Photo (c) Premaphotos/naturepl.com
Cover design by Nicki Averill Design

Contents

Colour plate pages fall between pp. 196 and 197

Contributors

Karen C. Abbott (Lead Author)
Department of Ecology and Evolution
University of Chicago
Chicago, IL 60637 USA

Anurag A. Agrawal
Department of Ecology and
Evolutionary Biology *and* Department
of Entomology
Cornell University
Ithaca, NY 14853 USA

Matthew P. Ayres
Department of Biological Sciences
Dartmouth College
Hanover, NH 03755 USA

Pedro Barbosa
Department of Entomology
University of Maryland
College Park, MD 20742 USA

Andrea Battisti
Department of Environmental
Agronomy and Crop Production
Università di Padova
Padova, Italy 35122

Spencer T. Behmer (Lead Author)
Department of Entomology
Texas A&M University
College Station, TX USA 77843

Christer Björkman
Department of Ecology
Swedish University of Agricultural
Sciences
750 07 Uppsala, Sweden

Ottar N. Bjørnstad
Departments of Entomology
and Biology
Penn State University
University Park, PA 16802 USA

Yasmin J. Cardoza (Lead Author)
Department of Entomology
North Carolina State University
Raleigh, NC 27695-7613 USA

Yves Carrière
Department of Entomology
University of Arizona
Tucson, AZ 85721-0036 USA

Walter P. Carson
Department of Biological Sciences
University of Pittsburgh
Pittsburgh, PA 15260 USA

Lee A. Dyer (Lead Author)
Department of Biology
University of Nevada
Reno, NV 89557 USA

Fritzi S. Grevstad
Olympic Natural Resources Center
Long Beach, WA 98631 USA

Kyle J. Haynes
Blandy Experimental Farm
Boyce VA 22620 USA

Dan A. Herms
Department of Entomology
Ohio Agricultural Research and
Development Center
The Ohio State University
Wooster, OH 44691 USA

Richard W. Hofstetter
Northern Arizona University
School of Forestry
Flagstaff, AZ USA 86011

Anthony Joern
Division of Biology
Kansas State University
Manhattan, KS 66506-4901 USA

André Kessler (Lead Author)
Ecology and Evolutionary Biology
Cornell University
Ithaca, NY 14853 USA

Maartje J. Klapwijk
Department of Ecology
Swedish University of Agricultural
Sciences
750 07 Uppsala, Sweden

Julia Koricheva (Lead Author)
School of Biological Sciences
Royal Holloway University of
London
Egham, Surrey, TW20 0EX, UK

Conrad C. Labandeira (Lead Author)
Smithsonian Institution
Washington, DC 20013-7012
USA

Douglas A. Landis
Department of Entomology
Michigan State University
East Lansing, MI USA 48824

Stig Larsson
Department of Plant and Forest
Protection
Swedish University of Agricultural
Sciences
750 07 Uppsala, Sweden

Egbert G. Leigh, Jr.
Smithsonian Tropical Research
Institute
Apartado 0843-03092
Balboa, Ancon, Republic
of Panama

Deborah K. Letourneau (Lead Author)
Environmental Studies Department
University of California
Santa Cruz, CA 95064 USA

Andrew M. Liebhold (Lead Author)
USDA Forest Service
Northern Research Station
Morgantown, WV 26505 USA

Eric M. Lind (Lead Author)
Department of Ecology, Evolution,
and Behavior
University of Minnesota
St. Paul, MN 55108 USA

Ann M. Lynch (Lead Author)
U.S. Forest Service
Rocky Mountain Research Station
Tucson, AZ 85721 USA

Peter B. McEvoy (Lead Author)
Department of Botany and Plant
Pathology
Oregon State University
Corvallis, OR 97331-4501 USA

Raul F. Medina (Lead Author)
Department of Entomology
Texas A&M University
College Station, TX USA 77843

Erik H. Poelman
Laboratory of Entomology
Wageningen University
6700 EH Wageningen,
The Netherlands

Katja Poveda
Georg August Universität
Dep. für Nutzpflanzenwissenschaften
Abteilung Agrarökologie
37073 Göttingen, Germany

Michael J. Raupp (Lead Author)
Department of Entomology
University of Maryland
College Park, MD 20742 USA

Shon S. Schooler
CSIRO Entomology – Indooroopilly
Indooroopilly QLD 4068, Australia

Timothy D. Schowalter (Lead Author)
Entomology Department
Louisiana State University
Baton Rouge, LA 70803 USA

J. Gwen Shlichta (Lead Author)
Université de Neuchâtel
Institut de biologie
Laboratoire d'entomologie évolutive
CH-2000 Neuchatel, Switzerland

Paula M. Shrewsbury
Department of Entomology
University of Maryland
College Park, Maryland, USA

Angela M. Smilanich
Desert Research Institute
University of Nevada, Reno, NV
89512 USA

Bruce E. Tabashnik (Lead Author)
Department of Entomology
University of Arizona
Tucson, AZ 85721-0036 USA
brucet@ag.arizona.edu

Fernando E. Vega
USDA, ARS
PSI, Sustainable Perennial Crops
Laboratory
Beltsville, MD 20705 USA

Benjamin P. Werling
Michigan State University
Department of Entomology
East Lansing, MI 48824 USA

J. Megan Woltz (Lead Author)
Department of Entomology
Michigan State University
East Lansing, MI 48824 USA

Louie H. Yang (Lead Author)
Department of Entomology
University of California
Davis, CA 95616 USA

Acknowledgments

The editors and authors of the book would like to express their great thanks to those who reviewed and provided such important comments and suggestions. It is clear that the quality of the chapters was significantly improved. These reviewers include Karen Abbott, Arianne Cease, Sheena Cotter, Les Ehler, Barbara Ekbom, Michael Engel, Nancy Gillette, Larry Hanks, Richard Hofstetter, Pekka Kaitaniemi, Richard Karban, Tero Klemola, Kwang Pum Lee, Andrew Liebhold, Timothy Paine, Dylan Parry, Nigel Straw, Robert Srygley, Art Woods, Michael Stastny, Toomas Tammaru, Jennifer Thaler, Nora Underwood, Stephen Wratten, Saskya van Nouhuys, W. Jan A. Volney, and an anonymous reviewer. Please keep in mind that if you enjoyed reading the chapters in this book, the credit should go to these wonderful and selfless people, listed above.

Preface

It has been more than 25 years since the publication of *Insect Outbreaks* (Barbosa and Schultz 1987). Over the last two decades, significant advances have been made in our understanding of certain aspects of outbreak dynamics and outbreak species. Thus, we have updated that original effort in order to present some new insights, concepts, and hypotheses and revisit older concepts. As it was with the original *Insect Outbreaks* book, in this new effort, *Insect Outbreaks Revisited*, chapters are included that are designed to expose the reader to novel and recently proposed ideas and perspectives, as well as old concepts needing reappraisal. Further, authors rigorously discuss dogma of relevance to insect outbreaks, dogma that is both tested and untested. The chapters in this edition also effectively stimulate the interest of expert or novice, and hopefully lead to a broader understanding of insect outbreaks.

This book is divided into sections which represent the level of inquiry that is taken in the chapters included in the section. Thus, chapters in the "Physiological and Life History Perspectives" section discuss the importance of nutrition, the ratios of elemental nutrients, changes in the quality and quantity of food, physiological processes such as immune responses, and thus how understanding these issues and analyses may lead to novel ways of investigating the causes and consequences of insect outbreaks. Similarly, chapters in the "Population Dynamics and Multispecies Interactions" section provide insights into regional spatial synchrony of outbreaks, perspectives on how symbionts may play a significant role in some outbreaks, the mechanisms in plants that are triggered by herbivory and their impact on outbreak dynamics, and dendrochronological analyses involving the dating of past outbreaks through the study of tree ring growth, which illustrates the lessons that can be learned from such an analysis. The latter provides a long-term perspective for understanding outbreak dynamics in forested ecosystems. Chapters in the "Population, Community, and Ecosystem Ecology" section provide a comparison of the life history traits of outbreaking and non-outbreaking species, an overview of the consequences of insect outbreaks to other members of a community, and a related chapter on the influences of outbreaks on ecosystem services. Further, there is a chapter providing an analysis of whether outbreaks are primarily a temperate or tropical phenomenon. In the "Genetics and Evolution" section, chapters explore the role of genetic differentiation in outbreaks occurring in agroecosystems, and an examination of the fossil record to determine the degree to which detectable macroevolutionary patterns of outbreaks exist and if there is any evidence of phylogenetic constraints to those relationships. Finally, chapters in the "Applied Perspectives" section use what we know about the behavior, ecology, and evolution of pest species and the habitats in which they live, as well as and how future climate change might alter critical interactions and the dynamics of outbreaks. The authors of these chapters speculate on whether there are valuable lessons to be learned from the management or mismanagement of pest species.

Reference

Barbosa, P. and Schultz, J. 1987. *Insect Outbreaks*. Academic Press, New York.

Part I

Physiological and Life History Perspectives

1

Insect Herbivore Outbreaks Viewed through a Physiological Framework: Insights from Orthoptera

Spencer T. Behmer and Anthony Joern

For they covered the face of the whole earth, so that the land was darkened; and they did eat every herb of the land, and all the fruit of the trees which the hail had left: and there remained not any green thing in the trees, or in the herbs of the field, through all the land of Egypt.

Exodus 10:15 (King James Version)

The Cloud was hailing grasshoppers. The cloud was grasshoppers. Their bodies hid the sun and made darkness. Their thin, large wings gleamed and glittered. The rasping whirring of their wings filled the whole air and they hit the ground and the house with the noise of a hailstorm.

On the Banks of Plum Creek (by Laura Ingalls Wilder)

Insect Outbreaks Revisited, First Edition. Edited by Pedro Barbosa, Deborah K. Letourneau and Anurag A. Agrawal.
© 2012 Blackwell Publishing Ltd. Published 2012 by Blackwell Publishing Ltd.

1.1   Introduction

Insect herbivore outbreaks, particularly orthopteran outbreaks, have plagued humans throughout recorded history. The Egyptian locust swarm described in the Old Testament is perhaps the most famous orthopteran outbreak story. Two species, the African desert locust (*Schistocerca gregaria* Forskål) and the migratory locust (*Locusta migratoria* (Linnaeus)), still outbreak regularly throughout large expanses of Africa and the Middle East. The most likely villain in the biblical swarm was the African desert locust, based on the broad array of the food plants described in the story. In contrast to the desert locust, the migratory locust is a specialist that feeds only on grasses. However, despite its restricted diet, the migratory locust has a larger geographic range, extending from all of northern and central Africa across to eastern China. It too has greatly impacted human society throughout historical time, especially in China. Parenthetically, the Chinese character for locust is composed of two parts, insect (虫) and emperor (皇); this character combination indicates the power of locusts – it was an insect capable of threatening an emperor's supremacy. In China's 5000-year history, 842 locust plagues have been recorded, with the earliest ones being described in the *Book of Songs* (770–476 BCE). How locust outbreaks endangered regimes and changed the destiny of China is also described in two other important ancient Chinese books – *Zizhi Tongjian* (which covers Chinese history from 403 BCE to 959 CE, including 16 dynasties) and *Ch'ien Han Shu* (which covers Chinese history from 206 BCE to 25 CE).

Although the recorded histories of Australia and the Americas are more recent, orthopteran outbreaks have a long history on these continents as well. The first recorded outbreak of the Australian plague locust (*Chortoicetes terminifera* (Walker)) was in 1844, followed by outbreaks from the 1870s onward (including multiple outbreaks in the early 2000s, most of which were controlled by the Australian Plague Locust Commission (Hunter 2004)). In the United States, massive outbreaks of the Rocky Mountain locust (*Melanoplus spretus* (Walsh)) were recorded in the 1870s. The largest of the swarms covered a "swath equal to the combined areas of Connecticut, Delaware, Maine, Maryland, Massachusetts, New Hampshire, New Jersey, New York, Pennsylvania, Rhode Island and Vermont" (Riley *et al.* 1880), and nearly derailed westward expansion. Charles Valentine Riley, now considered one of the founding fathers of entomology in the United States, was appointed by the US government to investigate these outbreaks. His work led him to request further federal assistance, which the government provided in the form of the US Entomological Commission; this agency quickly morphed into the US Department of Agriculture that still operates today. The last known Rocky Mountain locust swarm occurred in the very early 1900s; why it disappeared remains a mystery, although some interesting hypotheses have been proposed (Lockwood 2005). The Mormon cricket (*Anabrus simplex* (Haldeman)) is another orthopteran species renowned for its outbreaks. Populations of Mormon crickets usually occur at low densities throughout most of their range in western North America, but population explosions that exceed more than 1 million individuals, marching in roving bands at densities of more than 100 individuals/m^2, are not uncommon. In 1848 a Mormon cricket outbreak nearly thwarted the settlement of Salt Lake City, Utah, by Mormon pioneers. Although the story is controversial,

Mormon folklore recounts the miracle of the gulls. Legend claims that legions of seagulls, sent by God, appeared on June 9, 1848. These seagulls saved the settler's crops by eating all the crickets. South America and Central America also have orthopterans that show outbreak dynamics, the most notable being *Schistocerca cancellata* (Serville) and *Schistocerca piceifrons* (Walker), respectively.

Given the devastation and immense suffering inflicted on humans by orthopteran outbreaks, it is pressing to understand the causal factors that contribute to their outbreaks. With the exception of Mormon crickets (see Sword 2005), the orthopterans described above exhibit phase polyphenism – defined by Hardie and Lees (1985, 473) as "occurrence of two or more distinct phenotypes which can be induced in individuals of the same genotype by extrinsic factors." The African desert locust and African migratory locust are easily two of the best-known species to practice phase polyphenism. However, many orthopterans that do not exhibit phase polyphenism can also undergo outbreaks, as has been the case for many grasshopper species in the western United States (Branson *et al.* 2006).

In this chapter we concentrate primarily on orthopterans, but our aim is to understand factors that contribute to insect herbivore outbreaks more generally. We also discuss other types of insects, particularly lepidopterans, to make our points. Because insect outbreaks cannot happen without an initial increase in population size, we begin by focusing on individuals while considering factors, especially nutritional ones, that contribute to increased performance. We next explore how behavior and performance (e.g., survival, growth, and reproduction) of individual insect herbivores change as population densities increase. Shifting gears, we then discuss how ecological paradigms, particularly the "plant stress hypothesis," have influenced how we view insect herbivore outbreaks. We conclude the chapter by calling for an integrative approach that translates individual responses into group-level phenomena, couched within the contexts of their communities and ecosystems.

1.2   Which conditions favor the individual, and can lead to insect herbivore outbreaks?

Insect outbreaks are often cyclical and require a confluence of events to occur. Critical is the initial phase of an outbreak – insect herbivores must have access to sufficient food, and that food must be of good quality to ensure survival, rapid growth, and high reproductive output. Historically, plant quality has been defined in terms of its nitrogen content (e.g., McNeil and Southwood 1978, Mattson 1980, Scriber and Slansky 1981), but more recently there has been a shift away from a single currency approach. We now recognize that organisms, including insect herbivores, require a suite of nutrients and perform best when they acquire these nutrients in particular blends (Raubenheimer and Simpson 1999, Behmer 2009, Raubenheimer *et al.* 2009). Insect herbivores require upwards of 30 different nutrients, including protein (amino acids), digestible carbohydrates (e.g., simple sugars and starches), fatty acids, sterols, vitamins, minerals, and water (Chapman 1998, Schoonhoven *et al.* 2005). Plants contain all the nutrients that insect herbivores need, but securing these nutrients in the appropriate amounts and ratios is often challenging because plant nutrient content can be highly variable depending on plant type, age, and growing conditions (Mattson 1980, Scriber and Slansky 1981, Slansky and Rodriguez 1987, Bernays and Chapman 1994).

1.2.1 Factors that influence performance in immature stages

Two particularly important macronutrients for insect herbivores are protein and digestible carbohydrates. Plant proteins provide amino acids (the major source of nitrogen) used to construct insect proteins that serve structural purposes, as enzymes, for transport and storage, and as receptor molecules. In contrast, digestible carbohydrates are used primarily for energy, but they can also be converted to fat and stored, and their carbon skeleton can contribute to the production of amino acids. It has long been known that insufficient protein and carbohydrates can limit insect growth and performance.

Only recently, though, have we begun to appreciate the extent to which insect herbivores can regulate the intake of these two nutrients, and that they regulate them independently of one another. The most thoroughly explored insect with respect to protein–carbohydrate regulation is the African migratory locust (the gregarious phase). Laboratory experiments using artificial diets with fixed protein–carbohydrate ratios have shown that African migratory locusts regulate their protein–carbohydrate intake under a number of different conditions: (1) when presented with two nutritionally suboptimal but complementary foods (Chambers *et al.* 1995, Chambers *et al.* 1997, 1998), (2) as the relative frequency of two nutritionally complementary foods changes (Behmer *et al.* 2001), (3) as the physical space between nutritionally complementary foods increases (Behmer *et al.* 2003), and (4) in the presence of plant secondary metabolites (Behmer *et al.* 2002).

A key mechanism that allows *Locusta* to regulate their protein–carbohydrate intake involves taste receptors in hundreds of sensilla on and around the mouthparts. Each sensillum houses a small set of neurons, some of which are sensitive to amino acids and others to sugars (the other neurons detect water, salt and deterrent chemicals (Chapman 1998)). These neurons operate independently, and the sensitivity of the neurons for amino acids and sugars are inversely correlated with the levels of amino acids and sugars in the hemolymph, respectively (Simpson *et al.* 1990, Simpson *et al.* 1991, Simpson and Simpson 1992, Simpson and Raubenheimer 1993). Thus, if a locust is starved for protein, the amino-acid neurons are more easily stimulated when high-protein foods are encountered. Likewise, if hemolymph levels of sugar decline, sugar-sensitive neurons are stimulated when high-sugar foods are encountered. Self-selected protein and carbohydrate intake points have been identified in a number of insect herbivores, and the functional significance of these self-selected protein–carbohydrate ratio is revealed through no-choice experiments; the self-selected protein–carbohydrate intake point consistently aligns with the p:c ratio of foods that provide the best performance (Behmer and Joern 2008).

Regulation of other biomolecules, elements, and minerals is less well studied, which represents a serious limitation to understanding how nutrition contributes to outbreaks. Simpson *et al.* (1990) showed that a suite of 8 amino acids can stimulate amino acid neurons in locusts. One of these amino acids, proline, often elicits increased feeding in caterpillars (Heron 1965, Cook 1977, Bently *et al.* 1982) and grasshoppers (Cook 1977, Haglund 1980, Mattson and Haack 1987), and this may be functionally significant because free proline concentrations, particularly under drought conditions, are positively associated with concentrations of soluble N in plant tissues (Mattson and Haack 1987). Interestingly, adults of two grasshopper species show a sex-specific response to proline, with females, but not males, preferring proline-rich foods (Behmer and Joern 1994). Perhaps this difference reflects sex-specific nutrient demands; because they invest more in reproduction, females should need more protein than do males. Another amino acid, phenylalanine, is essential and needed in large amount for cuticle production by immature insects. In adults it is less important. Using choice-test experiments with fifth-instar and adult *Phoetaliotes nebrascensis* (Thomas) grasshoppers, Behmer and Joern (1993) showed that nymphs, but not adults, selected diets high in phenylalanine. This result, like the one for proline, suggested that an insect herbivore's nutritional requirements directly influence diet selection.

All insect herbivores also require dietary sources of sterol, but many plant sterols are unusable by insect herbivores (Behmer and Nes 2003). For grasshoppers, ingesting too much unsuitable sterol negatively affects survival (Behmer and Elias 1999a, 2000). However, grasshoppers can limit their intake of unsuitable sterols through a combination of post-ingestive feedbacks and learning (Behmer and Elias 1999b, Behmer *et al.* 1999). In natural settings, plant sterol content probably has little impact on insect herbivore populations; in agriculture, however, using plants with modified sterol profiles may be an effective way to manage and control economically important insect herbivores (Behmer and Nes 2003).

Sodium, which is involved in electrochemical functions, including message transmission in nerves, cellular signaling, and energy metabolism, is an important element for insect herbivores. Sodium typically occurs at low concentrations in plants, making it is easy to overlook its ecological importance. However, we know that female grasshoppers allocate large amounts of sodium to their offspring (Boswell *et al.* 2008), male butterflies exhibit puddling behavior as a mechanisms for collecting sodium that they later share with females during copulation (Arms *et al.* 1974), and that locust nymphs (*L. migratoria*) tightly regulate sodium intake when presented with pairs of foods that contain different salt concentrations (Trumper and Simpson 1993). Interestingly, salt regulation breaks down when locust nymphs are presented with foods that vary in their protein, carbohydrate and salt content. Here locusts prioritize protein and carbohydrate regulation, and ingest salt in amounts proportional to its concentration in the available foods (Trumper and Simpson 1993).

Historically, phosphorus has been considered a limiting nutrient in aquatic systems (Schindler 1977, Hecky and Kilham 1988, Karl *et al.* 1995). More recently there has been a growing appreciation for its role in insect nutrition. Phosphorus (P) comes mostly from nucleic acids (DNA, mRNA, tRNA, rRNA), which are on average about 9% P (Sterner and Elser 2002). Vacuoles in plants are important storage sites, and they often contain large amounts of P mostly as phosphate (PO_4^{3-}). Woods *et al.* (2004) explored allometric and phylogenetic variation in insect phosphorus content and found a negative relationship between body size and P content (measured as a %) within seven insect orders, although

Boswell *et al.* (2008) found that P content in different aged *S. americana* nymphs was constant across a range of different body sizes. However, Woods *et al.* (2004) found that recently derived insect orders had lower P content with the exception of the panorpids (Diptera + Lepidoptera), which had high P content. Unfortunately, few studies have explicitly addressed the effects of P concentrations on insect herbivores. One exception (Perkins *et al.* 2004) found that growth rates in the caterpillar *Manduca sexta* (L.) were higher, and developmental times shorter, with increasing levels of dietary P. Interestingly, caterpillars did not consistently exhibit compensatory feeding as dietary P levels decreased. Apple *et al.* (2009) also looked at caterpillar performance in response to food P levels, and they too found that that growth was enhanced as leaf P content (%) in lupines increased. Clearly more work on the role of dietary P levels on insect herbivore performance at both the individual and population level is needed. And if P truly is limiting for insect herbivores, we need to explore the extent to which different species regulate its intake using both pre- and post-ingestive mechanisms.

When insect densities are low, and plant resources abundant, individual herbivores should have ample opportunity to regulate nutrient intake by selectively feeding among different plants and plant parts, and, for many insect herbivores, diet mixing is an effective strategy for optimizing growth rates and performance (Bernays and Bright 1993, Hagele and Rowell-Rahier 1999, Singer 2001, Behmer *et al.* 2002). However, the opportunities to regulate nutrient intake through diet mixing may be constrained, either because their food choices are limited (Bernays and Chapman 1994), they are outcompeted by other insect herbivores, including conspecifics (Denno *et al.* 1995, Kaplan and Denno 2007), or they trade off foraging activity with risk from predators and parasitoids (Beckerman *et al.* 1997, Bernays 1997, Schmitz 1998, Danner and Joern 2003, Singer and Stireman 2003, Danner and Joern 2004, Schmitz 2008, Hawlena and Schmitz 2010). When dietary self-selection is constrained, insect herbivores can use compensatory mechanisms. In one of the most thorough studies exploring food macronutrient content in an insect herbivore, Raubenheimer and Simpson (1993) gave final-instar locust nymphs one of 25 artificial foods, containing one of five levels each of protein and digestible carbohydrate, and then measured food intake (providing estimates of protein and carbohydrate consumption) plus growth. Over the final stadium, locusts regulated their intake of both protein and carbohydrate, with nearly equal efficiency. Although locusts ate considerably different quantities of food on the different combinations of protein and carbohydrate, when average consumption points for each treatment were viewed as a whole, a striking pattern emerged – individuals ate particular foods in amounts that allowed them to reach the geometrically closest point their preferred protein–carbohydrate intake target.

Locusts also practiced post-ingestive compensation by differentially utilizing ingested nutrients, which allowed them to more closely approach their growth targets (defined as the quantity of nutrients needed for growth and storage tissues). For example, grasshoppers (Zanotto *et al.* 1993, Zanotto *et al.* 1997, Simpson and Raubenheimer 2001) and caterpillars (Telang *et al.* 2003) regulate their energy budgets by respiring carbohydrates or by converting them to lipids and storing them. Locusts utilize protein efficiently when it is at low to optimal concentrations (Simpson and Raubenheimer 2001). When dietary protein exceeds requirements, most of it is digested but the excess is eliminated either as uric acid or ammonium (Simpson and Raubenheimer 2001). When carbohydrates are limiting and protein in excess, using excess amino acids for gluconeogenesis may be an option (Thompson 2000, 2004). Not all insects can do this, and some do it better than others (Simpson *et al.* 2002, Raubenheimer and Simpson 2003).

Additional compensatory mechanisms are available to insect herbivores faced with ingesting large quantities of suboptimal food. For example, nutrient dilution is a common challenge for insect herbivores, and they can cope with this in two ways. First, they tend to greatly increase the amount of food they consume (Slansky and Wheeler 1991, 1992, Raubenheimer and Simpson 1993). Second, they can allocate more to gut tissues (Yang and Joern 1994a, Yang and Joern 1994b, Raubenheimer and Bassil 2007), serving two primary functions: (1) it allows a greater amount of food to be processed, and/or (2) it increases digestion efficiency because food can be retained for longer periods of time. Locusts can also differentially release key digestive enzymes when they eat foods with strongly imbalanced ratios of protein and carbohydrate (Clissold *et al.* 2010). Proteases with α-chymotrypsin-like activity are down-regulated when protein occurs in excess of carbohydrates, while carbohydrases with α-amylase-like activity are down-regulated when carbohydrates occur in excess of protein.

Temperature influences a number of life history traits. Lee and Roh (2010) recently explored how temperature interacts with food nutrients to affect growth rates in the generalist caterpillar *Spodoptera exigua* (Hübner). Using a factorial experiment with three temperatures (18°C, 26°C, and 34°C) and six different protein–carbohydrate ratios (ranging from heavily protein-biased to heavily carbohydrate-biased), they found a significant temperature-by-diet interaction. Differences in growth rates on the different temperatures were largest on diets with more balanced protein–carbohydrate ratios and smallest on the more imbalanced diets. Interestingly, growth rate was greatest at the highest temperature, but survival was greatest at the moderate temperature. Their results indicate developmental and physiological costs associated with fast growth. Interactions between temperature and food quality have also been examined in grasshoppers. Yang and Joern (1994b) showed that temperature had no effect on mass gain when food quality was good (3% N) or high (5% N), but temperature negatively affected growth when food quality was low. Miller *et al.* (2009a) showed that locusts select thermal regimes that result in rapid development and growth when allowed to choose, but at the expense of efficient nutrient utilization. Multiple studies clearly show the link between growth and development and temperature (Stamp 1990, Petersen *et al.* 2000, Levesque *et al.* 2002), but little is currently known about how thermal preferences and food availability or quality influence insect herbivores in the field, or how these factors interact to affect populations. Predators also play a role in affecting insect herbivore behavior ((Beckerman *et al.* 1997, Danner and Joern 2003, Hawlena and Schmitz 2010), including their effects on thermal preferences. With respect to grasshopper thermal preferences, Pitt (1999), for example, showed that predators matter. When birds are absent, grasshoppers sit high in vegetation, where temperatures are higher; when birds are present, grasshoppers are forced down into the vegetation, where temperatures are lower.

In most of the studies above, experiments were restricted to a single developmental stage, usually the final immature stage. This is done primarily to standardize for physiological condition, and final stage immatures are also relatively large in size, making them easier to handle. One issue associated with working on the last immature stage is that the nutritional conditions in earlier development are usually quite good, causing a potential complication – resources accumulated during earlier development might be mobilized to lessen the full effects of a particular diet treatment in later stages (Behmer and Grebenok 1998, Behmer and Elias 1999a). Recent studies using caterpillars have explored the lifetime effects of diet macronutrients (Lee 2010, Roeder 2010). These studies have shown that single-stage nutritional studies may underestimate the actual costs of compensatory feeding. In these studies, newly hatched neonates were placed on foods with different protein and carbohydrate levels and their performance was followed to pupation. Both studies revealed that compensatory mechanisms, when examined over an insect herbivore's lifetime, increasingly break down as the protein–carbohydrate ratio of the food became more imbalanced relative to the caterpillars' preferred protein–carbohydrate intake target. Roeder's study followed individuals through eclosion and revealed interesting gender differences. Females eclosed successfully across all diets except those that were heavily carbohydrate-biased. In contrast, males eclosed successfully on diets that had protein–carbohydrate ratios not far removed from their self-selected protein–carbohydrate ratio (Lee *et al.* 2006a), but success dropped off significantly in both directions as the protein–carbohydrate ratio of the experimental food became increasingly more imbalanced relative to the self-selected protein–carbohydrate ratio. Roeder speculated that this might be related to sex-specific differences in nitrogen utilization (Telang *et al.* 2000, Telang *et al.* 2002).

1.2.2 Factors that influence performance in the adult stage

To this point our focus has been on immature insect herbivores, assessing how nutrients influence feeding and performance. To more fully understand outbreak dynamics, we must also consider adult survival and reproduction and how these life-history traits are affected by plant quality. Here, species that provision offspring using recently gained resources (income breeders) are distinguished from those that provision with resources accumulated earlier (capital breeders).

Grasshoppers are examples of income breeders, and as such the diet quality they experience as adults affects demographic attributes. Joern and Behmer (1997, 1998) explored the effects of diet quality in three grasshopper species that represented different feeding guilds (two grass feeders and one mixed feeder) and distinct phylogenies (one gomphocerine and two melanoplines), and they observed variability in how these three species responded to foods with different protein–carbohydrate amounts and ratios. For example, adult survival was unaffected by diet quality in the mixed-feeding melanopline, but in the grass-feeding melanopline adult survival was longest on low-protein diets and decreased as dietary protein content increased (Joern and Behmer 1998).

For the grass-feeding gomphocerine species, survival depended on the protein–carbohydrate combination. On low-protein diets, survival increased as dietary carbohydrate content increased, but at moderate and high protein levels carbohydrate content became unimportant. Shorter adult lifetime can negatively affect reproduction by limiting the number of egg pods that can be produced (Sanchez *et al*. 1988, Branson 2006), but diet quality can also influence clutch size (eggs/pod). However, even here the effects of diet can be species-specific (Joern and Behmer 1997, 1998). In some cases protein does matter (the grass feeders), but in other instances carbohydrates are more important (the mixed feeder). It is important, though, to remember that food nutrient quality is not the only factor impacting reproduction. Predators can affect reproduction potential even when food quality is adequate by suppressing feeding rates through trait-mediated effects (Danner and Joern 2004).

On the other hand, studies exploring nutritional effects on reproduction in capital breeders require that individuals be fed throughout larval development, allowed to pupate, and then mated. Roeder (2010) has completed such a study, and then used his data on survival and reproduction to extrapolate to a population level. He found that population densities decreased significantly as the protein–carbohydrate content of the larval food became more imbalanced. This result suggests that caterpillar population outbreaks might be closely tied to the nutrient conditions of available foods, and that outbreaks are most likely to occur when conditions match those that are optimal for a given species.

1.2.3   What happens to the individual as population density increases?

The previous section focused on the conditions that lead to success at the individual level, in the build-up to population outbreaks. What happens when population density is high, and competition for resources increases? A growing literature shows that individual animals behave differently when part of a large group (Couzin and Krause 2003), and this increasingly seems to be the case for insect herbivores as well.

One fascinating example shows that being part of a crowd alters strategies of nutrient regulation. Although diet-choice studies show that solitary- and gregarious-phase locusts (*S. gregaria*) regulate their protein–carbohydrate intake to identical levels when allowed to self-select from suboptimal but complementary foods, a remarkable difference appears as the nutrient profile of available food changes (Simpson *et al*. 2002). First, gregarious nymphs consume more than solitary insects. Second, differences in intake become much larger as the protein–carbohydrate ratio of their food becomes more imbalanced. From a functional perspective, solitary nymphs minimize nutritional errors relative to their intake target. In doing so, they trade off the cost of processing nutrients ingested in excess of requirements against the cost of undereating required nutrients. In contrast, gregarious locusts use a strategy of nutrient maximization, in which they greatly overeat nutrients in excess of requirements to more closely approach their requirement for limiting nutrients. This shift in feeding behavior may correlate with contrasting nutritional environments, an idea that Simpson *et al*. (2002) refer to as the "nutritional heterogeneity hypothesis." The amount of nutritionally suboptimal food eaten should be tied to the probability that an equally and oppositely unbalanced food will be encountered. Solitary locusts are less active and more sedentary, and hence encounter a more limited range of host plant options, and under natural conditions there is a low probability that they will encounter foods with widely divergent nutritional content (van der Zee *et al.* 2002, Pener and Simpson 2009). Under such conditions, it makes sense for solitary locusts to be error minimizers if there are real physiological costs associated with long-term nutrient imbalances. In contrast, gregarious locusts are highly active and move great distances as both nymphs and adults over the course of a day, making it likely they will encounter a divergent range of food items and conspecifics. Under these conditions, they should take advantage of all food opportunities when possible.

More broadly, the regulatory rules associated with imbalanced foods may be a function of diet breadth, such that specialists (even in gregarious forms) are error minimizers and generalists are nutrient maximizers (Behmer 2009). In desert locusts, the solitary form is often effectively a specialist because it may spend significant time on a single host plant, whereas the gregarious form is a generalist because it encounters a wide array of plant species (Pener and Simpson 2009). But does this imply that only insect herbivores that practice nutrient maximization show outbreaks? Obviously the answer is no because grass-specialist locusts such as *L. migratoria* and *C. terminifera*, and tree specialists like the forest tent caterpillar (*Malacosoma disstria* (Hübner)), which are all error minimizers, often exhibit outbreak dynamics. The biology of generalists and specialists is quite different, which provides a context for asking more general questions about outbreaks of insect herbivores. For instance, do generalists or specialists have greater propensity to exhibit outbreaks? Do outbreaks by generalists and specialists occur with similar frequencies? And when an outbreak occurs, is its intensity a reflection of diet breadth? Is the duration of the outbreak associated with diet breadth? Finally, are there physiological similarities between generalists and specialists with outbreak dynamics, particularly in how they utilize ingested nutrients, that is associated with being able to outbreak? These questions link the nutritional ecology of species that exhibit outbreak dynamics to larger population processes.

One benefit of living in a large group is that large numbers can swamp predators' functional responses (Sword *et al.* 2005, Reynolds *et al.* 2009). On the other hand, living in a large group can for a number of reasons make individual members more susceptible to parasites and pathogens (McCallum *et al.* 2001, Moore 2002), which can lead to increased mortality (Anderson and May 1978) and decreased fecundity (Hurd 2001). Parasites and pathogens can also modify competitive interactions and predator–prey interactions (Hatcher *et al.* 2006). A significant literature indicates that withstanding infection is a function of host nutritional state (Chandra 1996, Lochmiller and Deerenberg 2000, Coop and Kyriazakis 2001, Lee *et al.* 2006b, Lee *et al.* 2008). Recent work by Lee *et al.* (2006b, 2008) using the caterpillar *S. littoralis* Boisduval suggests that resistance to pathogen attack and constitutive immune function are tied to dietary protein, not carbohydrate, and that individuals that self-select protein-rich diets survive viral diseases better. Interestingly, insects on high-protein diets also have more heavily melanized cuticles, and display higher antibacterial activity

(Lee *et al.* 2008). For insects in large groups that are often exposed to pathogens, a limited capacity to regulate their nutritional intake because of excessive competition for the nutritional resources needed to combat pathogens may be a contributing factor that leads to the population crashes.

The fate of eggs is a critical component of an insect outbreak that is sometimes difficult to assess. Grasshoppers lay eggs in the ground. When key environmental conditions align (proper soil moisture levels, or optimal temperatures), hatchling success can be high. But how does being part of a large group influence egg production and egg viability, and can plant quality modify density-dependent responses? Oogenesis in insects is typically nutrient-limited. Because grasshoppers are income breeders, the nutrients they allocate to eggs are acquired as adults (Wheeler 1996). Branson (2006) studied the interaction between plant quality and population density on reproduction using the grass-feeding grasshopper *Ageneotettix deorum* (Scudder), which undergoes regular population explosions in the western United States. Results suggested that increasing food quality lessened density-dependent effects, but this outcome may have been mediated through increased total plant material (as a function of fertilizer treatment). More work will be needed to clarify this relationship.

Laws (2009) explored interactions between density and parasitism on fecundity using the generalist grasshopper *Melanoplus dawsoni* (Scudder). Parasitism prevalence was similar across a range of densities, but parasitized grasshoppers in high-density treatments had significantly reduced fecundity relative to parasitized grasshoppers in low-density treatments. Here again there are potential negative costs associated with group living, which tie directly into resource availability. Pathogens can also target eggs that are waiting to hatch. Miller *et al.* (2009b) showed that hatchling locusts coming from crowded parents (i.e., high-density conditions) are more susceptible to fungal attack than are hatchlings from isolated parents (i.e., low-density conditions). The authors suggest that locusts developing at high densities, and are adapted for dispersal or migration, have fewer energetic or nutritional resources available for immune defense.

1.2.4   Density effects on group behavior

Many animal species that live in large groups (e.g., social insects, fish, birds, and ungulates) are capable of self-organization and often move as a group (Krause and Ruxton 2002, Couzin and Krause 2003). Such group behavior is often linked with foraging behavior and has important implications for ecological processes (Levin 1999). Recent work has demonstrated that self-organization and collective behavior also exist in insect herbivore at high densities, most notably Mormon crickets and desert locusts. These studies are noteworthy because they reveal underlying mechanisms that drive collective behavior, especially as it relates to directed mass movements.

Mormon crickets and desert locusts both regularly form large, cohesive migratory bands, consisting of millions of individuals moving in unison across the landscape. But what factors contribute to group formation, help maintain group cohesion, and influence its direction? For locusts, group formation during the switch from solitary to gregarious phase is tied to resource distribution patterns, particularly during the initial stage of an outbreak. Using computer simulations and laboratory experiments, Collett *et al.* (1998) showed that when resources (e.g., plants) are clumped, rather than uniformly dispersed, gregarization is induced. Under these conditions, phase change can occur rapidly and synchronously. A follow-up lab study (Despland *et al.* 2000) demonstrated that locusts are more active, experience more crowding, and become more gregarious when food is patchy. This same outcome is also observed under simulated field conditions (Despland and Simpson 2000b), and Babah and Sword (2004) found that plant distributions tend to be more aggregated in areas where the frequency of gregarization is high. Low-quality foods and patches that contained clumped nutritionally complementary foods promoted increased crowding and movement, which led to increased gregarization (Despland and Simpson 2000a).

For gregarious locusts, group formation is a function of *phase state* – after individuals come into contact and gregarize, they shift from being mutually repelled to being mutually attracted (Pener and Simpson 2009). In contrast, Mormon crickets do not exhibit behavior strictly consistent with phase-polyphenism (Sword 2005). Despite this key difference, locusts and Mormon crickets share common behaviors, particularly with respect to marching. The collective motion of locusts has been examined using models from theoretical physics, where individuals in a group are modeled as self-propelled particles (SPPs), with each "particle" modifying its behavior (speed and orientation) in response to its nearest neighbors (Toner and Tu 1998, Gregoire and Chate 2004). Using this approach, Buhl *et al.* (2006) demonstrated that marching in locusts is a product of density, and identified the critical density at which coordinated marching in locusts nymphs takes place (which they estimated to be about 20 locusts/m^2). Importantly, they also demonstrated dynamic instability, meaning that a group of locusts can switch direction without external inputs. But what happens when marching bands of locusts and Mormon crickets reach high density? And how does the marching band make a collective decision with respect to its orientation? In many large animal groups, only a small proportion of individuals are needed to influence the direction of a group (Couzin *et al.* 2005), and information about this decision can be transferred within groups in the absence of explicit signaling, and when group members are unaware of which individuals are making decisions. This seems to be the case with locusts and Mormon cricket marching bands. Their collective group movements are initiated by a small number of motivated individuals whose movements stimulate directed movement in individuals in their local vicinity. This creates a chain of events, where the other members of the collective group, using simple local orientation and movement rules, respond to their nearest neighbors. Scaled up, these local behaviors translate to into a cohesive marching band of insects.

Given that a just few individuals can determine the movement patterns of a massive group of insects, the next question to ask is what environmental factors influence these leaders? Wind and topography have been suggested to influence small-scale movement patterns in rangeland grasshoppers (Narisu *et al.* 1999,

2000), but radiotelemetric mark–recapture studies on Mormon crickets failed to find any correlation between migratory band movement direction and local wind direction (Lorch *et al.* 2005). Two key conclusions resulted from this study: (1) cues mediating directionality are likely to be group specific, and (2) landscape-scale environmental cues (e.g., large landmarks, and the position of the sun) likely have only a small effect. One emerging group-specific factor that appears to be important is the nutritional state of members of the group (Simpson *et al.* 2006, Srygley *et al.* 2009, Bazazi *et al.* 2008, in press). Large groups of insect herbivores can quickly deplete local resources, and hunger enhances the probability that the group will leave an area. Recently Simpson *et al.* (2006) demonstrated that hunger for two key nutrients – protein and salt – drives marching behavior in Mormon crickets. They also highlighted how the threat of cannibalism is a critical factor in maintaining marching behavior because the richest source of protein and salt for a hungry Mormon cricket is another Mormon cricket. By satiating crickets with protein and salt, cannibalism rates were reduced, and by satiating crickets with protein, walking was inhibited. They also demonstrated that crickets with reduced motility or mobility were at higher risk of being cannibalized. Rather vividly, these results describe a situation where Mormon crickets in the field are in effect on a forced march. That is, individuals are pulled forward by their hunger for protein and salt, while at the same time being pushed forward by trailing individuals that are, quite literally, nipping at their tarsal heels. The importance of cannibalism to understanding the dynamics of movement also was demonstrated recently in marching bands of locusts, which have a tendency to bite others (particularly in the abdomen) in the marching band. Through abdominal denervation, Basazi *et al.* (2008) showed that individual locusts that lost the ability to detect individuals approaching from behind decreased their probability of moving. In turn, this resulted in decreased overall group movement and ultimately led to significant increased cannibalism rates. Together, these studies suggest that cannibalism, specifically the threat of attack by trailing individuals, is a key factor in the onset of collective movement in Mormon crickets and locusts, and a key factor that keeps bands moving forward.

Group living includes both benefits (safety in numbers from predators) and costs (threat of cannibalism). Two other potential costs also have been identified. One relates to how resource availability, particularly protein limitation, might affect insect immune responses. Srygley *et al.* (2009) found that providing migrating Mormon crickets with a supplemental source of protein increased their phenoloxidase (PO) activity. This suggests that migrating Mormon crickets are compromised with respect to their PO activity, which is a critical early enzyme that triggers the production of melanin, part of the humoral encapsulation process by which insects fight off foreign invaders. Therefore, reduced PO activity makes them more susceptible to wounding, and a compromised cuticle decreases an insect's ability to physically block parasites and pathogen infections. A second potential cost is illustrated in social caterpillars, which can often live in large groups composed of hundreds to thousands of individuals. Dussutour *et al.* (2007) found that forest tent caterpillars (*M. disstria*), which show a capacity to regulate their nutrient intake when kept as individuals (Despland and

Noseworthy 2006), lose their "nutritional wisdom" when they are part of a group. This likely happens for two reasons. First, because a small number of individuals can dictate group behavior (Couzin *et al.* 2005), a poor choice by a focal individual can lead the group to a poor food. Second, as a result of trail-following behavior, the group as a whole can become trapped at the food (Dussutour *et al.* 2007). Such behavior suggests that when a 'decisive individual' (or a small group of decisive individuals) makes a poor choice, and the group responds based on rules governing local interactions, the performance of individuals within that group is compromised.

1.2.5 Population responses by insect herbivores to variably abundant nutritional resources

As reviewed above, we know much about how food intake rate and nutritional balance affect the behavior and performance of individuals. The clear importance of feeding to individuals suggests that nutrition should play a central role in understanding population dynamics and outbreaks of insect herbivores. But, is the natural variability in the availability of high-quality food sufficient to drive population processes in the face of many other critical factors that also affect the dynamics of natural populations? And, if the combined effects of food intake and nutrition are not the primary driver behind insect herbivore population dynamics, and especially outbreaks, what shared role and interactions with other factors might these factors play?

Insect ecologists have considered the role of variable food quality to insect population dynamics and outbreaks for some time, but have struggled to develop ecological models with enough mechanistic detail to explain population dynamics (Berryman 1987, 1999, Turchin 2003). Except at extreme densities, insect herbivores, and especially generalist feeders, typically encounter sufficient food, but that food may not be rich enough to support individuals. Depending on the nutritional landscape available to insect herbivores, and the rate at which food can be acquired, allocation budgets of resources to growth, developmental rate, survival, and egg production that drive population dynamics will also be highly variable as well. One challenge with many of the current paradigms, though, is that "diet quality" is often narrowly, or loosely, defined.

1.3 The plant-stress paradigm

Insect herbivore outbreaks are often correlated with weather patterns, leading to the presumption that food quality is a critical determinant of population outbreaks. Unless weather affects individual insect performance directly (e.g., temperature-dependent metabolism and physiological functions), it is best to identify mechanisms associated with variable weather patterns that could alter demographic responses of insect herbivore populations leading to outbreaks.

The plant stress hypothesis (PSH) states that moderate environmental stresses on plants decrease plant resistance to insect herbivory by altering biochemical source–sink relationships and foliar chemistry, thus providing better nutrition to insects (Mooney *et al.* 1991). Here, "environmental stress" is defined as the conditions that reduce plant performance below that achieved under optimal conditions, often measured as total accumulation of biomass. Obviously, optimal conditions vary by plant species, genotype, and tissue within a plant, and multiple environmental factors can contribute to the degree of stress experienced by the plant (e.g., water, light, and nutrient concentrations). PSH models incorporate relatively detailed physiological and biochemical responses by stressed plants to predict foliar nutritional quality available to insect herbivores (Mattson and Haack 1987, Jones and Coleman 1991, Louda and Collinge 1992, Lindroth *et al.* 1993, English-Loeb and Duffy 1997). Prominent discussions of PSH are those of White (1984, 1993) and Mattson and Haack (1987), who argued that primary nutritional environments of insect herbivores are typically inadequate. Rhoades (1983) extended PSH to include reduced production of both primary nutrients and chemical defenses under stress conditions. As the PSH matured, more options have been proposed. Some authors (White (1984, 1993) and Mattson and Haack (1987)) expect insect performance to increase with moderate plant stress; others suggest that insect herbivores do best when plant performance is near optimal (the plant vigor hypothesis; Price 1991), while still other authors note that different insect groups respond variably, but predictably, to the same plant responses to stress (the insect performance hypothesis; Larsson 1989). Jones and Coleman (1991) developed an integrated model that incorporated environmental stress explicitly, including the interplay between herbivory and plant stress. This model nicely summarizes the state of the art for plant stress perspectives, but it is less than ideal in some ways. Jones and Coleman's model includes a very large number of potential mechanistic paths that can vary in response to stress and interact, leading to an almost bewildering set of possibilities. As such, the approach is unable to predict outcomes without specific knowledge of responses at each level. Although the general theme underlying these approaches – plant quality to herbivores varies with level of stress, in turn affecting individual performance and population-level responses – is compelling, it has not proven sufficiently predictive for explaining population-level responses and especially outbreaks of insect herbivores.

The effects of multiple stresses (water, mineral N, and grasshopper herbivory) to the dominant grass *Bouteloua gracilis* Willd. ex Kunth Lag. ex Griffiths in a nutrient-poor grassland illustrates the need for more complex investigations of interactions (Joern and Mole 2005). Total plant biomass increased with the addition of water and mineral N fertilizer over moderate and ecologically relevant levels. Herbivory by a grass-feeding grasshopper altered total biomass only in dry years, and plants compensated for tissue loss in wetter years. Over three years of the study, foliar-N concentrations were 10–20% higher in low water treatments, and varied according to ambient levels of precipitation. In general, effects of individual stressors to *B. gracilis* showed strong, significant interactions under field conditions (Figure 1.1). Effect of grasshopper herbivory on total foliar-N and total nonstructural carbohydrates (TNC) was greatest in dry years, where it interacted with abiotic stresses. However, performance (developmental rates and survival) by two common grass-feeding grasshopper species differed from predictions of the PSH even though nutritional quality of plants changed. *Phoetaliotes nebrascensis* (Thomas) faced with different water and nitrogen levels showed no difference in survival, while *Ageneotettix deorum*

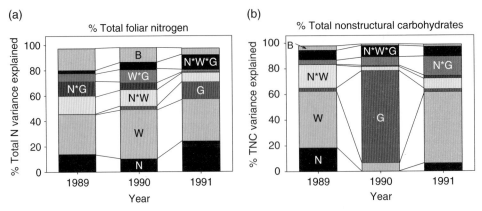

Figure 1.1 Variance explained of foliar-N and TNC in *Bouteloua gracilis* in response to stresses over three years in a field experiment (Joern & Mole 2005). Water addition, N fertilizer addition, and exposure to grasshopper herbivory were manipulated in a field experiment using *B. gracilis* in a nutrient-poor soil. Letters refer to variance explained by main effects of the experiment (N, nitrogen fertilization; W, water; and G, grasshopper herbivory) and statistical interactions (N*W, N*G, W*G). B represents the site effect. The percentage of the total variance was calculated as the variance associated with the treatment combination compared with the total variance of the experiment. Reprinted with kind permission from Springer Science+Business Media: *Journal of Chemical Ecology*, The plant stress hypothesis and variable responses by blue grama grass (*Bouteloua gracilis*) to water, mineral nitrogen and insect herbivory, volume 31, 2005, pages 2069–2090, Joern, A. and S. Mole.

survived equally well on water-stressed plants with no N-fertilzer, and watered plants survived best on water-stressed plants with no N-fertilizer as well as on watered plants with the highest levels of fertilizer applied. However, foliar-N levels were often low in the watered treatment with and without N-fertilizer. Two key take-home messages are clear from this study. First, responses to multiple, interacting stresses and variable weather conditions among years are complex and difficult to predict. Second, significant challenges still remain for developing general predictions of how plant stress affects insect herbivore population dynamics, particularly through its effect on variable plant tissue quality.

General empirical assessments of PSH are mixed, but mostly not strongly supportive (Waring and Cobb 1992, Watt 1992, Koricheva *et al.* 1998, Huberty and Denno 2004), and all crtitical evaluations of multiple studies identify important problems in its ability to reliably predict responses by insect herbivores to stressed plants. It is also the case that most studies focus on responses by woody plants, and few studies with grasses and forbs exist. Where multiple data sets are evaluated, vote-counting type assessment methods (Waring and Cobb 1992, Galway *et al.* 2004) are generally more favorable regarding the success of the PSH than are meta-analyses (Koricheva *et al.* 1998, Huberty and Denno 2004), especially for the role of plant water stress. In general these analyses conclude that insect herbivores responded to plant stress, but the directions and magnitudes of the responses were highly variable making the general notion unhelpful for predictive purposes. Refining the domain of taxa involved according to herbivore feeding guilds was helpful in meta-analysis, and we have found in our own work that recognizing feeding-guild membership often provides greater clarity to our understanding to a broad range of biological questions. Overall, rigorous comparisons among multiple studies on woody plants (Koricheva *et al.* 1998) indicate that the plant stress–insect performance interactions form a continuum, ranging from insects doing best on stressed plants to situations where they perform best on vigorous plants. However, current models do not identify underlying mechanisms necessary to inform us where on the continuum a particular herbivore–plant interaction lies.

Several reasons explain why potentially compelling integrative models of insect outbreaks did not provide the predictive frameworks promised. While the models were general and meant to reflect generalized plant responses to non-optimal conditions, each plant species, population, or genotype may respond to the same environmental stress in quantitatively different ways. Actual plant responses to stress were often not measured directly in many studies, so that ecologists could not accurately assess the nutritional environment actually encountered by herbivores. Many studies focused on the nature, magnitude, and duration of environmental stresses, not how the plant actually responded or what nutritional quality of the food was actually encountered by the herbivores. Moreover, general models often lumped all insect herbivores, failing to recognize different feeding constraints and/or physiological needs of different feeding guilds of insect herbivores, let alone that species-specific differences exist. Similarly, most tests of PSH focused on woody plants, whereas many insect herbivores feed on nonwoody forbs and grasses. Plant responses to stress may include two components that affect herbivore feeding – altered leaf characteristics affecting primary nutrition (e.g., protein, and soluble carbohydrates), and other secondary compounds that alter feeding levels independent of direct nutritional influences (e.g., alkaloids and tannins). Studies commonly focus on responses to primary nutrients (although often a very restricted subset), rarely assay secondary compounds, and even more rarely include both. Finally, most studies examine a single environmental stress factor at a time (Koricheva *et al.* 1998) so that the impact of multiple stresses common under field conditions and whether they are additive or interactive in their action cannot be assessed. Because we desire to develop general understanding of how insect herbivore populations respond to plant nutritional quality in nature and the role that environmental stress plays in such interactions, we strongly believe that we must take each of the above issues under consideration in developing new syntheses.

Plant stresses often result from variable weather conditions over geographic regions. Missing from many analyses is recognition of the appropriate regional scale and responses to weather, even though population cycles provide opportunities (and challenges) for understanding and predicting possible population processes leading to outbreaks. Regional climate cycles (El Nino, ENSO; North Atlantic Oscillation, NAO; and Pacific Decadal Oscillation, PDO) operate at about the same periodicity as observed for many irruptive herbivore populations and their food resources (Jaksic *et al.* 1996, Stenseth *et al.* 2003, McCabe *et al.* 2004, Sutton and Hodson 2005, Halkka *et al.* 2006). Operationally, these large-scale climatic cycles affect regional weather. They can alter precipitation and temperature patterns over multiple years, potentially affecting insect herbivores indirectly through their effects on plant growth and quality. The long-term dynamics of grasshopper populations at Konza Prairie (Kansas) are strongly correlated with weather, possibly acting through effects on plant quality (Jonas and Joern 2007). Temporal dynamics of densities observed in feeding guilds (Plate 1.1) show evidence of both intrinsic (density dependence, the slope of the relationship), and extrinsic influences (the variability along the diagonal relationship of the figure) over a 25-year period. Variation in species-specific and feeding-guild dynamics over this period correlated well with a regional North Atlantic Oscillation (NAO) index reflecting decadal time scales as well as with local growing season precipitation in the current and previous years. The positive phase of the NAO was associated with increased abundance of forb feeders and poor fliers, while mixed feeders and strong fliers were more common during the negative phase of the NAO cycle. Shifts in species composition relating to size and phenology were associated with local variation in heating degree days during the growing seasons of the previous year, variability of growing season precipitation in the current and previous years, fall temperature, and the decadal index. A regional Palmer Drought Index (PDI) also predicted grasshopper population responses reasonably well. Medium-bodied and late-hatching species were more common during years with high drought intensity values, while large-bodied and early-hatching species were more abundant during years with adequate precipitation.

However, host plant responses to weather are not always the appropriate underlying mechanism for understanding outbreak dynamics of insect herbivores, as demonstrated in the larch budmoth (*Zeiraphera diniana* (Guénée)). This species exhibits periodic 8–10-year population cycles throughout northern Europe and is well studied empirically (Turchin *et al.* 2003). Multiple hypotheses have been proposed to explain such regular population cycles, including: parasitoid–host interactions, delayed effects of plant quality, pathogen–host interactions, and maternal effects. Time-series analyses provided no evidence for maternal effects, and mortality effects from disease (the granulosis virus infection) were intermittent and could not explain long-term population patterns. Subsequent modeling based on mechanistic linkages complemented time-series analyses to examine the remaining two hypotheses. Here, the effect of plant quality was weak, whereas a model of larch budmoth–parasite interactions explained 90% of the variance in population growth rates. A combined model of food quality and parasitism suggested that the cycles are driven by parasitism but modified by an interaction with food quality. This example illustrates the complexity of even seemingly "obvious" population level responses to variation in environmental factors and the critical importance of considering underlying ecological mechanisms to explain population cycles.

1.4   Insect herbivore outbreaks – where do we go from here?

There is little doubt that our understanding of insect herbivore outbreaks has grown since the original incarnation of this book (Barbosa and Schultz 1987). However, the big picture is still blurry because we have yet to identify and properly rank the key factors that drive insect outbreaks, we know little about how many of these factors interact to drive insect outbreak dynamics, and it is difficult to classify how different insect groups respond. Although top-down forces are clearly an important factor when considering insect herbivore outbreaks (Dwyer *et al.* 2004), we have emphasized a bottom-up view throughout this chapter because insect herbivore population densities cannot readily break through the predator–pathogen ceiling without adequate resources. We see three key areas that need additional consideration to successfully build general models that better explain and predict insect outbreaks. We need to (1) develop a more comprehensive definition of plant quality and how it varies with environmental conditions; (2) more fully integrate physiological processes, which will allow us to identify generalized traits that are shared by insect herbivores that exhibit outbreak dynamics, while recognizing unique traits that contribute to outbreaks under different environmental conditions (e.g., how different are locusts from forest caterpillars?); and (3) develop collaborations across disciplines (e.g., physiology, ecology, behavior, ecosystem sciences) so that we can translate individual-based responses to a group level, couched within community and ecosystem contexts.

With respect to host plant quality, it is evident that from an insect herbivore's perspective (in fact from any organism's perspective) nutrients alter performance, affecting all life history stages (Behmer 2009). The emerging critical concept is nutrient balance rather than maximization of any single nutrient. But what are the key nutrients that need to be considered? A constant theme throughout this chapter has been the importance of two key macronutrients – protein and digestible carbohydrates – to insect herbivores. But what do we really know about variability in levels of protein and digestible carbohydrates in plants, especially their relative amounts? Unfortunately quantifying variation in plant protein and digestible carbohydrates can be arduous in field studies, and the co-occurrence of these two nutrients is rarely reported in field studies. An alternative, more feasible, approach is to consider these two nutrients as elements. Here protein is viewed in terms of nitrogen (N), while carbohydrates are viewed in terms of carbon (C). A critical question remains – how good are N and C are estimating plant protein and digestible carbohydrate levels, respectively? There is a strong case to be made that N is a good predictor of protein (Marschner 1995), but unfortunately C is a terrible predictor of digestible carbohydrates. For example, laboratory studies by Boswell (2009) measured C content in a range of artificial diets that contained variable amounts of digestible carbohydrates (sugar and starch) and cellulose. This work documented that C content was relatively constant (~43%) across diets that differed radically in their digestible carbohydrate (7–35%) and cellulose content (from 28–82%). In terms of characterizing the energy content of the plants available to insect herbivores, there is absolutely no substitute for quantifying digestible carbohydrates.

The nutritional requirements of insect herbivores extend beyond protein and digestible carbohydrates. But currently there is little information regarding how other biomolecules at low concentrations relative to protein and digestible carbohydrates might influence outbreak dynamics. As noted above, nucleic acids (DNA, mRNA, tRNA, and rRNA) contain significant amounts of plant P content, and some P is also found in storage vacuoles, typically as phosphate (PO_4^{3-}). There is a strong correlation between P (measured elementally) and the amount of nutrient compounds in plants containing P, so elemental measures of P in plants should be informative. And importantly, not all nutrients needed by insect herbivores come packaged as biomolecules. There is a suite of inorganic nutrient compounds, including sodium, potassium, calcium, magnesium, chloride, and phosphate, that primarily occur in plants as free ions. We know that many of these elements are essential for insect herbivores (Chapman 1998), and that their amounts can vary in plants (Marschner 1995), but we unfortunately know little about how "micronutrient" variation affects insect performance at either the individual or population level. Researchers must think multidimensionally when defining plant quality. This is important because quantifying key elements and digestible carbohydrates will allow us to construct a comprehensive view of the nutritional landscape encountered and occupied by insect herbivores. And, when combined with information on insect herbivore outbreak dynamics, it will allow us to identify which nutrients correlate with insect outbreaks.

Finally, our progress with respect to better understanding insect herbivore outbreaks and dynamics might be hindered by attempts to generalize across insect herbivores without taking into account feeding guild (e.g., generalist vs. specialists, grass feeder vs. tree feeder), and in some cases species-specific differences. Here we provide two examples from our own work on grasshoppers. First, our reproduction studies on three grasshopper species that regularly have population explosions in the grasslands of western Nebraska suggest that differential responses to food nutrients may be a function of feeding guild, phylogeny, and/ or phenology (Joern and Behmer 1997, 1998). Second, we have shown that closely related generalist species occupy unique nutritional feeding niches, which would provide a mechanism for facilitating coexistence (Behmer and Joern 2008). A key aspect of this latter work was to quantify, in the protein–carbohydrate nutritional landscape of grasses and forbs (Plate 1.2a), each species' specific protein–carbohydrate intake point (Plate 1.2b). This allowed us to visualize for each species its relative position in protein–carbohydrate nutrient space. Additionally, our approach allowed us to create a starting point for developing and testing predictions about which species are likely to be the strongest competitors, as well as predictions about the relative abundances of each species, and the extent to which they fluctuate year to year. For example, we predict that species that occupy a relatively central location in protein–carbohydrate nutritional space should have relative stable populations from year to year because they will be less susceptible to yearly differences that influence the availability of foods that allow them to most closely match their preferred protein–carbohydrate intake targets. And in good years, an increased amount of high-quality food will allow them to maximize performance, and thus increase their population size. However, we also offer one cautionary note about generalizing too broadly across species, especially for species that occupy a vast geographic range. Fielding and Defoliart (2008) recently documented differences in nutrient regulation in two populations (a subarctic one (from Alaska), and a temperate-zone one (from Idaho)) of a generalist grasshopper (*Melanoplus sanguinipes* (Fabricius)); they associated this difference with latitudinal gradients. They suggested that the observed differences might be related to physiological attributes associated with living in environments that impose different physical constraints.

Throughout this chapter we highlighted links between physiological processes and outbreaks of insect herbivores. We believe physiological approaches have and will continue to provide insight into insect herbivore outbreaks. However, significant advances will mostly come through collaborative efforts that extend across conceptual boundaries. We fully embrace the idea that multiple factors, including top-down effects and weather, contribute to insect herbivore outbreaks, but we feel that focusing on bottom-up factors provides the clearest starting point to lead to the biggest advances. One specific way forward is to ask explicitly: how do plant nutrients mediate interactions between insect herbivores and their predators and pathogens? With this in mind, the key is documenting variation in plant nutrient profiles, where we see an elemental approach as providing the best strategy for large-scale studies (with the exception of documenting digestible carbohydrate content in plants). Correlative studies between plant nutrient content and species abundance (either at the population or community level) will be the necessary first step to identify which nutrients are most influencing insect herbivores. Laboratory and field experiments that manipulate key nutrients to explore their individual and combined effects on insect herbivores underpin the critical next step. As recent work suggests, there is most likely a narrow range of nutrient combinations that favor outbreaks, although we expect that nutritional combinations that are optimal for one species are likely suboptimal for other species. It will also be important to build a temporal picture of the dynamics of plant quality and population abundances over time to capture the cyclical nature and contributions from time lags to understand how bottom-up factors drive subsequent generations of insect herbivores. With a comprehensive approach in place, we will also be better prepared to predict and deal with changes in insect herbivore dynamics that we are likely to see in response to climate change. And in contrast to the ancient Egyptians and Israelis, and more contemporarily, Laura Ingalls Wilder, a comprehensive approach will help demystify the causal factors that trigger insect herbivore outbreaks.

Acknowledgments

We want to thank Pedro Barbosa for the invitation to share our thoughts on insect outbreaks, and for his and Art Woods's helpful feedback and insightful comments on the chapter. We also thank Xiangfeng Jing (a PhD student in STB's lab) for providing background and information on the historical effects and significance of locusts in China. Many of the ideas and concepts in this chapter have come through a research project supported by a National Science Foundation (NSF) grant to AJ and STB (DEB-045622).

References

Anderson, R. M., and R. M. May. 1978. Regulation and stability of host–parasite population interactions. 1. Regulatory processes. *Journal of Animal Ecology* 47:219–247.

Apple, J. L., M. Wink, S. E. Wills, and J. G. Bishop. 2009. Successional change in phosphorus stoichiometry explains the inverse relationship between herbivory and lupin density on Mount St. Helens. *PLoS One* 4.

Arms, K., P. Feeny, and R. Lederhou. 1974. Sodium – stimulus for puddling behavior by tiger swallowtail butterflies, *Papilio glaucus. Science* 185:372–374.

Babah, M. A. O., and G. A. Sword. 2004. Linking locust gregarization to local resource distribution patterns across a large spatial scale. *Environmental Entomology* 33:1577–1583.

Barbosa, P., and J. C. Schultz. 1987. *Insect Outbreaks*. Academic Press, San Diego.

Bazazi, S., J. Buhl, J. J. Hale, M. L. Anstey, G. A. Sword, S. J. Simpson, and I. D. Couzin. 2008. Collective motion and cannibalism in locust migratory bands. *Current Biology* 18:735–739.

Bazazi, S., P. Ramanczuk, S. Thomas, L. Schimansky-Geier, J. J. Hale, G. A. Miller, G. Sword, S. J. Simpson, and I. D. Couzin. 2011. Nutritional state and collective motion: from individuals to mass migration. *Proceedings of the Royal Society B-Biological Sciences* 278:356–363.

Beckerman, A. P., M. Uriarte, and O. J. Schmitz. 1997. Experimental evidence for a behavior-mediated trophic cascade in a terrestrial food chain. *Proceedings of the National Academy of Sciences of the USA* 94:10735–10738.

Behmer, S. T. 2009. Insect herbivore nutrient regulation. *Annual Review of Entomology* 54:165–187.

Behmer, S. T., E. Cox, D. Raubenheimer, and S. J. Simpson. 2003. Food distance and its effect on nutrient balancing in a mobile insect herbivore. *Animal Behaviour* 66:665–675.

Behmer, S. T., and D. O. Elias. 1999a. The nutritional significance of sterol metabolic constraints in the generalist grasshopper *Schistocerca americana. Journal of Insect Physiology* 45:339–348.

Behmer, S. T., and D. O. Elias. 1999b. Phytosterol structure and its impact on feeding behaviour in the generalist grasshopper *Schistocerca americana. Physiological Entomology* 24:18–27.

Behmer, S. T., and D. O. Elias. 2000. Sterol metabolic constraints as a factor contributing to the maintenance of diet mixing in grasshoppers (Orthoptera: Acrididae). *Physiological and Biochemical Zoology* 73:219–230.

Behmer, S. T., D. O. Elias, and E. A. Bernays. 1999. Post-ingestive feedbacks and associative learning regulate the intake of unsuitable sterols in a generalist grasshopper. *Journal of Experimental Biology* 202:739–748.

Behmer, S. T., and R. J. Grebenok. 1998. Impact of dietary sterols on life-history traits of a caterpillar. *Physiological Entomology* 23:165–175.

Behmer, S. T., and A. Joern. 1993. Diet choice by a grass-feeding grasshopper based on the need for a limiting nutrient. *Functional Ecology* 7:522–527.

Behmer, S. T., and A. Joern. 1994. The influence of proline on diet selection – sex-specific feeding preferences by the grasshoppers *Ageneotettix deorum* and *Phoetaliotes nebrascensis* (Orthoptera: Acrididae). *Oecologia* 98:76–82.

Behmer, S. T., and A. Joern. 2008. Coexisting generalist herbivores occupy unique nutritional feeding niches. *Proceedings of the National Academy of Sciences of the USA* 105:1977–1982.

Behmer, S. T., and W. D. Nes. 2003. Insect sterol nutrition and physiology: a global overview. *Advances in Insect Physiology* 31:1–72.

Behmer, S. T., D. Raubenheimer, and S. J. Simpson. 2001. Frequency-dependent food selection in locusts: a geometric analysis of the role of nutrient balancing. *Animal Behaviour* 61:995–1005.

Behmer, S. T., S. J. Simpson, and D. Raubenheimer. 2002. Herbivore foraging in chemically heterogeneous environments: nutrients and secondary metabolites. *Ecology* 83:2489–2501.

Bently, M. D., D. E. Leonard, S. Leach, E. Reynolds, W. Stoddard, D. Tomkinson, G. Strunz, and M. Yatagai. 1982. Effects of some naturally occurring chemicals and extracts of non-host plants on feeding in spruce budworm larvae, *Chroristoneura fumiferana. Maine Agricultural Experimental Station Technical Bulletin 107*. Maine Agricultural Experimental Station, Orono.

Bernays, E. A. 1997. Feeding by lepidopteran larvae is dangerous. *Ecological Entomology* 22:121–123.

Bernays, E. A., and K. L. Bright. 1993. Mechanisms of dietary mixing in grasshoppers – a review. *Comparative Biochemistry and Physiology A-Physiology* 104:125–131.

Bernays, E. A., and R. F. Chapman. 1994. *Host-Plant Selection by Phytophagous Insects*. Chapman & Hall, New York.

Berryman, A. A. 1987. The theory and classification of outbreaks. Pages 3–30 *in* P. Barbosa and J. C. Schultz, editors. *Insect Outbreaks*. Academic Press, San Diego.

Berryman, A. A. 1999. *Principles of Population Dynamics*. Stanley Thornes Publishers, Ltd, Cheltenham, UK.

Boswell, A. W. 2009. *Insect Herbivore Stoichiometry: The Effect of Macronutrient Quantity, Ratio, and Quality*. M.S. Thesis, Texas A&M University, College Station, TX.

Boswell, A. W., T. Provin, and S. T. Behmer. 2008. The relationship between body mass and elemental composition in nymphs of the grasshopper *Schistocerca americana*. *Journal of Orthoptera Research* 17:307–313.

Branson, D. H. 2006. Life-history responses of *Ageneotettix deorum* (Scudder) (Orthoptera: Acrididae) to host plant availability and population density. *Journal of the Kansas Entomological Society* 79:146–155.

Branson, D. H., A. Joern, and G. A. Sword. 2006. Sustainable management of insect herbivores in grassland ecosystems: new perspectives in grasshopper control. *Bioscience* 56:743–755.

Buhl, J., D. J. T. Sumpter, I. D. Couzin, J. J. Hale, E. Despland, E. R. Miller, and S. J. Simpson. 2006. From disorder to order in marching locusts. *Science* 312:1402–1406.

Chambers, P. G., D. Raubenheimer, and S. J. Simpson. 1997. The selection of nutritionally balanced foods by *Locusta migratoria*: the interaction between food nutrients and added flavours. *Physiological Entomology* 22:199–206.

Chambers, P. G., D. Raubenheimer, and S. J. Simpson. 1998. The functional significance of switching interval in food mixing by *Locusta migratoria*. *Journal of Insect Physiology* 44:77–85.

Chambers, P. G., S. J. Simpson, and D. Raubenheimer. 1995. Behavioural mechanisms of nutrient balancing in *Locusta migratoria* nymphs. *Animal Behaviour* 50:1513–1523.

Chandra, R. K. 1996. Nutrition, immunity and infection: From basic knowledge of dietary manipulation of immune responses to practical application of ameliorating suffering and improving survival. *Proceedings of the National Academy of Sciences of the USA* 93:14304–14307.

Chapman, R. F. 1998. *The Insects: Structure and Function*. Cambridge University Press, Cambridge.

Clissold, F. J., B. J. Tedder, A. D. Conigrave, and S. J. Simpson. 2010. The gastrointestinal tract as a nutrient-balancing organ. *Proceedings of the Royal Society B-Biological Sciences* 277:1751–1759.

Collett, M., E. Despland, S. J. Simpson, and D. C. Krakauer. 1998. Spatial scales of desert locust gregarization. *Proceedings of the National Academy of Sciences of the USA* 95:13052–13055.

Cook, A. G. 1977. Nutrient chemicals as phagostimulants for *Locusta migratoria* (L). *Ecological Entomology* 2:113–121.

Coop, R. L., and I. Kyriazakis. 2001. Influence of host nutrition on the development and consequences of nematode parasitism in ruminants. *Trends in Parasitology* 17:325–330.

Couzin, I. D., and J. Krause. 2003. Self-organization and collective behavior in vertebrates. *Advances in the Study of Behavior* 32:1–75.

Couzin, I. D., J. Krause, N. R. Franks, and S. A. Levin. 2005. Effective leadership and decision–making in animal groups on the move. *Nature* 433:513–516.

Danner, B. J., and A. Joern. 2003. Resource-mediated impact of spider predation risk on performance in the grasshopper *Ageneotettix deorum* (Orthoptera: Acrididae). *Oecologia* 137:352–359.

Danner, B. J., and A. Joern. 2004. Development, growth, and egg production of *Ageneotettix deorum* (Orthoptera: Acrididae) in response to spider predation risk and elevated resource quality. *Ecological Entomology* 29:1–11.

Denno, R. F., M. S. Mcclure, and J. R. Ott. 1995. Interspecific interactions in phytophagous insects – competition reexamined and resurrected. *Annual Review of Entomology* 40:297–331.

Despland, E., M. Collett, and S. J. Simpson. 2000. Small-scale processes in desert locust swarm formation: how vegetation patterns influence gregarization. *Oikos* 88:652–662.

Despland, E., and M. Noseworthy. 2006. How well do specialist feeders regulate nutrient intake? Evidence from a gregarious tree-feeding caterpillar. *Journal of Experimental Biology* 209:1301–1309.

Despland, E., and S. J. Simpson. 2000a. The role of food distribution and nutritional quality in behavioural phase change in the desert locust. *Animal Behaviour* 59:643–652.

Despland, E., and S. J. Simpson. 2000b. Small-scale vegetation patterns in the parental environment influence the phase state of hatchlings of the desert locust. *Physiological Entomology* 25:74–81.

Dussutour, A., S. J. Simpson, E. Despland, and N. Colasurdo. 2007. When the group denies individual nutritional wisdom. *Animal Behaviour* 74:931–939.

Dwyer, G., J. Dushoff, and S. H. Yee. 2004. The combined effects of pathogens and predators on insect outbreaks. *Nature* 430:341–345.

English-Loeb, G. M., and S. S. Duffy. 1997. Drought stress in tomatoes: changes in plant chemistry and potential nonlinear consequences for insect herbivores. *Oikos* 79:456–468.

Fielding, D. J., and L. S. Defoliart. 2008. Discriminating tastes: self-selection of macronutrients in two populations of grasshoppers. *Physiological Entomology* 33:264–273.

Galway, K. E., R. P. Duncan, P. Syrett, R. M. Emberson, and A. W. Sheppard. 2004. Insect performance and host-plant stress: a review from a biological control perspective. In *Proceedings of the XI International Symposium on Biological Control of Weeds*.

Gregoire, G., and H. Chate. 2004. Onset of collective and cohesive motion. *Physical Review Letters* 92.

Hagele, B. F., and M. Rowell-Rahier. 1999. Dietary mixing in three generalist herbivores: nutrient complementation or toxin dilution? *Oecologia* 119:521–533.

Haglund, B. M. 1980. Proline and valine – cues which stimulate grasshopper herbivory during drought stress. *Nature* 288:697–698.

Halkka, A., L. Halkka, O. Halkka, and K. Roukka. 2006. Lagged effects of North Atlantic oscillation on spittlebug *Philaenus spumarius* (Homoptera) abundance and survival. *Global Change Biology* 12: 2250–2262.

Hardie, J., and A. D. Lees. 1985. Endocrine control of polymorphism and polyphenism. Pages 441–490 in G. A. Kerkut and L. I. Gilbert, editors. *Comprehensive Insect Physiology Biochemistry and Pharmacology*. Pergamon Press, Oxford.

Hatcher, M. J., J. T. A. Dick, and A. M. Dunn. 2006. How parasites affect interactions between competitors and predators. *Ecology Letters* 9:1253–1271.

Hawlena, D., and O. J. Schmitz. 2010. Herbivore physiological response to predation risk and implications for ecosystem nutrient dynamics. *Proceedings of the National Academy of Sciences of the USA* 107:15503–15507.

Hecky, R. E., and P. Kilham. 1988. Nutrient limitation of phytoplankton in fresh-water and marine environments – a review of recent-evidence on the effects of enrichment. *Limnology and Oceanography* 33:796–822.

Heron, R. J. 1965. Role of chemotactic stimuli in feeding behavior of spruce budworm larvae on white spruce. *Canadian Journal of Zoology* 43:247–269.

Huberty, A., and R. F. Denno. 2004. Plant water stress and its consequences for herbivorous insects: a new synthesis. *Ecology* 85:1385–1398.

Hunter, D. M. 2004. Advances in the control of locusts (Orthoptera: Acrididae) in eastern Australia: from crop protection to preventive control. *Australian Journal of Entomology* 43:293–303.

Hurd, H. 2001. Host fecundity reduction: a strategy for damage limitation? *Trends in Parasitology* 17:363–368.

Jaksic, F. M., S. I. Solva, P. L. Meserve, and J. R. Gutierrez. 1996. A long-term study of vertebrate predator responses to an El Nino (ENSO) disturbance in western South America. *Oikos* 78:341–354.

Joern, A., and S. T. Behmer. 1997. Importance of dietary nitrogen and carbohydrates to survival, growth, and reproduction in adults of the grasshopper *Ageneotettix deorum* (Orthoptera: Acrididae). *Oecologia* 112:201–208.

Joern, A., and S. T. Behmer. 1998. Impact of diet quality on demographic attributes in adult grasshoppers and the nitrogen limitation hypothesis. *Ecological Entomology* 23:174–184.

Joern, A., and S. Mole. 2005. The plant stress hypothesis and variable responses by blue grama grass (*Bouteloua gracilis*) to water, mineral nitrogen and insect herbivory. *Journal of Chemical Ecology* 31:2069–2090.

Jonas, J. L., and A. Joern. 2007. Grasshopper (Orthoptera: Acrididae) communities respond to fire, bison grazing and weather in North American tallgrass prairie: a long-term study. *Oecologia* 153:699–711.

Jones, C., and J. S. Coleman. 1991. Plant stress and insect herbivory: toward an integrated perspective. Pages 249–280 in H. A. Mooney, D. A. Winner, and E. J. Pell, editors. *Response of Plants to Multiple Stresses*. Academic Press, San Diego.

Kaplan, I., and R. F. Denno. 2007. Interspecific interactions in phytophagous insects revisited: a quantitative assessment of competition theory. *Ecology Letters* 10:977–994.

Karl, D. M., R. Letelier, D. Hebel, L. Tupas, J. Dore, J. Christian, and C. Winn. 1995. Ecosystem changes in the North Pacific Subtropical Gyre attributed to the 1991–92 El Nino. *Nature* 373:230–234.

Koricheva, J., S. Larsson, and E. Haukioja. 1998. Insect performance on experimentally stressed wood plants: a meta-analysis. *Annual Review of Entomology* 43:195–216.

Krause, J., and G. D. Ruxton. 2002. *Living in Groups*. Oxford University Press, Oxford.

Larsson, S. 1989. Stressful times for the plant stress–insect performance hypothesis. *Oikos* 56:277–283.

Laws, A. N. 2009. Density dependent reductions in grasshopper fecundity in response to nematode parasitism. *The Canadian Entomologist*. 141:415–421.

Lee, K. P. 2010. Sex-specific differences in nutrient regulation in a capital breeding caterpillar, *Spodoptera litura* (Fabricius). *Journal of Insect Physiology* 56:1685–1695.

Lee, K. P., S. T. Behmer, and S. J. Simpson. 2006a. Nutrient regulation in relation to diet breadth: a comparison of *Heliothis* sister species and a hybrid. *Journal of Experimental Biology* 209: 2076–2084.

Lee, K. P., J. S. Cory, K. Wilson, D. Raubenheimer, and S. J. Simpson. 2006b. Flexible diet choice offsets protein costs of pathogen resistance in a caterpillar. *Proceedings of the Royal Society B-Biological Sciences* 273:823–829.

Lee, K. P., and C. Roh. 2010. Temperature-by-nutrient interactions affecting growth rate in an insect ectotherm. *Entomologia Experimentalis et Applicata* 136:151–163.

Lee, K. P., S. J. Simpson, and K. Wilson. 2008. Dietary protein-quality influences melanization and immune function in an insect. *Functional Ecology* 22:1052–1061.

Levesque, K. R., M. Fortin, and Y. Mauffette. 2002. Temperature and food quality effects on growth, consumption and post-ingestive utilization efficiencies of the forest tent caterpillar *Malacosoma disstria* (Lepidoptera: Lasiocampidae). *Bulletin of Entomological Research* 92:127–136.

Levin, S. 1999. *Fragile Dominion: Complexity and the Commons*. Perseus Books, Cambridge.

Lindroth, R. L., K. K. Kinney, and C. L. Platz. 1993. Responses of deciduous trees to elevated atmospheric CO_2: productivity, phytochemistry and insect performance. *Ecology* 74:763–777.

Lochmiller, R. L., and C. Deerenberg. 2000. Trade-offs in evolutionary immunology: just what is the cost of immunity? *Oikos* 88:87–98.

Lockwood, J. A. 2005. *Locust: The Devasting Rise and Mysterious Disappearance of the Insect That Shaped the American Frontier*. Basic Books, New York.

Lorch, P. D., G. A. Sword, D. T. Gwynne, and G. L. Anderson. 2005. Radiotelemetry reveals differences in individual movement patterns between outbreak and non-outbreak Mormon cricket populations. *Ecological Entomology* 30:548–555.

Louda, S. M., and S. K. Collinge. 1992. Plant resistance to insect herbivores: a field test of the environmental stress hypothesis. *Ecology* 73:153–169.

Marschner, H. 1995. *Mineral Nutrition of Higher Plants*, 2nd ed. Academic Press, San Diego.

Mattson, W. J. 1980. Herbivory in relation to plant nitrogen-content. *Annual Review of Ecology and Systematics* 11:119–161.

Mattson, W. J., and R. A. Haack. 1987. The role of drought in outbreaks of plant-eating insects. *Bioscience* 37:110–118.

McCabe, G. J., M. A. Palecki, and J. L. Betancourt. 2004. Pacific and Atlantic Ocean influences on multidecadal drought frequency in the United States. *Proceedings of the National Academy of Science USA* 101:4136–4141.

McCallum, H., N. Barlow, and J. Hone. 2001. How should pathogen transmission be modelled? *Trends in Ecology & Evolution* 16:295–300.

McNeil, S., and T. R. E. Southwood. 1978. The role of nitrogen in the development of insect plant relationships. Pages 77–98 *in* J. B. Harborne, editor. *Biochemical Aspects of Plant and Animal Coevolution*. Academic Press, London.

Miller, G. A., F. J. Clissold, D. Mayntz, and S. J. Simpson. 2009a. Speed over efficiency: locusts select body temperatures that favour growth rate over efficient nutrient utilization. *Proceedings of the Royal Society B-Biological Sciences* 276:3581–3589.

Miller, G. A., J. K. Pell, and S. J. Simpson. 2009b. Crowded locusts produce hatchlings vulnerable to fungal attack. *Biology Letters* 5:845–848.

Moore, J. 2002. *Parasites and the Behavior of Animals*. Oxford University Press, Oxford.

Narisu, J. A. Lockwood, and S. P. Schell. 1999. A novel mark–recapture technique and its application to monitoring the direction and distance of local movements of rangeland grasshoppers (Orthoptera: Acrididae) in the context of pest management. *Journal of Applied Ecology* 36:604–617.

Narisu, J. A. Lockwood, and S. P. Schell. 2000. Rangeland grasshopper movement as a function of wind and topography: implications for pest management. *Journal of Orthoptera Research* 9:111–120.

Pener, M. P. and S. J. Simpson. 2009. Locust phase polyphenism: an update. *Advances in Insect Physiology* 36:1–272.

Perkins, M. C., H. A. Woods, J. F. Harrison, and J. J. Elser. 2004. Dietary phosphorus affects the growth of larval *Manduca sexta*. *Archives of Insect Biochemistry and Physiology* 55:153–168.

Petersen, C., H. A. Woods, and J. G. Kingsolver. 2000. Stage-specific effects of temperature and dietary protein on growth and survival of *Manduca sexta* caterpillars. *Physiological Entomology* 25:35–40.

Pitt, W. C. 1999. Effects of multiple vertebrate predators on grasshopper habitat selection: trade-offs due to predation risk, foraging, and thermoregulation. *Evolutionary Ecology* 13:499–515.

Price, P. W. 1991. The plant vigor hypothesis and herbivore attack. *Oikos* 62:244–251.

Raubenheimer, D., and K. Bassil. 2007. Separate effects of macronutrient concentration and balance on plastic gut responses in locusts. *Journal of Comparative Physiology B-Biochemical Systemic and Environmental Physiology* 177:849–855.

Raubenheimer, D., and S. J. Simpson. 1993. The geometry of compensatory feeding in the locust. *Animal Behaviour* 45:953–964.

Raubenheimer, D., and S. J. Simpson. 1999. Integrating nutrition: a geometrical approach. *Entomologia Experimentalis Et Applicata* 91:67–82.

Raubenheimer, D., and S. J. Simpson. 2003. Nutrient balancing in grasshoppers: behavioural and physiological correlates of dietary breadth. *Journal of Experimental Biology* 206:1669–1681.

Raubenheimer, D., S. J. Simpson, and D. Mayntz. 2009. Nutrition, ecology and nutritional ecology: toward an integrated framework. *Functional Ecology* 23:4–16.

Reynolds, A. M., G. A. Sword, S. J. Simpson, and D. R. Reynolds. 2009. Predator percolation, insect outbreaks, and phase polyphenism. *Current Biology* 19:20–24.

Riley, C. V., J. Packard, A. S., and C. Thomas. 1880. *Second Report of the United States Enotomological Commision*. Government Printing Office, Washington, DC.

Roeder, K. A. 2010. Dietary effects on the performance and body composition of the generalist insect herbivore *Heliothis virescens* (Lepidotera: Noctuidae). M.S. Thesis, Texas A&M University, College Station, TX.

Sanchez, N. E., J. A. Onsager, and W. P. Kemp. 1988. Fecundity of *Melanoplus sanguinipes* (F) in 2 crested wheatgrass pastures. *Canadian Entomologist* 120:29–37.

Schindler, D. W. 1977. Evolution of phosphorus limitation in lakes. *Science* 195:260–262.

Schmitz, O. J. 1998. Direct and indirect effects of predation and predation risk in old-field interaction webs. *American Naturalist* 151:327–342.

Schmitz, O. J. 2008. Herbivory from individuals to ecosystems. *Annual Review of Ecology Evolution and Systematics* 39:133–152.

Schoonhoven, L. M., J. J. A. v. Loon, and M. Dicke. 2005. *Insect–Plant Biology*. Oxford University Press, Oxford.

Scriber, J. M., and F. Slansky. 1981. The nutritional ecology of immature insects. *Annual Review of Entomology* 26:183–211.

Simpson, C. L., S. Chyb, and S. J. Simpson. 1990. Changes in chemoreceptor sensitivity in relation to dietary selection by adult *Locusta migratoria*. *Entomologia Experimentalis et Applicata* 56:259–268.

Simpson, S. J., S. James, M. S. J. Simmonds, and W. M. Blaney. 1991. Variation in chemosensitivity and the control of dietary selection behaviour in the locust. *Appetite* 17:141–154.

Simpson, S. J., and D. Raubenheimer. 1993. The central role of the hemolymph in the regulation of nutrient intake in insects. *Physiological Entomology* 18:395–403.

Simpson, S. J., and D. Raubenheimer. 2001. The geometric analysis of nutrient-allelochemical interactions: a case study using locusts. *Ecology* 82:422–439.

Simpson, S. J., D. Raubenheimer, S. T. Behmer, A. Whitworth, and G. A. Wright. 2002. A comparison of nutritional regulation in solitarious- and gregarious-phase nymphs of the desert locust *Schistocerca gregaria*. *Journal of Experimental Biology* 205:121–129.

Simpson, S. J., and C. L. Simpson. 1992. Mechanisms controlling modulation by hemolymph amino acids of gustatory responsiveness in the locust. *Journal of Experimental Biology* 168:269–287.

Simpson, S. J., G. A. Sword, P. D. Lorch, and I. D. Couzin. 2006. Cannibal crickets on a forced march for protein and salt. *Proceedings of the National Academy of Sciences of the USA* 103:4152–4156.

Singer, M. S. 2001. Determinants of polyphagy by a woolly bear caterpillar: a test of the physiological efficiency hypothesis. *Oikos* 93:194–204.

Singer, M. S., and J. O. Stireman. 2003. Does anti-parasitoid defense explain host-plant selection by a polyphagous caterpillar? *Oikos* 100:554–562.

Slansky, F., and J. G. Rodriguez. 1987. *Nutritional Ecology of Insects, Mites, Spiders, and Related Invertebrates*. Wiley, New York.

Slansky, F., and G. S. Wheeler. 1991. Food consumption and utilization responses to dietary dilution with cellulose and water by velvetbean caterpillars, *Anticarsia gemmatalis*. *Physiological Entomology* 16:99–116.

Slansky, F., and G. S. Wheeler. 1992. Caterpillars compensatory feeding response to diluted nutrients leads to toxic allelochemical dose. *Entomologia Experimentalis Et Applicata* 65:171–186.

Srygley, R. B., P. D. Lorch, S. J. Simpson, and G. A. Sword. 2009. Immediate protein dietary effects on movement and the generalised immunocompetence of migrating Mormon crickets *Anabrus simplex* (Orthoptera: Tettigoniidae). *Ecological Entomology* 34:663–668.

Stamp, N. E. 1990. Growth versus molting time of caterpillars as a function of temperature, nutrient concentration and the phenolic rutin. *Oecologia* 82:107–113.

Stenseth, N. C., G. Ottersen, J. W. Hurrell, *et al.* 2003. Studying climate effects on ecology through the use of climate indices: the North Atlantic Oscillation, El Nino Southern Oscillation, the North Atlantic Oscillation and beyond. *Proceedings ot the Royal Society of London B* 270:2087–2096.

Sterner, R. W., and J. J. Elser. 2002. *Ecological Stoichiometry: The Biology of Elements from Molecules to the Biosphere*. Princeton University Press, Princeton.

Sutton, R. T., and D. L. R. Hodson. 2005. Atlantic Ocean forcing of North American and European summer climate. *Science* 309:115–118.

Sword, G. A. 2005. Local population density and the activation of movement in migratory band-forming Mormon crickets. *Animal Behaviour* 69:437–444.

Sword, G. A., P. D. Lorch, and D. T. Gwynne. 2005. Migratory bands give crickets protection. *Nature* 433:703–703.

Telang, A., N. A. Buck, R. F. Chapman, and D. E. Wheeler. 2003. Sexual differences in postingestive processing of dietary protein and carbohydrate in caterpillars of two species. *Physiological and Biochemical Zoology* 76:247–255.

Telang, A., N. A. Buck, and D. E. Wheeler. 2002. Response of storage protein levels to variation in dietary protein levels. *Journal of Insect Physiology* 48:1021–1029.

Telang, A., R. F. Chapman, and D. E. Wheeler. 2000. Sexual differences in protein and carbohydrate utilization by larval tobacco budworm, *Heliothis virescens* (f.). *American Zoologist* 40:1230–1230.

Thompson, S. N. 2000. Pyruvate cycling and implications for regulation of gluconeogenesis in the insect, *Manduca sexta* L. *Biochemical and Biophysical Research Communications* 274:787–793.

Thompson, S. N. 2004. Dietary fat mediates hyperglycemia and the glucogenic response to increased protein consumption in an insect, *Manduca sexta* L. *Biochimica et Biophysica Acta* 1673:208–216.

Toner, J., and Y. H. Tu. 1998. Flocks, herds, and schools: A quantitative theory of flocking. *Physical Review E* 58:4828–4858.

Trumper, S., and S. J. Simpson. 1993. Regulation of salt intake by nymphs of *Locusta migratoria*. *Journal of Insect Physiology* 39:857–864.

Turchin, P. 2003. *Complex Population Dynamics: A Theoretical/Empirical Synthesis*. Princeton University Press, Princeton.

van der Zee, B., S. T. Behmer, and S. J. Simpson. 2002. Food mixing strategies in the desert locust: effects of phase, distance between foods, and food nutrient content. *Entomologia Experimentalis Et Applicata* 103:227–237.

Waring, G. L., and N. S. Cobb. 1992. The impact of plant stress on herbivore dynamics. Pages 167–226 *in* E. A. Bernays, editor. *Insect–Plant Interactions*. CRC Press, Boca Raton, FL.

Watt, A. D. 1992. The relevance of the stress hypothesis to insects feeding on tree foliage. Pages 73–85 *in* S. R. Leather, A. D. Watt, N. J. Mills, and K. F. A. Walters, editors. *Individuals, Populations and Patterns in Ecology*. Intercept Ltd, Andover, Hampshire, UK.

Wheeler, D. 1996. The role of nourishment in oogenesis. *Annual Review of Entomology* 41:407–431.

White, T. C. R. 1984. The abundance of invertebrate herbivores in relation to the availability of nitrogen stressed food plants. *Oecologia* 63:90–105.

White, T. C. R. 1993. *The Inadequate Environment: Nitrogen and the Abundance of Animals*. Springer-Verlag, Berlin.

Woods, H. A., W. F. Fagan, J. J. Elser, and J. F. Harrison. 2004. Allometric and phylogenetic variation in insect phosphorus content. *Functional Ecology* 18:103–109.

Yang, Y., and A. Joern. 1994a. Gut size changes in relation to variable food quality and body-size in grasshoppers. *Functional Ecology* 8:36–45.

Yang, Y. L., and A. Joern. 1994b. Influence of diet quality, developmental stage, and temperature on food residence time in the grasshopper *Melanoplus differentialis*. *Physiological Zoology* 67:598–616.

Zanotto, F. P., S. M. Gouveia, S. J. Simpson, D. Raubenheimer, and P. C. Calder. 1997. Nutritional homeostasis in locusts: is there a mechanism for increased energy expenditure during carbohydrate overfeeding? *Journal of Experimental Biology* 200:2437–2448.

Zanotto, F. P., S. J. Simpson, and D. Raubenheimer. 1993. The regulation of growth by locusts through postingestive compensation for variation in the levels of dietary protein and carbohydrate. *Physiological Entomology* 18:425–434.

2

The Dynamical Effects of Interactions between Inducible Plant Resistance and Food Limitation during Insect Outbreaks

Karen C. Abbott

2.1 Introduction

In a highly influential paper, Hairston *et al.* (1960) presented a series of broad observations and an accompanying set of arguments for what they believed those observations revealed about the regulation of species abundances. One of Hairston *et al.*'s observations in particular has left an enduring mark on the field of insect population ecology (but see Murdoch 1966, Ehrlich and Birch 1967 for critiques): plants are abundant and severe defoliation is rare, suggesting that herbivores are not regulated by bottom-up forces. Research on the population regulation of insect herbivores, therefore, has proceeded primarily along two paths. First, ecologists have sought to identify top-down regulatory forces, such as predators and pathogens. Second, researchers have proposed ways in which bottom-up regulation of insect herbivores can occur even without the widespread

Insect Outbreaks Revisited, First Edition. Edited by Pedro Barbosa, Deborah K. Letourneau and Anurag A. Agrawal.
© 2012 Blackwell Publishing Ltd. Published 2012 by Blackwell Publishing Ltd.

depletion of their food plants. In this chapter, I present results along this second line of inquiry that reveal ways in which insect outbreaks might result from plant–herbivore interactions. I focus primarily on foliage-feeding insects, which have been the subject of much of the work in this area.

Herbivore growth, survival, and reproduction can be affected by both the quantity of food available and also the quality of that food. Each of these aspects of the plant–herbivore interaction can potentially affect insect population dynamics, although this is not necessarily the case. Whether the demographic effects of plant quality or quantity result in outbreaks is an extremely difficult question to answer, and experimentally demonstrating that any particular mechanism is driving such a large-scale pattern may be impossible.

Our knowledge about how plant quantity and quality influence herbivore population dynamics therefore comes from integrating studies done at a hierarchy of scales. Information about how plant traits affect the performance of individual insects generally comes from small-scale controlled experiments. Evidence that plant effects scale up from individual herbivores to populations comes from larger scale experiments and from analyses of population-level observational data. Finally, theoretical population modeling, sometimes in conjunction with time series data, is used to make inferences about long-term population dynamics. Outbreaks are, of course, a population dynamical pattern, and so this latter scale may seem most relevant for evaluating the roles that plants play during insect outbreaks. However, the construction and evaluation of reasonable population dynamic models cannot proceed without the information gathered at those finer scales. Therefore, this chapter emphasizes the hierarchy of spatial and temporal scales, and summarizes results from both individual- and population-level studies that inform our understanding of when and how plant–herbivore interactions can drive insect outbreaks.

2.2 Inducible resistance and insect outbreaks

Even in the presence of lush green foliage, herbivores may not have ready access to high-quality, nutritious food (White 2005). When plants are subjected to herbivory, for instance, they can induce a wide range of responses that reduce their quality by deterring, harming, or even poisoning herbivores. These changes (reviewed in Karban and Myers 1989, Karban and Baldwin 1997) can involve altered nutritional content of tissues, changes to visual or chemical cues used by herbivores to locate plants, morphological feeding deterrents such as spines or trichomes, or an increase in secondary metabolites like proteinase inhibitors and tannins that interfere with digestion. With time, the induced response fades and the plant returns to its pre-herbivory state. The fluctuating nature of a plant's defense against herbivores has led some to hypothesize that insect outbreaks are driven by inducible resistance (Benz 1974, Haukioja *et al.* 1983, Rhoades 1985). Insect densities may build up while plants are relatively undefended, then crash when their intense feeding induces plant responses that are harmful to the insect's fitness.

The effects of inducible resistance on herbivore demographic rates have been measured in many different systems, often by comparing individuals raised on previously damaged plants against those raised on undamaged controls. This type of experiment reveals effects of inducible resistance that act at the level of individual herbivores, and such effects are sometimes quite strong. Herbivores raised on previously grazed plants have been shown to experience slower development and smaller body size (Haukioja 1980, Raupp and Denno 1984, Denno *et al.* 2000, Nykänen and Koricheva 2004), lower fecundity and fertility (Raupp and Denno 1984, Haukioja and Neuvonen 1985, Haukioja *et al.* 1985), and reduced survivorship (Haukioja 1980, Karban 1993).

In order for inducible resistance to drive herbivore outbreaks, several things must be true. First, herbivore population sizes must respond in some way to host plant quality. It is clear that inducible resistance can affect insect performance, but whether these effects result in overall changes in population size is a somewhat separate question (Fowler and Lawton 1985). Other factors could swamp out the influence of inducible resistance, or herbivores may compensate for some detrimental effects during subsequent stages of their development. Making conclusive statements about the effects of inducible resistance on insect population size therefore requires population-level investigations.

Second, the temporal dynamics of induction (i.e., how plant quality changes through time) must be such that the herbivore population experiences large amplitude fluctuations, the peaks of which represent outbreaks. Time lags between the induction of a plant response and its impact on herbivore density should promote such cycles (Lundberg *et al.* 1994, Underwood 1999). Time lags arise, for example, if the insect has non-overlapping generations or if the induced response lasts for multiple insect generations. Additionally, density-dependent induction, where the intensity of the plant's induced response is a function of herbivore density (Underwood 1999, Agrawal and Karban 2000, Lynch *et al.* 2006), results in a tightly coupled consumer–resource system that is prone to sustained fluctuations in population size (Underwood 2000).

2.2.1 Population-level responses to plant quality

Induced plant traits can alter both the strength of density dependence experienced by herbivores and their net population growth rate (Underwood 2000, Underwood and Rausher 2002, Agrawal 2004), as well as their oviposition behavior (Lynch *et al.* 2006). Clearly, the potential exists for plant quality to affect the size and distribution of herbivore populations.

Bioassays, where individual insects are fed plant material from induced versus uninduced plants, have informed much of what we know about how insect growth and fitness respond to inducible resistance. Bioassays are of course not intended to mimic the true feeding behavior of insects, and it is possible that herbivores in natural settings exhibit behaviors that reduce their susceptibility to negative plant effects. If herbivores evade induced plant responses by selectively feeding on undamaged plant tissues, for instance, then these insects might not experience significant demographic effects from induced responses. To examine this possibility, Underwood (2010) conducted experiments on beet armyworm

(*Spodoptera exigua*) larvae that were free to move around individual tomato plants. These insects suffered significant and nonlinear density-dependent effects mediated by the plant's induced response. Because this experiment was conducted on single plants, it serves as an important intermediate case between studies on individual herbivores and studies on whole populations. It provides compelling evidence that individual-level effects can scale up to larger groups, even when accounting for feeding behavior.

There are indeed several examples showing that host plant quality can influence the distribution and dynamics of larger herbivore populations. For instance, allocation to reproduction by the tree cholla cactus is a good predictor of spatial and temporal variation in populations of the herbivorous cactus bug, *Narnia pallidicornis Stål*, that prefers to feed on reproductive structures (Miller *et al.* 2006). Although cactus reproductive allocation does not appear to be an inducible trait (Miller *et al.* 2008), this nonetheless is clear evidence that insect herbivores can respond strongly to plant quality at the scale of the population. Similarly, the majority of temporal variance in the density of a wood-mining herbivorous insect, *Phytobia betulae* Kangas, can be explained by density-independent effects of host tree suitability (Ylioja *et al.* 1999).

Inducible changes in plant quality have also been shown to affect herbivores at the population scale. Although *Phytobia* population dynamics appeared to be mainly driven by density-independent tree effects, Ylioja *et al.* (1999) did find evidence that some of the remaining variation was density dependent. Based on what is known about the *Phytobia*–birch system, the authors concluded that this density dependence was most likely mediated by rapid induced responses.

Population-level effects of induced resistance have been demonstrated experimentally for several herbivores on cotton. Cell content-feeding spider mites, *Tetranychus urticae* Koch and *T. turkestani* Ugarov & Nikolskii, exhibited significantly lower population growth on plants that had been previously exposed to herbivory both in the lab (Karban and Carey 1984) and in the field (Karban 1986). Over the course of a growing season, induced cotton plants supported smaller populations of both spider mites and the phloem-feeding whitefly *Bemisia argentifolii* Bellows & Perring (Agrawal *et al.* 2000). Similarly, leaf-mining caterpillars, *Bucculatrix thurberiella* Busck, raised on previously damaged plants had significantly lower cumulative population sizes than those raised on undamaged plants (Karban 1993). This population-level effect appeared to be due to lower survivorship of individuals raised on induced plants.

2.2.2   Temporal induction dynamics driving insect outbreaks

Reductions in insect performance due to induced resistance can persist across multiple generations (Lynch *et al.* 2006), suggesting that the time scale of an herbivore's response to induction may be sufficient for induced resistance to drive herbivore population dynamics. Furthermore, in a meta-analysis of the insect herbivores of woody plants, Nykänen and Koricheva (2004) reported that induced responses had a much weaker effect on the insect generation imposing the damage than on the following generation. This indicates a time lag between induction and the bulk of the herbivore's population-level response to induction and thus provides an indirect clue that inducible resistance may drive outbreaks. Similarly, Haukioja (1980) documented a much slower decay of induced resistance in northern than in southern Fennoscandian birch trees. This might explain why only northern populations of the birch-eating autumnal moth, *Epirrita autumnata* (Borkhausen), exhibit outbreaks, although more recent work attributes the pattern to greater predator specialization in the north (Klemola *et al*. 2002).

Plant–herbivore theory gives us further insight into when and how outbreaks may result from inducible resistance. Early studies showed that if herbivorous insects have limited mobility or an Allee effect (positive density dependence at low density), then insect outbreaks can theoretically be driven by inducible resistance alone (Edelstein-Keshet 1986, Edelstein-Keshet and Rausher 1989). These results are valuable not only for demonstrating that inducible resistance-driven outbreaks are possible but also for suggesting that for many insect life histories, such outbreaks are not expected.

Edelstein-Keshet's foundational models operate in continuous time, meaning that both the insect population size and the level of induction can respond to one another instantaneously. Continuous-time models work well for describing populations with overlapping generations and no clear seasonality. For species with seasonal life histories and non-overlapping generations, such as many temperate insects, discrete-time models give a more appropriate representation of their population dynamics. This distinction has important implications that reach beyond the mathematical details of the models. Discrete-time systems include a built-in time lag because forces acting in one time step affect the population size in the following time step (Turchin 1990). Because of this time lag, species with non-overlapping generations might be more prone to outbreak (May 1974).

Indeed, discrete-time models for inducible resistance do appear to show outbreaks under much less limited conditions than Edelstein-Keshet's continuous-time models. A discrete-generation simulation model based on Edelstein-Keshet and Rausher's (1989) conceptual framework reveals that inducible resistance-driven outbreaks occur under a surprisingly broad range of assumptions about the nature of the inducible response and the insect's mobility and selectivity (Underwood 1999). As expected, outbreaks were even more likely when the time between herbivory and induction was longer.

Other discrete-time models with structures less similar to Edelstein-Keshet and Rausher's (1989) also readily exhibit outbreak dynamics. A discrete-time model for the interaction between larch budmoth and the quality of larch needles shows high-amplitude budmoth cycles with a realistic outbreak period (Turchin *et al*. 2003). Outbreaks likewise occurred in a discrete-time model in which the maximum insect survival rate was lower on damaged plants, provided that herbivory had a long-term impact on the plant population (Abbott *et al*. 2008).

Even for populations with overlapping generations, outbreaks might be more common in nature than is suggested by the deterministic models that are usually implemented for studying theoretical population dynamics. For instance, Lundberg *et al*. (1994) concluded that inducible resistance results in damped deterministic oscillations in herbivore abundance. If stochastic perturbations occur frequently enough, these oscillations will not be expected to fully dampen and the population will appear to undergo sustained oscillations. Likewise, when multiple herbivores feed on the same host plant, deterministic models predict stable, equilibrial herbivore dynamics (with either competitive exclusion or coexistence, depending on the specific model assumptions; Anderson *et al.* 2009). However, transient outbreaks can occur if the induction dynamics are slower than the herbivore population dynamics, opening the possibility that the observed outbreaks of some insects might be transient dynamics maintained by environmental stochasticity.

2.3   Food limitation and insect outbreaks

Unlike the work described in section 2.2 on inducible resistance, the hypothesis that insect population growth might be limited simply by food availability does not readily align with Hairston *et al.*'s (1960) observation that the world is green. If edible plant material is abundant, then how can herbivores be food limited? This question has led many to conclude that reduction in plant quality, but not quantity, is the only realistic hypothesis for how herbivore dynamics might be driven by plant–herbivore interactions. Nonetheless, plant productivity has a fundamental influence over populations at higher trophic levels (Power 1992) and instances of severe food limitation (i.e., regulation of herbivores by the quantity of plant material independent of its quality) have indeed been documented in insects (Monro 1967, White 1974). The challenge is to reconcile these observations with the expectation that severe defoliation is uncommon.

2.3.1   Individual- and population-level responses to food limitation

Reasonably, populations of starving insects suffer reduced population growth (e.g., Muthukrishnan and Delvi 1974). However, it is not obvious whether starvation will occur in most natural settings or whether inducible plant responses will prevent herbivores from ever consuming enough plant tissue for starvation to become a significant threat. Morris (1997) investigated this question in Colorado potato beetles, *Leptinotarsa decemlineata* Say, by raising the beetles on potato plants that had been exposed to a range of intensities of prior herbivory. By fitting models to disentangle the effects of food limitation from the effects of induced plant defenses in the data, Morris found that food limitation alone could explain the lower survivorship of beetles on more severely damaged plants. Although inducible defenses had some impact on another beetle trait (mass at pupation), the effects of food limitation were generally much stronger than the effects of inducible defenses on beetle performance. Because these were short-term experiments, this study does not directly address the role of food limitation in driving longer term insect dynamics like outbreaks. Nonetheless, the clear effect of food limitation on performance suggests that inquiries at longer time scales are warranted (Morris 1997).

Another case where food limitation may be important comes from the leaf-feeding beetles, *Trirhabda virgata*, *T. borealis*, and *T. canadensis*, which are outbreaking herbivores on goldenrod (Brown and Weis 1995, Herzig 1995). Effects of plant nutritional quality on *Trirhabda* performance appear to be overwhelmed by the much stronger effects of the quantity of food consumed by individuals (Brown and Weis 1995).

Food limitation appears to occur in other systems at the population scale as well. Harrison (1994) documented strong population-level food limitation in a manipulative experiment on tussock moth, *Orgyia vetusta*. Similarly, a statistical analysis of time series data for the insect herbivores of English oak showed that the spatial distribution of winter moth, *Operophtera brumata*, and the temporal dynamics of the green oak tortrix, *Tortrix viridana*, were predominantly explained by the quantity of food available (Hunter *et al.* 1997).

2.3.2   Food limitation driving insect outbreaks

Intense defoliation of ragwort by the cinnabar moth, *Tyria jacobaeae*, can lead to severe food limitation in subsequent generations (Dempster 1971). Whether outbreaks result, however, depends on other factors affecting plant dynamics. In habitats where ragwort recruitment is strongly limited by space availability, plant dynamics do not respond significantly to moth herbivory and cycles do not occur. In contrast, outbreaks are observed at sites with negligible space limitation and tightly coupled plant–herbivore dynamics (Bonsall *et al.* 2003).

Studies implicating food limitation in insect outbreaks are often grounded in theory, much like the work summarized in section 2.2.2 linking inducible resistance to herbivore dynamics. For example, models can be used to show that many features of *Trirhabda* population dynamics are consistent with the hypothesis that food limitation drives their outbreaks (Abbott and Dwyer 2007). Furthermore, *Trirhabda* outbreaks are unlikely to be caused by interactions with natural enemies (Messina 1983), and while the importance of inducible resistance in this system is relatively unknown, the clear effect of food limitation on performance (Brown and Weis 1995) makes it a strong candidate for the driver of *Trirhabda* outbreaks.

Models of consumer–resource interactions are notoriously unstable, in that the simplest models exhibit population cycles that may even expand in amplitude until extinction (Murdoch *et al.* 2003). For species involved in a tight relationship with their resources (or their consumers), it is actually their persistence, rather than their outbreaks, that usually begs explanation. From a theoretical perspective, therefore, we would expect the interaction between food-limited herbivores and their host plants to readily result in outbreaks. However, our knowledge of consumer–resource systems comes almost exclusively from predator–prey and host–parasitoid models, and these may be particularly inappropriate for plant–insect systems (Caughley and Lawton 1981, Abbott and Dwyer 2007). Whether food limitation results in outbreak dynamics that are realistic for insect herbivores, therefore, is not readily apparent from predator–prey and host–parasitoid theory.

An early argument against the importance of food limitation in driving insect outbreaks came from the visual inspection of time series data on four pine-feeding lepidopterans (Varley 1949). Some of the outbreaks ended at notably lower peak densities than others. Because larger insect densities were possible during other, larger outbreaks, this observation was interpreted as evidence that the smaller outbreaks must have ended before the food supply was exhausted. Varley also noted that in some years, all four insect species were at low densities, suggesting that there was surplus food not being exploited. Although these claims have intuitive appeal, and the conclusions may ultimately be correct, verbal arguments about complex nonlinear processes can be dangerously misleading. Without a theoretical framework, it is not straightforward to know what the temporal dynamics of a food-limited herbivore would look like and thus whether the pine lepidopteran data truly deviate from these expectations.

One problem with attempting to evaluate the potential for food limitation in herbivorous insects is that we do not actually know how green (and undefended) the world must be before we can conclude that food for herbivores is effectively unlimited and therefore incapable of driving herbivore dynamics. Collectively, herbivores consume 10–20% of annual plant production in typical communities (Bazzaz *et al.* 1987). We might, then, consider plants to be experiencing unusually intense defoliation when consumption significantly exceeds those levels. If we call a plant population "defoliated" anytime that total edible plant biomass is at least 25% below its maximum, then the host plants of some outbreaking insects are in fact defoliated as little as 10–20% of the time on average (Abbott and Dwyer 2007). That means that 80–90% of the time, the host plants even of notorious pests can have abundant foliage, just as Hairston *et al.* (1960) proclaim. Clearly, the actual severity of food limitation will depend on both food availability and herbivore density, but the historical argument contends that foliage is typically so abundant that no realistic herbivore population would find it inadequate. So, is lush foliage 80–90% of the time too "green" for food limitation to possibly be driving insect dynamics?

A suite of simple plant–herbivore models suggest that the answer to this question is no (Abbott and Dwyer 2007). In these models, herbivore population dynamics are driven entirely by the availability of food, such that per capita population growth increases with edible plant biomass up to some maximum. The models can show a range of dynamics depending on the parameter values used. For parameter values leading to insect outbreaks, host plant defoliation occurred as little as 10% of the time, in agreement with data. These models do not provide any evidence that insect outbreaks *are* caused by food limitation, but they clearly demonstrate that food-limited outbreaking insects inflict levels of defoliation that are completely consistent with our expectations for a green earth. This gives new validity to the hypothesis that food limitation can drive insect outbreaks.

Other plant–herbivore models involving food limitation also exhibit outbreaks. Allen *et al.* (1993) built a model for the apple twig borer, *Amphicerus bicaudatus*, on grape. Because it was built to describe the dynamics of a specific system, this model is considerably more complicated than those analyzed by Abbott and Dwyer (2007); for example, Allen *et al.* considered a growing season that was divided into three distinct phases. Despite these differences, Allen *et al.* also concluded that insect fluctuations could be attributed to a plant–herbivore interaction characterized by food limitation. Using a model to identify scenarios that result in successful biocontrol by food-limited insect herbivores, Buckley *et al.* (2005) found several traits that yield unstable population dynamics, such as weak plant density dependence, a large impact of herbivory, plant compensatory growth, and strong competition among insects. Taken together, these studies show that food-limited herbivores exhibit outbreaks under a broad range of biological assumptions.

2.4   Interactive effects of inducible resistance and food limitation

Effects of plant quality and plant quantity on insect dynamics have rarely been considered together. This is unfortunate, because herbivory should cause changes in both the quantity and quality of the plant material that will be encountered by later-feeding herbivores. From the studies reviewed above, it is clear that each of these effects can potentially drive dramatic fluctuations in insect density. However, it is generally poorly understood whether food limitation and inducible resistance when acting together are more or less likely to result in insect outbreaks than they are when acting singly.

The issue of whether plant quality and quantity can jointly drive outbreaks is not as intuitive as it may at first appear. We have two factors that occur simultaneously and can each cause outbreaks, so why can't we automatically conclude that their combined effect will also be herbivore outbreaks? One very general reason is that nonlinear processes, by virtue of their nonlinearity, do not usually "add up" in a straightforward way. With regard to the current context, we know that heterogeneity in a resource species' susceptibility to consumption can be stabilizing (May 1978, Abrams and Walters 1996, Verschoor *et al.* 2004). Variable levels of induction could potentially generate such heterogeneity within the plant population, opening the possibility that food limitation no longer drives herbivore cycles (but see Adler and Karban 1994). Therefore, understanding what dynamics actually result from the combined action of inducible resistance and food limitation requires research efforts that examine both factors together.

Few studies have measured both plant quantity and quality in a plant–herbivore interaction simultaneously. Morris (1997) found that Colorado potato beetles had lower mass at pupation due to the combined effects of induced resistance and reduced food availability on previously damaged plants. This suggests the possibility of an interaction between inducible resistance and food limitation at the population scale. More work is needed to understand whether this type of quantity-quality interaction is common to other species, and to assess the extent to which such interactions really do affect populations.

Several classic models for forest insect outbreaks include effects of both plant quality and quantity, although the potential for interactive effects on herbivore dynamics was not explicitly studied. For instance, Ludwig *et al.* (1978) presented a model for spruce budworm that included separate variables for tree size and foliage condition. In their model, foliage condition affected tree growth but not budworm density; however, the authors state that allowing foliage condition to directly affect the budworm resulted in no qualitative change in the model's behavior as long as the condition was never very poor. This suggests no interaction between quantity and quality, although it is not immediately apparent whether it is realistic to expect foliage condition not to drop too low when budworms are at peak density.

A second well-known forest insect model to include effects of both food availability and food quality is the simulation model for larch budmoth presented in Fischlin and Baltensweiler (1979). Although they concluded that plant–insect interactions were a likely contributor to budmoth outbreaks, their analysis did not allow a distinction to be made between the effects of food quantity and the effects of food quality. Furthermore, the model assumed a linear relationship between the abundance of needles and their raw fiber content (i.e., their quality), for which they cite empirical evidence from Omlin (1977). The simplicity of this relationship eliminates the possibility of a nonlinear quantity–quality interaction. The same is true of Turchin *et al.*'s (2003) larch budmoth model, in which the state variable representing plant quality is a linear function of larch needle length. Although the cycles in that model were attributed to induced changes in plant quality, the tight relationship between the quality and availability of needles means that the model's behavior could just as readily be attributed to the effects of food limitation. Nonetheless, the linear quantity–quality relationship precludes more complex interactive effects of inducible resistance and food limitation in the model.

Though justified empirically for larch (Omlin 1977), there is no clear reason to expect that a linear relationship between the level of induction and the intensity of food limitation is a general phenomenon. Interactive effects of plant quality and quantity may therefore occur. Since we know that plant quality and quantity can each individually drive herbivore outbreaks, the important question here is whether these interactive effects would amplify or damp outbreaks.

Population dynamic models that combine the effects of food limitation and inducible resistance on herbivore population dynamics reveal that insect outbreaks can occur under circumstances where neither food limitation alone, nor inducible resistance alone, causes outbreaks (Abbott *et al.* 2008). This effect is particularly striking in models for crop pests and other insects that feed on plants whose abundance and quality at the start of each growing season are determined by factors besides past herbivory. In these systems, neither food limitation nor inducible resistance can drive herbivore fluctuations when acting singly. Instead, these factors each maintain the herbivores at a stable equilibrium. However, insect outbreaks do occur in the models when both factors are present. This result plainly shows that the combined effects of plant quantity and plant quality on herbivore populations can deviate greatly from the sum of the individual effects.

This dramatic result is clearly somewhat specific to insects like crop pests, because we know from the studies summarized in this chapter that for plant–herbivore systems in which plant populations do respond to past herbivory, outbreaks can result from either food limitation alone or inducible resistance alone. Nonetheless, we see an interactive effect in these types of plant–herbivore systems as well. Although food limitation and inducible resistance can each drive outbreaks, they do not invariably do so. Outbreaks result in these models under a particular range of parameter values, representing a particular range of biological assumptions about the system. For systems where plants respond to past herbivory, Abbott *et al.* (2008) showed that food limitation and inducible resistance can jointly drive

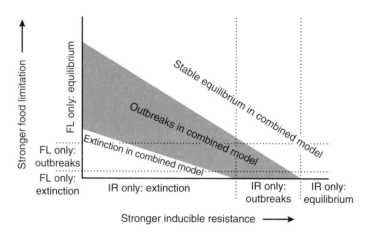

Figure 2.1 Synergistic effects of food limitation and inducible resistance. The labeled areas on the graph show the behaviors of a combined food limitation and inducible resistance model. Text along the *x*-axis shows the behavior of the model when inducible resistance (IR) is acting alone (i.e., the strength of food limitation is zero); similarly, text along the *y*-axis shows the behavior when food limitation (FL) acts alone. Outbreaks occur in the combined model for many parameter combinations that do not yield outbreaks in either single-factor model. Details of the models can be found in Abbott *et al.* (2008).

outbreaks for parameter values where neither factor alone would cause outbreaks (Figure 2.1). Again, this reveals that insect outbreaks occur under a new range of conditions when food limitation and inducible resistance act together.

These results suggest that plant-driven outbreaks might be much more common in nature than we would expect from studying plant effects one at a time. Whether this synergy occurs in real populations remains unknown, however, and further empirical work on both food limitation and inducible resistance at the population scale would be very informative.

2.5 A note on multiple drivers of outbreaks

The recognition that food limitation and inducible resistance occur simultaneously and may jointly drive outbreaks raises the broader point that in most natural systems, there are likely to be multiple factors driving population dynamics. Although regulation by natural enemies and regulation by host plants might appear to be competing hypotheses for what drives herbivore dynamics, it is probably the case that both factors act simultaneously in many systems. For example, in their study to evaluate the roles of foliage quality and parasitism in larch budmoth outbreaks, Turchin *et al.* (2003) ultimately concluded that the data were best described by a model that included the two factors together.

For some extinction-prone herbivore–predator systems, food limitation in addition to predation may be necessary to explain herbivore dynamics (May 1973). Ludwig *et al.* (1978) and McNamee *et al.* (1981) used graphical models to suggest that outbreaks are initiated when abundant foliage allows herbivores to temporarily escape suppression by enemies. The outbreak ends when the foliage has been consumed and the herbivore dies back to the initial low-density state maintained by natural enemy attack. Thus, food limitation might theoretically play a role in driving outbreaks even in those insects that have important associations with natural enemies.

2.6　Future directions

Clearly more work is needed to understand the interaction between food limitation and inducible resistance. In addition to interactive effects at the population scale (as in Abbott *et al.* 2008), these factors might interact within individual plants. Plant density can affect a plant's ability to induce herbivore resistance (Baldwin 1988, Karban *et al.* 1989), perhaps leading to an interactive effect of plant density and inducible resistance on herbivore population size (Karban 1993). This raises an interesting question about how to think about plant density. Many studies of food limitation consider the total amount of edible biomass available in the plant population, without regard to how many individual plants make up that biomass. However, if inducible resistance depends on the density of individuals rather than the density of total biomass, then herbivore populations might behave quite differently if they are exposed to high biomass that is due to a crowded plant population than if they are exposed to high biomass due to fewer, larger plants. The implications for this distinction on herbivore population dynamics would be interesting to explore.

Inducible resistance may have a strong impact on herbivore populations only if many local plants are induced. This leads naturally to the question of the spatial scale of induction relative to the scale of herbivore foraging. Some aspects of how space influences plant–herbivore interactions have been explored. Plant defenses may affect herbivore spatial spread (Morris and Dwyer 1997) and spatial distribution (Lewis 1994, Underwood *et al.* 2005). Plant–herbivore cycles due to food limitation can occur when the mechanism of change in local herbivore density is movement rather than reproduction (Nisbet *et al.* 1997). Generally speaking, however, there is a lot left to learn about the role of space and herbivore movement in plant-driven outbreaks.

The length and breadth of this book should immediately hint to the reader that researchers are far from narrowing in on a single causal mechanism for insect outbreaks. This is probably because no such single factor exists (Hunter and Price 1992, Walker and Jones 2001). A consideration of how plant–herbivore pairs fit within a larger ecological community should therefore be illuminating. The presence of a lower (Grover and Holt 1998) or higher (Lawton and McNeill 1979, Price *et al.* 1980, Ode 2006) trophic level is likely to influence plant–herbivore dynamics. Tritrophic models exist, but they typically include only two sets of direct interactions: say, plant–herbivore and herbivore-predator. Some plants release volatile compounds in response to herbivory that attract the herbivore's natural enemies (Turlings *et al.* 1995, DeMorales *et al.* 1998, Walling 2000). Others create specialized structures to house and protect species that prey on their herbivores (Marquis and Whelan 1996, Agrawal *et al*. 2000). Furthermore, some herbivores sequester plant defensive compounds that then affect the herbivore's natural enemies (Nishida 2002). These suggest an additional important interaction between plants and predators that could have a large dynamical effect.

Understanding how plant–herbivore interactions influence community structure is another promising avenue of research (Denno *et al*. 1995, 2000, Anderson *et al*. 2009). The complex and sometimes delayed nature of inducible responses means that herbivore species that feed at different times and on different plant structures may nonetheless experience strong interspecific competition (Lynch *et al*. 2006), suggesting that a simple application of niche theory to herbivore communities might be misleading (Kaplan and Denno 2007). Theory that accounts for the unique ways in which insect species compete for resources, in addition to being insightful in its own right (Anderson *et al*. 2009), may help identify the mechanisms underlying multispecies outbreaks (e.g., lepidopterans on *Pinus sylvestris*; Varley 1949).

Complementing the studies mentioned in this chapter, there is a literature on the interactions between vertebrate grazers and plants. The life histories of vertebrate and invertebrate herbivores are sufficiently different that it is unclear how readily ideas about vertebrate population dynamics translate to insect systems. However, models for food-limited mammalian herbivores show that inducible resistance can increase the likelihood of outbreaks (Feng *et al*. 2008, Liu *et al*. 2008), in general agreement with plant–insect herbivore theory (Abbott *et al*. 2008). Also in agreement with theory for insects (Abbott and Dwyer 2007), but in contrast to the conventional notion of what it means to live in a green world, food-limited vertebrate grazers do not necessarily cause a dramatic reduction in the populations of their host plants (van de Koppel *et al*. 1996). Perhaps some of the ideas for what drives cycles in vertebrate herbivores (e.g., Fox and Bryant 1984) can be carefully applied to outbreaking insect species as well. However, the high tolerance to herbivory that is exhibited by many grazer-adapted plants (McNaughton 1979; see also Kotanen and Rosenthal 2000), and the likely effect of tolerance on dynamics (Buckley *et al*. 2005), may further limit the applicability of vertebrate grazer models to many insect systems.

2.7 Concluding remarks

The empirical information presented in this chapter clearly shows the potential for plant-driven insect outbreaks to occur, since insects respond to changes in both plant quantity and quality at the population level. The theoretical studies clearly demonstrate that these types of responses can indeed result in outbreaks without violating our general expectations about the characteristics of natural populations. However, direct evidence that insect outbreaks *are* driven by plant–herbivore interactions, and especially by food limitation, is largely absent. This is surely due in part to the difficulty of conducting conclusive causal studies at large spatial and temporal scales. Of course, researchers investigating other hypotheses for insect outbreaks are confronted with the same challenge, so this cannot be the full explanation for why evidence of plant-driven outbreaks has lagged.

Another part of the explanation might simply be that such evidence does not exist because insect outbreaks are never caused by interactions with the host plant. This is certainly possible, but if it is true then a very important question is raised: why not? The work summarized above leads us to expect plant–herbivore cycles under a range of conditions that should be met by some populations of insects. It may instead be the case, therefore, that our evidence for plant-driven outbreaks is currently limited because relatively few researchers have been looking for it. Influential arguments by Varley (1949) and Hairston *et al.* (1960) for the unimportance of plants in herbivore dynamics likely discouraged many later ecologists from seriously considering plant effects or from accepting results of those who did. Now that we are armed with a much better idea of how plant–herbivore interactions scale up to the population level, it is clear that the story is not yet complete. The imperative now is to either bolster our evidence for plant-driven outbreaks or explain their surprising absence.

References

Abbott, K. C., and Dwyer, G. 2007. Food-limitation and insect outbreaks: complex dynamics in plant–herbivore interactions. *Journal of Animal Ecology* 76:1004–1014.

Abbott, K. C., Morris, W. F., and Gross, K. 2008. Simultaneous effects of food limitation and inducible resistance on herbivore population dynamics. *Theoretical Population Biology* 73:63–78.

Abrams, P. A., and Walters, C. J. 1996. Invulnerable prey and the paradox of enrichment. *Ecology* 77:1125–1133.

Adler, F. R., and Karban, R. 1994. Defended fortresses or moving targets? Another model of inducible defenses inspired by military metaphors. *American Naturalist* 144:813–832.

Agrawal, A. A. 2004. Plant defense and density dependence in the population growth of herbivores. *American Naturalist* 164:113–120.

Agrawal, A. A., and Karban, R. 2000. Specificity of constitutive and induced resistance: pigment glands influence mites and caterpillars on cotton plants. *Entomologia Experimentalis et Applicata* 96:39–49.

Agrawal, A. A., Karban, R., and Colfer, R. G. 2000. How leaf domatia and induced plant resistance affect herbivores, natural enemies and plant performance. *Oikos* 89:70–80.

Allen, L. J. S., Hannigan, M. K., and Strauss, M. J. 1993. Mathematical analysis of a model for a plant–herbivore system. *Bulletin of Mathematical Biology* 55:847–864.

Anderson, K. E., Inouye, B. D., and Underwood, N. 2009. Modeling herbivore competition mediated by inducible changes in plant quality. *Oikos* 118:1633–1646.

Baldwin, I. T. 1988. Damage-induced alkaloids in tobacco: pot-bound plants are not inducible. *Journal of Chemical Ecology* 14:1113–1120.

Bazzaz, F. A., Chiariello, N. R., Coley, P. D., and Pitelka, L. F. 1987. Allocating resources to reproduction and defense. *BioScience* 37:58–67.

Benz, G. 1974. Negative feedback by competition for food and space, and by cyclic induced changes in the nutritional base as regulatory principles in the population dynamics of the larch budmoth, *Zeiraphera diniana*. *Journal of Applied Entomology* 76:196–228.

Bonsall, M. B., van der Meijden, E., and Crawley, M. J. 2003. Contrasting dynamics in the same plant–herbivore interaction. *Proceedings of the National Academy of Sciences* 100:14932–14936.

Brown, D. G., and Weis, A. E. 1995. Direct and indirect effects of prior grazing of goldenrod upon the performance of a leaf beetle. *Ecology* 76:426–436.

Buckley, Y. M., Rees, M., Sheppard, A. W., and Smyth, M. J. 2005. Stable coexistence of an invasive plant and biocontrol agent: a parameterized coupled plant–herbivore model. *Journal of Applied Ecology* 42:70–79.

Caughley, G., and Lawton, J. H. 1981. Plant–herbivore systems. Pages 132–166 *in* R. M. May, editor. *Theoretical Ecology: Principles and Applications*. Sinauer Associates Inc., Sunderland, MA.

DeMorales, C. M., Lewis, W. J., Pare, P. W., Alborn, H. T., and Tumlinson, J. H. 1998. Herbivore-infested plants selectively attract parasitoids. *Nature* 393:570–573.

Dempster, J. P. 1971. The population ecology of the cinnabar moth, *Tyria jacobaeae* L. (Lepidoptera, Arctiidae). *Oecologia* 7:26–67.

Denno, R. F., McClure, M. S., and Ott, J. R. 1995. Interspecific interactions in phytophagous insects: competition reexamined and resurrected. *Annual Review of Entomology* 40:297–331.

Denno, R. F., Peterson, M. A., Gratton, C., Cheng, J., Langellotto, G. A., Huberty, A. F., and Finke, D. L. 2000. Feeding-induced changes in plant quality mediate interspecific competition between sap-feeding herbivores. *Ecology* 81:1814–1827.

Edelstein-Keshet, L. 1986. Mathematical theory for plant–herbivore systems. *Journal of Mathematical Biology* 24:25–58.

Edelstein-Keshet, L., and Rausher, M. D. 1989. The effects of inducible plant defenses on herbivore populations. I. Mobile herbivores in continuous time. *American Naturalist* 133:787–810.

Ehrlich, P. R., and Birch, L. C. 1967. The "balance of nature" and "population control." *American Naturalist* 101:97–107.

Fischlin, A., and Baltensweiler, W. 1979. Systems analysis of the larch bud moth system. Part 1. The larch–larch bud moth relationship. *Bull. Soc. Entomol. Suisse* 52:273–289.

Feng, Z., Liu, R., and DeAngelis, D. L. 2008. Plant–herbivore interactions mediated by plant toxicity. *Theoretical Population Biology* 73:449–459.

Fowler, S. V., and Lawton, J. H. 1985. Rapidly induced defenses and talking trees: the devil's advocate position. *American Naturalist* 126:181–195.

Fox, J. F., and Bryant, J. P. 1984. Instability of the snowshoe hare and woody plant interaction. *Oecologia* 63:128–135.

Grover, J. P., and Holt, R. D. 1998. Disentangling resource and apparent competition: realistic models for plant–herbivore communities. *Journal of Theoretical Biology* 191:353–376.

Hairston, N. G., Smith, F. E., and Slobodkin, L. B. 1960. Community structure, population control, and competition. *American Naturalist* 94:421–425.

Harrison, S. 1994. Resources and dispersal as factors limiting a population of the tussock moth (*Orgyia vetusta*), a flightless defoliator. *Oecologia* 99:27–34.

Haukioja, E. 1980. On the role of plant defenses in the fluctuations of herbivore populations. *Oikos* 35:202–213.

Haukioja, E., Kapiainen, P., Niemela, P., and Tuomi, J. 1983. Plant availability hypothesis and other explanations of herbivore cycles: complimentary or exclusive alternatives? *Oikos* 40:419–432.

Haukioja, E., and Neuvonen, S. 1985. The relationship between male size and reproductive potential in *Epirrita autumnata* (Lep., Geometridae). *Ecological Entomology* 10:267–270.

Haukioja, E., Suomela, J., and Neuvonen, S. 1985. Long-term inducible resistance in birch foliage: triggering cues and efficacy on a defoliator. *Oecologia* 65:363–369.

Herzig, A. 1995. Effects of population density on long-distance dispersal in the goldenrod beetle *Trirhabda virgata*. *Ecology* 76:2044–2054.

Hunter, M. D., and Price, P. W. 1992. Playing chutes and ladders: heterogeneity and the relative roles of bottom-up and top-down forces in natural communities. *Ecology* 73:724–732.

Hunter, M. D., Varley, G. C., and Gradwell, C. R. 1997. Estimating the relative roles of top-down and bottom-up forces on insect herbivore populations: a classic study revisited. *Proceedings of the National Academy of Sciences* 94:9176–9181.

Kaplan, I., and Denno, R. F. 2007. Interspecific interactions in phytophagous insects revisited: a quantitative assessment of competition theory. *Ecology Letters* 10:977–994.

Karban, R. 1986. Induced resistance against spider mites in cotton: field verification. *Entomologia Experimentalis et Applicata* 42:239–242.

Karban, R. 1993. Induced resistance and plant density of a native shrub, *Gossypium thurberi*, affect its herbivores. *Ecology* 74:1–8.

Karban, R., and Baldwin, I. T. 1997. *Induced Responses to Herbivory*. University of Chicago Press, Chicago, IL.

Karban, R., and Carey, J. T. 1984. Induced resistance of cotton seedlings to mites. *Science* 255:53–54.

Karban, R., Brody, A. K., and Schnathorst, W. C. 1989. Crowding and a plant's ability to defend itself against herbivores and diseases. *American Naturalist* 134:749–760.

Karban, R., and Myers, J. H. 1989. Induced plant responses to herbivory. *Annual Review of Ecology and Systematics* 20:331–348.

Klemola, T., Tunhuanpää, M., Korpimäki, E., and Ruohomäki, K. 2002. Specialist and generalist natural enemies as an explanation for geographical gradients in population cycles of northern herbivores. *Oikos* 99:83–94.

Kotanen, P. M., and Rosenthal, J. P. 2000. Tolerating herbivory: does the plant care if the herbivore has a backbone? *Evolutionary Ecology* 14:537–549.

Lawton, J. H., and McNeill, S. 1979. Between the devil and the deep blue sea: on the problem of being a herbivore. Pages 223–244 *in* R. M. Anderson, B. D. Turner, and L. R. Taylor, editors. *Population Dynamics*. Blackwell Scientific Publications, Oxford.

Lewis, M. A. 1994. Spatial coupling of plant and herbivore dynamics: the contribution of herbivore dispersal to transient and persistent waves of damage. *Theoretical Population Biology* 45:277–312.

Liu, R., Feng, Z., Zhu, H., and DeAngelis, D. L. 2008. Bifurcation analysis of a plant–herbivore model with toxin-determined functional response. *Journal of Differential Equations* 245:442–467.

Ludwig, D., Jones, D. D., and Holling, C. S. 1978. Qualitative analysis of insect outbreak systems: the spruce budworm and forest. *Journal of Animal Ecology* 47:315–332.

Lundberg, S., Jaremo, J., and Nilsson, P. 1994. Herbivory, inducible defence and population oscillations: a preliminary theoretical analysis. *Oikos* 71:537–540.

Lynch, M. E., Kaplan, I., Dively, G. P., and Denno, R. F. 2006. Host-plant-mediated competition via induced resistance: interactions between pest herbivores on potatoes. *Ecological Applications* 16:855–864.

Marquis, R. J., and Whelan, C. 1996. Plant morphology and recruitment of a third trophic level: subtle and little-recognized defenses? *Oikos* 75:330–334.

May, R. M. 1973. Time-delay versus stability in population models with two and three trophic levels. *Ecology* 54:315–325.

May, R. M. 1974. Biological populations with nonoverlapping generations: stable points, stable cycles, and chaos. *Science* 186:645–647.

May, R. M. 1978. Host–parasitoid systems in patchy environments: a phenomenological model. *Journal of Animal Ecology* 47:833–843.

McNamee, P. J., McLeod, J. M., and Holling, C. S. 1981. The structure and behavior of defoliating insect/forest systems. *Researches on Population Ecology* 23:280–298.

McNaughton, S. J. 1979. Grazing as an optimization process: grass–ungulate relationships in the Serengeti. *American Naturalist* 113:691–703.

Messina, F. J. 1983. Parasitism of two goldenrod beetles (Coleoptera: Chrysomelidae) by *Aplomyiopsis xylota* (Diptera: Tachinidae). *Environmental Entomology* 12:807–809.

Miller, T. E. X., Tyre, A. J., and Louda, S. M. 2006. Plant reproductive allocation predicts herbivore dynamics across spatial and temporal scales. *American Naturalist* 168:608–616.

Miller, T. E. X., Tenhumberg, B., and Louda, S. M. 2008. Herbivore-mediated ecological costs of reproduction shape the life history of an iteroparous plant. *American Naturalist* 171:141–149.

Monro, J. 1967. The exploitation and conservation of resource by populations of insects. *Journal of Animal Ecology* 36:531–547.

Morris, W. F. 1997. Disentangling effects of induced plant defenses and food quantity on herbivores by fitting nonlinear models. *American Naturalist* 150:299–327.

Morris, W. F., and Dwyer, G. 1997. Population consequences of constitutive and inducible plant resistance: herbivore spatial spread. *American Naturalist* 149:1071–1090.

Murdoch, W. W. 1966. "Community structure, population control, and competition" – a critique. *American Naturalist* 100:219–226.

Murdoch, W. W., Briggs, C. J., and Nisbet, R. M. 2003. *Consumer-Resource Dynamics*. Princeton University Press, Princeton.

Muthukrishnan, J., and Delvi, M. R. 1974. Effect of ration levels on food utilisation in the grasshopper *Poecilocerus pictus*. *Oecologia* 16:227–236.

Nisbet, R. M., Diehl, S., Wilson, W. G., Cooper, S. D., Donalson, D. D., and Kratz, K. 1997. Primary-productivity gradients and short-term population dynamics in open systems. *Ecological Monographs* 67:535–553.

Nishida, R. 2002. Sequestration of defensive substances from plants by Lepidoptera. *Annual Review of Entomology* 47:57–92.

Nykänen, H., and Koricheva, J. 2004. Damage-induced changes in woody plants and their effects on insect herbivore performance: a meta-analysis. *Oikos* 104:247–268.

Ode, P. J. 2006. Plant chemistry and natural enemy fitness: effects on herbivore and natural enemy interactions. *Annual Review of Entomology* 51:163–185.

Omlin, F. X. 1977. Zur populationsdynamischen Wirkung der durch Raupenfrass und Dungung veränderten Nahrungsbasis auf den Grauen Larchenwickler *Zeiraphera diniana* Gn. (Lep: Tortricidae). Dissertation No. 6064. ETH Zurich, Switzerland.

Power, M. E. 1992. Top-down and bottom-up forces in food webs: do plants have primacy? *Ecology* 73:733–746.

Price, P. W., Bouton, C. E., Gross, P., McPheron, B. A., Thompson, J. N., and Weis, A. E. 1980. Interactions among three trophic levels: influence of plants on interactions between insect herbivores and natural enemies. *Annual Review of Ecology and Systematics* 11:41–65.

Raupp, M. J., and Denno, R. F. 1984. The suitability of damaged willow leaves as food for the leaf beetle, *Phgiodera versicolora*. *Ecological Entomology* 9:443–448.

Rhoades, D. F. 1985. Offensive-defensive interactions between herbivores and plants: their relevance in herbivore population dynamics and ecological theory. *American Naturalist* 125:205–238.

Turchin, P. 1990. Rarity of density dependence or population regulation with lags? *Nature* 344: 660–663.

Turchin, P., Wood, S. N., Ellner, S. P., Kendall, B. E., Murdoch, W. W., Fischlin, A., Casas, J., McCauley, E., and Briggs, C. J. 2003. Dynamical effects of plant quality and parasitism on population cycles of larch budmoth. *Ecology* 84:1207–1214.

Turlings, T. C. J., Loughrin, J. H., McCall, P. J., Röse, U. S. R., Lewis, W. J., and Tumlinson, J. H. 1995. How caterpillar-damaged plants protect themselves by attracting parasitic wasps. *Proceedings of the National Academy of Sciences* USA 92:4169–4174.

Underwood, N. 1999. The influence of plant and herbivore characteristics on the interaction between induced resistance and herbivore population dynamics. *American Naturalist* 153:282–294.

Underwood, N. 2000. Density dependence in induced plant resistance to herbivore damage: threshold, strength, and genetic variation. *Oikos* 89:295–300.

Underwood, N. 2010. Density dependence in insect performance within individual plants: induced resistance to *Spodoptera exigua* in tomato. *Oikos* 119:1993–1999.

Underwood, N., and Rausher, M. 2002. Comparing the consequences of induced and constitutive plant resistance for herbivore population dynamics. *American Naturalist* 160:20–30.

Underwood, N., Anderson, K. E., and Inouye, B. D. 2005. Induced vs. constitutive resistance and the spatial distribution of insect herbivores among plants. *Ecology* 86:594–602.

van de Koppel, J., Huisman, J., van der Wal, R., and Olff, H. 1996. Patterns of herbivory along a productivity gradient: an empirical and theoretical investigation. *Ecology* 77:736–745.

Varley, G. C. 1949. Population changes in German forest pests. *Journal of Animal Ecology* 18:117–122.

Verschoor, A. M., Vos, M., and van der Stap, I. 2004. Inducible defences prevent strong population fluctuations in bi- and tritrophic food chains. *Ecology Letters* 7:1143–1148.

Walker, M., and Jones, T. H. 2001. Relative roles of top-down and bottom-up forces in terrestrial tritrophic plant–insect herbivore–natural enemy systems. *Oikos* 93:177–187.

Walling, L. L. 2000. The myriad plant responses to herbivores. *Journal of Plant Growth Regulation* 19:195–216.

White, R. R. 1974. Food plant defoliation and larval starvation of *Euphydryas editha*. *Oecologia* 14:307–315.

White, T. C. R. 2005. *Why Does the World Stay Green? Nutrition and Survival of Plant-Eaters.* CSIRO Publishing, Collingwood, Australia.

Ylioja, T., Roininen, H., Ayres, M. P., Rousi, M., and Price, P. W. 1999. Host-driven population dynamics in an herbivorous insect. *Proceedings of the National Academy of Sciences* USA 96:10735–10740.

3

Immune Responses and Their Potential Role in Insect Outbreaks

J. Gwen Shlichta and Angela M. Smilanich

3.1 Introduction

In this chapter, we focus on the insect immune response as a potential mechanism that may facilitate pest herbivores' escape from regulation by natural enemies, allowing them to reach outbreak levels. We start by providing some background on the insect immune response and how its components have been experimentally measured. We go on to discuss how the immune response varies, why it is relevant to understanding the potential mechanisms or conditions associated with outbreaks, and investigate some traits of outbreak species that may be associated with an increased immune response. Although many studies have intensely examined factors that influence the insect immune response, to date only a handful of studies have examined the potential role of immunocompetence in outbreaks (Kapari *et al.* 2006, Klemola *et al.* 2007a, Ruuhola *et al.* 2007, Yang *et al.* 2007, Klemola *et al.* 2008, Yang *et al.* 2008). For this reason, we draw on the studies available and present several potential hypotheses (Table 3.1) to stimulate

Insect Outbreaks Revisited, First Edition. Edited by Pedro Barbosa, Deborah K. Letourneau and Anurag A. Agrawal.
© 2012 Blackwell Publishing Ltd. Published 2012 by Blackwell Publishing Ltd.

Table 3.1 Summary of hypotheses and evidence for how insect immune responses interact with outbreak potential.

Hypotheses	Immunity measure	Hypothesis support	Comments	Species	References
Density-dependent prophylaxis – hosts living at high densities suffer a greater risk of disease and therefore invest more in immunity.	M	−	Larval mortality from NPV declined as density increased.	Mythimna separata*	Kunimi and Yamada 1990
	M	+	Resistance to granulosis virus increased as density increased.	Mythimna separata*	Kunimi and Yamada 1990
	M	−	Larvae were most susceptible to infection at high densities.	Mamestra brassicae	Goulson and Cory 1995
	M	+	Larval transmission of viruses declined as density increased.	Lymantria dispar*	D'Amico et al. 1996
	M	+	Larger egg masses had higher resistance to NPV.	Malacosoma californicum p.*	Rothman and Myers 1996
	HM, M, PO	+	Higher densities were more resistant to fungus.	Tenebrio moliator	Barnes and Siva-Jothy 2000
	EN, PO	+	Immunity measures increased with increasing gregariousness.	Spodoptera exempta*	Wilson et al. 2001
	AB, EN, HM, PO	+	Increased pathogen resistance under crowded conditions.	Schistocerca gregaria*	Wilson et al. 2002
	EN, HM, PO	−	Gregarious species had lower PO and HM.	Various Lepidoptera	Wilson et al. 2003
	EN, P	−	No difference in EN of moths from different sites.	Epirrita autumnata*	Klemola et al. 2006

Melanin – Outbreak species have higher cuticular melanin which is positively correlated with immune response.

M	+	Disease resistance increased as larval density decreased.	Lymantria dispar*	Reilly and Hajek 2008
M	–	Resistance to disease was not related to size of egg mass.	Malacosoma californicum p.*	Cory and Myers 2009
M	+	Melanic morphs five times more resistant to fungus.	Mythimna separata*	Mitsui and Kunimi 1988
M	+	Melanic morphs two times more resistant to NPV.	Mythimna separata*	Kunimi and Yamada 1990
M	–	Melanic morphs were more susceptible to NPV.	Mamestra brassicae	Goulson and Cory 1995
PO	+	Melanic morphs four times more resistant to NPV.	Spodoptera exempta	Reeson et al. 1998
HM, M, PO	+	Darker morphs have stronger immune defense.	Tenebrio moliator	Barnes and Siva-Jothy 2000
EN, HM	+	Positive association between size of wingspot and immunity.	Calopteryx splemdens	Rantala et al. 2000
EN, PO	+	Melanic larvae more resistant to fungal disease.	Spodoptera exempta*	Wilson et al. 2001
EN	–	Darker individuals showed lower encapsulation and melanization.	Hemideina maori	Robb et al. 2003
EN	+	Implanted pupae had darker melanization on wing as adults.	Pieris brassicae	Freitak et al. 2005
M, P	–	Parasitism was positively associated with melanism.	Operophtera brumata*	Hagen et al. 2006

(Continued)

Table 3.1 (Cont'd).

Hypotheses	Immunity measure	Hypothesis support	Comments	Species	References
	P	+	Higher parasitism in green larvae versus brown larvae.	*Macrolepidoptera*	Barbosa and Caldas 2007
	AB, PO	+	Causative relationship between melanism and immune response.	*Spodoptera littoralis**	Cotter *et al.* 2008
	AB, PO	+	High-protein diet increased melanization in cuticle and AB.	*Spodoptera littoralis**	Lee *et al.* 2008
	EN	+	Melanic morphs have stronger EN than pale morphs.	*Lymantria monacha*	Mikkola and Rantala 2010
Thermal effects – Continued increases in yearly temperature averages will lead to fluctuations in normal outbreak periodicity, due to altered immunity.	EN	+	Encapsulation of parasitoid highest during warmer months.	*Protopulvinaria pyriformis*	Blumberg 1991
	M	0	Virulence of fungus is strongly affected by temperature.	*Zonocerus variegatus*	Thomas and Jenkins 1997
	M	–	Melanization of silica beads decreased with increasing temperature.	*Anopheles gambiae*	Suwanchaichinda and Paskewitz 1998
	EN	+	Encapsulation of selected lines increased with temperature.	*Drosophila melanogaster*	Fellowes *et al.* 1999
	HM	+	Phagocytosis by hemocytes enhanced with increased temperature.	*Locusta migratoria*	Ouedraogo *et al.* 2002

M	+	High temperature reduced susceptibility to parasitism.	Acyrthosiphon pisum	Blanford et al. 2003
AB, AF	+	Heat-shocked larvae had enhanced immune response.	Galleria mellonella	Wojda and Jakubowicz 2007
Host plant – Species tend to outbreak on host plants that provide food quality conditions that are favorable for growth and immunity. EN	+	Host plant on which larvae grew fastest also had highest EN.	Pieris rapae	Benrey and Denno 1997
EN, PO	0	Nutritional conditions affected PO, but not EN of males.	Tenebrio moliator	Rantala et al. 2003
EN	+	Larvae encapsulate on host plant with highest antioxidants.	Parasemia plantaginis	Ojala et al. 2005
AB, EN, HM. PO	+	High protein content enhances NPV resistance and immunity.	Spodoptera littoralis*	Lee et al. 2006
EN	–	Hydolysable tannins have negative effects on EN.	Epirrita autumnata*	Haviola et al. 2007
EN, PO	–	Encapsulation was higher on low-quality foliage.	Epirrita autumnata*	Klemola et al. 2007b
EN, P	–	Larvae fed high- and low-quality diets did not differ in parasitism.	Epirrita autumnata*	Klemola et al. 2008
PO, AB	+	Larvae on high protein diet had higher AB activity.	Spodoptera littoralis*	Lee et al. 2008
AB, HM, M, PO	+	Increasing protein: carb diet increased resistance to infection.	Spodoptera exempta*	Povey et al. 2009

(Continued)

Table 3.1 (Cont'd).

Hypotheses	Immunity measure	Hypothesis support	Comments	Species	References
	EN	+	Iridoid glycosides have negative effects on EN.	*Junonia coenia*	Smilanich et al. 2009a
	EN	+	Differential effects of amides on generalist versus specialists.	*Eois nympha, S. frugiperda*	Richards et al. 2010
	EN	+	EN increased with increasing host plant quality.	*Manduca sexta*	Diamond and Kingsolver 2011
	EN	−	Encapsulation not associated with resistance to pathogen.	*Orgyia antiqua**	Sandre et al. 2011
Key natural enemies – Certain natural enemies can circumvent the immune response.	AB, EN, PO	+	Parasitoid benefits from melanization of respiratory tube.	*Teleogryllus oceanicus*	Bailey and Zuk 2008
	EN, PO	+	Parasitoid "hides" from host immune response.	*Trichoplusia ni*	Caron et al. 2008
Outbreaks are predicted to occur when these enemies are less abundant.	PO	−	Polydnavirus inhibits phenoloxidase cascade.	*Manduca sexta*	Beck and Strand 2007
	EN	+	Polydnavirus inhibits phenoloxidase cascade.	*Manduca sexta*	Lu et al. 2010
	HM	−	Polydnavirus disrupts hemocyte function.	*Pseudoplusia includens*	Strand et al. 2006

Transgenerational effects and natural selection – Periods of high parasitism may be followed by outbreaks due to either maternal effects or natural selection.				
—	+	Daphnia magna	Parental bacteria challenge increases resistance in progeny.	Little et al. 2003
HM	+	Leptinotarsa decemlineata	Higher HM in populations from different potato fields.	Ots et al. 2005
AB, PO, M	+	Tenebrio molitor	Parental immune challenge enhanced offspring immunity.	Moret 2006
—	+	Lymantria dispar*	Natural selection for host resistance effects insect outbreaks.	Elderd et al. 2008
M	+	Schistocera gregaria*	Transgenerational effect of crowding on pathogen resistance.	Miller et al. 2009
PO, AB	+	Trichoplusia ni*	Progeny of mothers exposed to bacteria exhibited priming.	Freitak et al. 2009
AB, PO	+	Tribolium castaneum	Mothers and fathers can transfer resistance to offspring.	Roth et al. 2010

Abbreviations: EN, encapsulation/melanization; HM, hemocytes; PO, phenoloxidase; AB, lysozyme activity/ antibacterial activity; AF, antifungal activity; M, mortality (or survival): P, Parasitism, NPV, nuclear polyhedrosis virus.
* Denotes outbreak or pest species.

further research. Our goal is to show that changes in the strength of insects' immune responses may affect the success of natural enemies (parasites, parasitoids, and pathogens) in the host population, which in turn may destabilize insect–enemy dynamics, leading to outbreaks.

The role of parasites, parasitoids, and pathogens in regulating insect populations has been researched extensively. One clear hypothesis that has emerged from the outbreak literature is that outbreaks result from insect populations escaping the control of their natural enemies (Morris 1963, Berryman 1996, Klemola *et al.* 2010). Research to explain the mechanisms behind the escape from natural enemies include those that are behavioral (Gross 1993, Gentry and Dyer 2002, Smilanich *et al.* 2011), morphological (Gentry and Dyer 2002, Barbosa and Caldas 2007), chemical (Dyer 1995, Gentry and Dyer 2002, Nishida 2002), and physiological (Carton *et al.* 2008, Smilanich *et al.* 2009a). While each of these mechanisms is an important adaptation for escape from natural enemies, immunity is a critical escape mechanism that has received little attention (Godfray 1994, Smilanich *et al.* 2009b).

A low incidence of parasitism has been found to be characteristic of growing defoliator and other outbreak populations (Berryman *et al.* 2002). Indeed, the low incidence of parasitism in populations reaching outbreak status has often been attributed to the inability of parasitoids to "catch up" with the population growth of the outbreak population. Nonetheless, this conclusion has not accounted for the disparity between realized parasitism and actual parasitism. Parasitism rates are often assessed as the number of parasitoids reared from infected hosts (realized parasitism). This assumes that parasitism rates are singularly dependent on the number of parasitoids and/or the incidence of parasitism, and does not assess those insect hosts that have successfully defended themselves against parasitism (Kapari *et al.* 2006).

In long-term studies monitoring larval density and parasitism of the autumnal moth, *Epirrita autumnata* Borkhausen (Lepidoptera: Geometridae), on birch in Fennoscandia, *E. autumnata* was found to have delayed density-dependent variation in larval parasitism rates ranging from 0% to 100% (Ruohomäki 1994, Tanhuanpää *et al.* 2002). The abundance of some parasitoid species (surveyed from trapping of adults) around the time of outbreaks was found to decline to almost zero within two years after outbreaks (Nuorteva 1971). This suggests that the larval immune response to parasitism during periods of high larval density may play a role in decreasing the number of parasitoids post outbreak. Furthermore, Klemola *et al.* (2008, 2010) showed decreases in *E. autumnata* density with increases in parasitism and vice versa. Variations in herbivore–pathogen or herbivore–parasitoid interactions may be due, in part, to the susceptibility of the host herbivore to disease or parasitism. For example, the extent of oak defoliation by the gypsy moth, *Lymantria dispar* (L.) (Lepidoptera: Lymantriidae), was correlated with increased survival when exposed to nuclear polyhedrosis virus (NPV) (Hunter and Schultz 1993). Hunter and Schultz (1993) suggest that *L. dispar* was inducing changes in the leaf quality of its host plant in order to increase gallotannin concentrations and inhibit NPV. Another possibility in their scenario is that declining leaf quality due to defoliation may positively impact the immune responses of *L. dispar* against NPV.

3.2   The insect immune response

The immune response is composed of a humoral and cellular response that works in concert to defend against parasitoids, parasites, and pathogens (reviewed by Carton *et al*. 2008, Strand 2008, Beckage 2008). The cellular response is directed by specialized hemocytes. In general, when a special type of hemocyte called "granular cells" contacts a foreign target, they lyse or degranulate, releasing material that promotes attachment of plasmatocytes. Multiple layers of plasmatocytes then form a capsule around the object, which is then melanized by the humoral response. A suite of proteins, including regulatory proteins, hemocyte response modulators, and melanization enzymes, are also involved in immunity, making it a complex and resource-costly process (Schmid-Hempel and Ebert 2003). The humoral response aids in the production of non-self-recognition proteins, enzymatic activity leading to melanization and clotting, and antimicrobial peptides. The combination of asphyxiation from encapsulation and the cytotoxic products produced during melanization is thought to contribute to killing the parasite (Strand 2008).

Experiments seeking to understand the variation in the immune response use a variety of methods. A common method for measuring encapsulation and melanization is injection or insertion of a synthetic object into the insect's hemocoel (Lavine and Beckage 1996, Klemola *et al*. 2007b). The immune response is quantified by measuring the color change and/or cell thickness around the object (Beckage 2008). Other measurements rely on quantifying the protein activity of the humoral response by determining the concentration of the enzyme, phenoloxidase (PO), which catalyzes the melanization cascade, or quantifying the activity of antibacterial lysozyme activity (Adamo 2004a). Still other measures include hemocyte counts (Ibrahim and Kim 2006), hemolymph protein concentration (Adamo 2004a), and gene expression (Freitak *et al*. 2009). Studies measuring the response of multiple immune parameters show that they do not always respond in the same way (Adamo 2004b) and can exhibit trade-offs (i.e., antibacterial efficiency trades off with PO activity) (Cotter *et al*. 2004b). For example, not only will a bacterial infection induce a specific component of the immune response that is different from a parasitoid infection, but also the response will be specific to the type of invading bacteria (Riddell *et al*. 2009). In relation to insect outbreaks, losing natural enemies that are capable of escaping the immune response may result in an outbreak. This idea will be further developed in the hypothesis section of this chapter.

3.3   Sources of variation in immune response associated with outbreaks

3.3.1   Host plants

Host plant diet is a major source of variation in the immune response (Smilanich *et al*. 2009a, Diamond and Kingsolver 2011). Since most aspects of herbivore outbreaks involve interactions with the host plant, it is reasonable to hypothesize that the immune response will play an integral role, from start to finish, in insect outbreak dynamics. The effects of host plants on immune responses can depend on plant chemistry, herbivore health, both herbivore and plant genotype, and the specific immune parameter being measured (e.g., PO activity, encapsulation of inert object, and lysozyme activity). Certain plant secondary metabolites can alter the effectiveness of immune responses. For example, ingestion of diets containing carotenoids enhances immune function due to their free radical–scavenging properties (de Roode *et al.* 2008, Babin *et al.* 2010). In contrast, ingestion of other secondary metabolites can negatively affect the immune response (Haviola *et al.* 2007, Smilanich *et al.* 2009a). Macronutrients can also affect the immune response. Protein and carbohydrates not only boost immune parameters (Lee *et al.* 2008, Srygley *et al.* 2009, Cotter *et al.* 2011), as does plant and diet quality (Yang *et al.* 2008, Bukovinszky *et al.* 2009, Diamond and Kingsolver 2011) but also are preferred by immune-challenged herbivores (Lee *et al.* 2006, Povey *et al.* 2009). In most cases, high plant quality (i.e., low secondary metabolite concentration and high nitrogen content) translates to increased immunity (but see Klemola *et al.* 2007b).

3.3.2 Heritability

Different immune responses have been shown to be heritable, indicating that if immune responses play a role in outbreak formation or decline, there is a genetic component to the potential for a population or species to outbreak (Kraaijeveld and Godfray 1997, Schmid-Hempel and Ebert 2003, Carton *et al.* 2005). Early work investigating the genetic basis of immune defense mostly focused on *Drosophila melanogaster*, and has since expanded considerably to include many other invertebrate species (Cotter and Wilson 2002, Ojala *et al.* 2005, Moret 2006, Rantala and Roff 2006, 2007, Freitak *et al.* 2009). A significant heritability for encapsulation rate was found in *E. autumnata*, feeding on high- and low-quality host plants (Klemola *et al.* 2007b). Knowing the heritability and genetic basis of the immune response helps to understand and illuminate insect outbreaks in an evolutionary context. Later in the chapter we propose and discuss the hypothesis that the heritability of the immune response may play a prominent role in outbreak periodicity.

3.4 Traits or conditions associated with outbreak species

Only a small fraction of insect species reach outbreak proportions (Mattson and Addy 1975, Hunter 1995). Most species maintain a relatively low, stable population size and do not become noticeable defoliators (Mason 1987). Although there have been numerous studies trying to find traits or factors to explain why some species outbreak, we are still left with some rather unsatisfying explanations that in some way contribute to population regulation, from weather and host plant chemistry, to life history traits and natural enemies. However, it remains unclear why some species in the same habitat, under the same environmental conditions, are outbreak species and others are not. Furthermore, the questions of why some populations of species outbreak in certain geographical areas, but not others, are still the subject of much debate (Ruohomäki *et al.* 1997, Bjornstad *et al.* 2010).

Several studies have used a comparative approach between outbreak and non-outbreak species to examine important characteristics that set them apart (Mason 1987, Wallner 1987, Hunter 1995, Cappuccino *et al.* 1995, Price 2003). Hunter (1991) examined the Canadian Forest Insect Survey (CFIS) data supplemented with other sources to compare life history traits of outbreak and non-outbreak species of macrolepidoptera. The findings that are most relevant to our focus on outbreaks and immune response include gregarious feeding behavior (Larsson *et al.* 1993, Cappuccino *et al.* 1995), overwintering as eggs in clusters or masses, and a wide diet breadth (a mean of 20 plant genera) (Hunter 1991). Although some advantages of gregarious behavior and laying eggs in clusters have been proposed (reviewed in Hunter 1991), we suggest that increased density of herbivores, which may be a result of egg clustering and gregarious behavior, may additionally lead to an increased immune defense against natural enemies. As discussed above, host plant attributes can alter the insect immune response; thus, polyphagy may be a trait that allows for adaptive responses in some outbreak species. The importance of these traits will be explored in the next section, where we propose several hypotheses for how the immune response plays an integral role in insect outbreaks (Table 3.1).

3.5 Hypotheses on insect outbreaks and the immune response

Hypothesis 1: Density-dependent prophylaxis (Wilson and Cotter 2009). This hypothesis proposes that hosts living at high densities suffer a greater risk of disease and therefore invest more in immunity.

Insects across a broad range of taxa are able to assess conspecific density, presumably as an indication of a potentially deteriorating quality or quantity of food, or exposure to pathogens or parasites (Carroll and Dingle 1996, Wilson and Cotter 2008). This is most often associated with insects that are able to migrate when conditions deteriorate (Denno *et al.* 1991, Carroll and Dingle 1996), or are capable of switching from a gregarious to solitary phase (Wilson and Cotter 2008). Given that not all insects are able to migrate when density increases (especially those in the larval stage or with limited mobility), such species may risk increased exposure to pathogens, parasites, and other natural enemies. With increasing density, the risk of transmission of disease increases due to the higher probability of interaction with potentially infected conspecifics (Steinhaus 1958, McCallum *et al.* 2001). Species that can modify their level of disease resistance to match potential risk of exposure would minimize the costly investment in disease resistance (Kraaijeveld and Godfray 1997, Wilson and Reeson 1998, Moret and Schmid-Hempel 2000). Many of the insects that exhibit density-dependent prophylaxis (DDP) are outbreak or pest species, suggesting that gregarious outbreak species may be more likely to allocate resources to immune defense as densities increase.

Resistance against parasites or pathogens may be costly to maintain and express (Sheldon and Verhulst 1996); therefore, a trade-off may exist between immune function and other traits. For example, Kraaijeveld and Godfray (1997) found a trade-off between competitive ability and immune function in *Drosophila* males. Natural selection should thus favor insects that are able to assess larval density and allocate more resources as needed for resistance against natural enemies (Goulson and Cory 1995, Wilson and Reeson 1998). This phenotypically plastic response of allocation of resources to defense has been demonstrated in numerous insect species (Wilson *et al.* 2001, Cotter *et al.* 2004a, Ruiz-Gonzàlez *et al.* 2009). In a comparative analysis using 15 studies of temperate Lepidoptera, Hochberg (1991) found a positive correlation between gregarious feeding behavior and age-related resistance to viral infection.

Several studies indicate that there may be a connection between DDP and species that have been known to be outbreak pests (D'Amico *et al.* 1996, Rothman and Myers 1996, Wilson and Reeson 1998) (Table 3.1). For example, egg masses of the northern tent caterpillar, *Malacosoma californicum pluviale* Dyar (Lepidoptera: Lasiocampidae), range in size from 130 to 375 eggs per mass. Rothman and Myers (1996) orally exposed caterpillar siblings from a range of egg masses of different sizes to NPV in their second larval instar. They found that individuals that came from larger egg masses had significantly higher resistance to NPV than those from smaller egg masses. Although these differences associated with larval density may be attributed to resource availability, genetic differences, or maternal effects, density appears to play a role in intensity of resistance to viral infection across studies (Table 3.1).

The majority of DDP research indicates that there is a positive association between density and immune response (Wilson and Cotter 2009). There are still many aspects of DDP that warrant further investigation, such as how insects assess density, the trade-offs associated with increasing immune response, the specificity of immune responses, and the mechanism behind allocating resources for immune defense. Although several levels of density have been tested experimentally, whether there is a critical density threshold that facilitates an increase in immune response remains unclear. A large number of the studies examining DDP were conducted in the laboratory (Kunimi and Yamada 1990, Barnes and Siva-Jothy 2000, Cotter *et al.* 2004a). These studies may not allow for the most biologically relevant assessment of immune defense. Field studies on outbreak species, such as that of Klemola *et al.* (2007a) which assessed the immune response of *E. autumnata*, allowed for direct assessments of immune response at different densities. Examining density-dependent resistance under more natural conditions will allow us to gain a better understanding of how immune responses influence the population dynamics of outbreak species.

Hypothesis 2: Melanism and disease resistance. Outbreak species have higher cuticular melanin which is positively correlated with immune response

Phase polymorphism, a phenomenon closely associated with DDP, occurs when species change to darker, more melanized forms under high-density conditions (Wilson and Cotter 2009). Changes in color form are often correlated with other traits such as increased activity, respiration rate (Shibizaki and Ito 1969), desiccation resistance (Parkash *et al.* 2008), thermotolerance (Parkash *et al.* 2010), or flight activity (Parker and Gatehouse 1985). Phase polymorphism has been shown in a diverse array of taxa, including many lepidopteran agricultural pests or outbreak species (reviewed by Goulson and Cory 1995). It has been suggested that melanism may be a potential indicator of high levels of immunocompetence in insects (Majerus 1998, Barnes and Siva-Jothy 2000, Wilson *et al.* 2001). The mechanism behind the association of melanism and immune defense remains unclear, although some studies have suggested possible mechanisms.

Cuticular melanin and the melanin involved in the immune response are sometimes correlated, thus successful outbreak species are more likely to have stages with increased melanin (Mikkola and Rantala 2010). Melanin is a product of the phenoloxidase (PO) cascade and is found in the cuticle, hemolymph, and midgut (Cotter *et al.* 2008). It plays a critical role in immune defense during recognition of foreign objects and encapsulation of large invaders (Wilson *et al.* 2001, Cotter *et al.* 2004b). It also improves the cuticle's ability to function as a physical barrier to the penetration of fungus, parasites, and pathogens (St. Leger *et al.* 1988, Hajek and St. Leger 1994, Wilson *et al.* 2001). An association between cuticular melanin and the insect immune defense against pathogens and parasites has been demonstrated in numerous species (Hung and Boucias 1992, Beckage and Kanost 1993, Wilson *et al.* 2001, Cotter *et al.* 2004b). Furthermore, melanin is toxic to microorganisms and has potential antimicrobial activity (Ourth and Renis 1993). Melanin is known to bind to a range of proteins and act as an inhibitor to many of the lytic enzymes produced by microorganisms (Bull 1970, Doering *et al.* 1999). The PO cascade involves a suite of enzymes that oxidize tyrosine to quinones and their polymerization product melanin (Nappi and Vass 1993). These same enzymes are also involved in cellular encapsulation, humoral encapsulation, and nodule formation (Hung and Boucias 1992, Beckage and Kanost 1993).

Although many of the studies mentioned showed increased melanism under crowded conditions, it is important to note that melanism is not just a by-product of increased density. Controlling for the effect of larval density, several studies have examined the association of melanism with disease resistance. At least four studies demonstrated that more melanic individuals had higher resistance against disease (NPV) or entomopathogenic fungi (Kunimi and Yamada 1990, Reeson *et al.* 1998, Barnes and Siva Jothy 2000). One similar study did not support these findings; Goulson and Cory (1995) found that melanic *Mamestra brassicae* (L.) (Lepidoptera: Noctuidae) larvae were less resistant to NPV compared to their less melanic conspecifics. The association of melanism and defense may be a highly specific interaction between herbivore host and natural enemy, such that in some cases the natural enemy may have evolved to overcome the herbivore host's defenses.

Conflicting data have been found on whether melanism is always associated with immune response, suggesting that it may be one of a number of factors that is important in immunocompetence. Using field data to quantify parasitoid attack rates in winter moth larvae, *Operophtera brumata* L. (Lepidoptera: Geometridae), Hagen *et al.* (2006) found that parasitoids were the greatest source of mortality (26%). However, cuticular melanin was positively associated with parasitoid attack (Hagen *et al.* 2006). Given that no research has definitively demonstrated that melanism is associated with increased immune response, there is a need for further research examining the role that melanin plays in immuno-competence, especially in outbreak species. The expression of melanism has been shown to be affected by a variety of environmental factors, including tempera-ture and humidity (Goulson 1994), light (Faure 1943), and population density (Goulson and Cory 1995). There can be a high degree of phenotypic variation between individuals and a significant effect of genetic family on melanism, indi-cating that melanism is heritable (Cotter *et al.* 2002, 2004a, 2008).

Hypothesis 3: Thermal effects on the immune response. Continued increases in yearly temperature averages will lead to fluctuations in normal outbreak perio-dicity, due to altered immunity.

This hypothesis centers on climate change and the projected increases in global mean temperatures. According to Mann *et al.* (2008), average temperature will increase by 4°C over the next 100 years. Undoubtedly, these temperature changes will impact insect populations and may disrupt regulation. The immune response will also be impacted by these changes since it is affected by temperature. However, the effects of temperature on the immune response are quite variable depending on the taxa, making a simple prediction between temperature and outbreaks impossi-ble. Thomas and Blanford (2003) state that temperature can affect parasite viru-lence, host resistance, and host recovery. Thus, since multiple variables are affected by temperature, a complex relationship with many possible outcomes exists. For example, heightened temperature can enhance either natural enemy performance (Fellowes *et al.* 1999, Thomas and Blanford 2003) or the immune response (Fellowes *et al.* 1999, Wojda and Jakubowicz 2007), while in other cases it depresses natural enemy performance (Blanford *et al.* 2003, Thomas and Blanford 2003) and the immune response (Suwanchaichinda and Paskewitz 1998). Thus, the outcome of increased temperature will depend on the response of the enemy and the response of the host, especially where temperatures reach the boundary of normality based on their evolutionary history (Thomas and Blanford 2003). One pattern that has emerged from parasitoid–host data shows that increased variability in climate pat-terns will disrupt host tracking by parasitoids (Stireman *et al.* 2005), possibly lead-ing to greater frequency in outbreaks. Whether this effect is due to increased immunity at higher temperatures is undetermined, yet holds implications for insect outbreaks. Greater immune defense at high temperatures could mean that parasi-toids will not be as effective at controlling and maintaining insect populations.

With selection history in mind, dramatically altering temperature will at the very least lead to destabilization of normal insect–enemy interactions, which may lead to increased outbreak frequency (Parmesan 2006, Ims *et al.* 2008). Johnson *et al.* (2010) showed that outbreak periodicity shifts of the larch budmoth (*Zeiraphera diniana* Guénée), which exhibited regular 8–10-year outbreaks since 800 CE, is partially explained by increases in mean winter temperatures. Using a traveling wave model, they showed that increases in temperature at the optimal elevation for larch budmoth population growth lead to destabilization of outbreak periodicity (Johnson *et al.* 2010). In this case, the net effect of increased temperature is clear, but whether the immune response contributes to the destabilization is unclear.

Hypothesis 4: Host plant quality. Species tend to outbreak on host plants that provide food quality conditions that are favorable for growth and immunity. In most cases, these conditions will involve high nutrient quality and low levels of toxins.

As mentioned earlier, high-quality host plants enhance immune parameters, either directly (i.e., increased nitrogen for melanization precursors) or indirectly (i.e., increased body fat). However, the term "high-quality" should be defined in order to allow one to make a meaningful prediction. Here, we follow suit with most insect ecology literature and define quality based upon herbivore performance (e.g., development time, pupal mass, and fecundity) and nutrient profiles. In terms of nutrients, host plants that are high in nitrogen and water content are considered high-quality hosts since nitrogen is a limiting resource for insects, and insects risk desiccation from low water content (Scriber and Slansky 1981, Behmer 2009).

Most of the examples given earlier in the chapter deal with the macronutrient, protein, showing that high protein content enhances immunity (Lee *et al.* 2006, 2008, Povey *et al.* 2009, but see Cotter *et al.* 2011). Since the immune response relies heavily on enzymatic reactions that require amino acid precursors, it is fitting that immune-stressed herbivores would not only have enhanced immunity on high-protein diets but also prefer these diets to a nutritionally balanced diet. However, regardless of whether an herbivore is immune challenged, decades of research show that herbivores will regulate their nutrient intake, favoring a high protein-to-carbohydrate ratio (Behmer 2009). With this evidence, it is easy to speculate that a characteristic of outbreak species is the ability to achieve high protein content in their diets, either by consuming host plants with high protein: carbohydrate ratios and/or by superior nutrient regulation. For example, the gregarious tent caterpillar, *Malacosoma disstria* (Hübner) (Lepidoptera: Lasiocampidae), is an outbreaking species that feeds preferentially on aspen and sugar maple. Research investigating the nutrient regulation of this forest pest shows that it does not regulate its diet to favor protein (Despland and Noseworthy 2006). Instead, its phenology is such that the beginning of its life cycle is in sync with new leaf flush on its host plants; this is a time when leaf nitrogen content will be high, which should favor a strong immune response.

Another reason why high-quality host plants may favor the occurrence of insect outbreaks is due to the documented trade-off between immunity and growth. Since the immune response is costly in terms of resources, it is predicted to trade off with other metabolic functions such as growth and fitness (Zuk and Stoehr 2002, Diamond and Kingsolver 2011). In other words, herbivores that are resource limited are not able to invest in both high growth and high immune capacity. For example, using path analysis, Diamond and Kingsolver (2011) found that host plant resource quality had a significant indirect positive effect on the encapsulation response via enhanced body condition (measured by growth rates) of *Manduca sexta* (LINNAUS) (Lepidoptera: Sphingidae) larvae. However, individuals with higher immune capacity exhibited slower growth. If outbreaking species are more likely to be found on high-quality host plants, then they may invest more in immune capacity than growth, which would favor escape from natural enemies. However, in some cases high-quality plants do not enhance immune capacity. Sandre *et al.* (2011) found that despite host plant–dependent resistance of *Orgyia antiqua* (Lepidoptera: Lymantriidae) caterpillars to the entomopathogenic fungus, *Metarhizium anisopliae* (Metsch.) Sorokin, this resistance was not correlated with host plant quality or the encapsulation response, indicating a direct effect of host plant on the pathogen.

Not surprisingly, the effects of secondary chemistry on immune parameters are much more variable and dependent upon the action that the compound takes in the herbivore's body. For example, certain secondary metabolites, such as carotenoids, can increase the effectiveness of the immune response by ameliorating the autoreactivity of the melanization cascade (de Roode *et al.* 2008, Babin *et al.* 2010). However, ingestion of hydrolysable tannins and imides, and high concentrations of iridoid glycosides, reduced encapsulation and melanization (Haviola *et al.* 2007, Smilanich *et al.* 2009a, Richards *et al.* 2010), while still other secondary metabolites have no effect on the immune capacity (Smilanich *et al.* 2011). These differing results reflect the enormous diversity of secondary metabolites and the role of evolutionary history between herbivore and host plants. In short, overall host plant quality will most likely be a better predictor of immune capacity than specific secondary metabolites.

Hypothesis 5: The key natural enemy hypothesis. Even though the immune response is an effective defense, certain natural enemies can circumvent the immune response. Outbreaks are predicted to occur when these enemies are less abundant.

Since the immune response is one of the most effective defenses against parasitic enemies (Smilanich *et al.* 2009b, Godfray 1994), it is evolutionarily fitting that these enemies will have evolved counter-adaptations to cope with or suppress the immune response. Indeed, the best example of a counter-adaptation to the insect immune response is exhibited by hymenopteran parasitoids in the families Braconidae and Ichneumonidae (Webb and Strand 2005). Species in these two families harbor polydnaviruses, which have become integrated into the wasp's genome and are passed vertically through the germ line to offspring (Strand 2009). The virus replicates in the reproductive tract of the female wasp and is injected into the host during oviposition (Beckage 2008). Once inside the host's hemocoel, the virus infects immune-functioning cells, enzymes, and tissues such as hemocytes, phenoloxidase, and fat body, and thereby suppresses the immune response (Strand 2009). Another example of counter-adaptation to the immune response is found in certain species of tachinid flies. Bailey and Zuk (2008) found a positive correlation between the phenoloxidase activity of the field cricket, *Teleogryllus oceanicus* (Le Guillou) (Orthoptera: Gryllidae), and the melanization of the respiratory funnel in the attacking tachinid fly, *Ormia ochracea*. Since the respiratory funnel is the means by which many tachinid flies receive oxygen, a stronger funnel that is less likely to break is beneficial. In this way, these flies may be co-opting the immune response for their own benefit as the funnel is strengthened by the melanization process. Other species of tachinids have evolved a behavioral counter-adaptation to the immune response. The broad generalist tachinid, *Compsilura concinnata*, hides from the host immune response by developing between the peritrophic membrane and the midgut, where the immune response has limited access (Caron *et al.* 2008). Similarly, other tachinids reside in certain tissues, such as fat bodies, to avoid the immune response (Salt 1968).

In insect populations, these natural enemies that are capable of suppressing or circumventing the immune response may be key sources of mortality. When populations of these key natural enemies are low, it may lead to outbreak situations. Moreover, insect species with the fewest key natural enemies will be the most likely to outbreak. Recent models by Bjørnstad *et al.* (2010) and Dwyer *et al.* (2004) demonstrate that population cycles of outbreaking insects are determined by a complement of natural enemies. In particular, Bjørnstad *et al.* (2010), show that the periodicity of outbreaks in *L. dispar* populations is governed by both generalist and specialist natural enemies. Generalist predators maintain the population up to a certain carrying capacity at which point an outbreak occurs. The outbreaking population is brought back to pre-outbreak size by a specialist pathogen that is capable of escaping the immune response. Similarly, Dwyer *et al.* (2004) use a host–pathogen–predator model to show that generalist predators maintain the population whereas specialist pathogens maintain the cycles. Although these data are not a perfect demonstration of our hypothesis, they support the idea that different types of natural enemies each play a role in outbreak cycles. Both examples hinge on the premise that the natural enemy is capable of escaping host defenses. In general, whether or not key natural enemies are avoiding or overcoming the immune response, and whether they are specialists or generalists, have yet to be tested.

Hypothesis 6: Transgenerational maternal effects and natural selection. Periods of high parasitism may be followed by outbreaks due to the following: (1) transgenerational maternal effects where individuals that survived parasitism attack have a heightened immune response and produce offspring that are immune primed, or (2) individuals surviving the attack are genetically selected for higher resistance, and progeny of these individuals will also have higher resistance via inheritance.

Data sets monitoring caterpillar density and numbers, and monitoring parasitism, support this hypothesis (Karban and Valpine 2010, Schott *et al.* 2010). These data sets show peak periods in natural enemy populations followed by peaks or outbreaks in insect populations. In addition, host–pathogen models predict this same trend where periods of high parasitism are followed by outbreaks (Elderd *et al.* 2008). Although the empirical data sets focus on caterpillars and parasitoids, the models are general for outbreaking species. An exception to this trend is seen in the data set with *E. autumnata* (Klemola *et al.* 2007a, 2008, 2010). This data set started in the 1970s and shows periods of high parasitism matching periods of low caterpillar density. In addition, Klemola *et al.* (2008) found no relationship between *E. autumnata* immunity and parasitism status, suggesting an alternative mechanism regulating population outbreaks for this caterpillar. Thus, our hypothesis may be most relevant for populations that rely heavily on the immune response to resist pathogens and parasites.

Transgenerational maternal effect

A spike in natural enemy population (pathogen, parasitoid) leads to many individuals in the population attacked and presented with an immune challenge. Epigenetic mechanisms may occur such that the progeny of immune-challenged and surviving parents are immune-primed against another attack. This may occur through mechanisms such as maternal effects, where the physiological result of the mother's developmental environment is passed on to offspring. Other genetic mechanisms are also possible, such as chromatin marking, where the structure of the DNA molecule is altered by an environmental stimulus (Jablonka and Lamb 2010). Transgenerational priming of the immune response has been shown in insects, although the exact mechanism is unclear (Little *et al.* 2003, Moret 2006, Freitak *et al.* 2009, Roth *et al.* 2010). The best evidence to date is that of Freitak *et al.* (2009), in which *Trichoplusia ni* (Hübner) (Lepidoptera: Noctuidae) progeny whose mothers were exposed to dietary bacteria exhibited transgenerational priming of enzyme activity, protein expression, and transcript abundance of immune-functioning genes. While these results support maternal effects of the immune response, the offspring's response did not exactly mirror the mother's response. Studies have shown that the maternal effects are fleeting, and without the stimulus will not be maintained in the population (Jablonka and Lamb 2010). Thus, if natural enemy populations are not as high during the F1 generation, the maternal effect is not as prominent in the population, and the outbreak subsides in the F2 generation.

Natural selection

Periods of insect outbreak following periods of high parasitism may be due to evolutionary processes such as natural selection. Regardless of maternal effects, individuals that have strong immune responses will be more likely to survive an attack, and, if heritable, these genes will be passed to the next generation. Using both a model and experimental approach, Elderd *et al.* (2008) demonstrate that resistance to natural enemies plays a prime role in population fluxes of *L. dispar* and that natural selection for resistance to these enemies drives population cycles. In this scenario, it is more difficult to explain the end of an outbreak if the progeny are strongly resistant; however, a wealth of additional factors, both physiological and ecological, can result in population crashes.

3.6 Conclusions

Outbreak species share traits and behaviors that implicate the immune response as having the potential for playing a significant role in insect outbreaks. Their gregarious behavior and egg clustering, which increase the density of conspecifics in a given area, may increase their likelihood of density-dependent prophylaxis and melanism (which in turn strengthens their resistance to natural enemies). Outbreak species are often polyphagous, allowing for the potential of differential host plant selection, sometimes resulting in the selection of diets which maximize immune defense when needed. As research on the insect immune response continues to accumulate, the rather coarse picture that we have drawn of the role of the immune response will be refined such that exact mechanisms of immunity on outbreaks can be defined and described. There are many avenues to explore on this topic, and the six hypotheses that we presented can be used as guidelines for future research (Table 3.1). These hypotheses are not mutually exclusive and most likely work in concert to influence outbreaks. Most of the examples pooled in this chapter come from research on forest insect pests, yet the hypotheses can likely be generalized to agricultural pests.

References

Adamo, S. A. 2004a. Estimating disease resistance in insects: phenoloxidase and lysozyme-like activity and disease resistance in the cricket *Gryllus texensis. Journal of Insect Physiology* 50:209–216.

Adamo, S. A. 2004b. How should behavioural ecologists interpret measurements of immunity? *Animal Behaviour* 68:1443–1449.

Babin, A., C. Biard, and Y. Moret. 2010. Dietary supplementation with carotenoids improves immunity without increasing its cost in a crustacean. *American Naturalist* 176:234–241.

Bailey, N. W., and M. Zuk. 2008. Changes in immune effort of male field crickets infested with mobile parasitoid larvae. *Journal of Insect Physiology* 54:96–104.

Barbosa, P., and A. Caldas. 2007. Do larvae of species in macrolepidopteran assemblages share traits that influence susceptibility to parasitism? *Environmental Entomology* 36:329–336.

Barnes, A. I., and M. T. Siva-Jothy. 2000. Density-dependent prophylaxis in the mealworm beetle *Tenebrio molitor* L-(Coleoptera: Tenebrionidae): cuticular melanization is an indicator of investment in immunity. *Proceedings of the Royal Society of London Series B-Biological Sciences* 267:177–182.

Beckage, N. E. 2008. *Insect Immunology*. Academic Press, Oxford.

Beckage, N. E., and M. R. Kanost. 1993. Effects of parasitism by the braconid wasp *Cotesia-congregata* on host hemolymph-proteins of the tobacco hornworm, *Manduca sexta. Insect Biochemistry and Molecular Biology* 23:643–653.

Behmer, S. T. 2009. Insect herbivore nutrient regulation. *Annual Review of Entomology* 54:165–187.

Benrey, B., and R. F. Denno 1997. The slow-growth high-mortality hypothesis: a test using the cabbage butterfly. *Ecology* 4:987–999.

Berryman, A. A. 1996. What causes population cycles of forest Lepidoptera? *Trends in Ecology and Evolution* 11:28–32.

Berryman, A. A., M. Lima, and B. A. Hawkins. 2002. Population regulation, emergent properties, and a requiem for density dependence. *Oikos* 99:600–606.

Bjornstad, O. N., C. Robinet, and A. M. Liebhold. 2010. Geographic variation in North American gypsy moth cycles: subharmonics, generalist predators, and spatial coupling. *Ecology* 91:106–118.

Blanford, S., M. B. Thomas, C. Pugh, and J. K. Pell. 2003. Temperature checks the Red Queen? Resistance and virulence in a fluctuating environment. *Ecology Letters* 6:2–5.

Blumberg, D. 1991. Seasonal-variations in the encapsulation of eggs of the Encyrtid parasitoid *Metaphycus stanleyi* by the pyriform scale, *Protopulvinaria pyriformis. Entomologia Experimentalis et Applicata* 3:231–237.

Bukovinszky, T., E. H. Poelman, R. Gols, G. Prekatsakis, L. E. M. Vet, J. A. Harvey, and M. Dicke. 2009. Consequences of constitutive and induced variation in plant nutritional quality for immune defence of a herbivore against parasitism. *Oecologia* 160:299–308.

Bull, A. T. 1970. Inhibition of polysaccharases by melanin–enzyme inhibition in relation to mycolysis. *Archives of Biochemistry and Biophysics* 137:345.

Cappuccino, N., H. Damman, and J-F. Dubuc. 1995. Spatial behavior and temporal dynamics of outbreak and nonoutbreak species. Pages 65–82 in N. Cappuccino and P. W. Price, editors. *New Approaches and Synthesis*. Academic Press, San Diego.

Caron, V., A. F. Janmaat, J. D. Ericsson, and J. H. Myers. 2008. Avoidance of the host immune response by a generalist parasitoid, *Compsilura concinnata* Meigen. *Ecological Entomology* 33:517–522.

Carroll, S. P., and H. Dingle. 1996. The biology of post-invasion events. *Biological Conservation* 78:207–214.

Carton, Y., A. J. Nappi, and M. Poirie. 2005. Genetics of anti-parasite resistance in invertebrates. *Developmental and Comparative Immunology* 29:9–32.

Carton, Y., M. Poirie, and A. J. Nappi. 2008. Insect immune resistance to parasitoids. *Insect Science* 15:67–87.

Cory, J. S., and J. H. Myers. 2009. Within and between population variation in disease resistance in cyclic populations of western tent caterpillars: a test of the disease defence hypothesis. *Journal of Animal Ecology* 3:646–655.

Cotter, S. C., and K. Wilson. 2002. Heritability of immune function in the caterpillar *Spodoptera littoralis*. *Heredity* 88:229–234.

Cotter, S. C., L. E. B. Kruuk, and K. Wilson. 2004a. Density-dependent prophylaxis and condition-dependent immune function in Lepidopteran larvae: a multivariate approach. *Journal of Animal Ecology* 73: 283–293.

Cotter, S. C., L. E. B. Kruuk, and K. Wilson. 2004b. Costs of resistance: genetic correlations and potential trade-offs in an insect immune system. *Journal of Evolutionary Biology* 17:421–429.

Cotter, S. C., J. P. Myatt, C. M. H. Benskin, and K. Wilson. 2008. Selection for cuticular melanism reveals immune function and life-history trade-offs in *Spodoptera littoralis*. *Journal of Evolutionary Biology* 21:1744–1754.

Cotter, S. C., S. J. Simpson, D. Raubenheimer, and K. Wilson. 2011. Macronutrient balance mediates trade-offs between immune function and life history traits. *Functional Ecology* 25:186–198.

D'Amico, V., J. S. Elkinton, G. Dwyer, J. P. Burand, and J. P. Buonaccorsi. 1996. Virus transmission in gypsy moths is not a simple mass action process. *Ecology* 77:201–206.

de Roode, J. C., A. B. Pedersen, M. D. Hunter, and S. Altizer. 2008. Host plant species affects virulence in monarch butterfly parasites. *Journal of Animal Ecology* 77:120–126.

Denno, R. F., G. K. Roderick, K. L. Olmstead, and H. G. Dobel. 1991. Density-related migration in planthoppers (homoptera, delphacidae) – the role of habitat persistence. *American Naturalist* 138: 1513–1541.

Despland, E., and M. Noseworthy. 2006. How well do specialist feeders regulate nutrient intake? Evidence from a gregarious tree-feeding caterpillar. *Journal of Experimental Biology* 209:1301–1309.

Diamond, S. E., and J. G. Kingsolver. 2011. Host plant quality, selection history and trade-offs shape the immune responses of *Manduca sexta*. *Proceedings of the Royal Society B-Biological Sciences* 278:289–297.

Doering, T. L., J. D. Nosanchuk, W. K. Roberts, and A. Casadevall. 1999. Melanin as a potential crypto-coccal defence against microbicidal proteins. *Medical Mycology* 37:175–181.

Dwyer, G., J. Dushoff, and S. H. Yee. 2004. The combined effects of pathogens and predators on insect outbreaks. *Nature* 430:341–345.

Dyer, L. A. 1995. Tasty generalists and nasty specialists? A comparative study of antipredator mechanisms in tropical lepidopteran larvae. *Ecology* 76:1483–1496.

Elderd, B. D., J. Dushoff, and G. Dwyer. 2008. Host–pathogen interactions, insect outbreaks, and natural selection for disease resistance. *American Naturalist* 172:829–842.

Faure, J. C. 1943. Phase variation in the armyworm, *Laphygma exempta* (Walk.). *Scientific Bulletin of the Department of Agriculture and Forestry of the Union of South Africa* 234:2–17.

Fellowes, M. D. E., A. R. Kraaijeveld, and H. C. J. Godfray. 1999. Cross-resistance following artificial selection for increased defense against parasitoids in *Drosophila melanogaster*. *Evolution* 53:966–972.

Freitak, D., A. Vanatoa, I. Ots, and M.J. Rantala. 2005. Formation of melanin-based wing patterns is influenced by condition and immune challenge in *Pieris brassicae*. *Entomologia Experimentalis et Applicata* 116:237–243.

Freitak, D., D. G. Heckel, and H. Vogel. 2009. Dietary-dependent trans-generational immune priming in an insect herbivore. *Proceedings of the Royal Society Biological Sciences* 276:2617–2624.

Futuyma, D. J., S. L. Leipertz, and C. Mitter. 1981. Selective factors affecting clonal variation in the fall cankerworm *alsophila-pometaria* (Lepidoptera, Geometridae). *Heredity* 47:161–172.

Gentry, G., and L. A. Dyer. 2002. On the conditional nature of neotropical caterpillar defenses against their natural enemies. *Ecology* 83:3108–3119.

Godfray, H. C. J. 1994. *Parasitoids: Behavioral and Evolutionary Ecology*. Princeton University Press, Princeton.

Goulson, D. 1994. Determination of larval melanization in the moth, *Mamestra brassicae*, and the role of melanin in thermoregulation. *Heredity* 73:471–479.

Goulson, D., and J. S. Cory. 1995. Responses of *Mamestra brassicae* (Lepidoptera, Noctuidae) to crowding interactions with disease resistance, color phase and growth. *Oecologia* 104:416–423.

Gross, P. 1993. Insect behavioral and morphological defenses against parasitoids. *Annual Review of Entomology* 38:251–273.

Hagen, S. B., O. Sorlibraten, R. A. Ims, and N. G. Yoccoz. 2006. Density-dependent melanism in winter moth larvae (Lepidoptera: Geometridae): a countermeasure against parasitoids? *Environmental Entomology* 35:1249–1253.

Hajek, A. E., and R. J. St. Leger. 1994. Interactions between fungal pathogens and insect hosts. *Annual Review of Entomology* 39:293–322.

Haviola, S., L. Kapari, V. Ossipov, M. J. Rantala, T. Ruuhola, and E. Haukioja. 2007. Foliar phenolics are differently associated with *Epirrita autumnata* growth and immunocompetence. *Journal of Chemical Ecology* 33:1013–1023.

Hochberg, M. E. 1991. Nonlinear transmission rates and the dynamics of infectious-disease. *Journal of Theoretical Biology* 153:301–321.

Hung, S. Y., and D. G. Boucias. 1992. Influence of *Beauveria bassiana* on the cellular defense response of the beet armyworm, *Spodoptera exigua*. *Journal of Invertebrate Pathology* 60:152–158.

Hunter, A. F. 1991. Traits that distinguish outbreaking and nonoutbreaking Macrolepidoptera feeding on northern hardwood trees. *Oikos* 60:275–282.

Hunter, M. D., A. D. Watt, and M. Docherty. 1991. Outbreaks of the winter moth on Sitka spruce in Scotland are not influenced by nutrient deficiencies of trees, tree budburst, or pupal predation. *Oecologia* 86:62–69.

Hunter, M. D., and J. C. Schultz. 1993. Induced plant defenses breached? Phytochemical induction protects an herbivore from disease. *Oecologia* 94:195–203.

Hunter, A. F. 1995. Ecology, life history, and phylogeny of outbreak and nonoutbreak species. Pages 41–64 in N. Cappuccino and P. W. Price, editors. *New Approaches and Synthesis*. Academic Press, San Diego.

Ibrahim, A. M. A., and Y. Kim. 2006. Parasitism by *Cotesia plutellae* alters the hemocyte population and immunological function of the diamondback moth, *Plutella xylostella*. *Journal of Insect Physiology* 52:943–950.

Ims, R. A., J. A. Henden, and S. T. Killengreen. 2008. Collapsing population cycles. *Trends in Ecology and Evolution* 23:79–86.

Jablonka, E., and M. J. Lamb. 2010. Transgenerational epigenetic inheritance. Pages 137–174. *Evolution: The Extended Synthesis*. MIT Press, Cambridge, MA.

Johnson, D. M., U. Buntgen, D. C. Frank, K. Kausrud, K. J. Haynes, A. M. Liebhold, J. Esper, and N. C. Stenseth. 2010. Climatic warming disrupts recurrent Alpine insect outbreaks. *Proceedings of the National Academy of Sciences of the USA* 107:20576–20581.

Kapari, L., E. Haukioja, M. J. Rantala, and T. Ruuhola. 2006. Defoliating insect immune defense interacts with induced plant defense during a population outbreak. *Ecology* 87:291–296.

Karban, R., and P. de Valpine. 2010. Population dynamics of an Arctiid caterpillar–tachinid parasitoid system using state-space models. *Journal of Animal Ecology* 79:650–661.

Klemola, T., N. Klemola, T. Andersson, and K. Ruohomäki. 2007a. Does immune function influence population fluctuations and level of parasitism in the cyclic geometrid moth? *Population Ecology* 49:165–178.

Klemola, N., T. Klemola, M. J. Rantala, and T. Ruuhola. 2007b. Natural host-plant quality affects immune defence of an insect herbivore. *Entomologia Experimentalis et Applicata* 123:167–176.

Klemola, N., L. Kapari, and T. Klemola. 2008. Host plant quality and defence against parasitoids: no relationship between levels of parasitism and a geometrid defoliator immunoassay. *Oikos* 117:926–934.

Klemola, N., T. Andersson, K. Ruohomäki, and T. Klemola. 2010. Experimental test of parasitism hypothesis for population cycles of a forest lepidopteran. *Ecology* 91:2506–2513.

Kraaijeveld, A. R., and H. C. J. Godfray. 1997. Trade-off between parasitoid resistance and larval competitive ability in *Drosophila melanogaster*. *Nature* 389:278–280.

Kunimi, Y., and E. Yamada. 1990. Relationship of larval phase and susceptibility of the armyworm, *Pseudaletia-separata* walker (Lepidoptera, Noctuidae) to a nuclear polyhedrosis-virus and a granulosis-virus. *Applied Entomology and Zoology* 25:289–297.

Larsson, S., C. Bjorkman, N. Kidd, and A. C. Neil. 1993. Outbreaks in diproid sawflies: why some species and not others. Pages 453–483 in M. R. Wagner, K. F. Raffa, editors. *Sawfly Life History Adaptations to Wood Plants*. Academic Press, San Diego.

Lavine, M. D., and N. E. Beckage. 1996. Temporal pattern of parasitism-induced immunosuppression in *Manduca sexta* larvae parasitized by *Cotesia congregata*. *Journal of Insect Physiology* 42:41–51.

Lee, K. P., J. S. Cory, K. Wilson, D. Raubenheimer, and S. J. Simpson. 2006. Flexible diet choice offsets protein costs of pathogen resistance in a caterpillar. *Proceedings of the Royal Society B-Biological Sciences* 273:823–829.

Lee, K. P., and K. Wilson. 2006. Melanism in a larval Lepidoptera: repeatability and heritability of a dynamic trait. *Ecological Entomology* 31:196–205.

Lee, K. P., S. J. Simpson, and K. Wilson. 2008. Dietary protein-quality influences melanization and immune function in an insect. *Functional Ecology* 22:1052–1061.

Little, T. J., B. O'Connor, N. Colegrave, K. Watt, and A. F. Read. 2003. Maternal transfer of strain-specific immunity in an invertebrate. *Current Biology* 13:489–492.

Majerus, E. N. 1998. *Melanism: Evolution in Action.* Oxford University Press, Oxford.

Mann, M. E., Z. H. Zhang, M. K. Hughes, R. S. Bradley, S. K. Miller, S. Rutherford, and F. B. Ni. 2008. Proxy-based reconstructions of hemispheric and global surface temperature variations over the past two millennia. *Proceedings of the National Academy of Sciences of the USA* 105:13252–13257.

Mattson, W. J., and N. D. Addy. 1975. Phytophagous insects as regulators of forest primary production. *Science* 190:515–522.

Mason, R. R. 1987. Nonoutbreak species of forest Lepidoptera. Pages 31–52 in P. Barbosa and J. C. Schultz, editors. *Insect Outbreaks.* Academic Press, San Diego.

McCallum, H., N. Barlow, and J. Hone. 2001. How should pathogen transmission be modelled? *Trends in Ecology and Evolution* 16:295–300.

Mikkola, K., and M. J. Rantala. 2010. Immune defence, a possible nonvisual selective factor behind the industrial melanism of moths (Lepidoptera). *Biological Journal of the Linnean Society* 99:831–838.

Miller, G. A., J. K. Pell, and S. J. Simpson. 2009. Crowded locusts produce hatchlings vulnerable to fungal attack. *Biology Letters* 5:845–848.

Moret, Y. 2006. Trans-generational immune priming: specific enhancement of the antimicrobial immune response in the mealworm beetle, *Tenebrio molitor. Proceedings of the Royal Society B-Biological Sciences* 273:1399–1405.

Moret, Y., and P. Schmid-Hempel. 2000. Survival for immunity: the price of immune system activation for bumblebee workers. *Science* 290:1166–1168.

Morris, R. F., ed. 1963. The dynamics of epidemic spruce budworm populations. *Memoirs of the Entomological Society of Canada* 31:1–332.

Nappi, A. J., and E. Vass. 1993. Melanogenesis and the generation of cytotoxic molecules during insect cellular immune-reactions. *Pigment Cell Research* 6:117–126.

Nishida, R. 2002. Sequestration of defensive substances from plants by Lepidoptera. *Annual Review of Entomology* 47:57–92.

Nuorteva. P. 1971. Decline of the parasite population of the geometrid moth, *Oporinia autumnata* (Bkh.) during the second year after a calamity on birches. *Ann. Entomol. Fenn.* 37:96.

Ojala, K., R. Julkunen-Tiito, L. Lindstrom, and J. Mappes. 2005. Diet affects the immune defence and life-history traits of an Arctiid moth *Parasemia plantaginis. Evolutionary Ecology Research* 7:1153–1170.

Ouedraogo, R.M., A. Kamp, M.S. Goettel, J. Brodeur, and M.J. Bidochka. 2002. Attenuation of fungal infection in thermoregulating Locusta migratoria is accompanied by changes in hemolymph proteins. *Journal of Invertebrate Pathology* 81:19–24.

Ourth, D. D., and H. E. Renis. 1993. Antiviral melanization reaction of *Heliothis virescens* hemolymph against DNA and RNA viruses in-vitro. *Comparative Biochemistry and Physiology B-Biochemistry & Molecular Biology* 105:719–723.

Parkash, R., S. Rajpurohit, and S. Ramniwas. 2008. Changes in body melanisation and desiccation resistance in highland vs. lowland populations of D-melanogaster. *Journal of Insect Physiology* 54:1050–1056.

Parkash, R., B. Kalra, and V. Sharma. 2010. Impact of body melanisation on contrasting levels of desiccation resistance in a circumtropical and a generalist *Drosophila* species. *Evolutionary Ecology* 24:207–225.

Parker, W. E., and A. G. Gatehouse. 1985. The effect of larval rearing conditions on flight performance in females of the african armyworm, *Spodoptera exempta* (Walker) (Lepidoptera, Noctuidae). *Bulletin of Entomological Research* 75:35–47.

Parmesan, C. 2006. Ecological and evolutionary responses to recent climate change. *Annual Review of Ecology Evolution and Systematics* 37:637–669.

Povey, S., S. C. Cotter, S. J. Simpson, K. P. Lee, and K. Wilson. 2009. Can the protein costs of bacterial resistance be offset by altered feeding behaviour? *Journal of Animal Ecology* 78:437–446.

Price, P. W. 2003. *Macroevolutionary Theory on Macroecological Patterns.* Cambridge University Press, Cambridge.

Rantala, M. J., and D. A. Roff. 2006. Analysis of the importance of genotypic variation, metabolic rate, morphology, sex and development time on immune function in the cricket, *Gryllus firmus. Journal of Evolutionary Biology* 19:834–843.

Rantala, M. J., and D. A. Roff. 2007. Inbreeding and extreme outbreeding cause sex differences in immune defence and life history traits in *Epirrita autumnata. Heredity* 98:329–336.

Reeson, A. F., K. Wilson, A. Gunn, R. S. Hails, and D. Goulson. 1998. Baculovirus resistance in the Noctuid *Spodoptera exempta* is phenotypically plastic and responds to population density. *Proceedings of the Royal Society of London Series B-Biological Sciences* 265:1787–1791.

Reilly, J. R., and A. E. Hajek. 2008. Density-dependent resistance of the gypsy moth *Lymantria dispar* to its nucleopolyhedrovirus, and the consequences for population dynamics. *Oecologia* 154: 691–701.

Richards, L. A., L. A. Dyer, A. M. Smilanich, and C. D. Dodson. 2010. Synergistic effects of amides from two piper species on generalist and specialist herbivores. *Journal of Chemical Ecology* 36:1105–1113.

Riddell, C., S. Adams, P. Schmid-Hempel, and E. B. Mallon. 2009. Differential expression of immune defences is associated with specific host–parasite interactions in insects. *Plos One* 4:e7621. doi:10.1371/journal.pone.0007621

Robb, T., M. R. Forbes, and I. G. Jamieson. 2003. Greater cuticular melanism is not associated with greater immunogenic response in adults of the polymorphic mountain stone weta, *Hemideina maori. Ecological Entomology* 28:738–746.

Roth, O., G. Joop, H. Eggert, J. Hilbert, J. Daniel, P. Schmid-Hempel, and J. Kurtz. 2010. Paternally derived immune priming for offspring in the red flour beetle, *Tribolium castaneum. Journal of Animal Ecology* 79:403–413.

Rothman, L. D., and J. H. Myers. 1996. Is fecundity correlated with resistance to viral disease in the western tent caterpillar? *Ecological Entomology* 21:396–398.

Ruiz-Gonzalez, M. X., Y. Moret, and M. J. F. Brown. 2009. Rapid induction of immune density-dependent prophylaxis in adult social insects. *Biology Letters* 5:781–783.

Ruohomäki, K. 1994. Larval parasitism in outbreaking and non-outbreaking populations of *Epirrita autumnata* (Lepidoptera geometridae). *Entomologica Fennica* 5:27–34.

Ruohomäki, K., T. Virtanen, P. Kaitaniemi, and T. Tammaru. 1997. Old mountain birches at high altitudes are prone to outbreaks of *Epirrita autumnata* (Lepidoptera: Geometridae). *Environmental Entomology* 26:1096–1104.

Ruuhola, T., J. P. Salminen, S. Haviola, S. Y. Yang, and M. J. Rantala. 2007. Immunological memory of mountain birches: effects of phenolics on performance of the autumnal moth depend on herbivory history of trees. *Journal of Chemical Ecology* 33:1160–1176.

Salt, G. 1968. Resistance of insect parasitoids to defence reactions of their hosts. *Biological Reviews of the Cambridge Philosophical Society* 43:200–232.

Sandre, S. L., T. Tammaru, and H. M. T. Hokkanen. 2011. Pathogen resistance in the moth *Orgyia antiqua*: direct influence of host plant dominates over the effects of individual condition. *Bulletin of Entomological Research* 101:107–114.

Schmid-Hempel, P., and D. Ebert. 2003. On the *Evolutionary Ecology* of specific immune defense. *Trends in Ecology and Evolution* 18:27–33.

Schott, T., S. B. Hagen, R. A. Ims, and N. G. Yoccoz. 2010. Are population outbreaks in sub-arctic geometrids terminated by larval parasitoids? *Journal of Animal Ecology* 79:701–708.

Scriber, J. M., and J. Slansky. 1981. The nutritional ecology of immature insects. *Annual Review of Entomology* 26:183–211.

Sheldon, B. C., and S. Verhulst. 1996. Ecological immunology: costly parasite defences and trade-offs in evolutionary ecology. *Trends in Ecology and Evolution* 11:317–321.

Shibizaki, A., and Y. Ito. 1969. Respiratory rates of green and black larvae of the armyworm, Leucania seperata (Lepidoptera: Noctuidae). *Appl. Entomol. Zool.* 4:100–101.

Smilanich, A. M., L. A. Dyer, J. Q. Chambers, and M. D. Bowers. 2009a. Immunological cost of chemical defence and the evolution of herbivore diet breadth. *Ecology Letters* 12:612–621.

Smilanich, A. M., L. A. Dyer, and G. L. Gentry. 2009b. The insect immune response and other putative defenses as effective predictors of parasitism. *Ecology* 90:1434–1440.

Smilanich, A. M., P. A. Mason, L. Sprung, T. R. Chase, and M. S. Singer. 2011. Complex effects of parasitoids on pharmacophagy and diet choice of a polyphagous caterpillar. *Oecologia* 165:995–1005.

Srygley, R. B., P. D. Lorch, S. J. Simpson, and G. A. Sword. 2009. Immediate protein dietary effects on movement and the generalised immunocompetence of migrating Mormon crickets *Anabrus simplex* (Orthoptera: Tettigoniidae). *Ecological Entomology* 34:663–668.

Stireman, J. O., L. A. Dyer, D. H. Janzen, M. S. Singer, J. T. Li, R. J. Marquis, R. E. Ricklefs, G. L. Gentry, W. Hallwachs, P. D. Coley, J. A. Barone, H. F. Greeney, H. Connahs, P. Barbosa, H. C. Morais, and I. R. Diniz. 2005. Climatic unpredictability and parasitism of caterpillars: Implications of global warming. *Proceedings of the National Academy of Sciences of the USA* 102:17384–17387.

St. Leger, R. J., R. M. Cooper, and A. K. Charnnley. 1988. The effect of melanization of *Manduca sexta* cuticle on growth and infection by *Metarhizium anisopliae*. *Journal of Invertebrate Pathology* 52:459–470.

Strand, M. R. 2008. The insect cellular immune response. *Insect Science* 15:1–14.

Strand, M. R. 2009. The interactions between polydnavirus-carrying parasitoids and their lepidopteran hosts. In: Goldsmith, M.R., Marec, F. (Eds.), *Molecular Biology and Genetics of the Lepidoptera*, pp. 321–336. CRC Press, Boca Raton, FL.

Steinhaus, E. A. 1958. Crowding as a possible stress factor in insect disease. *Ecology* 3:503–514.

Suwanchaichinda, C., and S. M. Paskewitz. 1998. Effects of larval nutrition, adult body size, and adult temperature on the ability of *Anopheles gambiae* (Diptera: Culicidae) to melanize Sephadex beads. *Journal of Medical Entomology* 35:157–161.

Tanhuanpää, M., K. Ruohomäki, P. Turchin, M. Ayres, H. Bylund, P. Kaitaniemi, T. Tammaru, and E. Haukioja. 2002. Population cycles of the autumnal moth in Fennoscandia. Pages 142–154 in A. Berryman, *Population Cycles: The Case for Trophic Interactions*. Oxford Press, New York.

Thomas, M. B., and N. E. Jenkins. 1997. Effects of temperature on growth of *Metarhizium flavoviride* and virulence to the variegated grasshopper, *Zonocerus variegatus*. *Mycological Research* 101:1469–1474.

Thomas, M. B., and S. Blanford. 2003. Thermal biology in insect–parasite interactions. *Trends in Ecology and Evolution* 18:344–350.

Wallner, W. E. 1987. Factors affecting insect population dynamics – differences between outbreak and non-outbreak species. *Annual Review of Entomology* 32:317–340.

Webb, B. A., and M. R. Strand 2005. The biology and genomics of polydnaviruses.In: Gilbert, L. I., Iatrou, K., Gill, S. S. (Eds.), *Comprehensive Molecular Insect Science*, vol. 5, pp. 260–323. Elsevier Press, San Diego.

Wilson, K., and A. F. Reeson. 1998. Density-dependent prophylaxis: Evidence from Lepidoptera–baculovirus interactions? *Ecological Entomology* 23:100–101.

Wilson, K., S. C. Cotter, A. F. Reeson, and J. K. Pell. 2001. Melanism and disease resistance in insects. *Ecology Letters* 4:637–649.

Wilson, K., M.B. Thomas, S. Blanford, M. Doggett, S.J. Simpson, and S.L. Moore. 2002. Coping with crowds: density-dependent disease resistance in desert locusts. *Proceedings of the National Academy of Science of the USA* 99:5471–5475.

Wilson, K., R. Knell, M. Boots, and J. Koch-Osborne. 2003. Group living and investment in immune defence: an interspecific analysis. *Journal of Animal Ecology* 72:133–143.

Wilson, K., and S. C. Cotter. 2008. Density-dependent prophylaxis in insects. Pages 137–176 in T. N. Ananthakrishnan & D. W. Whitman, editors. *Insects and Phenotypic Plasticity: Mechanisms and Consequences*. Science Publishers Inc., Plymouth, UK.

Wilson, K., and S. C. Cotter. 2009. Density-dependent prophylaxis in insects. Pages 191–232 in T. N. Ananthakrishnan and D. W. Whitman, editors. *Phenotypic Plasticity of Insects: Mechanisms and Consequences*. Science Publishers Inc., Plymouth, UK.

Wojda, I., and T. Jakubowicz. 2007. Humoral immune response upon mild heat-shock conditions in *Galleria mellonella* larvae. *Journal of Insect Physiology* 53:1134–1144.

Yang, S. Y., T. Ruuhola, and M. J. Rantala. 2007. Impact of starvation on immune defense and other life-history traits of an outbreaking geometrid, *Epirrita autumnata*: a possible causal trigger for the crash phase of population cycle. *Annales Zoologici Fennici* 44:89–96.

Yang, S., T. Ruuhola, S. Haviola, and M.J. Rantala. 2008. Effects of host plant shift on immune and other key life history traits of an eruptive Geometrid, *Epirrita autumnata* (Borkhausen). *Ecological Entomology* 33:510–516.

Zuk, M., and A. M. Stoehr. 2002. Immune defense and host life history. *American Naturalist* 60:S9–S22.

4

The Role of Ecological Stoichiometry in Outbreaks of Insect Herbivores

Eric M. Lind and Pedro Barbosa

4.1 Introduction

Insect herbivores are regulated by a combination of "top-down" and "bottom-up" forces, although the relative strength of these forces has long been debated (Andrewartha and Birch 1954, Hairston *et al.* 1960, Coley *et al.* 1985, Hunter and Price 1992, Hunter 2001). Outbreaks of insect herbivores suggest temporal variance in this regulation, and so provide an opportunity to examine this fundamental debate in plant–animal interactions from a different angle. It has been demonstrated that herbivore fitness and population expansion are tied to the quality of the host plant, variously defined (Awmack and Leather 2002). However, explorations of outbreaks have often confounded the lack of negative impacts from plants with positive impacts of plant nutrition when ascribing "quality" to a host plant.

At least two theoretical frameworks now exist for understanding the role of host plant nutritional quality in insect outbreaks. First, the geometric framework for nutrient regulation (Simpson and Raubenheimer 1993, Raubenheimer and Simpson 1997, Behmer *et al.* 2001) explores how actively foraging consumers balance competing priorities and deficiencies to build tissues from available food.

Insect Outbreaks Revisited, First Edition. Edited by Pedro Barbosa, Deborah K. Letourneau and Anurag A. Agrawal.
© 2012 Blackwell Publishing Ltd. Published 2012 by Blackwell Publishing Ltd.

How this consumption model aids understanding of insect outbreaks is discussed in detail elsewhere in this volume (Chapter 1, this volume), but in brief the geometric framework focuses on macronutrients and antinutrients (including plant defenses) and how herbivores ingest them optimally. A more general framework for exploring the role of plant quality in promoting outbreaks is provided by the rapidly developing field of ecological stoichiometry (Sterner and Elser 2002). Ecological stoichiometry has the potential for explanatory power across many scales, by focusing on ratios of elements within and between organisms, and connecting the abundance of these elements to their fundamental roles in cellular processes (Sterner and Elser 2002).

In this chapter, we explore the applicability of ecological stoichiometry to the initiation, dynamics, and consequences of insect outbreaks. By following the ratios and roles of elements across multiple levels, from genes to individuals to ecosystems, we hope to add another dimension to ecologists' understanding of the processes behind insect outbreaks at multiple scales.

4.2   Ecological and biological stoichiometry

Stoichiometry is a word of Greek derivation that means a "measuring of the elements." In the science of chemistry, stoichiometry is the process of recording, in ratios, the proportions of elements in compounds. *Ecological stoichiometry* (Sterner and Elser 2002) uses elemental ratios as a common currency to document and explain ecological interactions within and between organisms rather than reactions between compounds. The empirical approach of ecological stoichiometry is thus to quantify elemental ratios across ecological interactions.

An expanded framework is made possible by combining the measurements of elemental ratios with understanding of the roles of elements at the cellular level and the simple principle of conservation of matter. Nutrient ratios within organisms are self-similar in that whole-body nutrient ratios reflect cellular nutrient ratios. *Biological stoichiometry* seeks to understand the consequences of cellular nutrient allocation and availability on whole-organism behavior, ecology, and evolution (Elser *et al.* 2000b).

Elemental nutrients are characteristically associated with (though by no means exclusively found in) different essential cellular processes (Elser *et al.* 1996, 2003, Sterner and Elser 2002). Carbon (C) is the main component of energy sources and storage molecules such as sugars and lipids, and also of structural compounds (lignin in plants, and chitin in insects) and some chemical defenses in plant tissues. Nitrogen (N) is critical to cellular enzymatic processes in the form of amino acids used to make proteins, and is also present in insect chitin exoskeletons. Phosphorus (P) is incorporated into the nucleic acid molecules DNA and RNA, importantly including the ribosomal RNA (rRNA), which are essential to the machinery of cellular division and thus organismal growth.

Biological and ecological stoichiometry theories emphasize different aspects of the roles of the nutrients in organismal and ecosystem ecology (Sterner and Elser 2002). Fundamentally, ratios of nutrients are a common currency that can be measured across many scales, leading to novel ways of investigating the causes and consequences of insect outbreaks. This simple accounting can reveal important insights into the dynamics of these interactions. For example, one fundamental pattern emphasized by quantifying stoichiometric ratios is the strong difference in C:N and C:P across trophic levels: between terrestrial insect herbivores and their plant hosts (Fagan *et al.* 2002), between insect predators and their prey (Fagan and Denno 2004), and among detritivores (Martinson *et al.* 2008). While plant–herbivore theory has long considered the role of host plant quality in herbivore performance (Awmack and Leather 2002), and C:N ratios in particular (Mattson 1980), ecological stoichiometry provides an expanded framework for understanding the connection between nutrient limitation and life history in herbivorous insects.

4.3   Nutrient ratios and insect herbivory

The most basic application of ecological stoichiometry to insect outbreaks is to focus on the degree of limitation or mismatch between herbivores and their host plant food. This is far from a novel approach, as plant tissue concentrations of nutrients, especially nitrogen with respect to carbon (C:N ratio), have long been thought to be a factor in rates of insect herbivory (Mattson 1980) and outbreaks (Broadbeck and Strong 1987). TCR White (1984) proposed that available N in plant tissues led to increased insect herbivore outbreaks in large part due to the higher resource levels, which increased survival of early instar larvae. White focused on outbreaks of psyllids and other sap-sucking insects, and hypothesized that "stressed" plants increased the nitrogen available to these herbivores as amino acids were mobilized to combat that stress (White 1984). While hardly universally accepted (Larsson 1989, Price 1997), this plant stress hypothesis was among the first to formalize the idea that intraspecific variation in plant nutrient quality, specifically in terms of relative amino acid content, could explain the high variance in population growth typical of outbreaking insect herbivore species.

One of the competing paradigms attempting to explain the mechanisms that lead to differences in insect herbivory is the plant vigor hypothesis, formalized by Price (1991). Based on his research with gall-forming sawflies on willow (*Salix* spp.), Price advocated the idea that "vigorous" plant tissues (generally described as longer shoots) were more supportive of herbivory (Price 1991). While no explicit mechanistic nutrient mechanism was proposed, vigorous growth in plant tissues can reflect a high nutrient availability to the plant, and thus implicate this nutrient surplus in resulting herbivore performance (individual growth and survival, leading to greater population growth).

Although conceived, at least in part, as an alternative to the plant stress hypothesis, the plant vigor hypothesis is its complement in that both hypotheses implicitly favor a strong role for nutrient ratios as reflective of positive nutrient content as an explanation for herbivore population dynamics. These ideas stand together in contrast to the focus on evolutionary and ecological explanations of insect herbivory through plant defensive chemistry and growth tradeoffs that dominate much of the field (Coley *et al.* 1985, Herms and Mattson 1992, Strauss and Agrawal 1999). However, because ecological stoichiometric ratios capture all potential uses of elements in an interaction, it is not straightforward to separate nutrient from antinutrient forms. For instance, the same C:N ratio can reflect vastly different concentrations of molecules with identical nitrogen content, but with positive (e.g., amino acids) or negative (e.g., alkaloids) impacts on an herbivore.

Nonetheless, because insect herbivores use N-rich proteins for structural molecules (chitin) as well as for enzymatic activity, they require a C:N ratio significantly lower than in the plants they eat. Thus, there is a fundamental elemental ratio mismatch, in which insect herbivores have a higher per mass N requirements than their host plants can provide. This mismatch results in strong selection for feeding behaviors and physiology to maximize efficiency of use of what little N exists in any food source. For instance, the cabbage white *Pieris rapae* (L.) (Lepidoptera: Pieridae) increases its consumption rate with decreasing %N (increasing C:N), and can maintain similar growth trajectories as caterpillars grown on low C:N plants (Slansky and Feeny 1977). When increased consumption is directly deleterious to larval development, as when nutrition is accompanied by toxic compounds, the behavioral drive to minimize C:N can result in increased mortality. Slansky and Wheeler (1992) fed velvetbean caterpillars (*Anticarsia gemmatalis* (Hübner) [Lepidoptera: Noctuidae]) artifical diet with a range of nutrient and caffeine concentrations, and found the caterpillars on low-nutrient diet increased their feeding rate 2.5-fold over the high nutrient consumers, resulting in a self-administered lethal dose of caffeine.

Nitrogen ratios can influence the effectiveness of plant defenses in more subtle ways. Broadway and Duffey (1988) varied the proteins used in combination with a plant-derived proteinase inhibitor (PI) known to severely impact digestive ability of the armyworm caterpillar *Spodoptera exigua* (Hübner) (Lepidoptera: Noctuidae). Their manipulations of artificial diet revealed caterpillars fed proteins richest in the amino acids arginine and lysine were least affected by the PI (Broadway and Duffey 1988). Importantly, these two amino acids have N-rich side chains, and thus low C:N protein overwhelmed negative effects of the PI while high C:N protein led to severe reductions in caterpillar growth. Search for lower C:N tissues may contribute to the documented response of increased consumption by herbivores when confronted with PIs (Winterer and Bergelson 2001), which in turn may have selected for more diverse chemical responses in plants (Steppuhn and Baldwin 2007).

The limitations due to C:N may influence evolutionary patterns of herbivory as well as behavioral and physiological traits. In Lepidoptera, outbreaking species are known to be more likely to occur in species that overwinter as eggs and feed as larvae on early-season tissue (Hunter 1991). Many deciduous and evergreen trees exhibit predictable seasonal patterns in which a sharp peak in nutrient content corresponds to early-season leaf expansion and growth (Chapin and Kedrowski 1983). Thus, stoichiometric constraints may contribute to selection for outbreaking species of caterpillars to consume foliage at the time of year in which the leaves have the lowest C:N ratios, as the plants are growing themselves.

Potassium and Phosphorus

Compared to the historical attention given to the influence of C:N on insect herbivore outbreaks, other nutrients such as P and potassium (K) have been relatively neglected. Yet these (and other) elements have fundamental roles in insect physiology and ecology that allow predictions of how the stoichiometry of these elements might influence herbivore outbreaks.

The role of C:K ratios in herbivore performance and efficiency is not well-studied. Physiologically K acts in cell signaling but perhaps more importantly to herbivory, is critical in the regulation of the alkalinity of the gut lumen and malpighian tubules, and absorption of electrolytes and organic solutes (Azuma *et al.* 1995, Beyenbach *et al.* 2010). The high pH of the insect herbivore gut is created by pumping K^+ into the lumen, creating an electric current promoting H^+ flow back out of the lumen (Azuma *et al.* 1995). A severe increase in C:K could thus theoretically limit the ability of an insect herbivore to create the necessary alkaline digestive environment for disabling phenolics, tannins, and other acid accompaniments to plant protein (Felton and Duffey 1991). Extremely high C:K can limit plant growth, however (Sterner and Elser 2002), and whether plant C:K can be low enough to support plant growth but high enough to limit herbivore physiology is an open question.

Phospshorus is also a limiting element in plant productivity (Tilman 1982, Vitousek *et al.* 2010). In contrast to the focus on C:N in terrestrial systems, in aquatic systems research on herbivory has focused on the influence of C:P on primary producers and their zooplankton grazers (Sterner and Hessen 1994, Elser *et al.* 2000a). In these systems, C:P has been found to limit herbivore individual developmental growth rates (Urabe *et al.* 1997), limit population growth rates (DeMott *et al.* 2001, Makino *et al.* 2002), and explain lake-to-lake variance in population densities (Hessen 1992) of zooplankton. Given these examples of C:P stoichiometry in aquatic plant–animal relationships and its role in rapid population increases (Makino *et al.* 2002), it is surprising that more has not been done to investigate the importance of P in terrestrial herbivore outbreaks.

The basic physiology of how insect herbivores use P has been established experimentally. Woods *et al.* (2002) reared *Manduca sexta* (L.) (Lepidoptera: Sphingidae) on control and manipulated leaves (decreasing leaf C:P by standing petioles in high-P water). These caterpillars were then analyzed for P content in the whole body, gut portions, and extracted hemolymph (Woods *et al.* 2002). By calculating a mass balance of P from the leaves, caterpillars, and frass, they showed that at naturally low P levels the caterpillar incorporated into tissues on average 85% of leaf P, but that with higher levels of P this efficiency decreased as more P was excreted (Woods *et al.* 2002). Interestingly, the *M. sexta* larvae were found to store excess P in the hemolymph, which the authors conclude is a buffer mechanism against both free inorganic phosphorus and future P deficits (Woods *et al.* 2002). Somewhat in contrast, Meehan and Lindroth (2009) found P excretion rates of larval Orgyia *leucostigma* (J. E. Smith) (Lepidoptera: Lymantriidae) were independent of leaf P, but this may have been due to compensatory feeding by the caterpillar. As amino acid receptors are thought to most often regulate

consumer intake, this compensatory feeding may have been stimulated by correlated N increases due to constant N:P levels in the food (Meehan and Lindroth
2009). Woods *et al.* (2002) tested for and found no compensatory feeing behavior when manipulating C:P independently.

Perkins *et al.* (2004) also manipulated the C:P of both artificial diet and natural leaves, and demonstrated faster development and higher final mass for
M. sexta when reared on low C:P food. This effect was strongest when larvae had
been reared from hatching on the low C:P food, indicating the potential importance of early-development nutrition (Perkins *et al.* 2004). This is analogous to
the importance of such neonate effects emphasized by White (1984) in his work
on N-limited whitefly outbreaks.

An important insight of ecological stoichiometry is that because optimal nutrient
ratios reflect efficient cellular performance, not only deficits but also surpluses of
elements can be damaging to herbivore fitness (Boersma and Elser 2006). For
instance, while a limiting element in its own right, excess P can also be damaging
to insect herbivore fitness by altering N:P and C:N:P ratios. If such ratios are
homeostatic and insect herbivores cannot regulate P through buffering or excretion, C and N levels required to maintain optimal C:N:P may be unavailable,
leading to decreased performance (Boersma and Elser 2006). In diet studies on
the grasshopper *Melanoplus bivittatus* (Say) (Orthoptera: Acrididae), P was
found to have little impact on nymph development rate or ultimate mass in comparison to C:N ratios in the form of protein–carbohydrate manipulations (Loaiza
et al. 2008). Indeed low C:N:P was damaging, as grasshoppers fed otherwise
optimal diet took longer to complete development (Loaiza *et al.* 2008). A similar
effect can result if maintenance of N:P balance in plant tissues causes nutrient
uptake by the plant. Zehnder and Hunter (2009) found population growth rates
of *Aphis nerii* (Hemiptera: Aphididae) were lowered by excess P added to the
soil, as this increased plant uptake of N to deleterious concentrations in the host
tissues. Similarly, Huberty and Denno (2006) found that P uptake of *Spartina*
plants is weakly related to P fertilization, but positively influenced by N fertilization. The response of two *Proklesia* (Hemiptera: Delphacidae) planthopper species likewise depended on plant C:N, even under P fertilization (Huberty and
Denno 2006).

4.4   The growth rate hypothesis and insect outbreaks

Investigating herbivore outbreaks from first principles, ecological stoichiometry
suggests an emphasis on C:P because population and individual growth are ultimately driven by cellular division, which is in turn limited by the amount of
P-rich ribosomal RNA. The growth rate hypothesis (GRH) formally proposes
that organismal C:P should vary with growth rates, reflecting differential allocation to the molecular machinery of growth (Elser *et al.* 2000a). Research supporting the GRH in herbivores has come mainly from freshwater and marine
aquatic systems (Elser *et al.* 2003, Hessen *et al.* 2007). Experimental evidence
also reveals that the GRH holds best when P is limiting in the environment
(Makino and Cotner 2004, Shimizu and Urabe 2008).

Relatively few studies have specifically considered the role of C:P or P limitation in the growth rates of terrestrial insects. Janssen (1994) found strong effects of P on growth rates of the outbreaking caterpillar African armyworm, *Spodoptera exempta* (Walker) (Lepidoptera: Noctuidae) using a P-deficient plant variety in leaf section experiments. Of measured leaf variables (N, P, K, and water content), only P was found to be significantly explanatory of larval growth across leaf types (Janssen 1994). Elser *et al.* (2006) reared five *Drosophila* species from varying food sources and found strong positive correlations between relative growth rate of developing individuals and %P of their tissues. Kay *et al.* (2006) found strong differences in stoichiometry among life stages when manipulating the ratio of sucrose to phosphorus (C:P) in the diet of the pavement ant *Tetramorium caespitum*, but found no specific relationship between individual or colony growth and C:P.

In order for the GRH to be informative of eruptive population dynamics of outbreaking insects, growth rates of individuals must correlate with population growth rates. The larval stage is often a key factor influencing the population dynamics of Lepidoptera (Dempster 1983), and growth rate is a commonly used metric of performance and fitness in caterpillars (e.g., Coley *et al.* 2006). Shorter development times can lead to higher rates of population growth in herbivores, which can also translate into increased long-term abundance (Lind and Barbosa 2010).

A finding that outbreaking insects have lower C:P and N:P than non-outbreaking insects would thus validate the importance of the GRH to outbreak dynamics. Using a co-occurring group of macrolepidoptera as a test case, we evaluated the elemental stoichiometry of 24 species of caterpillars from seven families, of which seven species were known to exhibit outbreaking behavior in some portion of their range. Outbreak species had marginally higher C:N (two-tailed t-test p=0.07) and C:P (p=0.14), meaning lower whole-body content of both nutrients in the outbreaking species (Lind and Barbosa, unpublished data). There was no difference in the N:P ratios (p=0.4). While this may be a weak test in terms of replication and diversity of species, these results are nonetheless suggestive of lower rather than higher C:P, in contradiction to the GRH.

Data are still lacking to properly evaluate the GRH and its applicability to terrestrial insects, in general, and outbreaks, in particular. Specifically, a comparative approach, correcting for phylogenetic relatedness could be employed to test whether across a broad sampling of insect taxa, those species known to exhibit outbreaks have lower C:P than related non-outbreak species. Yet the GRH is not the only way in which ecological stoichiometry can influence insect outbreaks. The variance of host plant stoichiometry in type, space, and time is also likely to play a role in the initiation of outbreaks.

4.5   Variance in host plant stoichiometry and insect outbreaks

Herbivores face a dual challenge in building tissues by consuming plants as food. First, herbivores must overcome the fundamental limitations of much higher ratios of carbon to other important nutrients in plants than their bodies

(Fagan *et al*. 2002). Within this constraint, however, herbivores face plants which vary in time and space in their own C:N:P ratios. This variance can be classified into distinct categories: the variance among plant lineages and species, the variance among plant individuals, and the variance within plant individuals. These different variance components can explain different aspects of insect outbreaks.

At the evolutionary scale, leaf tissue N:P appears to tend toward a central two-thirds power law relationship (i.e., log[N] ~ 2/3log[P] in leaf tissue; Reich *et al*. 2010). This lends support to the tenets of ecological stoichiometry, that cellular machinery (in this case, of photosynthesis and growth) is composed of molecules whose elemental signature can be observed in the whole-body tissues (Sterner and Elser 2002). However, there is massive variance around this central tendency within taxonomic, functional, and life history group (Reich *et al*. 2010). Foliar stoichiometry has also been demonstrated to change in predictable ways across global temperature gradients (Kerkhoff *et al*. 2005). Specifically, Kerkhoff *et al*. (2005) found that plants in colder latitudes have significantly higher P in their tissues, independent of mass. Nitrogen, however, showed no such variance across temperature and latitude (Kerkhoff *et al*. 2005). Given the role suggested by the GRH for P, it is intriguing to note that the most dramatic and expansive outbreaks of forest insect herbivores occur at high latitudes. Tropical forest outbreaks typically are localized within a forest or along edges and in gaps or in agricultural or reforestation settings, and are qualitatively different (see Chapter 11, this volume).

Plant lineages can have distinct elemental stoichiometry as well. In an analysis of nearly 1300 species, Kerkhoff *et al*. (2006) found strong explanatory power of phylogenetic relatedness on N:P tissue stoichiometry at multiple levels through family, genus, and species. This is not surprising if stoichiometric ratios are a reflection of heritable aspects of plant life history. Still, from an herbivore perspective it raises the possibility that certain lineages will provide a better stoichiometric match, all else being equal. Certainly plant species can differ strongly in their C:N:P stoichiometry. For instance, gypsy moth *Lymantria dispar* (L.) (Lepidoptera: Lymantriidae) larval performance and pupal weight are highest on the host plant species with the highest foliar %N and %P, although these are not necessarily the preferred hosts in the field (Barbosa and Greenblatt 1979). In broad surveys of foliar nutrients across global sites, Townsend *et al*. (2007) found as much variance in N:P among species within a single forest, and among species within single families, as exists across a global data set of nutrient values. More research is needed on the ecological implications of this type of variation.

Beyond interspecific variation in stoichiometric ratios, environmental conditions can shift nutrient content of plants from site to site. This is an important source of variance that could help explain where outbreaks occur. For instance, leaf P is known to vary directly with the available soil P, while N and N-based defensive compounds vary much less (Wright *et al*. 2010). Studies of tropical forests have consistently shown relationships between foliar nutrient concentration and soil nutrients (Vitousek *et al*. 1995), and temperate soil chronosequences can exhibit similar relationships (Richardson *et al*. 2004). Microsite conditions such as differences in sun versus shade also can change foliar stoichiometry, although studies differ in their conclusions as to whether sun exposure decreases leaf C:N (Osier and Jennings 2007) or increases it (Mooney *et al.* 2009). Finally, whereas plant "stress" (variously defined) has been implicated in increasing nutrient availability to outbreaking insects, this response is likely guild-dependent (Huberty and Denno 2004) and may have little impact on foliar nutrient concentrations (Bosu and Wagner 2007).

Perhaps most germane to ecological studies of outbreaking herbivores is the variance between individuals of a plant species, even within a site, in tissue stoichiometry. This variance has been well documented in terms of tissue nutrient availability and even in terms of nutrient ratios, and specifically invoked to explain responses of outbreaking and other herbivores (Bryant *et al.* 1991b, Bryant *et al.* 1993, Hunter and Schultz 1995, Kaitaniemi *et al.* 1998, Raffa *et al.* 1998). Though these prior studies focused almost exclusively on the availability of N in the form of nutrients and plant defensive chemistry, more recent studies of intraspecific variation in stoichiometry have attempted to connect herbivory with C:P. In a desert ecosystem, the density of the weevil *Sabinia setosa* (Coleoptera: Curcurlionidae) exponentially decreases with linearly increasing C:P, but is unrelated to a major C-based defensive compound, suggesting P itself positively influences population dynamics of the insect (Schade *et al.* 2003). Leaf phosphorus content has also been found to predict winter moth caterpillar outbreaks in a monospecific host plant stand, whereas other variables did not (Hunter *et al.* 1991). Similarly, the spatial distribution of high densities of herbivores can correspond with the distribution of plants with low C:P ratios in nearly monospecific patches (Fagan *et al.* 2004, Apple *et al.* 2009). However, in an explicit test of P limitation in the lace bug *Corythuca arcuata* (Hemiptera: Tingidae) feeding on oaks under different burning regimes, other environmental factors overwhelmed any stoichiometric effect (Kay *et al.* 2007).

At the smallest scale, plants can vary in elemental stoichiometry across individual plant parts in the same individual. Obvious differences can be caused by the structure and function of tissues. For example, woody stem tissue and photsythesizing leaf tissue will have very different elemental ratios. Yet the stoichiometry of plant organs (roots, stems, leaves, and reproductive structures) appears to be strongly correlated within species (Kerkhoff *et al.* 2006). Thus, plants with high-nutrient leaves also have higher levels of nutrients in stem and root tissue, although whether this reflects nutrient uptake, transport, or storage is unclear (Kerkhoff *et al.* 2006). For outbreaking insects this suggests that plants with a generalized high-nutrient (low C:N and C:P) profile should support, for instance, outbreaks on leaves and reproductive structures simultaneously without a trade-off in food quality from the herbivore's perspective. Within tissue types, studies of leaf quality focusing on defensive plant chemistry have found extreme variation from leaf to leaf within a given individual tree (Roslin *et al.* 2006, Gripenberg *et al.* 2007). Little data exist on the variance among leaves within individuals in nutrient ratios, though this would be a fruitful area of investigation for researchers seeking to understand the sometimes high local variance in herbivore success.

4.6   Outbreak population dynamics models incorporating stoichiometry

Variance in plant stoichiometry plays a key role in modeling how nutrient ratios could exhibit the density-dependent, delayed feedbacks necessary for generating the cycling population dynamics observed in outbreaking insects. Andersen *et al.* (2004) proposed incorporating plant stoichiometry as a measure of food quality into cyclic population models which had been mostly concerned with food quantity or top-down control. Building on classical Lotka–Volterra population models with carrying capacities for autotrophs and consumers to initiate feedback cycles, Andersen *et al.* (2004) showed how herbivory can at first have a positive impact on succeeding generations of herbivores, and then switch to a negative influence through time. Other theoretical models have also demonstrated how low levels of herbivory can promote plant productivity at low levels through the recycling of limiting nutrients (de Mazancourt and Loreau 2000).

A simplified version of the role for elemental stoichiometry in insect outbreaks is presented in Figure 4.1. With the loss of nutrients due to low levels of herbivory (Figure 4.1a), autotrophs may concentrate nutrients in remaining tissue or uptake more nutrients, lowering C:N and C:P ratios in remaining available tissue (Figure 4.1b). This can continue as long as there are nutrients in the system available to the autotroph, and as long as there is enough autotroph tissue of sufficient nutrient quality to support growing herbivore populations. As nutrients are concentrated into herbivore tissue, plant tissue C:N and C:P ratios will rise, causing more stoichiometrically homeostatic large consumer populations to collapse (Figure 4.1c). Nutrients from frass and dead herbivores are deposited in the local soil pool, where they may be diluted (leached from the system entirely). The plant–herbivore cycle restarts as nutrients become re-available from inputs of decaying autotroph and consumer matter and are taken up into plant tissue (Figure 4.1d).

Explicitly incorporating stoichiometry into resource–consumer models has led to other insights. Because considering multiple nutrients adds dimension to resource niches, explicitly considering stoichiometry can help explain coexistence of consumer species on seemingly homogeneous resources (Loladze *et al.* 2004). Stiefs *et al.* (2010) found that although the conversion efficiency of consumers is critical to the cycling behavior of their populations, intraspecific competition of autotrophs can introduce stability to the system when nutrient limitation is in effect.

Importantly, the insights of these population dynamics models gained by including stoichiometry appear to match large-scale patterns quite well. In a global meta-analysis, Hillebrand *et al.* (2009) demonstrated that considering stoichiometric mismatch is imperative for explaining variance in both individual and population levels of herbivory. While the focus of these studies was plant biomass removed rather than herbivore population density, for macroinvertebrates the degree of stoichiometric mismatch was far more explanatory of population-level herbivory than body size or temperature effects (Hillebrand *et al.* 2009). Nutrient-influenced population models also emphasize that ecological

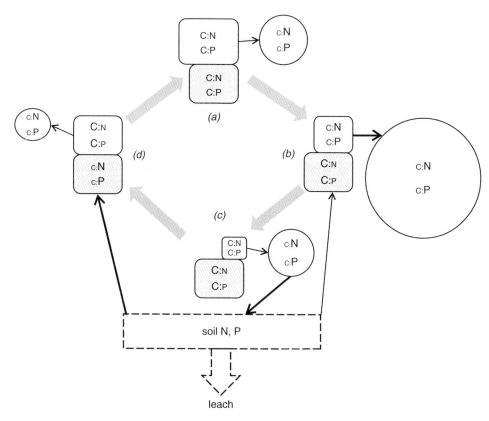

Figure 4.1 A conceptualization of how nutrient stoichiometry may influence insect herbivore outbreaks, based on Andersen *et al.* (2004). Plants (squares) are divided into available (e.g., leaf, unfilled squares) and unavailable (e.g., root, filled squares) tissues. Both plants and herbivores (circles) are drawn in proportion to their population density, with the relative stoichiometry of their tissues indicated by the relative size of the element abbreviations. Nutrient flows of interest at each focal step are drawn as solid arrows. Gray arrows represent movement between focal time steps in a plant–herbivore nutrient-based population cycle. (a) With the loss of nutrients due to low levels of herbivory, (b) autotrophs may concentrate nutrients in remaining tissue or uptake more nutrients, lowering C:N and C:P ratios in remaining available tissue. (c) As nutrients are concentrated into herbivore tissue, plant tissue C:N and C:P ratios will rise, causing more stoichiometrically homeostatic large consumer populations to collapse. Nutrients from frass and dead herbivores are deposited in the local soil pool, where they can be diluted (leached from the system entirely). (d) The plant–herbivore cycle restarts as nutrients become reavailable from inputs of decaying autotroph and consumer matter and are taken up into plant tissue.

stoichiometry of herbivory is not a one-way process of nutrient acquisition, but rather a temporary reallocation of existing nutrient molecules among various states of organic and inorganic matter in a local environment. It is, thus, as important to consider the impact of outbreaks on the stoichiometry of the local environment, as the reverse.

4.7   Impact of insect outbreaks on plant and environmental stoichiometry

A critical feedback in the dynamics of insect population outbreaks is the impact of herbivory, and especially defoliation, on plant resource stoichiometry (Figure 4.1). If plants respond to herbivore outbreaks by shifting ratios of carbon to other nutrients, this could have an impact on subsequent herbivory. This dynamic is well-studied empirically, although not necessarily in an explicitly stoichiometric context. Outbreaks may also have direct impacts on the biogeochemistry of nutrients in the environments on the local and even watershed scales, which can result in feedbacks to the plant tissue through time as well.

The impact of outbreaking herbivores on remaining or regrown plant tissues has a long history of investigation, especially as it relates to delayed induced resistance (DIR), in which plants respond to repeated or intense herbivory by chemical changes including increasing defensive compounds and lowering nitrogen levels (Haukioja and Niemela 1977, Haukioja 1991, Bryant *et al.* 1993, Nykanen and Koricheva 2004). Although these and other studies focused mostly on defensive compounds, many studies quantified changes in N as part of an investigation into trade-offs between growth and defense in attacked plants (e.g., Bryant *et al.* 1991a). In a model system for these investigations, mountain birch under defoliation from the outbreaking moth caterpillar *Epirrita autumnata* (Lepidoptera: Geometridae), leaves growing back after outbreaks exhibit the high defensive chemistry typical of young leaves, but the high C:N concentrations of mature leaves (Kaitaniemi *et al.* 1998). Thus the stoichiometric challenge for the herbivore is compounded by diluted nutrient resources being even better protected.

Bryant *et al.* (1993) examined whether nutrient addition could change the DIR dynamic. By adding N and P factorially to experimentally defoliated leaves, they tested whether the same nutrient limitations in leaf response to outbreaks would occur. In fact, N changed little between heavily attacked and unattacked trees under any fertilization treatment, while P was consistently higher in leaves when heavy defoliation was followed by fertilization of any kind, including N alone (Bryant *et al.* 1993).

In the unusual case of the outbreaking aquatic moth caterpillar *Acentria ephemerella* (Denis and Schiffermüller) (Lepidoptera: Crambidae), the pond-weed *Potamogeton* species that serve as its host respond to defoliation by shifting C:N:P stoichiometry among constituent parts, translocating N and P away from leaves and to resting buds which remain unattacked (Miler and Straile 2010). Even moderate herbivory can cause such redistributions in attacked plants. Frost and Hunter (2008) found oak seedlings raised C:N in leaves and increased storage pools of N after only two days of attack by tussock moth caterpillars *O. leucostigma*. Studies of whole-tree responses likewise show physiological responses including suppression of nutrient uptake by trees undergoing outbreaks by gypsy moth *L. dispar* caterpillars (Kosola *et al.* 2001).

Other systems do not necessarily follow the reduced nutrient pattern. Norway spruce leaves of defoliated plants had higher soluble nitrogen following three years of defoliation by the outbreaking sawfly *Pristiphora abietina* (Christ)

(Hymenoptera: Tenthredinidae) than those which had not been attacked (Schafellner *et al.* 1999). In fact, few generalizations can be drawn across plant and herbivore type in terms of chemical response to outbreaks, perhaps the most important being that under damage, C becomes concentrated in plant tissues, raising C:N ratios by default (Nykanen and Koricheva 2004).

Herbivore outbreaks represent a concentration of nutrients into animal tissue that under normal circumstances reside in plant tissue. The ultimate fate of the nutrients may be the same, in that they will be recycled through soil absorption and decomposition, yet the different form in which they fall can have consequences for the local environment. One change is that C and N fall as frass, a highly labile form of these elements as compared with senescent leaf tissue, which has immediate impacts on the soil biota and belowground processes (Bardgett and Wardle 2003). Folivore outbreaks can as much as double the normal input rates of C and N (le Mellec and Michalzik 2008). Like the leaf tissue from which it derives, however, frass can differ in C:N stoichiometry even among individual caterpillars on the same plant species (Madritch *et al.* 2007). Phosphorus is less well studied in terms of both its prevalence in frass and its resultant impact on soil processes. However, Meehan and Lindroth (2009) demonstrated in the laboratory that P did occur in measurable concentrations in *O. leucostigma* caterpillar frass, and was not related to leaf P.

In addition to direct inputs, outbreaking herbivores can change nutrient cycling through indirect inputs. For instance, *Ips* and *Dendroctonus* (Coleoptera: Curculionidae) bark beetle outbreaks can cause premature leaf drop in ponderosa pine forests, resulting in foliage with a C:N ratio half that of unattacked foliage (Morehouse *et al.* 2008). The resulting soils under the outbreak populations had higher nitrogen mineralization rates and ammonium accumulation (Morehouse *et al.* 2008). Similar results were found in the soils mixed forests invaded by the hemlock woolly adelgid *Adelges tsugae* (Annand) (Hemiptera: Adelgidae) (Orwig *et al.* 2008). These empirical demonstrations of increased nutrient availability to plants accord well with the theoretical models of outbreak feedbacks to plant production (Figure 4.1; Andersen *et al.* 2004).

Importantly, N and P mobilized from plant tissue into herbivore tissue and then deposited through feces or death do not necessarily stay in the local ecosystem but can be leached out as dissolved ions (Frost and Hunter 2004, 2007). Caterpillar outbreaks can contribute to measurable, multiyear changes in forest stream chemistry as first nitrates, and then other minerals, are leached (Lewis and Likens 2007). The export of N following defoliation events can even result in same-year limitations to plant growth at a landscape scale (McNeil *et al.* 2007).

4.8 Ecological stoichiometry and natural enemy regulation of outbreaks

While little is definitive about the dynamics of ecological stoichiometry in insect outbreaks, less still is known about the importance of stoichiometry to higher trophic levels. In general, arthropod predators and parasitoids are better matched stoichiometrically to their food resources than herbivores are to plants, since C:N:P of herbivore tissue is close to that of predatory insects (Fagan *et al.* 2002). Evidence exists that hyperparasitoids face even smaller C:N differences from their food than primary parasitoids (Harvey *et al.* 2009), though their life histories are challenging in other ways. Herbivores should be more predictable stoichiometric resources since animals are thought to be much more homeostatic in their resource ratios than plants, with fewer options for nutrient reallocation. However, the homeostasis assumption has not proven robust (Bertram *et al.* 2008), with terrestrial insects found to be especially variable in constitutive P (Persson *et al.* 2010).

Outbreaking herbivores could thus, in theory, present a link between the favorable stoichiometry of host plants and successful reproduction in high-nutrient predators and parasitoids. If both abundance and stoichiometric quality of herbivores increase simultaneously during outbreaks, parasitoid (and parasite) enemies may be the ultimate beneficiaries. Even more promising for understanding outbreak dynamics would be the decoupling of these rates. If abundant herbivores were significantly variable in stoichiometry, it could explain some of the delayed control action of parasitoid functional response.

4.9 Conclusions

Ecological stoichiometry holds promise for using novel approaches to exploring fundamental questions about insect outbreaks. Specific predictions of ecological stoichiometry remain largely untested. Under the GRH, insects with lower C:P would be expected to grow more rapidly under P limitation, potentially leading to outbreak dynamics. Alternatively, fluctuating C:N:P stoichiometry in host plants through space and time may help initiate outbreaks as herbivores become temporarily less distant from the elemental components of their food sources. Population dynamics cycles could be directly influenced by the shifts in local nutrient pools between autotrophs, consumers, and soil, with transport out of local environments through leaching an important feedback process. Finally, despite the bottom-up focus of this chapter, ecological stoichiometry predicts the regulation of herbivores by higher trophic levels also should be influenced by the dynamics of nutrient ratios.

References

Andersen, T., Elser, J. J., and Hessen, D. O. 2004. Stoichiometry and population dynamics. *Ecology Letters* 7:884–900.

Andrewartha, H. G., and Birch, L. C. 1954. *The Distribution and Abundance of Animals*. University of Chicago Press, Chicago.

Apple, J. L., Wink, M., Wills, S. E., and Bishop, J. G. 2009. Successional change in phosphorus stoichiometry explains the inverse relationship between herbivory and lupin density on Mount St. Helens. *PLoS One* 4.

Awmack, C. S., and Leather, S. R. 2002. Host plant quality and fecundity in herbivorous insects. *Annual Review of Entomology* 47:817–844.

Azuma, M., Harvey, W. R., and Wieczorek, H. 1995. Stoichiometry of K+/H+ antiport helps to explain extracellular pH 11 in a model epithelium. *Febs Letters* 361:153–156.

Barbosa, P., and Greenblatt, J. 1979. Suitability, digestibility, and assimilation of various host plants of the gypsy moth, *Lymantria dispar* L. Lepidoptera, Lymantriidae. *Oecologia* 43:111–119.

Bardgett, R. D., and Wardle, D. A. 2003. Herbivore-mediated linkages between aboveground and below-ground communities. *Ecology* 84:2258–2268.

Behmer, S. T., Raubenheimer, D., and Simpson, S. J. 2001. Frequency-dependent food selection in locusts: a geometric analysis of the role of nutrient balancing. *Animal Behaviour* 61:995–1005.

Bertram, S. M., Bowen, M., Kyle, M., and Schade, J. 2008. Extensive natural intraspecific variation in stoi-chiometric C:N:P composition in two terrestrial insect species. *Journal of Insect Science* 8:1–7.

Beyenbach, K. W., Skaer, H., and Dow, J. A. T. 2010. The developmental, molecular, and transport biology of malpighian tubules. *Annual Review of Entomology* 55:351–374.

Boersma, M., and Elser, J. J. 2006. Too much of a good thing: on stoichiometrically balanced diets and maximal growth. *Ecology* 87:1325–1330.

Bosu, P. P., and Wagner, M. R. 2007. Effects of induced water stress on leaf trichome density and foliar nutrients of three elm *Ulmus* species: implications for resistance to the elm leaf beetle. *Environmental Entomology* 36:595–601.

Broadbeck, B., and Strong, D. R. 1987. Amino acid nutrition of herbivorous insects and stress to host plants. In: *Insect Outbreaks*, edited by P. Barbosa and J. C. Schultz. Academic Press, San Diego.

Broadway, R. M., and Duffey, S. S. 1988. The effect of plant protein-quality on insect digestive physiology and the toxicity of plant proteinase-inhibitors. *Journal of Insect Physiology* 34:1111–1117.

Bryant, J. P., Heitkonig, I., Kuropat, P., and Owensmith, N. 1991a. Effects of severe defoliation on the long-term resistance to insect attack and on leaf chemistry in 6 woody species of the southern African savanna. *American Naturalist* 137:50–63.

Bryant, J. P., Provenza, F. D., Pastor, J., Reichardt, P. B., Clausen, T. P., and Dutoit, J. T. 1991b. Interactions between woody plants and browsing mammals mediated by secondary metabolites. *Annual Review of Ecology, Evolution, and Systematics* 22:431–446.

Bryant, J. P., Reichardt, P. B., Clausen, T. P., and Werner, R. A. 1993. Effects of mineral-nutrition on delayed inducible resistance in Alaska paper birch. *Ecology* 74:2072–2084.

Chapin, F. S., and Kedrowski, R. A. 1983. Seasonal-changes in nitrogen and phosphorus fractions and autumn retranslocation in evergreen and deciduous taiga trees. *Ecology* 64:376–391.

Coley, P. D., Bateman, M. L., and Kursar, T. A. 2006. The effects of plant quality on caterpillar growth and defense against natural enemies. *Oikos* 115:219–228.

Coley, P. D., Bryant, J. P., and Chapin, F. S. 1985. Resource availability and plant antiherbivore defense. *Science* 230:895–899.

DeMott, W. R., Gulati, R. D., and Van Donk, E. 2001. Effects of dietary phosphorus deficiency on the abundance, phosphorus balance, and growth of *Daphnia cucullata* in three hypereutrophic Dutch lakes. *Limnology and Oceanography* 46:1871–1880.

Dempster, J. P. 1983. The natural control of populations of butterflies and moths. *Biological Reviews of the Cambridge Philosophical Society* 58:461–481.

Elser, J. J., Acharya, K., Kyle, M., Cotner, J., Makino, W., Markow, T., et al. 2003. Growth rate-stoichiometry couplings in diverse biota. *Ecology Letters* 6:936–943.

Elser, J. J., Dobberfuhl, D. R., MacKay, N. A., and Schampel, J. H. 1996. Organism size, life history, and N:P stoichiometry. *Bioscience* 46:674–684.

Elser, J. J., Fagan, W. F., Denno, R. F., Dobberfuhl, D. R., Folarin, A., Huberty, A., et al. 2000a. Nutritional constraints in terrestrial and freshwater food webs. *Nature* 408:578–580.

Elser, J. J., Watts, T., Bitler, B., and Markow, T. A. 2006. Ontogenetic coupling of growth rate with RNA and P contents in five species of *Drosophila*. *Functional Ecology* 20:846–856.

Elser, J. J., Sterner, R. W., Gorokhova, E., Fagan, W. F., Markow, T. A., Cotner, J. B., et al. 2000b. Biological stoichiometry from genes to ecosystems. *Ecology Letters* 3:540–550.

Fagan, W. F., Siemann, E., Mitter, C., Denno, R. F., Huberty, A. F., Woods, H. A., et al. 2002. Nitrogen in insects: implications for trophic complexity and species diversification. *American Naturalist* 160:784–802.

Fagan, W. F., and Denno, R. F. 2004. Stoichiometry of actual vs. potential predator–prey interactions: insights into nitrogen limitation for arthropod predators. *Ecology Letters* 7:876–883.

Fagan, W. F., Bishop, J. G., and Schade, J. D. 2004. Spatially structured herbivory and primary succession at Mount St Helens: field surveys and experimental growth studies suggest a role for nutrients. *Ecological Entomology* 29:398–409.

Felton, G. W., and Duffey, S. S. 1991. Reassessment of the role of gut alkalinity and detergency in insect herbivory. *Journal of Chemical Ecology* 17:1821–1836.

Frost, C. J., and Hunter, M. D. 2004. Insect canopy herbivory and frass deposition affect soil nutrient dynamics and export in oak mesocosms. *Ecology* 85:3335–3347.

Frost, C. J., and Hunter, M. D. 2007. Recycling of nitrogen in herbivore feces: plant recovery, herbivore assimilation, soil retention, and leaching losses. *Oecologia* 151:42–53.

Frost, C. J., and Hunter, M. D. 2008. Herbivore-induced shifts in carbon and nitrogen allocation in red oak seedlings. *New Phytologist* 178:835–845.

Gripenberg, S., Salminen, J. P., and Roslin, T. 2007. A tree in the eyes of a moth – temporal variation in oak leaf quality and leaf-miner performance. *Oikos* 116:592–600.

Hairston, N. G., Smith, F. E., and Slobodkin, L. B. 1960. Community structure, population control, and competition. *American Naturalist* 94:421–425.

Harvey, J. A., Wagenaar, R., and Bezemer, T. M. 2009. Interactions to the fifth trophic level: secondary and tertiary parasitoid wasps show extraordinary efficiency in utilizing host resources. *Journal of Animal Ecology* 78:686–692.

Haukioja, E. 1991. Induction of defenses in trees. *Annual Review of Entomology* 36:25–42.

Haukioja, E., and Niemela, P. 1977. Retarded growth of a geometrid larva after mechanical damage to leaves of its host tree. *Annales zoologici Fennici* 14:48.

Herms, D. A., and Mattson, W. J. 1992. The dilemma of plants – to grow or defend. *Quarterly Review of Biology* 67:283–335.

Hessen, D. O. 1992. Nutrient element limitation of zooplankton production. *American Naturalist* 140:799–814.

Hessen, D. O., Jensen, T. C., Kyle, M., and Elser, J. J. 2007. RNA responses to N- and P-limitation; reciprocal regulation of stoichiometry and growth rate in *Brachionus*. *Functional Ecology* 21:956–962.

Hillebrand, H., Borer, E. T., Bracken, M. E. S., Cardinale, B. J., Cebrian, J., Cleland, E. E., et al. 2009. Herbivore metabolism and stoichiometry each constrain herbivory at different organizational scales across ecosystems. *Ecology Letters* 12:516–527.

Huberty, A. F., and Denno, R. F. 2004. Plant water stress and its consequences for herbivorous insects: a new synthesis. *Ecology* 85:1383–1398.

Huberty, A. F., and Denno, R. F. 2006. Consequences of nitrogen and phosphorus limitation for the performance of two planthoppers with divergent life-history strategies. *Oecologia* 149:444–455.

Hunter, A. F. 1991. Traits that distinguish outbreaking and nonoutbreaking macrolepidoptera feeding on northern hardwood trees. *Oikos* 60:275–282.

Hunter, M. D. 2001. Multiple approaches to estimating the relative importance of top-down and bottom-up forces on insect populations: experiments, life tables, and time-series analysis. *Basic and Applied Ecology* 2:295–309.

Hunter, M. D., and Price, P. W. 1992. Playing chutes and ladders – heterogeneity and the relative roles of bottom-up and top-down forces in natural communities. *Ecology* 73:724–732.

Hunter, M. D., and Schultz, J. C. 1995. Fertilization mitigates chemical induction and herbivore responses within damaged oak trees. *Ecology* 76:1226–1232.

Hunter, M. D., Watt, A. D., and Docherty, M. 1991. Outbreaks of the winter moth on Sitka spruce in Scotland are not influenced by nutrient deficiencies of trees, tree budburst, or pupal predation. *Oecologia* 86:62–69.

Janssen, J. A. M. 1994. Impact of the mineral-composition and water-content of excised maize leaf sections on fitness of the African armyworm, *Spodoptera exempta* Lepidoptera, Noctuidae. *Bulletin of Entomological Research* 84:233–245.

Kaitaniemi, P., Ruohomaki, K., Ossipov, V., Haukioja, E., and Pihlaja, K. 1998. Delayed induced changes in the biochemical composition of host plant leaves during an insect outbreak. *Oecologia* 116:182–190.

Kay, A. D., Rostampour, S., and Sterner, R. W. 2006. Ant stoichiometry: elemental homeostasis in stage-structured colonies. *Functional Ecology* 20:1037–1044.

Kay, A., Schade, J., Ogdahl, M., Wesserle, E. O., and Hobbie, S. 2007. Fire effects on insect herbivores in an oak savanna: the role of light and nutrients. *Ecological Entomology* 32:754–761.

Kerkhoff, A. J., Enquist, B. J., Elser, J. J., and Fagan, W. F. 2005. Plant allometry, stoichiometry and the temperature-dependence of primary productivity. *Global Ecology and Biogeography* 14:585–598.

Kerkhoff, A. J., Fagan, W. F., Elser, J. J., and Enquist, B. J. 2006. Phylogenetic and growth form variation in the scaling of nitrogen and phosphorus in the seed plants. *American Naturalist* 168:E103–E122.

Kosola, K. R., Dickmann, D. I., Paul, E. A., and Parry, D. 2001. Repeated insect defoliation effects on growth, nitrogen acquisition, carbohydrates, and root demography of poplars. *Oecologia* 129:65–74.

Larsson, S. 1989. Stressful times for the plant stress – insect performance hypothesis. *Oikos* 56:277–283.

Lewis, G. P., and Likens, G. E. 2007. Changes in stream chemistry associated with insect defoliation in a Pennsylvania hemlock–hardwoods forest. *Forest Ecology and Management* 238:199–211.

Lind, E. M., and Barbosa, P. 2010. Life history traits predict relative abundance in an assemblage of forest caterpillars. *Ecology* 91:3274–3283.

Loaiza, V., Jonas, J. L., and Joern, A. 2008. Does dietary P affect feeding and performance in the mixed-feeding grasshopper, Acrididae *Melanoplus bivitattus*? *Environmental Entomology* 37:333–339.

Loladze, I., Kuang, Y., Elser, J. J., and Fagan, W. F. 2004. Competition and stoichiometry: coexistence of two predators on one prey. *Theoretical Population Biology* 65:1–15.

Madritch, M. D., and Cardinale, B. J. 2007. Impacts of tree species diversity on litter decomposition in northern temperate forests of Wisconsin, USA: a multi-site experiment along a latitudinal gradient. *Plant and Soil* 292:147–159.

Makino, W., and Cotner, J. B. 2004. Elemental stoichiometry of a heterotrophic bacterial community in a freshwater lake: implications for growth- and resource-dependent variations. *Aquatic Microbial Ecology* 34:33–41.

Makino, W., Urabe, J., Elser, J. J., and Yoshimizu, C. 2002. Evidence of phosphorus-limited individual and population growth of Daphnia in a Canadian shield lake. *Oikos* 96:197–205.

Martinson, H. M., Schneider, K., Gilbert, J., Hines, J. E., Hamback, P. A., and Fagan, W. F. 2008. Detritivory: stoichiometry of a neglected trophic level. *Ecological Research* 23:487–491.

Mattson, W. 1980. Herbivory in relation to plant nitrogen content. *Annual Review of Ecology and Systematics* 11:119–161.

de Mazancourt, C., and Loreau, M. 2000. Grazing optimization, nutrient cycling, and spatial heterogeneity of plant–herbivore interactions: should a palatable plant evolve? *Evolution* 54:81–92.

McNeil, B. E., de Beurs, K. M., Eshleman, K. N., Foster, J. R., and Townsend, P. A. 2007. Maintenance of ecosystem nitrogen limitation by ephemeral forest disturbance: an assessment using MODIS, Hyperion, and Landsat ETM. *Geophysical Research Letters*, 34.

Meehan, T. D., and Lindroth, R. L. 2009. Scaling of individual phosphorus flux by caterpillars of the whitemarked tussock moth, *Orygia leucostigma*. *Journal of Insect Science* 9.

le Mellec, A., and Michalzik, B. 2008. Impact of a pine lappet *Dendrolimus pini* mass outbreak on C and N fluxes to the forest floor and soil microbial properties in a Scots pine forest in Germany. *Canadian Journal of Forest Research–Revue Canadienne De Recherche Forestiere* 38:1829–1841.

Miler, O., and Straile, D. 2010. How to cope with a superior enemy? Plant defence strategies in response to annual herbivore outbreaks. *Journal of Ecology* 98:900–907.

Mooney, E. H., Tiedeken, E. J., Muth, N. Z., and Niesenbaum, R. A. 2009. Differential induced response to generalist and specialist herbivores by *Lindera benzoin* Lauraceae in sun and shade. *Oikos* 118:1181–1189.

Morehouse, K., Johns, T., Kaye, J., and Kaye, A. 2008. Carbon and nitrogen cycling immediately following bark beetle outbreaks in southwestern ponderosa pine forests. *Forest Ecology and Management* 255:2698–2708.

Nykanen, H., and Koricheva, J. 2004. Damage-induced changes in woody plants and their effects on insect herbivore performance: a meta-analysis. *Oikos* 104:247–268.

Orwig, D. A., Cobb, R. C., D'Amato, A. W., Kizlinski, M. L., and Foster, D. R. 2008. Multi-year ecosystem response to hemlock woolly adelgid infestation in southern New England forests. *Canadian Journal of Forest Research–Revue Canadienne De Recherche Forestiere* 38:834–843.

Osier, T. L., and Jennings, S. M. 2007. Variability in host plant quality for the larvae of a polyphagous insect folivore in midseason: the impact of light on three deciduous sapling species. *Entomologia Experimentalis Et Applicata* 123:159–166.

Perkins, M. C., Woods, H. A., Harrison, J. F., and Elser, J. J. 2004. Dietary phosphorus affects the growth of larval *Manduca sexta*. *Archives of Insect Biochemistry and Physiology* 55:153–168.

Persson, J., Fink, P., Goto, A., Hood, J. M., Jonas, J., and Kato, S. 2010. To be or not to be what you eat: regulation of stoichiometric homeostasis among autotrophs and heterotrophs. *Oikos* 119:741–751.

Price, P. W. 1991. The plant vigor hypothesis and herbivore attack. *Oikos* 62:244–251.

Price, P. W. 1997. *Insect Ecology*. John Wiley and Sons, Inc., New York.

Raffa, K. F., Krause, S. C., and Reich, P. B. 1998. Long-term effects of defoliation on red pine suitability to insects feeding on diverse plant tissues. *Ecology* 79:2352–2364.

Raubenheimer, D., and Simpson, S. J. 1997. Integrative models of nutrient balancing: application to insects and vertebrates. *Nutrition Research Reviews* 10:151–179.

Reich, P. B., Oleksyn, J., Wright, I. J., Niklas, K. J., Hedin, L., and Elser, J. J. 2010. Evidence of a general 2/3-power law of scaling leaf nitrogen to phosphorus among major plant groups and biomes. *Proceedings of the Royal Society B-Biological Sciences* 277:877–883.

Richardson, S. J., Peltzer, D. A., Allen, R. B., McGlone, M. S., and Parfitt, R. L. 2004. Rapid development of phosphorus limitation in temperate rainforest along the Franz Josef soil chronosequence. *Oecologia* 139:267–276.

Roslin, T., Gripenberg, S., Salminen, J. P., Karonen, M., O'Hara, R. B., Pihlaja, K., *et al.* 2006. Seeing the trees for the leaves – oaks as mosaics for a host-specific moth. *Oikos* 113:106–120.

Schade, J., Kyle, M., Hobbie, S. E., Fagan, W. F., and Elser, J. J. 2003. Stoichiometric tracking of soil nutrients by a desert insect herbivore. *Ecology Letters* 6:96–101.

Schafellner, C., Berger, R., Dermutz, A., Fuhrer, E., and Mattanovich, J. 1999. Relationship between foliar chemistry and susceptibility of Norway spruce Pinaceae to Pristiphora abietina Hymenoptera: Tenthredinidae. *Canadian Entomologist* 131:373–385.

Shimizu, Y., and Urabe, J. 2008. Regulation of phosphorus stoichiometry and growth rate of consumers: theoretical and experimental analyses with Daphnia. *Oecologia* 155:21–31.

Simpson, S. J., and Raubenheimer, D. 1993. A multilevel analysis of feeding-behavior – the geometry of nutritional decisions. *Philosophical Transactions of the Royal Society of London Series B-Biological Sciences* 342:381–402.

Slansky, F., and Feeny, P. 1977. Stabilization of rate of nitrogen accumulation by larvae of cabbage butterfly on wild and cultivated food plants. *Ecological Monographs* 47:209–228.

Slansky, F., and Wheeler, G. S. 1992. Caterpillars compensatory feeding response to diluted nutrients leads to toxic allelochemical dose. *Entomologia Experimentalis Et Applicata* 65:171–186.

Steppuhn, A., and Baldwin, I. T. 2007. Resistance management in a native plant: nicotine prevents herbivores from compensating for plant protease inhibitors. *Ecology Letters* 10:499–511.

Sterner, R. W., and Elser, J. J. 2002. *Ecological Stoichiometry: The Biology of Elements from Molecules to the Biosphere*. Princeton University Press, Princeton.

Sterner, R. W., and Hessen, D. O. 1994. Algal nutrient limitation and the nutrition of aquatic herbivores. *Annual Review of Ecology and Systematics* 25:1–29.

Stiefs, D., van Voorn, G. A. K., Kooi, B. W., Feudel, U., and Gross, T. 2010. Food quality in producer–grazer models: a generalized analysis. *American Naturalist* 176:367–380.

Strauss, S. Y., and Agrawal, A. A. 1999. The ecology and evolution of plant tolerance to herbivory. *Trends in Ecology and Evolution* 14:179–185.

Tilman, D. 1982. *Resource Competition and Community Structure*. Princeton University Press, Princeton.

Urabe, J., Clasen, J., and Sterner, R. W. 1997. Phosphorus limitation of *Daphnia* growth: is it real? *Limnology and Oceanography* 42:1436–1443.

Vitousek, P. M., Turner, D. R., and Kitayama, K. 1995. Foliar nutrients during long-term soil development in Hawaiian montane rain-forest. *Ecology* 76:712–720.

Vitousek, P. M., Porder, S., Houlton, B. Z., and Chadwick, O. A. 2010. Terrestrial phosphorus limitation: mechanisms, implications, and nitrogen-phosphorus interactions. *Ecological Applications* 20:5–15.

White, T. C. R. 1984. The abundance of invertebrate herbivores in relation to the availability of nitrogen in stressed food plants. *Oecologia* 63:90–105.

Winterer, J., and Bergelson, J. 2001. Diamondback moth compensatory consumption of protease inhibitor-transformed plants. *Molecular Ecology* 10:1069–1074.

Woods, H. A., Perkins, M. C., Elser, J. J., and Harrison, J. F. 2002. Absorption and storage of phosphorus by larval *Manduca sexta*. *Journal of Insect Physiology* 48:555–564.

Wright, D. M., Jordan, G. J., Lee, W. G., Duncan, R. P., Forsyth, D. M., and Coomes, D. A. 2010. Do leaves of plants on phosphorus-impoverished soils contain high concentrations of phenolic defence compounds? *Functional Ecology* 24:52–61.

Zehnder, C. B., and Hunter, M. D. 2009. More is not necessarily better: the impact of limiting and excessive nutrients on herbivore population growth rates. *Ecological Entomology* 34:535–543.

Part II

Population Dynamics and Multispecies Interactions

5

Plant-Induced Responses and Herbivore Population Dynamics

André Kessler, Katja Poveda, and Erik H. Poelman

Insect Outbreaks Revisited, First Edition. Edited by Pedro Barbosa, Deborah K. Letourneau and Anurag A. Agrawal.
© 2012 Blackwell Publishing Ltd. Published 2012 by Blackwell Publishing Ltd.

5.1   Introduction

The search for the mechanisms underlying outbreak events is arguably one of the most interesting and most debated aspects of the study of plant–insect interactions. Recurrent outbreak events can generally be described as herbivore population oscillations comprised of a consistent pattern of population increase; a peak in population density, which normally causes major productivity loss in the host plants; and an herbivore population breakdown (Berryman 1987, Myers 1988, Ginzburg and Taneyhill 1994). The importance of extrinsic versus intrinsic factors in creating critical population densities at the onset of an outbreak, as well as causing recurrent population cycles, has been the subject of debate because these factors are frequently intercorrelated and can theoretically result in similar patterns of population dynamics (Mattson *et al.* 1991, Turchin *et al.* 1991, Ginzburg and Taneyhill 1994, Berryman 1995). Among the factors that can intrinsically produce cyclic population dynamics, herbivore density-dependent effects such as (1) qualitative and quantitative changes in the host plants, (2) herbivore–natural enemy interactions, and (3) maternal effects have received much attention (Berryman 1987, Haukioja and Neuvonen 1987, Myers 1988, Ginzburg and Taneyhill 1994, Myers *et al.* 1998, Turchin *et al.* 2003).

Plant defensive chemistry has long been considered as a major factor potentially affecting the density-dependent or delayed density-dependent population patterns (Edwards and Wratten 1983, Bergelson *et al.* 1986, Silkstone 1987, Underwood *et al.* 2005). However, its role in affecting outbreaks is still unclear in most cases, likely because plants have multiple chemical defense strategies, including direct and indirect defenses; constitutive, rapidly induced, or delayed induced resistance; as well as tolerance responses (Karban and Baldwin 1997, Kessler and Baldwin 2002). Thus, it may be difficult to link single-plant defense traits to herbivore population dynamics (Turchin 2003, Donaldson and Lindroth 2008). Moreover, different herbivore species may respond differently to plant resistance and tolerance, making generalizations problematic. Plant responses to herbivory can be hypothesized to be important at two stages in herbivore population cycles: at (1) low densities, affecting population growth to densities critical to create outbreaks; or (2) outbreak densities, when they may contribute to the crash of the outbreak-level herbivore population, because most plants in an attacked population are defoliated, destroyed, or of low quality as a result of massive herbivore attack. For the latter case, inducible resistance has been hypothesized as a potential mechanism causing herbivore population crashes, through decreased survival and fecundity in a delayed density-dependent manner (Haukioja and Neuvonen 1987, Abbott *et al.* 2008). At low herbivore densities, the distribution of resistant genotypes within a plant population and the altered distribution of low- and high-quality food plants as a result of induction of resistance and/or tolerance by herbivores are potential plant-mediated factors affecting herbivore population growth. We focus this chapter on the potential effect of induced plant responses to herbivory as factors influencing herbivore population dynamics at low densities and discuss their role in affecting the probability of outbreaks and the suppression of outbreak-level densities.

5.1.1 The multiple faces of plant-induced responses to herbivory

Most plant species change their metabolism and thus their chemical phenotype when attacked by herbivores (Karban and Baldwin 1997). These changes include the production of compounds or physical structures that can directly repel and/or reduce the performance of herbivores and thus increase the resistance of the plant to subsequent attacks (so-called direct resistance), or cause herbivores to move away from the plants they damaged. Here, we refer to (induced) resistance traits as herbivore-induced traits whose expression decreases the performance and/or survival of the attacking herbivores (insect's perspective). Respectively, (induced) defenses are induced responses that through repellence or the reduction in herbivore performance have a positive effect on plant fitness (from the plant's perspective) (Karban and Baldwin 1997). In addition to mediating direct resistance, some metabolic changes after herbivory can function as cues that may influence the interaction of the plant with other organisms. For example, plants produce herbivore-induced volatile organic compounds (HIVOCs) when attacked by herbivores, which can attract natural enemies (predators and parasitoid) of the herbivores to the plant and thus reduce herbivore survival and abundance (so-called indirect resistance) (Dicke and Baldwin 2010, Kessler and Heil 2011). Plant-mediated multitrophic interactions can be very complex in nature. This complexity may be partially caused by high specificity of plant responses to different herbivore species with differential effects on subsequent species with which they interact (Kessler and Halitschke 2007). Moreover, because herbivore attack also affects plant primary metabolism, induced responses can include changes in photosynthetic rates and resource allocation from and to storage, which allow plants to tolerate certain amounts of tissue loss without negative impacts on relative fitness, and which also can affect plant palatability to herbivores through altered availability of basic nutrients (Mattson 1991, Agrawal 2000, Stowe *et al.* 2000, Poveda *et al.* 2010).

To a certain extent, all of the above described changes in primary and secondary metabolism could be simultaneously expressed in the same plant and concertedly influence the interactions with other organisms (Kessler and Baldwin 2002). Thus, in order to assess the impact of plant-induced responses on population outbreak dynamics we need to account for direct and indirect (interaction-mediated) effects of induced responses on arthropod performance and fecundity. Moreover, we need to differentiate between rapidly induced and delayed induced plant responses, which can have dramatically different effects on herbivore population growth and outbreak dynamics (Haukioja and Neuvonen 1987). Here we try to merge some basic principles that can be derived from studies on rapidly induced plant responses and their effects on arthropod populations with the general theory of herbivore outbreaks. Because there are virtually no studies that test the effect of inducibility (phenotypic plasticity in resistance and tolerance) as a trait on the probability of herbivore populations reaching outbreak densities, this attempt is meant to generate new integrative hypotheses and cannot function as a thorough proof of principle.

5.1.2 Plant-induced responses in the light of the general theory of outbreaks

The general theory of outbreaks is based on a simple logistic growth model that considers the population dynamics of a single species (Berryman 1987). Accordingly, a general model for the population growth of a species could be described as

$$r = dN/Ndt$$

where r is the specific growth rate and N the density of the population. Assuming that (1) populations don't grow when no organisms are present ($dN/dt = 0$ when $N = 0$) and (2) population growth is limited ($dN/dt = 0$ when $N >> 0$, or $r = dN/dt = 0$ when N reaches some relatively large population size at carrying capacity K), then

$$r = a_0 + a_1 N$$

where $a_0 > 0$ and describes the specific maximal growth rate of the population at the limit ($N \rightarrow 0$), and a_1 is a coefficient describing the negative effects ($a_1 < 0$) of the interactions among members of the population and results from a population density-dependent change in growth rate (Berryman 1987). For $r=0$, it then follows that

$$K = -a_0 / a_1$$

Berryman (1987) argues that, following the above logic, outbreaks can occur only when the equilibrium density K is increased (e.g., from K_1 to K_2) through the increase of the maximum specific growth rate of the population, a_0, or the decrease of the negative interaction among the members of the population, a_1 (Figure 5.1). Such an increase of K presumes the existence of a threshold population density (outbreak threshold) that separates low-density dynamics from outbreak dynamics (Berryman 1982) and at which the probability of an outbreak becomes more likely. What ecological factors can influence the creation of an outbreak threshold and thus influence the probability of herbivore populations to reach outbreak densities?

The values of a_0 and a_1, and consequentially the specific population growth rate r, are functions of genetic properties of the population and all physical and biotic properties of the environment that can influence the survival, reproduction, and dispersal of a species (Berryman 1987). Plant resistance traits as well as plant-mediated interaction with the arthropod community (e.g., plant-mediated predator attraction, competition, cross-resistance, and induced susceptibility) are among those environmental properties that can affect insect herbivore fitness, spatial population dynamics, and thus the probability for insect populations to reach outbreak levels. More specifically, rapidly induced plant responses to herbivory can be hypothesized to influence a_1 and thus the carrying capacity K, through known mechanisms such as herbivore-induced increased resistance and increased predator attraction that affect herbivore fitness. If the expression of such induced direct and indirect resistance traits has a more negative effect on

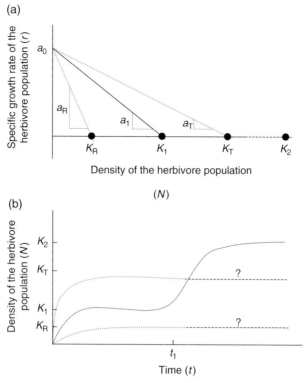

Figure 5.1 Induced plant responses as factors affecting herbivore population growth. (a) Specific growth rate of the herbivore population (r) as a function of population density (N). (b) Herbivore population density (N) as a function of time (t). Following a simple logistic equation, the specific growth rate of the herbivore population declines linearly as population density N increases producing the equilibrium point K_1 (continuous line). Plant-induced responses are hypothesized to shift the equilibrium point to a new equilibrium $K_R < K_1$ in cases of induced resistance and to $K_T > K_1$ in cases of tolerance. a_1 represents the negative interaction among members of the population and is largest in a_R and smallest in a_T. An outbreak occurs when K_1 increases to K_2 at time t_1.

the population growth than would be expected from mere density-dependent exhaustion of consumable plant material ($a_1 < a_R$ in Figure 5.1), an outbreak threshold would be reached more slowly or not at all (K_R in Figure 5.1).

Alternatively, herbivore-induced responses such as induced susceptibility or tolerance could result in greater herbivore survival rates and higher fecundity of the offspring ($a_1 > a_T$ in Figure 5.1) and thus increase the probability to reach outbreak threshold population densities (K_T in Figure 5.1). Support for the contention that plant-induced responses to herbivory have an effect on population dynamics, potentially leading up to critical outbreak levels, comes from recent studies on the mechanisms of plant trait–mediated effects on herbivore fitness. However, such mechanisms have rarely been considered as important factors in outbreak dynamics. Part of the problem, when trying to establish a connection between induced responses to herbivory and an insect population reaching outbreak thresholds, is that the heritability of inducibility (phenotypic plasticity) is relatively difficult to study. Moreover, induced responses may affect different herbivore species differently so that it is important not only how damage by a certain herbivore affects resistance against the herbivore but also how herbivory by different herbivore species or plant-mediated tritrophic interactions affect the performance of the attacker on the responding plant. Thus, strong ecological context dependency of herbivore-induced responses on arthropod performance can be expected. We use the above simplified theoretical framework to ask if plant responses to herbivory can (directly or indirectly) influence herbivore fitness and population growth and thus the probability of reaching population densities that can be considered outbreaks. We review studies that demonstrate and/or suggest effects of rapid induced plant responses to herbivory on arthropod population growth and community composition and have implications for the understanding of outbreaks. We do so by focusing on the three major categories of responses: induced direct, indirect resistance, and tolerance.

5.2   Direct induced resistance and herbivory

5.2.1   Direct induced resistance influencing herbivore population growth

Direct induced resistance is usually measured as the reduction of performance (reduction of feeding, growth, survival, and reproduction) of herbivores on plants that were previously damaged compared to undamaged control plants. The intensity and magnitude of metabolic changes are functions of time, because metabolic changes may be dependent on time-consuming alterations of gene expression and the *de novo* biosynthesis of enzymes and secondary metabolites (Karban and Baldwin 1997). The intensity of plant responses to herbivory depend on the amount of damage (Mithofer *et al.* 2005) and the presence of herbivore-derived chemical elicitors (Alborn *et al.* 1997, Pohnert *et al.* 1999, Schmelz *et al.* 2006, Alborn *et al.* 2007), and a response may be detectable only when certain damage thresholds are reached (Kessler and Halitschke 2007). Moreover, the decay of induced resistance response can be shorter (rapid induced resistance) or longer (long-term inducible resistance) than the generation time of the attacking insect. Long-term induced resistance was hypothesized to play a role in the decline of insect populations in a delayed density-dependent manner (Haukioja and Neuvonen 1987). In contrast, the rapid induction of plant metabolic changes was hypothesized to have rather stabilizing effects on the herbivore population due to negative feedback loops that affect herbivore population growth before reaching outbreak thresholds (Haukioja and Neuvonen 1987).

Rapid induction of plant metabolic changes may increase mortality directly (Zavala *et al.* 2004, van Dam *et al.* 2005), or increase feeding time and slow development (Kessler and Baldwin 2004, van Dam *et al.* 2005, Gomez *et al.* 2008, Poelman *et al.* 2010) of herbivores, which could cause prolonged exposure to potential mortality factors, such as natural enemies (Kessler and Baldwin 2004). Depending on how herbivores can cope with plant responses, effects on herbivore population growth can vary but may be generally negative as a consequence of the negative fitness effects on the herbivores or increased herbivore dispersion (e.g., increasingly negative impact of interactions on population growth $a_1 \rightarrow a_R$, Figure 5.1). Whether or not an herbivore is dispersing away from a damaged plant or avoids an already damaged plant when searching for food depends on the magnitude and impact of previous damage, on the expression of plant responses, and on herbivore properties that allow coping with plant responses (Kessler and Baldwin 2002). In addition, herbivore and plant life history traits make certain herbivore species relatively more likely to move (i.e., to be mobile) than others or make the movement from one plant to another more or less likely (i.e., to exhibit a dense or sparse plant distribution).

Assuming negative effects of rapid induced resistance on performance and dispersion, one can hypothesize that the probability of reaching density levels close to an outbreak threshold is smaller in plant systems with induced resistance phenotypes as compared to those with constitutive resistance. This is if induced resistance affects herbivore performance at plant damage levels significantly below those that would cause direct resource limitation of the herbivores in noninducible plants. Herbivores able to escape the increasingly bad food quality of previously damaged plants can also escape the associated negative fitness consequences as long as enough undamaged plants remain in the population and increased movement is not correlated with increased mortality. Thus, the influence of induced resistance as a plant trait on herbivore population growth also depends on its effect on herbivore dispersion.

Induce resistance to herbivores with significant effects on herbivore fitness has been shown in a number of study systems (Haukioja and Neuvonen 1987, Karban and Baldwin 1997, Osier and Lindroth 2004, Donaldson and Lindroth 2008). Moreover, we are beginning to understand how induced resistance affects the dispersion of herbivores and thus the distribution of herbivory. Early studies suggested that locally induced resistance could cause a more equal distribution of herbivory within individual plants with modular or clonal structure (Edwards and Wratten 1983, Karban and Baldwin 1997). In white clover, *Trifolium repens*, induced resistance causes a more equal distribution of herbivory among the clonal runners connected to a mother plant (Gomez *et al.* 2008). Accordingly, Edwards and Wratten (1983) proposed that induced resistance should cause an "overdispersion" of herbivory within plants, hence a more even than random or clumped distribution, because herbivores would move away from areas of previous damage. Resulting increased dispersal of damage among plants would cause the frequency of highly resistant plant phenotypes to increase faster with the number of herbivores than would be expected from resource-limiting density-dependent effects alone. This way induced resistance could slow herbivore population growth and so delay or prevent outbreak threshold levels (Figure 5.1). Nevertheless, the increased dispersal could be to the benefit of the individual herbivore by, for example, partitioning the food resource with conspecifics, or reducing its exposure to natural enemies that may use induced plant cues for host or prey search (Kessler and Baldwin 2001).

Only few studies have manipulated rapid induction while documenting the distribution of herbivores and their damage among plants, and these studies found both even dispersion and aggregation of herbivory as a function of induced resistance (Bergelson *et al.* 1986, Silkstone 1987, Underwood *et al.* 2005) with significant effects on insect herbivore fitness (Stamp 1980, Denno and Benrey 1997). More rigorous tests of the spatial dynamics of herbivores as a function of inducibility of resistance suggest that herbivory dispersion patterns are dependent on the time lag of induction and/or the induction threshold. Underwood *et al.* (2005) used a spatially explicit model to demonstrate that increasing lags and thresholds for the production of induced resistance cause increased aggregation of herbivores and their damage. However, without any lag or threshold (or very short lag times and low thresholds), induced resistance can lead to an even distribution of herbivores with the above-discussed potential consequences on population dynamics. Genotypic diversity in constitutive resistance within a plant population also does influence the distribution of herbivory (Underwood 2004) and would have to be integrated with phenotypically plastic responses to understand the distribution of herbivory in plant populations and its consequences for outbreak dynamics.

In many simulation models like those noted above, herbivores are assumed to move randomly through the plant population without having information about the induced resistance level of the plants before tasting (Underwood *et al.* 2005). The ability of the herbivore to use herbivore-inducible plant cues, such as HIVOCs, to evaluate the quality of the current host plant in comparison to the potential future host plants seems to be an important factor in determining if induced resistance results in clumped, or even, distribution of herbivory. Both are possible, due to herbivores accumulating on or avoiding a damaged plant. Chrysomelid (Schutz *et al.* 1997) and scolytid beetles (Raffa 2001) do aggregate on damaged plants using HIVOC cues induced by the feeding of conspecifics. Higher conspicuity of HIVOC-emitting attacked plants is thought to be the reason for why *Spodoptera frugiperda* (J. E. Smith) caterpillars prefer HIVOCs from damaged plants over less volatile-producing undamaged plants although the caterpillars grow slower on damaged, more resistant plants (Carroll *et al.* 2006). In contrast, *Manduca quinquemaculata* (Haworth) (Kessler and Baldwin 2001) and *Heliothis virescens* (Fabricius) (De Moraes *et al.* 2001) female moths use herbivore-induced VOCs to avoid herbivore-attacked and low-quality plants for oviposition.

Current herbivore dispersion models also do not account for plants having information about future herbivory (Underwood *et al.* 2005). There is an increasing number of examples that plants respond to volatile cues emitted from damaged neighbors, and so can be primed against subsequent herbivory or directly induce resistance against future herbivory (Baldwin *et al.* 2006, Heil and Karban 2010). Such interactions can potentially increase the proportion of resistant plants in the population far beyond the proportion of plants actually damaged, potentially resulting in accelerated inhibition of herbivore population growth (e.g., increasingly negative impact of interactions on population growth $a_1 \rightarrow a_R$, Figure 5.1). However, no data are available to support such claims, which provide ample opportunities for future research.

In conclusion, herbivore-induced direct resistance can reduce the fitness of feeding and newly arriving insects by reducing plant quality and the distribution thereof in an accelerated density-dependent manner, supporting earlier hypotheses that rapid induced resistance has a stabilizing effect on herbivore population growth (Haukioja and Neuvonen 1987). Factors that likely contribute to the effect of induced resistance on herbivore population growth include plant genotypic diversity, the herbivore's ability to assess host plant quality (random vs. directed movement), and the ability of plants to induce resistance in response to cues from neighboring damaged plants.

5.2.2 Induced resistance-mediated herbivore community interactions

Feeding damage of one herbivore species can induce plant resistance to conspecifics as well as to different herbivore species (Kessler and Baldwin 2004, Van Zandt and Agrawal 2004, Lynch *et al*. 2006). For example, the mirid *Tupiocoris notatus* (Distant) induces cross-resistance of *Nicotiana attenuata* Torr. ex S.Watson plants to the more damaging *Manduca* hornworms (Plate 5.1, Kessler and Baldwin 2004); and *Asclepias syriaca* L. damaged by early, stem-feeding weevil *Rhyssomatus lineaticollis* (Say) are more resistant to a number of leaf chewers, including monarch butterfly larvae, *Danaus plexippus* (L.), that occur later in the season (Van Zandt and Agrawal 2004). Such induced cross-resistance effects may not only occur among leaf feeders but also, due to systemically induced responses, affect herbivores on other tissues and mediate above-below ground interactions (Poveda *et al*. 2003, van Dam *et al*. 2005, Soler *et al*. 2007), or affect plant–pollinator interaction (Poveda *et al*. 2003, Adler *et al*. 2006, Kessler *et al*. 2011). The effects of cross-resistance on outbreak probability are largely unknown, but, like the intraspecific induced resistance effects, it can be hypothesized to stabilize population growth of the affected herbivore species (Haukioja and Neuvonen 1987).

On the other hand, induced responses to herbivory can also facilitate choice or enhance performance of other herbivore species, particularly when specialist herbivores inhibit the production of defensive secondary metabolites (Kahl *et al*. 2000, Musser *et al*. 2005, Poelman *et al*. 2008, Poelman *et al*. 2010). In such cases initial damage by a particular species could increase performance, survival, and thus population growth of a subsequently arriving herbivore species and so affect the probability of reaching outbreak densities positively. For example, tomato plants damaged by the aphid *Macrosiphum euphorbiae* (Thomas) were preferred for oviposition and supported a better larval growth of *Spodoptera exigua* (Hübner) compared to undamaged control and *S. exigua*–damaged plants (Rodriguez-Saona *et al*. 2005). In other examples, artificial silencing of downstream components of the jasmonic acid (JA) wound-signaling cascade in *N. attenuata* recruits novel generalists to the plant, suggesting that wound-induced JA responses can have a profound effect on herbivore host selection (Kessler *et al*. 2004, Paschold *et al*. 2007). Moreover, field studies suggest that plants can be constrained in responding to entire herbivore communities, because a response to the first attacker will result in an altered attack by the future herbivore community in both composition and abundance of attackers (Viswanathan *et al*. 2005, Bukovinszky *et al*. 2010, Poelman *et al*. 2010).

In the light of such cross-resistance and induced susceptibility examples, it seems possible that population growth and thus outbreak probability of certain species can be affected by other herbivores in the community. Accordingly, outbreaking species can be hypothesized to have disproportionately high effects on the rest of the arthropod community. However, we are not aware of studies that specifically address these hypotheses (but see Chapter 10, "The Ecological Consequences of Insect Outbreaks").

5.3 Plant tolerance and herbivory

Plant tolerance is mediated through traits that reduce the impact of herbivory on plant fitness with compensatory growth (Tiffin 2000, Nunez-Farfan *et al.* 2007) and is thus commonly quantified as the degree to which plant fitness is affected by herbivore damage relative to fitness in undamaged plants (Strauss and Agrawal 1999). Tolerance to insect herbivory seems ubiquitous, and it has been shown in a number of plant–herbivore systems (Fornoni and Nunez-Farfan 2000, Valverde *et al.* 2003, Kessler and Baldwin 2004, Kaplan *et al.* 2008) including plant species prone to being affected by outbreak events (Mattson *et al.* 1991, Kaitaniemi *et al.* 1999, Stevens *et al.* 2007). Although we have a relatively good understanding of chemical, physical, and biotic mechanisms underlying plant resistance (Kessler and Baldwin 2002), the mechanisms responsible for tolerance are considerably less well understood. Strauss and Agrawal (1999) reviewed the literature on the basic mechanisms that could be involved in plants being able to tolerate herbivory. They include (1) increased photosynthesis, (2) higher relative growth rates, (3) increased branching due to a release of apical dominance, after damage, and (4) a higher storage of carbon belowground, which can be allocated aboveground after damage. In addition, it has been proposed that tolerance in trees can be related to the amount of resources stored before damage in the tree's stem or allocated to the stem after damage depending on the amount of nutrients in the soil (Stevens *et al.* 2008). There is a growing consensus that tolerance is an adaptive plant trait under natural selection by herbivores in both herbaceous species and trees (Mattson *et al.* 1991, Stowe *et al.* 2000, Lennartsson and Oostermeijer 2001, Weinig *et al.* 2003, Soundararajan *et al.* 2004). It has been suggested that tolerance traits are evolutionarily older than resistance traits (Fornoni *et al.* 2004) because tolerance to damage imposed by both frost and desiccation were crucial steps during evolution into terrestrial environments (Oliver *et al.* 2000, Agrawal *et al.* 2004) before insect herbivores started exerting selection on plants (Fornoni *et al.* 2004). Assuming that tolerance reduces natural selection on resistance traits, it can be hypothesized that tolerance could slow down the evolution of herbivore counterresistant responses (Stinchcombe 2002, Garrido Espinosa and Fornoni 2006). It remains unclear how far tolerance can affect the herbivore population growth, but there is some evidence that tolerance can affect herbivore fitness measures, such as growth and fecundity.

5.3.1 Effect of plant tolerance on herbivore performance

Given that tolerant plants can compensate for tissue loss and that the compensation mechanisms could lead to equal or even better resources for herbivores (i.e., higher photosynthetic rates or nutrient uptake rates could increase the plant's tissue quality), it has been suggested that those plants have at least no negative effects on herbivore development or fitness (Tiffin 2000, Stinchcombe 2002). In the only study testing this hypothesis, Garrido Espinosa and Fornoni (2006) investigated how genotypes of *Datura stramonium* L., varying in their degree of tolerance, affect the development of the Chrysomelidae beetle, *Lema trilineata* (Olivier). They found no effects of plant tolerance on herbivore development, suggesting that host tolerance can relax the selective pressure on herbivores as was proposed earlier (Stinchcombe 2002). Although not specifically tested, indirect evidence in aspen trees (*Populus tremuloides* Michx.) suggests a similar pattern (Stevens *et al*. 2008). Aspen trees have been shown to exhibit tolerance responses to outbreaking herbivores such as forest tent caterpillars (*Malacosoma disstria* (Hübner)) but phytochemical differences between genotypes before and after an outbreak event seemed not to affect herbivore performance (Donaldson and Lindroth 2008). Although evidence is very scarce, we expect that herbivore dynamics would be very different in populations of predominantly tolerant or resistant plants (Tiffin 2000, Figure 5.1). Some studies demonstrate a preference of insect herbivores for regrown tissue after mammalian herbivore damage in comparison to undamaged tissue (Hjalten and Price 1996, Roininen *et al*. 1997). For example, herbivory by aphids, psyllids, miners, and leaf chewers was, in general, higher on regrown birch tissue in the spring after plants had been damaged by moose in the previous season, as compared to undamaged control plants (Danell *et al*. 1985). In the same study, leaf quality was higher in terms of nutrient content, size, and biomass. In conclusion, a combination of neutral effects of tolerant plants on herbivore performance and a possible preference of herbivores for the regrowth of tolerant plants would possibly lead to decreased limitation of herbivore population growth (e.g., decreasingly negative impact of interactions on population growth $a_1 \to a_T$, Figure 5.1). Even when we consider the apparently common simultaneous expression of tolerance and resistance traits in plants (Leimu and Koricheva 2006, Nunez-Farfan *et al*. 2007, Stevens *et al*. 2007), it could be hypothesized that reaching outbreak threshold population sizes would be more probable to occur in predominantly tolerant plant populations compared to less tolerant populations (Figure 5.1). The fact that tolerant plants compensate for the damage caused by herbivores suggests that herbivores are less resource limited when on regrowth plant tissue, which at the same time would diminish the negative effects of the interactions among members of the populations (a_1 in Figure 5.1). Similarly, the recruitment of seeds from tolerant plant phenotypes into the next generation would be much higher than from less tolerant phenotypes, increasing the size of the tolerant population and so providing resources for a growing herbivore population.

5.3.2 Specific induction of plant tolerance and arthropod community effects

Plants can specifically express tolerance responses to different herbivore species, suggesting similar plant-mediated community interaction effects as were hypothesized for plant resistance traits. When studying the effect of different herbivores on the tolerance response of scarlet gilia (*Ipomopsis aggregata* (Pursh V.E. Grant)), a species known to overcompensate in response to mammalian browsing, Juenger and Bergelson (1998) found that clipping as well as insect herbivory had a negative effect on plant fitness, instead of the expected positive effect. Similarly, wild tobacco can compensate for the damage caused by *T. notatus* while it suffers a reduction in fitness in response to *Manduca sexta* (L.) damage (Plate 5.1; Kessler and Baldwin 2004). Studies like this suggest that an herbivores' feeding type could be the main determinant for whether a plant expresses a tolerance response. However, most studies on plant tolerance suggest that compensatory responses are rather unspecific with regard to the damaging agent. Mechanical damage often has been enough to elicit the plant's tolerance response (Fornoni and Nunez-Farfan 2000, Hochwender *et al.* 2000). Examples of tolerance responses have been reported for mammalian browsers or grazers (Danell and Hussdanell 1985, Paige and Whitham 1987, Lennartsson 1997), natural communities of insect herbivores (Mauricio *et al.* 1997, Valverde *et al.* 2003), insect chewers (Pilson 2000, Siemens *et al.* 2003), piercing and sucking insects (Kessler and Baldwin 2004, Kaplan *et al.* 2008), tuber borers (Poveda *et al.* 2010), and even clipping or mowing (Huhta *et al.* 2000, Lennartsson and Oostermeijer 2001).

It has been proposed that plants should be under natural selection to compensate for herbivory if (1) herbivory is continuously very strong and size selective, (2) there is only one bout of herbivory, (3) the risk of herbivory occurs before flowering, (4) most of the resource acquisition occurs before herbivory, and (5) the abiotic conditions such as nutrient availability are not limiting (Maschinski and Whitham 1989, Strauss and Agrawal 1999, Agrawal 2000). Given that many insect outbreaks are cyclical (Berryman 1987) and that they could definitively have a very strong effect on plant fitness, be restricted to one bout of herbivory, be in accordance with the right plant phenological state, and could occur in regions where abiotic factors do not limit growth, it could be concluded that outbreaking insect species should strongly select for tolerance traits in plants as suggested by Mattson *et al.* (1991) and Stevens *et al.* (2007, 2008). We are not aware of any study that has empirically tested the effect of frequency and predictability of insect herbivore attack on selection of tolerance or resistance traits. Moreover, it remains unknown how natural selection by single or multiple herbivores on tolerance traits affects herbivore population dynamics, which again should strongly depend on the interaction between plant resistance and tolerance.

5.3.3   Resistance and tolerance as integrative factors affecting herbivore population dynamics

Based on assumptions of the resource allocation principle (Karban and Baldwin 1997), it has been proposed that there should be trade-offs between resistance and tolerance traits in plants (Vandermeijden *et al.* 1988, Simms and Triplett 1994, Fineblum and Rausher 1995). However, recent reviews and empirical evidence (Leimu and Koricheva 2006, Nunez-Farfan *et al.* 2007, Stevens *et al.* 2007) suggest that mixed patterns of defenses are commonly found in nature. Resistance and tolerance to herbivores can be positively or negatively correlated, or may not be correlated, depending on the study system. However, the relationship between tolerance and resistance in a given system may have significant impacts on herbivore population dynamics and thus the probability of outbreaks. This is particularly true when we consider that (1) the probability of an insect herbivore outbreak is a function of plant-mediated effects on herbivore fitness (e.g., fecundity), (2) plants with higher tolerance increase the fitness of individual herbivores, and (3) plants with higher resistance decrease the herbivore's fitness. Accordingly, herbivore outbreak probability should increase with increased predominance of plant tolerance phenotypes with relatively low resistance and decrease with the predominance of high-resistance phenotypes in the plant population. These conclusions are based on limited empirical evidence and hypothetical effects of plant tolerance and resistance on the probability of insect populations reaching outbreak threshold levels (Figure 5.1) and are extrapolated from relatively simple interactions. Nevertheless, the review does highlight potential focus areas of future research. Moreover, drawing these conclusions represents a bi-trophic view on plant–insect interactions and does not consider plant-mediated interactions between multiple trophic levels, which we want to highlight in Section 5.4.

5.4 Plant indirect resistance affecting arthropod community interactions

Predator and parasitoid populations can dampen the growth of herbivore populations and thus have been hypothesized as one of the major forces driving herbivore population dynamics (Hunter 1992), potentially exceeding the negative effects of direct plant resistance (Kessler and Baldwin 2001, Turchin *et al.* 2003). Therefore, plants have been widely studied for traits that may maximize local abundance of natural enemies of herbivores, such as predators and parasitoids. Those traits include resource provisioning to predators such as food (food bodies and extrafloral nectaries) and shelter, which can dramatically reduce herbivory (Bronstein *et al.* 2006). Moreover, by damaging leaf tissue, herbivores elicit changes in the plants' production of HIVOCs. Altered volatile emissions can function as information cues for natural enemies of herbivores to locate their prey or host (Dicke and Baldwin 2010, Kessler and Heil 2011). Through resource and information-mediated indirect resistance, natural enemies may alter the interactions of plants with herbivore communities and affect herbivore population growth patterns in three ways: via (1) direct top-down control of herbivores; (2) trait-mediated effects when predators and parasitoids (2a) make herbivores move to neighboring plants, or (2b) affect the behavior of herbivores in a way that alters herbivore–plant interactions (e.g., shifts in feeding site or pattern); and (3) preying on herbivores that through their feeding on plants mediate effects on herbivore abundance and species composition. Because of the ubiquity of HIVOCs as cues for herbivory, they may also play a major role in mediating associational resistance effects between plants within a community and can through the above mechanisms affect herbivore performance on neighboring plants (Barbosa *et al.* 2009). To evaluate whether rapidly induced indirect resistance can negatively affect herbivore population growth and so reduce the probability that a population will reach outbreak threshold levels, we need to answer two general questions: (1) are herbivore-induced indirect resistance traits, such as HIVOCs more efficient in attracting predators and parasitoids than would be expected from random prey search behavior; and (2) are natural enemies of herbivores attracted to plant cues and able to significantly reduce herbivore fitness and so affect population growth (e.g., increasingly negative impact of interactions on population growth $a_1 \rightarrow a_R$ and reduced probability to reach outbreak threshold population densities $K_1 \rightarrow K_R$; Figure 5.1)?

5.4.1 Top-down control of herbivore populations through plant-induced responses

HIVOCs can provide information useful to natural enemy foraging or host searching in that it allows for fine-tuned responses of predators and parasitoids to herbivore damaged plants, which potentially results in a more efficient control of herbivore populations (Havill and Raffa 2000, Kessler and Baldwin 2001). Parasitoid wasps, in particular, are well known for using plant-derived cues to locate their herbivore hosts. Different species and herbivore instars differentially affect HIVOC emissions, resulting in relatively specific volatile blends associated with each type of herbivore (Dicke and Baldwin 2010, Mumm and Dicke 2010). Hence, changes in plant volatile emission may be a reliable source of information to parasitoids and may allow them to detect a suitable host (Vet and Dicke 1992, Mumm and Dicke 2010). Variation in HIVOCs has been shown to result in different levels of parasitism and predation in the field (Poelman *et al.* 2009, Bezemer *et al.* 2010). However, the distance from which natural enemies are attracted in response to volatiles, or whether HIVOC-guided predators and parasitoids are more efficient than randomly searching natural enemies under field conditions, is largely unknown. It is also unclear if volatile-mediated attraction alters the population growth and community composition of herbivores (but see Kessler and Baldwin 2001, 2004).

Similarly, it is not well understood how induced direct and indirect resistance traits interact with each other in affecting arthropod performance and population dynamics. For example, HIVOC release, and direct resistance by cyanogenesis, were negatively correlated in a comparison of different lines of lima bean *Phaseolus lunatus* (L.) (Ballhorn *et al.* 2008). Moreover, parasitoid responses to HIVOCs can result in positive feedback loops reducing herbivore populations, but simultaneous changes in plant quality may activate negative feedback loops to parasitoid populations by reducing the fecundity of parasitoids on herbivore-induced plants (Havill and Raffa 2000). In contrast, HIVOC emission and defensive secondary metabolite production function synergistically in wild tobacco *N. attenuata* and can, in concert, dramatically increase herbivore mortality (Kessler and Baldwin 2001, 2004; Plate 5.1). Because of the interaction between direct and indirect resistance on an arthropod community level, it may be difficult to predict how induced indirect resistance traits influence herbivore population growth patterns (Kessler and Heil 2011). Novel whole-community approaches promise to provide more predictive analyses of the effects of herbivore-induced responses on herbivore population dynamics and emphasize the role of non-additive effects when studying herbivore population dynamics in a community context.

5.4.2   Natural enemies exploring multi-herbivore communities

HIVOC-mediated interactions have generally been studied in simple trophic cascades (Dicke 2009). However, ecosystem complexity resulting from both plant and herbivore community diversity has been found to affect parasitoid foraging in multiple dimensions.

First, in field habitats volatile release from plants that neighbor the parasitoids' host-induced plants may compromise host localization (Gols *et al.* 2005, Bukovinszky *et al.* 2007). Although parasitoids may be attracted by plant volatiles induced by their hosts, they also may be attracted to HIVOCs induced by nonhosts. Naïve parasitoids may not be able to discriminate plants under attack by host or nonhost herbivores (Geervliet *et al.* 1996). Second, parasitoids need to deal with multi-herbivore communities on the same plant that may release volatiles that distort cues related to the host herbivore. Plants under attack by two herbivore species have been found to either become more attractive to parasitoids or may become less attractive than plants damaged by either of the herbivore species alone. The effect of multi-herbivore communities on parasitoid responses to volatiles depends on the composition of the community, and are typically non-additive (Dicke *et al.* 2009).

Third, when arriving on a plant under attack by multiple herbivore species, parasitoids may encounter nonhost herbivores alongside their herbivore hosts. Depending on the distribution of the herbivores over the plant, parasitoids may encounter hosts only, nonhosts only, and mixed patches of hosts and nonhosts. Encounters with hosts may stimulate parasitoids to increase their searching time spent on the plant. However, each encounter with nonhost herbivores may discourage parasitoids to search for hosts, which results in fewer attacks of host herbivores and may permanently impair the parasitoid's association of the HIVOC cue with their host herbivore (Shiojiri *et al.* 2001, 2002). The reduced attacks on host herbivores may provide them with enemy-free space.

Moreover, the ubiquity of the HIVOC emissions allows a large number of organisms to utilize the information associated with the cue. Herbivores use plant volatiles to locate their host plant and may be deterred or even attracted to plants that have been induced by other herbivores (Poelman *et al.* 2008). Furthermore, plants interact differently with their pollinators when responding to herbivory (Kessler *et al.* 2011) and plants themselves may respond to volatiles that are derived from their own or neighboring plants (Barbosa *et al.* 2009, Heil and Karban 2010). Therefore, plant volatile responses to herbivory have been considered as major factors influencing arthropod community dynamics because changes in the volatile headspaces make plants more apparent to all members of the community (Poelman *et al.* 2010). However, the community-wide effects of HIVOC releases may outweigh the benefit of attracting natural enemies to the plant (Kessler *et al.* 2011), and may make HIVOC-mediated indirect resistance inherently less likely to affect population dynamics than traits that mediate resistance more directly (Kessler and Heil 2011). Thus, when considering multi-herbivore communities, natural enemies may play a less effective role influencing herbivore dynamics and may have little influence on stabilizing herbivore populations at relatively low density levels (e.g., moving $K_1 \rightarrow K_R$). Far more research is needed in this area to evaluate the role of arthropod community complexity in affecting plant-mediated organismal interactions.

5.4.3 Natural enemies affecting herbivore community composition

Natural enemies themselves may have a major impact on the composition of multi-attacker communities and thus on the population dynamics of each herbivore species comprising the community (Davic 2003, Utsumi *et al.* 2010). Functioning as keystone predators, they may shift arthropod community composition by selectively preying on particular herbivore species, therefore limiting population growth of one species (Kessler and Baldwin 2004; Plate 5.1) while potentially accelerating growth of the other.

The sheer presence of natural enemies may alter herbivore phenotypes, including changes in morphology, life history, and behavior, which affect the interaction of herbivores and plants (Utsumi *et al.* 2010). Predator-induced changes in herbivore behavior may include the avoidance of plants by herbivores when the plants are colonized by predators (Stamp and Bowers 1996), or alter the feeding rates or location of herbivores (Griffin and Thaler 2006). Similarly, natural enemies may cause significant immigration by herbivores to neighboring plants (Schmitz 1998, Barbosa *et al.* 2009). The complexity as well as the obvious context dependency of herbivore-induced attraction of predators as an indirect resistance trait make the impact on herbivore population dynamics difficult to predict and reemphasize the necessity of taking a whole-community approach when assessing the effects of rapidly induced plant response on herbivore population growth patterns.

5.5 Conclusion

With this review, we identified major characteristics of herbivore-induced plant responses that have been shown to influence herbivore performance and thus may affect population dynamics and community composition at low densities. The available data primarily derive from relatively simplified plant-mediated bi- and tritrophic systems with little information on the effect of plant-induced responses on herbivore population dynamics, which necessitates an extrapolation of underlying principles to derive hypotheses describing the role of induced responses on outbreak dynamics. Nevertheless, the significant progress of our understanding of plant–insect interactions in the past few decades allows some interesting new conclusions and the generation of new hypotheses.

Among the basic principles underlying plant-mediated interactions with herbivores, we find that rapidly induced plant responses to herbivore damage are complex and interact to affect herbivore population dynamics and community composition (Kessler and Halitschke 2007). The induction of those responses is specific to the attacking herbivore, and the ecological consequences of interactions with the herbivore community are context dependent. These principles apply to all three basic categories of herbivore-induced plant responses that allow plants to cope with herbivory: induced direct and indirect resistance, and tolerance.

Rapidly induced direct resistance affects herbivore fitness and dispersal in a large number of plant systems (Karban and Baldwin 1997) with the potential consequence of limiting herbivore population growth to densities below outbreak threshold levels (e.g., decreasing the probability of reaching outbreak threshold densities, $K_1 \rightarrow K_R$; Figure 5.1). Moreover, through specifically induced cross-resistance or induced susceptibility effects, damage by one herbivore species can influence the performance of other herbivores, which illustrates the importance of using whole-community approaches when evaluating the effects of induced plant responses on herbivore population growth.

Similar to induced resistance, plant tolerance responses can be specifically induced by particular herbivores. They allow plants to compensate or overcompensate for lost tissue or the allocation of resources into resistance compound production. If the resulting regrown tissue is assumed to have a similar herbivore resistance as the tissue on undamaged plants, prolonged and/or increased availability of the food resource can be hypothesized to decrease the negative effects of the interactions among members of the herbivore population ($a_1 \rightarrow a_1$; Figure 5.1). However, it seems crucial to consider the interaction of plant tolerance and resistance traits to affect herbivore performance in order to understand both the evolutionary processes acting on those traits and their effect on herbivore population dynamics.

While the population effects of plant-induced indirect resistance have been suggested for some systems (Kessler and Baldwin 2001, Wooley *et al.* 2007), the direct effects that induced responses have on herbivore populations (e.g., induced resistance and tolerance) seem generally more important in affecting herbivore population dynamics. Moreover, the ubiquity of HIVOC emission and the resulting complexity of interactions can overwrite the indirect resistance function of those traits, thus further reducing the probability to detect significant herbivore population effects of plant-mediated predator attraction. However, new statistical and experimental methods (Poelman *et al.* 2010) may allow assessing community-wide effects of herbivore-induced VOC emission in complex multitrophic systems and thus enable predictions on herbivore population dynamics and plant fitness. Such community-wide assessments of the impact of plant responses are also capable of addressing the characteristic context dependency of the effects associated with plant-induced responses. Because plant phenotypic changes are induced by interacting organisms, the context in which they are affecting a particular herbivore species is determined by the other interacting organisms in the community (Barbosa *et al.* 2009). Modern genetic and molecular tools are increasingly available for nonmodel systems and can help to establish causal links between herbivore-induced plant responses and herbivore population patterns. Therefore, comparative studies and within-species comparisons, using nonresponding and responding genotypes, as well as phyllogentically controlled multispecies comparisons are likely the first valuable steps toward this goal.

Acknowledgments

We are grateful to Pedro Barbosa, Stephen Ellner, Anurag Agrawal, Robert Raguso, and three anonymous, very constructive reviewers for helpful discussions and comments during the preparation of this work. Because of the brevity and the focus of this chapter, we have not been able to refer to all the relevant literature, and we apologize to those authors whose work we have not had space to cite. The preparation of this chapter was supported with funds from the National Science Foundation (IOS-0950225) and Cornell University.

References

Abbott, K. C., W. F. Morris, and K. Gross. 2008. Simultaneous effects of food limitation and inducible resistance on herbivore population dynamics. *Theoretical Population Biology* 73:63–78.

Adler, L. S., M. Wink, M. Distl, and A. J. Lentz. 2006. Leaf herbivory and nutrients increase nectar alkaloids. *Ecology Letters* 9:960–967.

Agrawal, A. A. 2000. Overcompensation of plants in response to herbivory and the by-product benefits of mutualism. *Trends in Plant Science* 5:309–313.

Agrawal, A. A., J. K. Conner, and J. R. Stinchcombe. 2004. Evolution of plant resistance and tolerance to frost damage. *Ecology Letters* 7:1199–1208.

Alborn, H. T., T. V. Hansen, T. H. Jones, D. C. Bennett, J. H. Tumlinson, E. A. Schmelz, and P. E. A. Teal. 2007. Disulfooxy fatty acids from the American bird grasshopper *Schistocerca americana*, elicitors of plant volatiles. *Proceedings of the National Academy of Sciences of the United States of America* 104:12976–12981.

Alborn, T., T. C. J. Turlings, T. H. Jones, G. Stenhagen, J. H. Loughrin, and J. H. Tumlinson. 1997. An elicitor of plant volatiles from beet armyworm oral secretion. *Science* 276:945–949.

Baldwin, I. T., R. Halitschke, A. Paschold, C. C. von Dahl, and C. A. Preston. 2006. Volatile signaling in plant–plant interactions: "talking trees" in the genomics era. *Science* 311:812–815.

Ballhorn, D. J., S. Kautz, U. Lion, and M. Heil. 2008. Trade-offs between direct and indirect defences of lima bean (*Phaseolus lunatus* L.). *Journal of Ecology* 96:971–980.

Barbosa, P., J. Hines, I. Kaplan, H. Martinson, A. Szczepaniec, and Z. Szendrei. 2009. Associational resistance and associational susceptibility: having right or wrong neighbors. *Annual Review of Ecology Evolution and Systematics* 40:1–20.

Bergelson, J., S. Fowler, and S. Hartley. 1986. The effects of foliage damage on casebearing moth larvae, *Coleophora-serratella*, feeding on birch. *Ecological Entomology* 11:241–250.

Berryman, A. A. 1982. Biological-control, thresholds, and pest outbreaks. *Environmental Entomology* 11:544–549.

Berryman, A. A. 1987. The theory and classification of outbreaks. Pages 3–30 *in* P. Barbosa and S. J. C., editors. *Insect Outbreaks*. Academic Press, San Diego.

Berryman, A. A. 1995. Population-cycles – a critique of the maternal and allometric hypotheses. *Journal of Animal Ecology* 64:290–293.

Bezemer, T. M., J. A. Harvey, A. F. D. Kamp, R. Wagenaar, R. Gols, O. Kostenko, T. Fortuna, T. Engelkes, L. E. M. Vet, W. H. Van der Putten, and R. Soler. 2010. Behaviour of male and female parasitoids in the field: influence of patch size, host density, and habitat complexity. *Ecological Entomology* 35:341–351.

Bronstein, J. L., R. Alarcon, and M. Geber. 2006. The evolution of plant–insect mutualisms. *New Phytologist* 172:412–428.

Bukovinszky, T., R. Gols, L. Hemerik, J. C. Van Lenteren, and L. E. M. Vet. 2007. Time allocation of a parasitoid foraging in heterogeneous vegetation: implications for host–parasitoid interactions. *Journal of Animal Ecology* 76:845–853.

Bukovinszky, T., R. Gols, A. Kamp, F. de Oliveira-Domingues, P. A. Hamback, Y. Jongema, T. M. Bezemer, M. Dicke, N. M. van Dam, and J. A. Harvey. 2010. Combined effects of patch size and plant nutritional quality on local densities of insect herbivores. *Basic and Applied Ecology* 11:396–405.

Carroll, M. J., E. A. Schmelz, R. L. Meagher, and P. E. A. Teal. 2006. Attraction of *Spodoptera frugiperda* larvae to volatiles from herbivore-damaged maize seedlings. *Journal of Chemical Ecology* 32:1911–1924.

Danell, K., and K. Hussdanell. 1985. Feeding by insects and hares on birches earlier affected by moose browsing. *Oikos* 44:75–81.

Danell, K., K. Hussdanell, and R. Bergstrom. 1985. Interactions between browsing moose and 2 species of birch in Sweden. *Ecology* 66:1867–1878.

Davic, R. D. 2003. Linking keystone species and functional groups: a new operational definition of the keystone species concept – response. *Conservation Ecology* 7.

De Moraes, C. M., M. C. Mescher, and J. H. Tumlinson. 2001. Caterpillar-induced nocturnal plant volatiles repel nonspecific females. *Nature* 410:577–580.

Denno, R. F., and B. Benrey. 1997. Aggregation facilitates larval growth in the neotropical nymphalid butterfly *Chlosyne janais*. *Ecological Entomology* 22:133–141.

Dicke, M. 2009. Behavioural and community ecology of plants that cry for help. *Plant Cell and Environment* 32:654–665.

Dicke, M., and I. T. Baldwin. 2010. The evolutionary context for herbivore-induced plant volatiles: beyond the 'cry for help'. *Trends in Plant Science* 15:167–175.

Dicke, M., J. J. A. van Loon, and R. Soler. 2009. Chemical complexity of volatiles from plants induced by multiple attack. *Nature Chemical Biology* 5:317–324.

Donaldson, J. R., and R. L. Lindroth. 2008. Effects of variable phytochemistry and budbreak phenology on defoliation of aspen during a forest tent caterpillar outbreak. *Agricultural and Forest Entomology* 10:399–410.

Edwards, P. J., and S. D. Wratten. 1983. Wound induced defenses in plants and their consequences for patterns of insect grazing. *Oecologia* 59:88–93.

Espinosa, E. G., and J. Fornoni. 2006. Host tolerance does not impose selection on natural enemies. *New Phytologist* 170:609–614.

Fineblum, W. L., and M. D. Rausher. 1995. Tradeoff between resistance and tolerance to herbivore damage in a morning glory. *Nature* 377:517–520.

Fornoni, J., and J. Nunez-Farfan. 2000. Evolutionary ecology of *Datura stramonium*: Genetic variation and costs for tolerance to defoliation. *Evolution* 54:789–797.

Fornoni, J., J. Nunez-Farfan, P. L. Valverde, and M. D. Rausher. 2004. Evolution of mixed strategies of plant defense allocation against natural enemies. *Evolution* 58:1685–1695.

Garrido Espinosa, E., and J. Fornoni. 2006. Host tolerance does not impose selection on natural enemies. *New Phytologist* 170:609–614.

Geervliet, J. B. F., L. E. M. Vet, and M. Dicke. 1996. Innate responses of the parasitoids *Cotesia glomerata* and *C. rubecula* (Hymenoptera: Braconidae) to volatiles from different plant–herbivore complexes. *Journal of Insect Behavior* 9:525–538.

Ginzburg, L. R., and D. E. Taneyhill. 1994. Population-cycles of forest lepidoptera – a maternal effect hypothesis. *Journal of Animal Ecology* 63:79–92.

Gols, R., T. Bukovinszky, L. Hemerik, J. A. Harvey, J. C. Van Lenteren, and L. E. M. Vet. 2005. Reduced foraging efficiency of a parasitoid under habitat complexity: implications for population stability and species coexistence. *Journal of Animal Ecology* 74:1059–1068.

Gomez, S., Y. Onoda, V. Ossipov, and J. F. Stuefer. 2008. Systemic induced resistance: a risk-spreading strategy in clonal plant networks? *New Phytologist* 179:1142–1153.

Griffin, C. A. M., and J. S. Thaler. 2006. Insect predators affect plant resistance via density- and trait-mediated indirect interactions. *Ecology Letters* 9:335–343.

Haukioja, E., and S. Neuvonen. 1987. Insect population dynamics and induction of plant resistance: The testing of hypotheses. Pages 411–428 *in* P. Barbosa and J. C. Schultz, editors. *Insect Outbreaks*. Academic Press, San Diego.

Havill, N. P., and K. F. Raffa. 2000. Compound effects of induced plant responses on insect herbivores and parasitoids: implications for tritrophic interactions. *Ecological Entomology* 25:171–179.

Heil, M., and R. Karban. 2010. Explaining evolution of plant communication by airborne signals. *Trends in Ecology & Evolution* 25:137–144.

Hjalten, J., and P. W. Price. 1996. The effect of pruning on willow growth and sawfly population densities. *Oikos* 77:549–555.

Hochwender, C. G., R. J. Marquis, and K. A. Stowe. 2000. The potential for and constraints on the evolution of compensatory ability in *Asclepias syriaca*. *Oecologia* 122:361–370.

Huhta, A. P., T. Lennartsson, J. Tuomi, P. Rautio, and K. Laine. 2000. Tolerance of *Gentianella campestris* in relation to damage intensity: an interplay between apical dominance and herbivory. *Evolutionary Ecology* 14:373–392.

Hunter, M. D. 1992. Interactions within herbivore communities mediated by the host plant the keystone herbivore concept. Hunter, M. D., T. Ohgushi and P. W. Price (Ed.). *Effects of Resource Distribution on Animal–Plant Interactions*: 287–325. Academic Press, San Diego.

Juenger, T., and J. Bergelson. 1998. Pairwise versus diffuse natural selection and the multiple herbivores of scarlet gilia, *Ipomopsis aggregata*. *Evolution* 52:1583–1592.

Kahl, J., D. H. Siemens, R. J. Aerts, R. Gabler, F. Kuhnemann, C. A. Preston, and I. T. Baldwin. 2000. Herbivore-induced ethylene suppresses a direct defense but not a putative indirect defense against an adapted herbivore. *Planta* 210:336–342.

Kaitaniemi, P., S. Neuvonen, and T. Nyyssonen. 1999. Effects of cumulative defoliations on growth, reproduction, and insect resistance in mountain birch. *Ecology* 80:524–532.

Kaplan, I., G. P. Dively, and R. F. Denno. 2008. Variation in tolerance and resistance to the lealhopper *Empoasca fabae* (Hemiptera : Cicadellidae) among potato cultivars: Implications for action thresholds. *Journal of Economic Entomology* 101:959–968.

Karban, R., and I. T. Baldwin. 1997. *Induced Responses to Herbivory*. Chicago University Press, Chicago.

Kessler, A. and I. T. Baldwin. 2001. Defensive function of herbivore-induced plant volatile emissions in nature. *Science* 291:2141–2144.

Kessler, A. and I. T. Baldwin. 2002. Plant responses to insect herbivory: the emerging molecular analysis. *Annual Review of Plant Biology* 53:299–328.

Kessler, A. and I. T. Baldwin. 2004. Herbivore-induced plant vaccination. Part I. The orchestration of plant defenses in nature and their fitness consequences in the wild tobacco *Nicotiana attenuata*. *Plant Journal* 38:639–649.

Kessler, A., and R. Halitschke. 2007. Specificity and complexity: the impact of herbivore-induced plant responses on arthropod community structure. Current Opinion in Plant Biology 10:409–414.

Kessler, A., R. Halitschke, and K. Poveda. 2011. Herbivory-mediated pollinator limitation: Negative impacts of induced volatiles on plant–pollinator interactions. *Ecology* (in press).

Kessler, A., R. Halitschke, and I. T. Baldwin. 2004. Silencing the jasmonate cascade: Induced plant defenses and insect populations. *Science* 305:665–668.

Kessler, A., and M. Heil. 2011. The multiple faces of indirect defences and their agents of natural selection. *Functional Ecology* 25:348–357.

Leimu, R., and J. Koricheva. 2006. A meta-analysis of tradeoffs between plant tolerance and resistance to herbivores: combining the evidence from ecological and agricultural studies. *Oikos* 112:1–9.

Lennartsson, T. 1997. Seasonal differentiation – a conservative reproductive barrier in two grassland Gentianella (Gentianaceae) species. *Plant Systematics and Evolution* 208:45–69.

Lennartsson, T., and J. G. B. Oostermeijer. 2001. Demographic variation and population viability in *Gentianella campestris*: effects of grassland management and environmental stochasticity. *Journal of Ecology* 89:451–463.

Lynch, M. E., I. Kaplan, G. P. Dively, and R. F. Denno. 2006. Host-plant-mediated competition via induced resistance: Interactions between pest herbivores on potatoes. *Ecological Applications* 16:855–864.

Maschinski, J., and T. G. Whitham. 1989. The continuum of plant-responses to herbivory – the influence of plant-association, nutrient availability, and timing. *American Naturalist* 134:1–19.

Mattson, W., D. Herms, J. Witter, and D. Allen. 1991. Woody plant grazing systems: North American outbreak folivores and their host plants. Pages 53–84 *in* Y. Baranchikov, W. Mattson, F. Hain, and T. Payne, editors. Forest Insect Guilds: Patterns of Interaction with Host Trees. USDA Forest Service, Abakan.

Mauricio, R., M. D. Rausher, and D. S. Burdick. 1997. Variation in the defense strategies of plants: Are resistance and tolerance mutually exclusive? *Ecology* 78:1301–1311.

Mithofer, A., G. Wanner, and W. Boland. 2005. Effects of feeding *Spodoptera littoralis* on lima bean leaves. II. Continuous mechanical wounding resembling insect feeding is sufficient to elicit herbivory-related volatile emission. *Plant Physiology* 137:1160–1168.

Mumm, R., and M. Dicke. 2010. Variation in natural plant products and the attraction of bodyguards involved in indirect plant defense. *Canadian Journal of Zoology – Revue Canadienne De Zoologie* 88:628–667.

Musser, R. O., D. F. Cipollini, S. M. Hum-Musser, S. A. Williams, J. K. Brown, and G. W. Felton. 2005. Evidence that the caterpillar salivary enzyme glucose oxidase provides herbivore offense in Solanaceous plants. *Archives of Insect Biochemistry and Physiology* 58:128–137.

Myers, J. H. 1988. Can a general hypothesis explain population-cycles of forest lepidoptera. *Advances in Ecological Research* 18:179–242.

Myers, J. H., G. Boettner, and J. Elkinton. 1998. Maternal effects in gypsy moth: Only sex ratio varies with population density. *Ecology* 79:305–314.

Nunez-Farfan, J., J. Fornoni, and P. L. Valverde. 2007. The evolution of resistance and tolerance to herbivores. *Annual Review of Ecology Evolution and Systematics* 38:541–566.

Oliver, M. J., Z. Tuba, and B. D. Mishler. 2000. The evolution of vegetative desiccation tolerance in land plants. *Plant Ecology* 151:85–100.

Osier, T. L., and R. L. Lindroth. 2004. Long-term effects of defoliation on quaking aspen in relation to genotype and nutrient availability: plant growth, phytochemistry and insect performance. *Oecologia* 139:55–65.

Paige, K. N., and T. G. Whitham. 1987. Overcompensation in response to mammalian herbivory – the advantage of being eaten. *American Naturalist* 129:407–416.

Paschold, A., R. Halitschke, and I. T. Baldwin. 2007. Co(i)-ordinating defenses: NaCOI1 mediates herbivore-induced resistance in *Nicotiana attenuata* and reveals the role of herbivore movement in avoiding defenses. *Plant Journal* 51:79–91.

Pilson, D. 2000. The evolution of plant response to herbivory: simultaneously considering resistance and tolerance in *Brassica rapa*. *Evolutionary Ecology* 14:457–489.

Poelman, E. H., C. Broekgaarden, J. J. A. Van Loon, and M. Dicke. 2008. Early season herbivore differentially affects plant defence responses to subsequently colonizing herbivores and their abundance in the field. *Molecular Ecology* 17:3352–3365.

Poelman, E. H., A. M. O. Oduor, C. Broekgaarden, C. A. Hordijk, J. J. Jansen, J. J. A. v. Loon, N. M. v. Dam, L. E. M. Vet, and M. Dicke. 2009. Field parasitism rates of caterpillars on *Brassica oleracea* plants are reliably predicted by differential attraction of Cotesia parasitoids. *Functional Ecology* 23:951–962.

Poelman, E. H., J. J. A. Van Loon, N. M. Van Dam, L. E. M. Vet, and M. Dicke. 2010. Herbivore-induced plant responses in *Brassica oleracea* prevail over effects of constitutive resistance and result in enhanced herbivore attack. *Ecological Entomology* 35:240–247.

Pohnert, G., V. Jung, E. Haukioja, K. Lempa, and W. Boland. 1999. New fatty acid amides from regurgitant of lepidopteran (Noctuidae, Geometridae) caterpillars. *Tetrahedron* 55:11275–11280.

Poveda, K., M. I. Gómez Jimenez, and A. Kessler. 2010. The enemy as ally: herbivore-induced increase in crop yield. *Ecological Applications* 20:1787–1793.

Poveda, K., I. Steffan-Dewenter, S. Scheu, and T. Tscharntke. 2003. Effects of below- and above-ground herbivores on plant growth, flower visitation and seed set. *Oecologia* 135:601–605.

Raffa, K. F. 2001. Mixed Messages across Multiple Trophic Levels: The Ecology of Bark Beetle Chemical Communication Systems. Birkhauser Verlag AG, Basel.

Rodriguez-Saona, C., J. A. Chalmers, S. Raj, and J. S. Thaler. 2005. Induced plant responses to multiple damagers: differential effects on an herbivore and its parasitoid. *Oecologia* 143:566–577.

Roininen, H., P. W. Price, and J. P. Bryant. 1997. Response of galling insects to natural browsing by mammals in Alaska. *Oikos* 80:481–486.

Schmelz, E. A., M. J. Carroll, S. LeClere, S. M. Phipps, J. Meredith, P. S. Chourey, H. T. Alborn, and P. E. A. Teal. 2006. Fragments of ATP synthase mediate plant perception of insect attack. *Proceedings of the National Academy of Sciences of the United States of America* 103:8894–8899.

Schmitz, O. J. 1998. Direct and indirect effects of predation and predation risk in old-field interaction webs. *American Naturalist* 151:327–342.

Schutz, S., B. Weissbecker, A. Klein, and H. E. Hummel. 1997. Host plant selection of the Colorado potato beetle as influenced by damage induced volatiles of the potato plant. *Naturwissenschaften* 84:212–217.

Shiojiri, K., J. Takabayashi, S. Yano, and A. Takafuji. 2001. Infochemically mediated tritrophic interaction webs on cabbage plants. *Population Ecology* 43:23–29.

Shiojiri, K., J. Takabayashi, S. Yano, and A. Takafuji. 2002. Oviposition preferences of herbivores are affected by tritrophic interaction webs. *Ecology Letters* 5:186–192.

Siemens, D. H., H. Lischke, N. Maggiulli, S. Schurch, and B. A. Roy. 2003. Cost of resistance and tolerance under competition: the defense-stress benefit hypothesis. *Evolutionary Ecology* 17:247–263.

Silkstone, B. E. 1987. The consequences of leaf damage for subsequent insect grazing on birch (*Betula spp*) – a field experiment. *Oecologia* 74:149–152.

Simms, E. L., and J. Triplett. 1994. Costs and benefits of plant-responses to disease – resistance and tolerance. *Evolution* 48:1973–1985.

Soler, R., T. M. Bezemer, A. M. Cortesero, W. H. Van der Putten, L. E. M. Vet, and J. A. Harvey. 2007. Impact of foliar herbivory on the development of a root-feeding insect and its parasitoid. *Oecologia* 152:257–264.

Soundararajan, R. P., P. Kadirvel, K. Gunathilagaraj, and M. Maheswaran. 2004. Mapping of quantitative trait loci associated with resistance to brown planthopper in rice by means of a doubled haploid population. *Crop Science* 44:2214–2220.

Stamp, N. E. 1980. Egg deposition patterns in butterflies – why do some species cluster their eggs rather than deposit them singly. *American Naturalist* 115:367–380.

Stamp, N. E., and M. D. Bowers. 1996. Consequences for plantain chemistry and growth when herbivores are attacked by predators. *Ecology* 77:535–549.

Stevens, M. T., E. L. Kruger, and R. L. Lindroth. 2008. Variation in tolerance to herbivory is mediated by differences in biomass allocation in aspen. *Functional Ecology* 22:40–47.

Stevens, M. T., D. M. Waller, and R. L. Lindroth. 2007. Resistance and tolerance in *Populus tremuloides*: genetic variation, costs, and environmental dependency. *Evolutionary Ecology* 21:829–847.

Stinchcombe, J. R. 2002. Can tolerance traits impose selection on herbivores? *Evolutionary Ecology* 16:595–602.

Stowe, K. A., R. J. Marquis, C. G. Hochwender, and E. L. Simms. 2000. The evolutionary ecology of tolerance to consumer damage. *Annual Review of Ecology and Systematics* 31:565–595.

Strauss, S. Y. and A. A. Agrawal. 1999. The ecology and evolution of plant tolerance to herbivory. *Trends in Ecology & Evolution* 14:179–185.

Tiffin, P. 2000. Mechanisms of tolerance to herbivore damage: what do we know? *Evolutionary Ecology* 14:523–536.

Turchin, P., P. L. Lorio, A. D. Taylor, and R. F. Billings. 1991. Why do populations of southern pine beetles (Coleoptera, Scolytidae) fluctuate. *Environmental Entomology* 20:401–409.

Turchin, P., S. N. Wood, S. P. Ellner, B. E. Kendall, W. W. Murdoch, A. Fischlin, J. Casas, E. McCauley, and C. J. Briggs. 2003. Dynamical effects of plant quality and parasitism on population cycles of larch budmoth. *Ecology* 84:1207–1214.

Underwood, N. 2004. Variance and skew of the distribution of plant quality influence herbivore population dynamics. *Ecology* 85:686–693.

Underwood, N., K. Anderson, and B. D. Inouye. 2005. Induced vs. constitutive resistance and the spatial distribution of insect herbivores among plants. *Ecology* 86:594–602.

Utsumi, S., Y. Ando, and T. Miki. 2010. Linkages among trait-mediated indirect effects: a new framework for the indirect interaction web. *Population Ecology* 52:485–497.

Valverde, P. L., J. Fornoni, and J. Nunez-Farfan. 2003. Evolutionary ecology of *Datura stramonium*: equal plant fitness benefits of growth and resistance against herbivory. *Journal of Evolutionary Biology* 16:127–137.

van Dam, N. M., C. E. Raaijmakers, and W. H. van der Putten. 2005. Root herbivory reduces growth and survival of the shoot feeding specialist *Pieris rapae* on *Brassica nigra*. *Entomologia Experimentalis et Applicata* 115:161–170.

Van Zandt, P. A., and A. A. Agrawal. 2004. Community-wide impacts of herbivore-induced plant responses in milkweed (*Asclepias syriaca*). *Ecology* 85:2616–2629.

Vandermeijden, E., M. Wijn, and H. J. Verkaar. 1988. Defense and regrowth, alternative plant strategies in the struggle against herbivores. *Oikos* 51:355–363.

Vet, L. E. M., and M. Dicke. 1992. Ecology of infochemical use by natural enemies in a tritrophic context. *Annual Review of Entomology* 37:141–172.

Viswanathan, D. V., A. J. T. Narwani, and J. S. Thaler. 2005. Specificity in induced plant responses shapes patterns of herbivore occurrence on *Solanum dulcamara*. *Ecology* 86:886–896.

Weinig, C., J. R. Stinchcombe, and J. Schmitt. 2003. Evolutionary genetics of resistance and tolerance to natural herbivory in *Arabidopsis thaliana*. *Evolution* 57:1270–1280.

Wooley, S. C., J. R. Donaldson, A. C. Gusse, R. L. Lindroth, and M. T. Stevens. 2007. Extrafloral nectaries in aspen (*Populus tremuloides*): heritable genetic variation and herbivore-induced expression (vol. 100, pg. 1337, 2007). *Annals of Botany* 100:1607–1607.

Zavala, J. A., A. G. Patankar, K. Gase, D. Q. Hui, and I. T. Baldwin. 2004. Manipulation of endogenous trypsin proteinase inhibitor production in *Nicotiana attenuata* demonstrates their function as antiherbivore defense. *Plant Physiology* 134:1181–1190.

6

Spatial Synchrony of Insect Outbreaks

Andrew M. Liebhold, Kyle J. Haynes, and Ottar N. Bjørnstad

6.1 Introduction

The concept of "spatial synchrony" refers to the tendency of the densities of spatially disjunct populations to be correlated in time (Bjørnstad *et al.* 1999a, Liebhold *et al.* 2004). Outbreaking forest insects offer many of the classic examples of this phenomenon (Figure 6.1). The spatial extent of synchrony of outbreaks is probably one of the most important – yet most underappreciated – characteristics that cause certain insect species to be classified as noxious pests. Locally eruptive population behavior alone would rarely qualify a species for "outbreak" status. Rather, regionalization of eruptions is what elevates ecological and socioeconomic impacts of certain species to high levels of concern, which gives them pest status. Consider for example the contrast between the fall web worm, *Hyphantria cunea*, and the gypsy moth, *Lymantria dispar* (Figure 6.2). The former can reach very high localized densities and denude branches of leaves, but in North America this defoliation is typically limited to isolated colonies on different branches on different trees over several years. Due to the highly localized nature of its damage, this species is not considered a major pest in the eastern United States. The gypsy moth in contrast is considered a highly noxious species because its defoliation is extremely synchronized, extending continuously over thousands of hectares in a single year.

Outbreaks of forest insects can have vast impacts on ecosystem functions. These effects include nitrogen leaching, carbon sequestration, and alteration of fire regimes (McCullough *et al.* 1998, Lovett *et al.* 2002, Schowalter 2011). There is a

Insect Outbreaks Revisited, First Edition. Edited by Pedro Barbosa, Deborah K. Letourneau and Anurag A. Agrawal.
© 2012 Blackwell Publishing Ltd. Published 2012 by Blackwell Publishing Ltd.

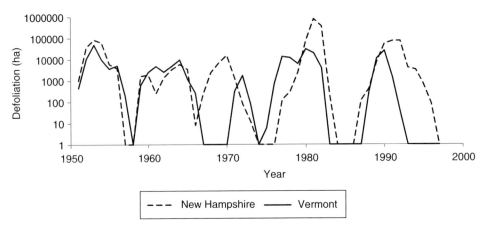

Figure 6.1 Spatial synchrony in gypsy moth populations. Time series of annual area defoliated by the gypsy moth, *Lymantria dispar*, in two adjacent states in the United States, 1951–2008. Defoliated area serves as a proxy for regional population density. Note the general temporal coincidence of both peaks and troughs in the time series.

(a)   (b)

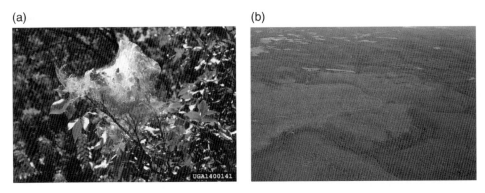

Figure 6.2 Contrast between regionally synchronized and unsynchronized outbreaks. (a) defoliation caused by *Hyphantria cunea* (photo: Linda Haugen, USDA Forest Service, Bugwood.org); due to the highly localized nature of its damage, this species is not considered a major pest in the eastern United States. (b) Extensive defoliation caused by *Lymantria dispar* (photo: Karl Mierzejewski, Centre Co., PA); due to the synchronous occurrence of high densities over large areas, this species is considered a major pest species in the eastern United States.

critical spatial component to all of these impacts. For example, the consequences of outbreak-caused elevated nitrogen leaching on stream chemistry are much more pronounced when outbreaks occur simultaneously over an entire watershed (Eshleman *et al.* 2004). There is also a strong (though poorly studied) relationship between spatial synchrony of insect outbreaks and socioeconomic impacts. For example, widespread forest defoliation and tree mortality are more likely to result in a loss of scenic value and tourism revenues (Lynch and Twery 1992). Similar spatial properties characterize the economic effects of outbreak synchrony on agricultural and timber values. Forest disturbances, such as bark beetle outbreaks, that extend over large areas often result in a depression of timber prices created by excessive supply generated from salvage harvest (Holmes 1991). Prices of agricultural commodities often rise as a result of widespread insect outbreaks that cause drops in supply and/or cause increases in production costs (Hoddle *et al.* 2003).

6.2   Quantifying synchrony

Spatial synchrony and the resultant regionalization of outbreak dynamics have been measured in many ways. The simplest and most common approach is to use the Pearson product–moment correlation among pairs of spatially disjunct time series of local abundance (Bjørnstad *et al.* 1999a, Buonaccorsi *et al.* 2001, de Valpine *et al.* 2010). Very often, however, data on local insect abundance are not available. In such cases, records of outbreak incidence and severity (e.g., defoliation) over time are used as proxies for abundance. While the resulting time series of outbreak incidence may fail to capture the detailed variations in abundance, particularly those that occur at sub-outbreak levels, such data tend to be reasonably correlated with true population densities and thus provide adequate proxies for assessing synchrony (e.g., Bjørnstad *et al.* 2002).

Spatial synchrony cannot be understood or quantified without reference to spatial scale because synchrony typically depends on the distance between locations (Liebhold *et al.* 2004, Fox *et al.* 2011). The maximum geographical extent over which synchrony extends varies among species and locations (Hanski and Woiwod 1993, Peltonen *et al.* 2002), but often extends over several hundred kilometers in forest insects (Peltonen *et al.* 2002). Quantifying the relationship between synchrony and distance is valuable because it can provide critical information, not only about the consequences of synchrony but also about its causes (Bjørnstad *et al.* 1999a, b). For example, differentiation among possible mechanisms driving observed synchrony can be accomplished by comparing the scale of synchrony in data simulated using alternate mechanistic models. However, it is often difficult to infer process from pattern because very different processes can lead to similar patterns of spatial dynamics.

When many georeferenced time series are available, it is useful to calculate all pairwise correlation coefficients between them and examine how the correlation depends on spatial distance (Figure 6.3). In most biological systems, cross-correlations decline with increasing distance between sampling locations (Liebhold *et al.* 2004, Fox *et al.* 2011, but see Ranta and Kaitala 1997, Bjørnstad and Bascompte 2001 for exceptions to this rule). The spatial scale of synchrony is usually measured as either the distance at which cross-correlations decline to zero or to the region-wide average correlation (Bjørnstad *et al.* 1999a, b). Several related statistical methods are available for characterizing the relationship between synchrony and distance. Many of these, including Koenig's (1999) modified correlogram and the nonparametric spatial covariance function (Bjørnstad and Falck 2001), are implemented in the NCF package of the R statistical program (R Development Core Team 2010).

When sampling to quantify synchrony, it is important to consider the spatial extent of the sample, that is, the size of the region over which synchrony is investigated, and the spatial extent and spacing of the individual sample units

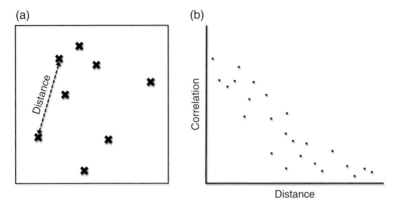

Figure 6.3 Illustration of a hypothetical time series correlogram. (a) Irregular spatial configuration of sample locations for eight population time series. (b) Time series correlogram of hypothetical data displays the correlation coefficient calculated between each pairwise combination of time series versus the distance between sample locations.

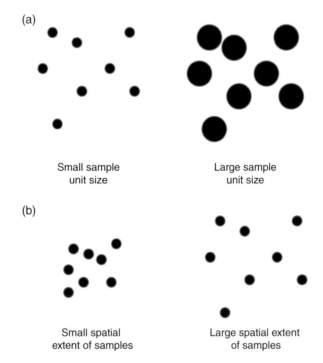

Figure 6.4 Two aspects of scale in sampling spatial synchrony. Black circles represent the sample area for time series. (a) Varying the sample unit size for each sample. (b) Varying the spatial extent among sample locations.

(Figure 6.4). The spatial extent of the sample unit can greatly influence the quantification of synchrony. Fluctuations in abundance in small sampling areas may be dominated by local stochasticities in dynamics as well as observational errors when densities are low that may distort both patterns of synchrony. As the sample unit size increases, patterns of synchrony may become clearer, though sometimes this clarity comes at the cost of diluting the synchrony versus distance signal (Bjørnstad and Falck 2001). This phenomenon is not limited to insect outbreaks or even to ecological data, and is a widely recognized phenomenon referred to as the "modifiable areal unit problem" in geography in which the shape and/or size of the units on which data are mapped can change the resulting correlations or statistical models generated from the data (Openshaw 1984, Wu 2004). If the spatial extent of the study is small, one may fail to quantify the scale of synchrony because it might be larger than the greatest distances considered. Similarly, if the spatial extent is large but the number of sample units is small, one may fail to quantify the scale because it may be smaller than the shortest distance among widely spaced samples.

6.3   Causes of spatial synchrony

Despite the ubiquity of spatial synchrony in forest insect outbreaks, its causes are often difficult to identify because of the general "inverse problem" of inferring process from pattern – very different processes can lead to similar dynamic patterns (Nelson *et al.* 2004). There are multiple potential causes of spatial synchrony: dispersal, regional stochasticity, and trophic interactions (Bjørnstad *et al.* 1999a, Liebhold *et al.* 2004). Moreover, all of these forces act simultaneously, but with varying importance in different insect species.

6.3.1   Dispersal

Mathematical models show that the density-independent exchanges of even a small number of individuals among spatially disjunct populations can lead to synchrony (Barbour 1990, Bjørnstad and Bolker 2000, Liebhold *et al.* 2004). In highly mobile species, dispersal might be expected to lead to synchronous dynamics across large regions. However, population synchrony has been observed over large distances even among totally isolated populations (Grenfell *et al.* 1998, Ims and Andreassen 2000). Moreover, a comparison of spatial synchrony among six forest insect species revealed that there was little relationship between a species' dispersal ability and the distances over which populations are synchronous (Peltonen *et al.* 2002). An extreme example is provided by the gypsy moth in North America; adult females are completely flightless, yet defoliation records indicate that populations are synchronous across distances of 600–900 km (Peltonen *et al.* 2002, Haynes *et al.* 2009) (Figure 6.1).

A further complication is that the ability of dispersal to induce synchrony appears to be highly dependent on the relationship between dispersal and density. Most theoretical studies indicating dispersal as a synchronizing force assume a density-independent dispersal rate. However, Ims and Andreassen (2005) demonstrated experimentally and theoretically that dispersal is not an important synchronizing agent when dispersal is negatively density dependent (i.e., when the dispersal rate is highest from low-density populations). In contrast to positive density-dependent dispersal, which would tend to force the densities of subdivided populations toward a regional average, negative density-dependent dispersal would tend to inflate differences in density among populations as individuals would move from low-density populations and settle in high-density populations. Negative density-dependent dispersal is common in a variety of taxa, perhaps due to the difficulty of finding mates or avoidance of inbreeding in low-density populations (Hanski 1999, Ims and Hjermann 2001, Clobert *et al.* 2004).

6.3.2   Regional stochasticity

Moran (1953) showed that stochastic density-independent factors – such as precipitation or temperature anomalies that are correlated across great distances (see Koenig 2002, Peltonen *et al.* 2002) – can drive population synchrony across large regions. Specifically, Moran showed that the correlation of two populations exhibiting identical log-linear population dynamics should equal the correlation in the stochastic factor affecting both populations. In the following decades, the ability of this process, known as the "Moran effect" or "Moran's theorem" (Royama 1992), to synchronize populations across regional scales has been explored extensively in both theoretical and empirical studies (e.g., Grenfell *et al.* 1998, Bjørnstad *et al.* 1999b, Hudson and Cattadori 1999, Koenig 2002, Peltonen *et al.* 2002). Regionally correlated random factors, particularly weather, have been widely accepted as a major driver of observed patterns of population synchrony (reviews in Bjørnstad *et al.* 1999a, Liebhold *et al.* 2004).

Moran (1953) showed that the regionally stochastic effect may exert a relatively weak effect on change in abundance compared to density-dependent effects but still have a strongly synchronizing effect. Because these synchronizing stochastic effects may be small, it often is difficult or impossible to isolate their identity, though there are a few examples where information on this has been inferred. For example, outbreaks of the spruce budworm in eastern Canada may be synchronized by spatial correlation in the recruitment rate of eggs resulting from dependence of the recruitment rate on weather conditions (Royama *et al.* 2005).

Moran's (1953) original work assumed that all subpopulations were governed by identical density-dependent processes. Peltonen *et al.* (2004) and Liebhold *et al.* (2006) used the same second-order autoregressive population model used by Moran (1953) to show that population synchrony generated through synchronous variability in weather may be weakened by geographic variability in the strength of density dependence. Furthermore, considerable geographic variation in density dependence and population behaviors (e.g., outbreak frequency and periodicity) has been observed in a variety of species (Johnson *et al.* 2006, Cooke and Lorenzetti 2006, Peltonen *et al.* 2004). In the two-year cycle spruce budworm, *Choristoneura biennis*, for example, regional synchrony of outbreaks appears to be reduced due to geographic variation in population dynamics associated with local differences in forest stand characteristics (Zhang and Alfaro 2003).

Moran (1953) also assumed that the dynamics of synchronized populations followed a second-order log-linear model, but such assumptions of linearity are not realistic for most forest insects (Turchin 2003). Nonlinearities in density-dependent population growth can greatly influence population synchrony.

Whereas the synchrony of populations governed by linear dynamics equals the synchrony of environmental perturbations (Moran 1953), nonlinear density dependence can cause population synchrony to be substantially greater or lower depending on the type of population behavior exhibited (e.g., cycles or chaos) (Grenfell *et al.* 1998, Ranta *et al.* 1998, Bjørnstad 2000). Nonlinearities producing cyclic dynamics enhance population synchrony through nonlinear phase locking (Bjørnstad 2000). In this process, exchange of very few individuals between populations will bring populations into synchrony (Bjørnstad *et al.* 1999a, Jansen 1999). Moreover, phase locking may produce synchrony over distances much greater than the dispersal distances of individuals (Jansen 1999).

6.3.3 Trophic interactions

Probably the best understood examples of food web interactions leading to population synchrony involve microtine rodent populations synchronized by nomadic predators; avian predators concentrate in patches of high prey density moving prey densities in these patches to levels more similar to the regional average (Ydenberg 1987, Ims and Steen 1990). Many natural enemy species are highly mobile, and orient toward areas of locally high prey population densities, resulting in spatial density dependence in their effects on host populations (Walde and Murdoch 1988). Such spatial density dependence is sometimes capable of stabilizing host populations (Hassell and May 1973, Murdoch and Stewart-Oaten 1989). Although synchrony in forest insect populations has not been directly linked to mobile predator populations, some findings suggest that this possibility warrants further consideration. In particular, Barber *et al.* (2008) found that cuckoos, which are known to exploit insect outbreaks, concentrate in areas of gypsy moth outbreaks. Parasitoids of forest insects are known to aggregate into areas of high-host density (Liebhold and Elkinton 1989, Gould *et al.* 1990, Parry *et al.* 1997); however, the dispersal capacity of parasitoids may be too low to synchronize hosts over regional scales though they may lead to local synchrony. Though it has received less attention, mobile insect pathogens could similarly serve to synchronize host populations. However, in the case of entomopathogenic viruses, movement of these pathogens is likely insufficient to influence the synchrony of forest insects across large regions (Abbott and Dwyer 2008).

   In recent years, there has been growing recognition of a broader role of food web interactions in population synchrony. Traditionally, Moran effects are thought to be caused by the direct action of weather on the survival or reproduction of individual species. It is becoming increasingly clear that exogenous effects can indirectly synchronize the fluctuations of a given species by increasing the synchrony of resource or consumer species (Cattadori *et al.* 2005, Haynes *et al.* 2009). In the northeastern United States, large-scale resource pulses provided by mast seeding of oaks are a major driver of population cycles of the white-footed mouse (Wolff 1996, McShea 2000, Elias *et al.* 2004, Clotfelter *et al.* 2007, Schmidt and Ostfeld 2007), which prey opportunistically on gypsy moths (Smith 1985; Yahner and Smith 1991). Synchrony in white-footed mouse populations resulting from mast seeding of their primary winter food source, red-oak acorns, has been hypothesized to help synchronize gypsy moth populations (Ostfeld and

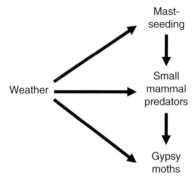

Figure 6.5 Interactions by which stochasticity in weather contributes to synchrony among gypsy moth populations (Haynes *et al.* 2009). Weather may cause direct synchronized stochastic effects (e.g. extreme cold temperatures may cause mortality in certain life stages). In addition, weather may act indirectly to cause synchronized stochasticity in gypsy moth populations by altering mast seeding or small mammal predator populations.

Jones 1996, Liebhold *et al.* 2000). Haynes *et al.* (2009) recently found that red-oak acorns, white-footed mice, and gypsy moths display similar levels of spatial synchrony over comparable distances. They also found that a mechanistic model of interactions among these species (as well as a specialist viral pathogen of the gypsy moth) indicated that regional synchronization of mast seeding through climatic forcing could indirectly synchronize the fluctuations of the white-footed mouse and gypsy moth (Figure 6.5). Similar mast-induced synchrony that trickles down the trophic chain has been documented in the apple fruit moth, *Argyresthia conjugella*, and its parasitoid, *Microgaster politus* (Satake *et al.* 2004).

It is often difficult to sort out the causes from the consequences of herbivore outbreak synchrony on higher trophic levels. Little information is available about the effects of herbivore spatial synchrony on populations at higher trophic levels (but see Satake *et al.* 2004). We can perhaps infer from the literature on mast seeding in plants (Silvertown 1980, Kelly and Sork 2002) that spatial synchrony in prey insect oscillations can satiate predator and parasitoid populations and thereby limit their ability to regulate hosts via spatial density dependence. Thus, spatial density dependence in the response of predators and parasitoids may contribute to the synchronization of host populations, but spatial synchrony in host populations may adversely affect the same spatial density dependence of higher trophic level impacts.

6.4 The ubiquity of synchrony and its implications

The presence of spatial synchrony in population dynamics in nearly every major taxa (Liebhold *et al.* 2004) begs the question: why is population synchrony so ubiquitous? The most likely explanation is that weather appears to co-vary through time in a similar fashion at continental scales across all parts of the Earth (Koenig 2002). All species may disperse at some – albeit often quite limited – observable scale. The lack of relationship between dispersal ability and spatial extent of population synchrony, even among forest insects with similar population dynamics (Peltonen *et al.* 2002), suggests dispersal may not explain the pervasiveness of population synchrony. Although weather may be the dominant force underlying population synchrony, the relationship between weather and population synchrony is probably almost never as simple as predicted by Moran's (1953) theorem. Spatial heterogeneities and nonlinearities in population dynamics both can modify the synchronizing effects of weather on populations. Furthermore, the synchrony of a target species can be synchronized by spatially correlated weather both directly (e.g., through weather-induced mortality) and indirectly due to propagation of synchrony through trophic chains (Satake *et al.* 2004, Cattadori *et al.* 2005, Haynes *et al.* 2009) (Figure 6.5).

Understanding the causes of outbreak synchrony has important implications for how insect populations are best managed. Specifically, if synchrony is primarily caused by dispersal, then suppression of outbreaks can be expected to have a positive impact on preventing outbreaks in neighboring areas. The concept of dispersal-driven synchrony is closely related to the "epicenter hypothesis," namely, that outbreaks start in specific areas and that outbreaks in neighboring areas are triggered by immigration from epicenters; this concept has been advanced to describe the spatial dynamics of several different forest insects such as the gypsy moth and spruce budworm, but evidence for the concept is sparse (Liebhold and McManus 1991, Royama 1992, Johnson *et al.* 2004). Implicit to the epicenter hypothesis is the existence of multiple stable equilibria and massive levels of dispersal, neither of which are well documented in most forest insect species. Given the questionable status of the epicenter hypothesis and the overall dominant role of weather (acting either directly or indirectly via trophic interactions) as the driving force behind spatial synchrony in insect outbreaks, there seems to be little justification for pro-actively suppressing "epicenters" in order to prevent outbreaks from spreading to adjoining areas.

While the ubiquity of spatial synchrony in insect outbreaks is remarkable, care should be taken in characterizing the extent of synchrony. Examples of what we feel are exaggerations of spatial synchrony include studies by Myers (1998) and Hawkins and Holyoak (1998), who reported synchrony of outbreaks extending over thousands of km, sometimes among continents. Synchrony in interannual variation in weather generally does not extend over such massive distances (Koenig 2002), and thus it is difficult to comprehend how a Moran effect could explain such extreme spatial extent of synchrony. Johnson *et al.* (2005) provide a detailed description of the failure of Myers (1998) and Hawkins and Holyoak (1998) to account for spatial autocorrelation in time series and how this may have negated the statistical analyses upon which they based their conclusions. We believe that a similar problematic application of statistical tests of correlation that neglect inherent temporal autocorrelation can be found in Selås *et al.* (2004), who proposed that forest insect outbreaks may be synchronized indirectly by low sunspot activity, which would result in global synchrony. The authors reasoned that UV-B radiation from the sun peaks in periods of low sunspot activity, causing trees to respond by increasing the production of UV-B protective phenolics at the expense of production of secondary compounds for herbivore resistance. This is a creative idea but we believe that, like Hawkins and Holyoak (1998) and Myers (1998), Selås *et al.* (2004) failed to account for temporal autocorrelation in insect and sunspot data and consequently reached a spurious conclusion.

The classic perspective on population synchrony is that it is a characteristic but not a cause of population fluctuations. This is perhaps most clearly embodied in the Moran theorem that effectively establishes that the mechanisms responsible for synchronizing populations may be completely independent of the mechanisms responsible for regulating populations (Moran 1953). With respect to outbreak ecology, this perspective implies that synchrony may be important because it exacerbates the economic burden and strains the logistical abilities of management agencies to mitigate impacts. An alternative perspective, however, is that synchrony plays a much more integral, causal role in outbreak ecology (Bjørnstad *et al.* 2008). The idea is that synchrony results in regionalization of *population* fluctuations and this regionalization dilutes the regulating abilities of mobile natural enemies that would otherwise provide control of local eruptions. Synchrony, then, may be a critical cause of outbreaks.

Acknowledgments

This work was supported in part by the National Research Initiative of the USDA Cooperative State Research, Education and Extension Service Grants to O.N. Bjørnstad and A.M. Liebhold (2002, 2006). Some of the work in preparing this chapter was conducted as a part of the "Modeling Forest Insect Dynamics" Working Group at the National Institute for Mathematical and Biological Synthesis, sponsored by the National Science Foundation, the US Department of Homeland Security, and the US Department of Agriculture through NSF Award No. EF-0832858, with additional support from The University of Tennessee, Knoxville.

References

Abbott, K.C., and G. Dwyer. 2008. Using mechanistic models to understand synchrony in forest insect populations: the North American gypsy moth as a case study. *The American Naturalist* 172:613–624.

Barber, N.A., R.J. Marquis, and W.P. Tori. 2008. Invasive prey impacts the abundance and distribution of native predators. *Ecology* 89:2678–2683.

Barbour, D.A. 1990. Synchronous fluctuations in spatially separated populations of cyclic forest insects. In *Population Dynamics of Forest Insects*, ed. A.D. Watt, S.R. Leather, M.D. Hunter, N.A.C. Kidd, pp. 339–46. Andover, UK: Intercept.

Bjørnstad, O.N. 2000. Cycles and synchrony: two historical 'experiments' and one experience. *Journal of Animal Ecology* 69:869–873.

Bjørnstad, O.N., R.A. Ims, and X. Lambin. 1999a Spatial population dynamics: analyzing patterns and processes of population synchrony. *Trends in Ecology and Evolution* 14:427–432.

Bjørnstad, O.N., N. C. Stenseth, and T. Saitoh. 1999b. Synchrony and scaling in dynamics of voles and mice in northern Japan. *Ecology* 80:622–637.

Bjørnstad, O.N., and B. Bolker. 2000. Canonical functions for dispersal-induced synchrony. *Proceedings of Royal Society London-B* 267:1787–1794.

Bjørnstad, O.N., and J. Bascompte. 2001. Synchrony and second order spatial correlation in host–parasitoid systems. *Journal of Animal Ecology* 70:924–933.

Bjørnstad, O.N., and Falck, W. 2001. Nonparametric spatial covariance functions: estimation and testing. *Environmental and Ecological Statistics* 8:53–70.

Bjørnstad, O.N, M. Peltonen, A.M. Liebhold, and W. Baltensweiler. 2002. Waves of larch budmoth outbreaks in the European Alps. *Science* 298:1020–23.

Bjørnstad, O.N., A.M. Liebhold, and D.M. Johnson. 2008. Transient synchronization following invasion: revisiting Moran's model and a case study. *Population Ecology* 50:379–389.

Buonaccorsi, J.P., J.S. Elkinton, S.R. Evans, and A.M. Liebhold. 2001. Measuring and testing for spatial synchrony. *Ecology* 82:1668–1679.

Cattadori, I.M., D.T. Haydon, and P.J. Hudson. 2005. Parasites and climate synchronize red grouse populations. *Nature* 433:737–741.

Clobert, J., R.A. Ims, and F. Rousset. 2004. Causes, mechanisms and consequences of dispersal. In *Ecology, Genetics and Evolution of Metapopulations* (ed. I. Hanski & O. Gaggiotti), pp. 307–336. Academic Press, London.

Clotfelter, E.D., A.B. Pedersen, J.A. Cranford, N. Ram, E.A. Snajdr, V. Nolan, and E.D. Ketterson. 2007. Acorn mast drives long-term dynamics of rodent and songbird populations. *Oecologia* 154:493–503.

Cooke, B.J., and F. Lorenzetti. 2006. The dynamics of forest tent caterpillar outbreaks in Quebec, Canada. *Forest Ecology and Management* 226:110–121.

De Valpine, P., K. Scranton, and C.P. Ohmart. 2010. Synchrony of population dynamics of two vineyard arthropods occurs at multiple spatial and temporal scales. *Ecological Applications* 20:1926–1935.

Elias, S.P., J.W. Witham, and M.L. Hunter. 2004. *Peromyscus leucopus* abundance and acorn mast: population fluctuation patterns over 20 years. *Journal of Mammalogy* 85:743–747.

Eshleman, K.N., D.A. Fiscus, N.M. Castro, J.R. Webb, and A.T. Herlihy. 2004. Regionalization of disturbance-induced nitrogen leakage from mid-Appalachian forests using a linear systems model. *Hydrological Processes* 18:2713–2725.

Fox, J.W., D.A. Vasseur, S. Hausch, and J. Roberts. 2011. Phase locking, the Moran effect and distance decay of synchrony: experimental tests in a model system. *Ecology Letters* 14:163–168.

Grenfell, B.T., K. Wilson, B.F. Finkenstadt, T.N. Coulson, S. Murray, S.D. Albon, J.M. Pemberton, T.H. Clutton-Brock, and M.J. Crawley. 1998. Noise and determinism in synchronized sheep dynamics. *Nature* 394:674–677.

Gould, J.R., Elkinton, J.S., and Wallner, W.E. 1990. Density-dependent suppression of experimentally created gypsy moth, *Lymantria dispar* (Lepidoptera: Lymantriidae), populations by natural enemies. *Journal of Animal Ecology* 59:213–233.

Hanski, I. 1999. *Metapopulation Ecology*. Oxford University Press, Oxford.

Hanski, I., and I.P. Woiwod. 1993. Spatial synchrony in the dynamics of moth and aphid populations. *Journal of Animal Ecology* 62:656–668.

Hassell, M.P., and R.M. May. 1973. Stability in insect host–parasite models. *Journal of Animal Ecology* 43:567–594.

Hawkins, B.A., & Holyoak, M. 1998 Transcontinental crashes of insect populations? *American Naturalist* 152:480–484.

Haynes, K.J., A.M. Liebhold T.M. Fearer, G. Wang, G.W. Norman, and D.M. Johnson. 2009. Spatial synchrony propagates through a forest food web via consumer–resource interactions. *Ecology* 90:2974–2983.

Hoddle, M.S., K.M. Jetter, and J.G. Morse. 2003. The economic impact of *Scirtothrips perseae* Nakahara (Thysanoptera: Thripidae) on California avocado production. *Crop Protection* 22:485–493.

Holmes, T.P. 1991. Price and welfare effects of catastrophic forest damage from southern pine beetle epidemics. *Forest Science* 37:500–516.

Hudson, P.J., and I.M. Cattadori. 1999. The Moran effect: a cause of population synchrony. *Trends in Ecology and Evolution* 14:1–2.

Ims, R.A, and Andreassen, H.P. 2000. Spatial synchronization of vole population dynamics by predatory birds. *Nature* 408:194–196.

Ims, R.A, and Andreassen, H.P. 2005. Density-dependent dispersal and spatial population dynamics. *Proceedings of the Royal Society B-Biological Sciences* 272:913–918.

Ims, R. A., and D.Ø. Hjermann. 2001. Condition dependent dispersal. In *Dispersal*, ed. J. Clobert, J. D. Nichols, E. Dancin & A. Dhondt, pp. 203–216. Oxford University Press, Oxford.

Ims, R. A., and H. Steen. 1990. Geographical synchrony in microtine population-cycles – a theoretical evaluation of the role of nomadic avian predators. *Oikos* 57:381–387.

Jansen, V.A.A. 1999. Phase locking: another cause of synchronicity in predator–prey systems. *Trends in Ecology and Evolution* 14:278–279.

Johnson, D. M., O. N. Bjørnstad, and A. M. Liebhold. 2004. Landscape geometry and travelling waves in the larch budmoth. *Ecology Letters* 7:967–974.

Johnson, D.M., A.M. Liebhold, O.N. Bjørnstad, and M.L. McManus. 2005. Circumpolar variation in periodicity and synchrony among gypsy moth populations. *Journal of Animal Ecology* 74:882–892.

Johnson, D.M., A.M. Liebhold, and O.N. Bjørnstad. 2006. Geographical variation in the periodicity of gypsy moth outbreaks. *Ecography* 29:367–374.

Kelly, D., and V.L. Sork. 2002. Mast seeding in perennial plants: why, how, where? *Annual Review of Ecology and Systematics* 33:427–447.

Koenig, W. D. 1999. Spatial autocorrelation of ecological phenomena. *Trends in Ecology and Evolution* 14:2226.

Koenig, W. D. 2002. Global patterns of environmental synchrony and the Moran effect. *Ecography* 25:283–288.

Liebhold, A.M., and J.S. Elkinton. 1989. Elevated parasitism in artificially augmented populations of *Lymantria dispar* (Lepidoptera, Lymantriidae). *Environmental Entomology* 18:988–995.

Liebhold, A.M., and M.L. McManus. 1991 Does Larval Dispersal Cause the Spread of Gypsy Moth Outbreaks? *Northern Journal of Applied Forestry* 8:95–98.

Liebhold, A.M., J. Elkinton, D. Williams, and R.M. Muzika. 2000. What causes outbreaks of the gypsy moth in North America? *Population Ecology* 42:257–266.

Liebhold, A.M., W.D. Koenig, and O.N. Bjørnstad. 2004. Spatial synchrony in population dynamics. *Annual Review of Ecology, Evolution, and Systematics* 35:467–490.

Liebhold, A.M., D.M. Johnson, and O.N. Bjørnstad. 2006. Geographic variation in density-dependent dynamics impacts the synchronizing effect of dispersal and regional stochasticity. *Population Ecology* 48:131–113.

Lovett, G.M., L.M. Christenson P.M. Groffman, C.G. Jones, J.E. Hart, and M.J. Mitchell. 2002. Insect defoliation and nitrogen cycling in forests. *BioScience* 52:335–341.

Lynch, A.M., and M.J. Twery. 1992. Forest visual resources and pest management: potential applications of visualization technology. *Landscape and Urban Planning* 21:319–321.

McCullough, D.G., R.A. Werner, and D. Neumann. 1998. Fire and insects in northern and boreal forest ecosystems of North America. *Annual Review of Entomology* 43:107–127.

McShea, W. M. 2000. The influence of acorn crops on annual variation in rodent and bird populations. *Ecology* 81:228–238.

Moran, P.A. 1953. The statistical analysis of the Canadian lynx cycle. II. Synchronization and meteorology. *Australian Journal of Zoology* 1:291–298.

Murdoch, W.W., and A. Stewart-Oaten. 1989. Aggregation by parasitoids and predators: effects on equilibrium and stability. *American Naturalist* 134:288–310.

Myers, J.H. 1998 Synchrony in outbreaks of forest Lepidoptera: a possible example of the Moran effect. *Ecology* 79:1111–1117.

Nelson, W. A., E. McCauley, and J. Wimbert. 2004. Capturing dynamics with the correct rates: Inverse problems using semiparametric approaches. *Ecology* 85:889–903.

Openshaw, S. 1984. *Concepts and Techniques in Modern Geography Number 38: The Modifiable Areal Unit Problem*. GeoBooks, Norwick Norfolk.

Ostfeld, R.S., C.G. Jones, and J.O. Wolff. 1996. Of mice and mast: ecological connections in eastern deciduous forests. *BioScience* 46:323–330.

Parry D., J.R. Spence, and W.J.A. Volney. 1997. Responses of natural enemies to experimentally increased populations of the forest tent caterpillar, *Malacosoma disstria*. *Ecological Entomology* 22:97–108.

Peltonen, M., Liebhold, A.M., O.N. Bjornstad, and D.W. Williams. 2002 Spatial synchrony in forest insect outbreaks: roles of regional stochasticity and dispersal. *Ecology* 83:3120–3129.

R Development Core Team. 2010. *R: A Language and Environment for Statistical Computing*. R Foundation for Statistical Computing, Vienna. http://www.r-project.org

Ranta, E., V. Kaitala, and J. Lindstrom. 1997. Dynamics of Canadian lynx populations in space and time. *Ecography* 20:454–460.

Ranta, E., V. Kaitala, J. Lindstrom, and H. Linden. 1998. Synchrony in population dynamics. *Proceedings of the Royal Society B-Biological Sciences* 262:113–118.

Royama, T. 1992. *Analytical Population Dynamics*. Chapman and Hall, London.

Royama, T., W.E. MacKinnon, E.G. Kettela, N.E. Carter, and L.K. Hartling. 2005. Analysis of spruce budworm outbreak cycles in New Brunswick, Canada, since 1952. *Ecology* 86:1212–1224.

Satake, A., O.N. Bjørnstad, and S. Kobro. 2004. Masting and trophic cascades: interplay between rowan trees, apple fruit moth, and their parasitoid in southern Norway. *Oikos* 104:540–550.

Schmidt, K.A., and R.S. Ostfeld. 2007. Numerical and behavioral effects within a pulse-driven system: consequences for direct and indirect interactions among shared prey. *Ecology* 89:635–646.

Selås, V., O. Hogstad, S. Kobro, and T. Rafoss. 2004. Can sunspot activity and ultraviolet–B radiation explain cyclic outbreaks of forest moth pest species? *Proceedings of the Royal Society B-Biological Sciences* 271:1897–1901.

Silvertown, J.W. 1980. The evolutionary ecology of mast seeding in trees. *Biol. Journal of the Linnean Society* 14:235–250.

Smith, H.R. 1985. Wildlife and the gypsy moth. *Wildlife Society Bulletin* 13:166–174.

Turchin, P. 2003. *Complex Population Dynamics: A Theoretical/Empirical Synthesis*. Princeton University Press

Walde, S., and W.W. Murdoch. 1988. Spatial density dependence in parasitoids. *Annual Review of Entomology* 33:441–466.

Wolff, J.O. 1996. Population fluctuations of mast-eating rodents are correlated with production of acorns. *Journal of Mammalogy* 77:850–856.

Wu, J. 2004. Effects of changing scale on landscape pattern analysis: scaling relations. *Landscape Ecology* 19:125–138.

Yahner, R.H., and H.R. Smith. 1991. Small mammal abundance and habitat relationships on deciduous forested sites with different susceptibility to gypsy-moth defoliation. *Environmental Management* 15:113–120.

Ydenberg, R.C. 1987. Nomadic predators and geographical synchrony in microtine population cycles. *Oikos* 50:270–272.

Zhang, Q.B., and R.I. Alfaro. 2003. Spatial synchrony of the two-year cycle budworm outbreaks in central British Columbia, Canada. *Oikos* 102:146–154.

7

What Tree-Ring Reconstruction Tells Us about Conifer Defoliator Outbreaks

Ann M. Lynch

7.1 Introduction

Our ability to understand the dynamics of forest insect outbreaks is limited by the lack of long-term data describing the temporal and spatial trends of outbreaks, the size and long life span of host plants, and the impracticability of manipulative experiments at relevant temporal and spatial scales. Population responses can be studied across varying site and stand conditions, or for a few years under somewhat controlled circumstances, but it is difficult to study temporal variability for species that outbreak only two or three times a century. Fortunately, dendrochronology enables us to explore decadal- and century-scale outbreak dynamics at spatial scales ranging from within-tree to continental. Evidence of past insect defoliation can be identified, dated, and sometimes quantified using variation in the width and morphology of annual growth rings in trees (Plate 7.1). Mortality

Insect Outbreaks Revisited, First Edition. Edited by Pedro Barbosa, Deborah K. Letourneau and Anurag A. Agrawal.

events can be identified and dated by dating the last growth ring on dead trees, growth release events in survivors, and postdisturbance recruitment events. Insect defoliation can be distinguished from weather and other disturbance agents by comparing responses in host and nonhost trees.

A historical perspective increases our understanding of ecological disturbances and provides a reference for assessing modern disturbance events and ecosystem response to change. Dendrochronology enables the identification of outbreaks in times and places before the documentary period. This long-term perspective helps explain outbreak dynamics in forested ecosystems, and permits investigation of multiple factors interacting in space and time – time frames that encompass multiple events. It allows investigators to determine if observations made from contemporary events are unique to modern circumstances, perhaps influenced by anthropogenic factors, or are comparable to what occurred in the past. The role of insects as disturbance agents and regulators of forest ecosystems can be explored at the same scales at which the insect populations operate.

Tree-ring chronologies of insect outbreaks are proxy representations of population cycles. They excel at quantifying the temporal and spatial variability in events and at describing the nature of population oscillations – at determining how often outbreaks occur, the periodic nature of successive outbreaks, and the presence and strength of spatial autocorrelation. With other proxy or observational data, inferences can be made about associations with other disturbance regimes and climate. Tree-ring chronologies do less well at determining the precise timing of some types of events, particularly the onset of outbreaks that develop slowly, and at identifying causal relationships.

In this chapter I review the information gained through dendrochronologic study of coniferous forest defoliators, focusing on what has been learned through these studies. I provide brief summaries of species life cycles and biology to provide a context for the long-term record, and then summarize information learned from tree-ring studies. At the end of the chapter I review common themes emerging from dendrochronological study of insects, and the implications for our understanding of forest insect and disturbance ecology.

7.2   Methodological considerations

Dendrochronology is the science of dating wood material to the exact year and extracting from it information about events and environmental change. Tree growth reflects comprehensive integration of several processes and environmental conditions and is a reliable and robust indicator of environmental conditions (Pretzsch, 2009). Trees grow more slowly in periods of drought or other environmental stress than they do under favorable conditions, producing seasonal and annual variation in growth rings (Fritts, 1976). Because trees are long-lived organisms and because remnant material from dead trees may persist in the environment for long times and be incorporated into structures, wood material can be used to provide information about past environmental conditions.

Dendrochronology methodology is reliable, robust, and very well tested (Cook and Kairiudstis, 1990; Fritts, 1976; Schweingruber, 1996). Space does not permit a comprehensive review of the dendrochronology methodology used to develop insect outbreak chronologies. To obtain a good understanding of tree growth, how it is measured, and the influences of environmental factors on it, see Duff and Nolan (1953), Fritts (1976), Pretzsch (1996), Schweingruber (1996), and Telewski and Lynch (1990). Cook and Kairiudstis (1990), Schweingruber (1996), and Speer (2010) present the methodology of dendrochronology as it is used for a variety of disciplines. For methods related to detecting defoliator outbreaks, see Swetnam *et al.* (1985) and papers cited herein; for bark beetle outbreaks, see Berg *et al.* (2006) and Veblen *et al.* (1991), and papers cited therein.

Defoliation suppresses radial growth, and can affect the relative widths and physical characteristics of earlywood and latewood, wood density, cell structure, and chemistry (Cook and Kairudstis, 1990; Swetnam *et al.*, 1985). These effects are often visible in raw wood samples (Plate 7.1). Using crossdated material, quantitative techniques can identify and separate variation in ring width related to climate, insects, fire, and other disturbances. The insect outbreak chronologies that are developed are long-term proxy data sets of insect population behavior that can be subjected to statistical analysis (Swetnam *et al.*, 1985; Swetnam and Lynch, 1989, 1993; Speer, 2010; Volney, 1994). Dendrochronologic reconstruction of defoliator outbreak chronologies has been widely used, with refinements and adjustments for individual circumstances (Hadley and Veblen, 1993; Jardon *et al.*, 1994; Morin and Laprise, 1990; Nishimura and Laroque, 2010; Swetnam *et al.*, 1985, and papers cited herein).

Outbreak chronologies are usually analyzed by computing the intervals (length of the period between the beginnings of two successive outbreaks), duration (length of the period encompassing the beginning and end of outbreak-related growth suppression), and sometimes severity of growth suppression (determined by methods that vary considerably) on the basis of absolute numbers or the proportion of sampled sites involved in an outbreak at any given time, and for the presence and strength of periodic and synchronous patterns. Spectral analysis is used to identify cyclical components in chronologies by decomposing the time series into a spectrum of cycles of different lengths. Spatial autocorrelation analysis is used to quantify the degree of geographic dependency in outbreak chronologies and can be applied at any relevant scale. Significant autocorrelation values between time series suggest that observations in multiple areas are correlated with or even dependent on each other. Significant components found in insect chronologies may involve a single significant value (e.g., 36), a range of periods around a peak value (e.g., 34–38), or multiple periodicities (e.g., 25, 36, 48–51). For time series in which outbreaks have cyclical components, multiple components are somewhat analogous to variance components, in that components (significant periodicities) may account for certain proportions of the variance in temporal variation, say, 40% and 28% for the first and second periodicities. Multiple frequencies within the same system may be harmonic or nearly harmonic; this outcome appears to result from natural processes, but in some cases may be a mathematical artifact. Time series are analyzed for correlation with other time series, such as precipitation records or reconstructions. Investigators attempt to interpret causal relationships that produce periodic and synchronous patterns, perhaps as inherent to population dynamics or as entrained by climatic cycles, but while evidence may be strong, proof is usually lacking.

7.3 Reconstructions of outbreak histories

7.3.1 Spruce budworm

Spruce budworm (SBW) (*Choristoneura fumiferana* (Clemens) (Lepidoptera: Tortricidae)) and western spruce budworm (WSBW) (*Choristoneura occidentalis* (Freeman)) are closely related univoltine, wasteful early-season defoliators. Tree bud development and suitable weather in spring play important roles in bud-worm survival and population growth (Volney and Fleming, 2007). Outbreaks are generally extensive, often involving entire regions, and persist for many years. Balsam fir (*Abies balsamea*) and to a lesser extent white spruce (*Picea glauca*) are the most important SBW hosts, and Douglas-fir (*Pseudotsuga menziesii*) and true firs, especially white fir (*A. concolor*), are the most important WSBW hosts.

Tree-ring chronologies
Spruce budworm outbreaks are identified in the tree-rings primarily through numerous adjacent narrow rings (Plate 7.1) not observed in nonhost species. SBW reconstructions have been developed for many areas of eastern Canada and the United States (Table 7.1). One extraordinarily long chronology from Québec includes 13 outbreaks since the early 1500s. Outbreak initiation and duration vary considerably amongst trees and stands, and numerous missing rings occur during outbreaks (Blais, 1961; Morin and Laprise, 1990; Lemieux and Fillion, 2004). Chronologies show that intensity may decline during the course of the outbreak, but later resurge (Blais, 1983; Morin *et al.*, 1993). Radial growth is usually suppressed for 11–14 years per outbreak, though duration is variable (Table 7.1). Growth suppressions detected in the tree-ring series begins 2–4 years after noticeable defoliation (Blais 1983). Mortality continues for up to 4 years after defoliation has ceased (Blais 1981a), which implies that growth suppression may persist for that long as well.

Tree-ring evidence indicates that outbreak extent and severity can be influenced by precedent landscape- and region-scale wildfires and SBW outbreaks (Blais, 1954; Bergeron and Archambault, 1993; Bergeron *et al.*, 2001; Bouchard *et al.*, 2006; Morin *et al.*, 2007).

Associations with temperature and precipitation
Climate and SBW reconstructions longer than two centuries show that outbreaks tend to occur during periods of relatively dry, sunny, early-summer conditions (Blais, 1961, 1984; Morin *et al.*, 1993), which result in more rapid larval development and increased fecundity as well as increased staminate flower production (staminate flowers are easily accessible, rich sources of nutrients for small larvae) (Greenbank, 1956; Pilon and Blais, 1961). In Minnesota, near the prairie-forest boundary where conditions for balsam fir are marginal, outbreaks are weakly

Table 7.1 For spruce budworm, location, sample size, initial year and length of chronology, number of outbreaks in the chronology (NOB), intervals, average interval length or significant periodicities from time series analyses (periods are indicated by square brackets [ ]; a number followed by parenthetical numbers indicates a peak periodicity within a range of significant period values), and duration or mean duration±SD (DUR), as determined in dendrochronologically reconstructed chronologies. The interval just prior to the first twentieth-century outbreak is indicated in **bold**. Some statistics are computed from tabulated data or text references in publications. Information is presented in order of publication date, as dendrochronology and analysis methodologies have improved over time. See Blais (1983) for a figure depicting many of the eastern Canadian locations with respect to one another. Note that advances in methodology in the late 1980s improved the precision of subsequent chronologies.

General location	No. sites (no. trees)	First year	Length	NOB	Intervals or [periods]¹	DUR	Publication	Place name
Northwest Ontario	81 (~800)	1802	170	3	average 93		Blais 1954, 1983	Lac Seul area
South-central Ontario	6 (300)	1704	268	3	average 93		Blais 1983	Lake Nipogen
Eastern Ontario	86 (430)	1802	180	2	**100**, 37	<21	Blais 1983	Algoma
Southeast Québec	32 (160)	1812	140	2	none, 45		Blais 1961	Gaspé Peninsula
Southeast Québec	37	1870	120	3–4	73, 35, **38**		Blais 1961	Lower St. Lawrence
Southern Québec and New Brunswick		1940	20	1		11	Blais 1964	Laurentide Park
Southern Québec	87	1680	281	6	44, 60, 25, **76**, 37	3–11	Blais 1965	Laurentide Park
Southwest Québec	18	1761	211	3	**127**, 28, 28		Blais 1981a	Ottawa River
Western Québec	(20)	1764	215	4	**127**, 30, 27		Blais 1981b	Dumoin and Coulonge Rivers
Southeast Québec		~1750	250	6	42, 75, **24**, 29, 27		Blais 1983	Eastern Townships, Lower St. Lawrence
Northeast Québec and Newfoundland	15	1810	162	4	**130**, 27		Blais 1983	North Shore
Southern Québec	8	1790	200	4	**76**, 36, 26	11.8±2.8	Morin and Laprise (1990)	Lac Saint-Jean
Southwest Québec	30 (160)	1780	200	4	**109**, 11, 40	11–23	Morin et al. 1993	Lake Duparquet, Abitibi

Location	Sites	Start year	Length	No.	Outbreak years	Interval	Reference	Site
Southern Québec	8 buildings (153)	1672	283	5–6	44, 57, 24, 33	3–11	Krause 1997	Chicoutimi, east of Lac Saint-Jean
Southern Québec	1 (121)	1859	130	4	**37**, 38, 23	7.3 ± 2.2	Filion et al. 1998	South of Lac Saint-Jean (highland sites with no history of wildfire)
Mid-latitude Québec	28	1847	152	1		10	Simard and Payette 2001	Eastern Laurentian Highlands
Northeast British Columbia	7	1817	176	0–6	26	6–17	Burleigh et al. 2002	Fort Nelson area
Southwest Québec	32	1875	130	5	[25–38, 29–34]		Jardon et al. 2003	Subregional collection
Northern and central New Brunswick	4	1940		1		9	Krause et al. 2003	Kedgewick
Southern Québec	7 buildings 1 stand	1513 1816	406 186	11	65, 36, 32, 42, 53, 27, 36, **46**, 33, 28	14.5 ± 5.7 (7–25)	Boulanger and Arseneault 2004	South shore of St. Lawrence River
Southern Québec	1 (25)	1859	141	3	38, 23	4–9	Lemieux and Filion 2004	Mount Mégantic
Western Québec	3	1820	180	5–6	180 yrs / 5 = 36		Bouchard et al. 2006	Témiscamingue region
Northern Maine	37	1700	300	5	53, 46, **106**, 62		Fraver et al. 2007	Big Reed Forest Reserve
Northwest Minnesota	5	1916	90	3–4	35, 38		Rauchfuss et al. 2009	Itasca State Park
Islandic Québec[2]	14	1850	152	4–7	26, 14, 11, **10**, 13, 8, 38		Barrette et al. 2010	Anticosti Island
Eastern Québec	9	1750			ca. **35**, **35**, 30		Bouchard and Pothier 2010	North Shore

[1] **None** indicates that no outbreaks are inferred in the chronology prior to 1900.
[2] SBW and hemlock looper outbreaks are not distinguished in this chronology.

associated with wetter-than-normal spring conditions (Rauchfuss *et al.*, 2009), similar to WSBW (discussed later). A change in outbreak frequency, extent, and severity in the twentieth century appears to be associated with a change in polar atmospheric circulation patterns (discussed later in this chapter).

Periodicity and synchrony

Tree-ring studies provide consistent evidence that periodic SBW outbreaks cycle synchronously at landscape and regional scales every 25 to 45 years (Table 7.1; Cooke *et al.*, 2007; Jardon *et al.*, 2003; Volney and Fleming 2007), similar to cycle lengths produced in modeled population dynamics (Royama *et al.*, 2005). Cycle amplitude varies considerably, and synchrony decays at very long distances (Gray *et al.*, 2000; Volney and Fleming, 2007). Time series analyses from southwestern Québec indicate a tighter signal of 25–38 years, with a true periodic component at 29–34 years, as well as quasi-periodic or complexly periodic components (Jardon *et al.*, 2003). Synchrony in SBW chronologies is attributed to moth migration (Jardon *et al.*, 2003). Tree-ring studies show that large SBW outbreaks usually develop over several years, and that outbreaks sometimes exhibit patterned, almost epicentric expansion, particularly during extreme events in which outbreaks envelop adjacent areas where conditions are marginal for SBW (Blais 1954, 1981b; Morin and Lapris, 1990; Morin *et al.*, 1993; Burleigh, 2002; Fraver *et al.*, 2007; Jardon *et al.*, 2003). Multicentury tree-ring studies and multidecadal survey data both indicate that areas less suited to budworm incur less frequent and less intense outbreaks (Blais, 1961, 1983; Burleigh *et al.*, 2002; Candau and Fleming, 2005; Gray *et al.*, 2000; Rauchfuss *et al.*, 2009; Williams and Birdsey, 2003).

Twentieth-century patterns

Outbreaks in the twentieth century have been synchronous (extensive) and severe, but not to the extraordinary degree indicated in the literatures. Such outbreaks occurred in the earlier dendrochronology record, but infrequently. If the Blais (1983) summary of his work from across eastern Canada is viewed as proportions of the number of areas in the chronologies at a given time, 70% were involved in an outbreak in the first decade of 1800. In later work, all chronologies that include the early nineteenth century show an outbreak in the first decade of 1800 (Elliott, 1960; Morin *et al.*, 1993; Krause, 1997; Bouchard and Pothier, 2010; Boulanger and Arsenault, 2004; Fraver *et al.*, 2007), except for one chronology (Morin and Laprise, 1990) in which the stands would have been too young in the 1800s to be susceptible to SBW.

There is a common (but not universal) pattern of a lengthy period without outbreaks prior to the first twentieth-century outbreak, followed by a dramatic increase in frequency and severity in the twentieth century (Table 7.1). In some cases, the first inferred or documented outbreak in an area occurred in the last half of the twentieth century, even though host trees were present earlier (Blais, 1961; Simard and Payette, 2001). The generally accepted interpretation of increased synchrony and severity in twentieth century SBW outbreaks has been that it is caused by human influences, particularly those of fire suppression and logging (which favored balsam fir abundance) and pesticide use (which maintained the forest in a budworm-susceptible condition), and less to natural changes in forest structure and species composition. Later developments have favored climate as the driving factor, perhaps exacerbated by human land use practices mentioned earlier. Coincidental reconstruction of fire history, SBW outbreaks, and climate have shown that fire frequency in eastern Canada dramatically decreased circa 1850, which was associated with changes in polar atmospheric circulation patterns and summer isolation (Bergeron and Archambault, 1993; Bergeron *et al.*, 2001; Carcaillet *et al.*, 2010). This favored increased balsam fir abundance over extensive areas; these forests matured into budworm-susceptible conditions at the turn of the twentieth century (Blais, 1954; Bergeron and Leduc, 1998; Morin *et al.*, 2007). It is interesting to note that all 81 sample sites in the Blais (1954) western Ontario chronology, which began in 1850, were of fire origin.

7.3.2   Western spruce budworm

Tree-ring chronologies

Several WSBW chronologies developed from Douglas-fir and white fir encompass 3–4 centuries and include 8–14 major outbreaks (Table 7.2). Two extraordinarily long chronologies from northern New Mexico and southern Colorado are from six to nearly eight centuries years long and include 17–20 outbreaks. The WSBW chronologies are characterized by temporal variability, missing rings during outbreaks, and mid-outbreak reprieves in outbreak intensity. Outbreak initiation and end are usually asynchronous, and there is considerable variability between outbreaks at multiple temporal and spatial scales (Swetnam and Lynch, 1989, 1993; Swetnam *et al.*, 1985; Ryerson *et al.*, 2003; Campbell *et al.*, 2006), though outbreak initiation is markedly synchronous in the Blue Mountains of northeast Oregon (Swetnam *et al.*, 1995), and for occasional outbreaks elsewhere (Swetnam and Lynch, 1989; Ryerson *et al.*, 2003). Radial growth response typically lags detectable defoliation by 1–3 years (Belyea, 1952; Kulman, 1971; Brubaker, 1978; Alfaro *et al.*, 1982; Swetnam *et al.*, 1985; Wickman *et al.*, 1980). Trees and stands also differ in post-outbreak rates of recovery, indicating that outbreaks may persist in some trees or stands longer than in others, or that damage severity or post-outbreak weather conditions influence the recovery period (Brubaker and Greene, 1979; Mason *et al.*, 1997).

Return intervals are commonly 32–40 years, and significant periodicities are 30–43 years (Table 7.2); in northern New Mexico, the strongest periods are 25 and 50 years. The mean duration of growth suppression is consistently between 11 and 15 years, though some inferred outbreaks in the southwestern reconstructions lasted two to several decades (Swetnam and Lynch, 1989). Years without evidence of infestation somewhere are rare in all regional chronologies. Outbreaks are neither more nor less frequent in old-growth stands than in younger stands (Lynch and Swetnam, 1992). Duration of growth suppression is generally longer in old-growth stands, but it is not clear if this is because WSBW population density persists longer in old stands, or if tree recovery is protracted in older stands (Lynch and Swetnam, 1992).

Table 7.2 For western spruce budworm and two-year cycle budworm, chronology length (years), number of outbreaks in the chronology (NOB), mean return intervals, significant periodicities from time series analyses, and mean duration of outbreaks, as determined in dendrochronologically reconstructed chronologies. The most significant of multiple periods are indicated with **bold** type. Some statistics are computed from tabulated data or text references in publications.

Location	Length	NOB	Intervals	Periods	Publication
Western spruce budworm					
Western Montana	100	3			McCune 1983
Western Montana	169	8	32–35	13.9; 23.2 after 1910	Anderson et al. 1987
Southern Rocky Mountains	280	8	34.9	12.9	Swetnam and Lynch 1989
Northern New Mexico	389	9		**25** (20–33); 50 (30–11); 2 (2.0–2.1)	Swetnam and Lynch 1993
Northeast Oregon	300	8	21–47; average[a] 36±10	13–17; average[a] 15±2	Swetnam et al. 1995a
North-central Colorado	277	9		6.9	Weber and Schweingruber 1995
South-central Colorado	348	14	32–79; average 36±10	25, **37**, 83	Ryerson et al. 2003
South-central British Columbia	300	8		30, **43**, 70	Campbell et al. 2006
Two-year cycle budworm					
Southern British Columbia	310	7	34+1.1	3–23	Parish and Antos 2002
Central British Columbia	278	7	32	10–16	Zhang and Alfaro 2002, 2003

Includes Douglas-fir tussock moth (*Orgyia pseudotsugata* (McDunnough) (Lepidoptera: Lymantriidae)).

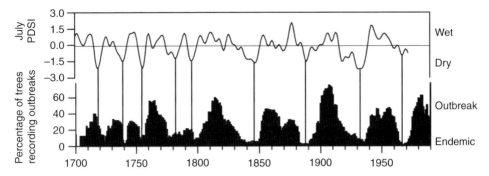

Figure 7.1 Western spruce budworm chronology (bottom) and Palmer Drought Severity Index (PDSI) (top) from the northern Blue Mountains, Oregon (T. Swetnam). Budworm outbreaks occur during relatively wet periods, and do not occur during droughts. Note that the PDSI cycle changes frequency about 1795, and the budworm cycle follows suit.

Associations with temperature and precipitation

WSBW outbreaks are positively associated with periods of increased spring and summer precipitation, and endemic periods are associated with drought conditions in the southwestern United States (Swetnam and Lynch, 1989, 1993; Ryerson *et al.*, 2003; Swetnam and Betancourt, 1998). These authors concluded that foliage production was abundant during pluvials, thereby supporting increased WSBW populations. Swetnam and Lynch (1993) also noted a weak pattern of cooler temperatures during periods of higher precipitation. Swetnam and Betancourt (1998) show a chronology from Oregon in which not only do WSBW outbreaks coincide with summer pluvials, but outbreak frequency decreases coincidentally with a decrease in pluvial frequency circa 1795 (Figure 7.1). This evidence strongly supports inferences that WSBW outbreaks in the southwestern United States occur during periods of increased moisture, and that the weakly or quasi-periodic population dynamics are entrained by climate.

WSBW outbreaks in British Columbia were associated with periods of average spring air temperatures after dry winters (Campbell *et al.*, 2006), a pattern more similar to those of SBW. Campbell *et al.* (2006) interpreted these circumstances as reflecting early spring conditions, which cued budworm emergence from hibernacula to coincide with the appropriate stage of bud development. The dissimilar findings from different regions are not incompatible, and regional differences should be expected.

Periodicity and synchrony

Some degree of synchrony is evident in the regional chronologies from each region, with significant periodicities of about 25, 30–43, and 70–80 (Table 7.2). Commonalities in periodicities from multiple regions, with multiple peaks in individual regions, indicate that WSBW population dynamics are at least inherently stable as periodic or quasi-periodic systems. Migration is an unlikely (but not impossible) mechanism to synchronize WSBW outbreaks because mountainous terrain and fire history cause host forests to be spatially discontinuous. Synchronization therefore requires exposure to one or more common factors able to harmonize the population oscillations. Given the demonstrated associations with climate, which is strongly periodic in western North America, it is implicated as the entraining factor.

Some WSBW cycles have unusually great amplitude, with a 70–80-year period (Table 7.2). This pattern is clearly evident in the south-central Colorado reconstruction (Ryerson *et al.*, 2003) which shows fairly regular outbreaks, with every second or third outbreak being more extensive or severe. Whether these cycles occur when the maxima of all cycles coincide by chance (Campbell *et al.*, 2006), are inherent in the population dynamics, or are driven by unknown exogenous factors is presently unknown. The temporal dynamics of root disease organisms should be examined as a possible factor affecting long-term cycles in western North American forests.

Twentieth-century patterns
The chronologies document changes in WSBW incidence in the twentieth century, including increased frequency, synchrony, duration, and severity. Of these, evidence is strongest for increased synchrony. Initiation of the 1940s–1950s and 1970s–1980s outbreaks across disjunct areas is notably synchronous in the Southwest, Oregon, and British Columbia (Campbell *et al.*, 2006), as well as Montana for the 1970s–1980s outbreak. Synchronous outbreaks occurred in the past in all of these regions, but the two most recent regional outbreaks involved at least as many, if not more, sites than any earlier outbreak in the Southwest or British Columbia. The chronologies from northeast Oregon are notably synchronous throughout, especially for outbreak initiation.

Evidence of greater outbreak frequency in the twentieth century relative to the nineteenth century is found in several chronologies (Swetnam and Lynch, 1989, 1993; Swetnam *et al.*, 1995), but not in others (Anderson *et al.*, 1987; Ryerson *et al.*, 2003). Conversely, interval length has varied considerably over time, and frequency during the twentieth century is within the reconstructed range of variability. There is consistent evidence of longer duration and increased severity, which is supported by well-documented high levels of tree mortality during the 1970s–1980s outbreaks. However, long, severe outbreaks occurred in the past in most of the chronologies.

The generally accepted interpretation of increased synchrony and severity in modern WSBW outbreaks pertains to the effects of fire exclusion, which fostered proliferation of shade-tolerant host species in multistoried canopies, thereby improving budworm habitat quality at large spatial scales (Fellin *et al.*, 1983; McCune, 1983; Swetnam and Lynch, 1989, 1993; Swetnam *et al.*, 1995). In much of western North America, fire exclusion began with livestock grazing, which reduced surface fuel amounts and connectivity. It became more comprehensive with changes in wildland fire suppression policies after devastating fires in the late nineteenth and early twentieth centuries, and improved fire-fighting technology increased the success of those policies (Swetnam, 1990; Keane *et al.*, 2002). Extensive forests that established after those fires would not have supported WSBW outbreaks for several decades, but later reached budworm-susceptible condition more or less simultaneously (Hadley and Veblen, 1993). A period of unusually low budworm activity can be seen in the early 1900s in several reconstructions (Swetnam and Lynch, 1989, 1993; Campbell *et al.*, 2006). Timing and duration of this period vary between regions, but the regional effects were to synchronize outbreaks in the latter half of the twentieth century over extensive areas. A second interpretation of increased synchrony and severity of modern WSBW outbreaks pertains to weather conditions suitable for population release. Ryerson *et al.* (2003) noted that if outbreaks are triggered by periods of increased moisture (as seen in most of the western chronologies), then synchrony of the 1970s–1980s outbreak might be associated with increases in precipitation, which is observed in the instrumental record and the dendrochronologic reconstructions. Probably both explanations are germane.

7.3.3 Two-year cycle spruce budworm

Biology
Two-year cycle spruce budworm (*Choristoneura biennis* (Freeman)) is semivoltine, overwintering as second and fourth instars. It is an early-season defoliator in both years of the life cycle, with heavy feeding on buds and new foliage in the second year, producing a distinctive tree-ring signal (Plate 7.1). Trees recover somewhat in alternate years, reducing the impact (Parish and Antos, 2002). It inhabits cold high-latitude high-elevation forests with Engelmann spruce (*P. engelmannii*), white spruce, and subalpine fir (*A. lasiocarpa*) as hosts.

Tree-ring chronologies
Our knowledge of the long-term patterns shown by this species was acquired by Parish and Antos (2002) and Zhang and Alfaro (2002, 2003) (Table 7.2). Suppression periods last 3–23 years (most last 7–16 years), about half of which may be a recovery period. Return intervals vary from 24 to 39 years, with averages of 32–34 years (Table 7.2). The initiation year, intensity, and extent of outbreaks vary spatially and temporally. Power spectra periods are significant at 6.8 and 43.6–45 years (Parish and Antos, 2002). The 45-year period and variability in outbreak extent and severity suggest quasi-periodic population dynamics influenced by a large-scale extrinsic factor, perhaps climate (Zhang and Alfaro, 2003). Some stands sampled by Parish and Antos (2002) were over 300 years old but did not incur an outbreak until almost 100 years into the chronology, indicating that stand structure and/or age are probably important features of susceptibility. In outbreak areas, tree growth over time is influenced more by two-year spruce budworm than it is by climate (Parish and Antos, 2002).

7.3.4   Larch sawfly

Biology
Larch sawfly (*Pristiphora erichsonii* Hartig (Hymenoptera: Diprionidae)) is usually univoltine, but pupation is sometimes prolonged to two winters. Males are rare, and most reproduction is parthenogenic. Adult emergence and egg-laying periods are nonsynchronous and long (up to 6 weeks), resulting in feeding throughout the summer. Populations do not develop in young stands. Due to the ability of larch (*Larix* spp.) to refoliate, and to sawfly preference for foliage on older twigs, impact is usually in terms of growth loss, but mortality can be severe in prolonged outbreaks (Drooz, 1971).

Tree-ring chronologies
Key attributes of tree-rings during larch sawfly outbreaks are the presence of pale latewood rings (which during the first year of the outbreaks may be of normal width), followed by radial growth suppression that is usually abrupt and synchronous amongst trees, and numerous (up to nine consecutive) missing or incomplete rings (Jardon *et al.*, 1994; Filion and Cournoyer, 1995; Girardin *et al.*, 2005; Nishimura and Laroque, 2010). Interval lengths vary considerably (6–56 years), but on average are between 25 and 35 years (Table 7.3). Data taken from the literature indicate that outbreaks generally last 6–10 years in individual stands, with a range of 2–18 years (Table 7.3). Once stands reach susceptible age, eastern larch (*L. laricina*) is defoliated on average 30–50% of the time (Girardin *et al.*, 2002; Jardon *et al.*, 1983). Both local and regional outbreaks are strongly synchronous (Girardin *et al.*, 2001, 2002; Nishimura and Laroque, 2010).

Dendrochronology studies indicate that larch sawfly is the most important regulator of eastern larch stand dynamics, influencing establishment dates and stand age structure (Girardin *et al.*, 2002; Busque and Arseneault, 2005). After severe outbreaks, larch establish from increased seed production on surviving trees (larch continues to reproduce during outbreaks) (Tailleux and Cloutier, 1993). After mild outbreaks, new stands establish from surviving advanced regeneration (Girardin *et al.*, 2002). In both cases, competition from shade-tolerant species is reduced.

Dendrochronologic investigations provided strong evidence that larch sawfly, once thought to be exotic, is native to North America. The evidence for this conclusion is the occurrence of numerous outbreaks since the early 1700s in Québec and Manitoba, established by Jardon *et al.* (1983) and Nishimura and Laroque (2010).

Twentieth-century patterns and associations with temperature
The 1894–1906 larch sawfly outbreak was probably the most severe and extensive disturbance event in eastern larch forests in 150 years (Arquillière *et al.*, 1990; Jardon *et al.*, 1994; Girardin *et al.*, 2002; Busque and Arseneault, 2005). Arquillière *et al.* (1990) and Nishimura and Laroque (2010) conclude that this mesocale event and other outbreaks between 1883 and 1911 are associated with a shift in sawfly disturbance regimes associated with warmer, improved conditions for larch growth and vigor. Investigators do not note periodic behavior associated with temperature or precipitation.

Table 7.3 For conifer-feeding defoliators other than SBW and WSBW: location, sample size, initial year and length (years) of chronology, number of outbreaks in the chronology (NOB), intervals or significant periodicities from time series analyses, and duration or mean duration±SD (DUR) of suppressed growth, as determined in dendrochronologically reconstructed chronologies. Some statistics are computed from tabulated data or text references in publications.

Insect and general location	Number of sites (trees)	First year	Length	NOB	Intervals or [periods]	DUR	Publication
Cedar shoot moth, *Dichelia cedricola* (Diakonoff) (Lepidoptera: Tortricidae)							
Turkey	4 (17)	1954	48	3	30, 14	12, 6	Carus and Avci 2005
Epinotia subsequana (Hw). (Lepidoptera: Tortricidae)							
Northwest Spain	2 (10)	1900	100			3	Camarero *et al.* 2003
Hemlock looper, *Lambdina fiscellaria* (Guénée) (Lepidoptera: Geometridae), with SBW							
Eastern Québec	14	1846	160	4–7	26, 14, 11, 10, 13, 8, 38		Barrette *et al.* 2010
Larch budmoth, *Zeiraphera diniana* Guénée (Lepidoptera: Tortricidae)							
Switzerland	35	1503	488	57	8.6 ± 1.7		Weber 1997
French Alps	7	1800	190+	22	8.9 ± 1.0		Rolland *et al.* 2001
Northwest Italy	1	1760	240	19	[9, 23–28, ave. 27]	3.6 ± 1.3	Nola *et al.* 2006
European Alps	(180)	832	1173		[8.1–9.3]		Esper *et al.* 2007
European Alps	(70)	1700	300		9, varies with elevation		Büntgen *et al.* 2009
Larch sawfly, *Pristiphora erichsonii* Hartig (Hymenoptera: Tenthredinidae)							
Northern Québec	10	1596	391	2	35	4, 7	Arquillière *et al.* 1990
Northern Québec	1 (26)	1680	300	10	27 ± 9 (6–49)	9 (5–15)	Jardon *et al.* 1994
Southwest Québec	12	1858	142	5			Giardin *et al.* 2002
Manitoba	4	1880	200	5	35 ± 18 (12–56)	7 ± 3 (6–10)	Girardin *et al.* 2005
Western Labrador	12	~1720	280	8	25 ± 9 (14–36)		Nishimura and Laroque 2010

(Continued)

Table 7.3 (Cont'd).

Pandora moth, *Coloradia pandora* (Blake) (Lepidoptera: Saturniidae)							
South-central Oregon	14	1379	622	22	27 (8–44) [18–24, 37–41]	10 (4–18)	Speer et al. 2001
Pine butterfly, *Neophasia menapia* (C. Felder and R. Felder) (Lepidoptera: Pieridae)							
Idaho	(100)	1916	30	1		2	Evenden 1940
Pine false webworm, *Acantholyda erythrocephala* (L.) (Hymenoptera: Pamphilidae)							
Northern New York	21	1950	50	1		5	Mayfield et al. 2005
Pine sawfly, *Diprion pini* L. (Hymenoptera: Diprionidae)							
France	2	1953	31	2			Laurent-Hervouët 1986b
Southern Finland	1	1966	36	1			Kurkela et al. 2005
Processionary caterpillars, *Thaumetopoea pityocampa* Denis & Schiffermüller, *T. wilkinsoni* Tams., and *T. ispartaensis* (Doğanlar and Avci) (Lepidoptera: Geometridae)							
France	2	1959	25	1			Laurent-Hervouët 1986a
Turkey	1	1981	23	3–4	12, 7	6 ± 0	Carus 2004
Turkey	5	1947	57	4	8, 31, 13	5, 8, 5	Avcı and Carus 2005
Turkey	1	1985	20	3	5–7	4–7	Carus 2010
Western false hemlock looper, *Nepytia freemanii* Munroe (Lepidoptera: Geometridae)							
British Columbia	(200)	1880	105	1		5	Alfaro and MacDonald 1988

7.3.5 Pandora moth

Biology

The Pandora moth (*Coloradia pandora* (Blake) (Lepidoptera: Saturniidae)) is a large-bodied semivoltine insect that overwinters as second or third instars at the base of needles and as pupae in the soil (which may diapause up to 5 years). Large larvae occur early- to midseason every other year, heavily defoliating old foliage but not buds or new shoots. Adults are strong flyers. Outbreaks spread from centers, though centers may coalesce. Outbreaks collapse rapidly after 6–8 years from a nucleopolyhedrosis virus infection (Schmid and Bennett, 1986). Mortality to hosts ponderosa pine (*P. ponderosa*) and lodgepole pine (*P. contorta*) is generally low, because defoliation occurs only in alternate years, allowing some recovery, and because buds are not consumed (Schmid and Bennett, 1986).

Tree-ring chronology

Speer *et al.* (2001) developed a 622-year chronology of Pandora moth from ponderosa pine in central Oregon (Table 7.3). Radial growth in the first suppression year is sometimes a white ring, is reduced about 50%, and has very thin latewood (Plate 7.1). The pattern of narrow rings in alternate years found in two-year cycle budworm is absent in Pandora moth material, or at least is not striking. This difference is not discussed in the literature, but is probably related to feeding behavior. Both are early-season defoliators, but two-year cycle budworm feeds on buds and new foliage, while Pandora moth feeds on old foliage and not on buds or new foliage.

Initiation of the suppression periods are typically represented by a cluster of dates rather than a single year, originating in small groups of stands, consistent with a pattern of population spread from multiple initial locations over several years. Intervals between outbreaks range from 8 to 48 years, but are sometimes very long – up to 156 years. Time series analyses showed significant periods at 18–24 years and 37–41 years. Both periods are evident in the regional chronology, in which approximately a third of outbreaks were more extensive. The regional time series exhibited a striking episodic pattern with quasi-cyclic recurring outbreaks. Mean duration is 10 years (4–18 years). Twentieth-century outbreaks were not more synchronous, severe, or longer in duration than outbreaks in previous centuries, but there was an unusual 60-year reduction in regional activity during about 1920–1980 (similar to WSBW). Temporal variability of outbreaks does not appear to be associated with climate (Speer and Jensen, 2003), but variable amplitude and strengths of the cycles over centuries suggest that an exogenous process entrains population fluctuations (Speer *et al.*, 2001). Nucleopolyhedrosis viruses persist in the environment for several years (Ghassemi *et al.*, 1983; Olofsson, 1988) and might restrict cycle frequency at the short end, but once the virus degrades then other factors might influence periodicity.

Reduced fire frequency is associated with Pandora moth outbreaks (Speer *et al.*, 2001), but it is not clear if this is because fine surface fuel accumulation is reduced because of needle consumption by larvae, thereby reducing fire frequency, or if Pandora moth survival is greater when fires are reduced.

7.3.6   Larch budmoth

Biology
Larch budmoth (*Zeiraphera diniana* Guénée) (Lepidoptera: Tortricidae)) (LBM) inhabits subalpine larch and larch-Swiss stone pine (*L. decidua* and *P. cembra*) forests of the European Alps. It is a univoltine early-season wasteful defoliator that pupates in the litter, though seasonality is variable (Baltensweiler and Fischlin, 1988). Larval eclosion is synchronized with host bud flush, and this coincidence is essential. Adults are strong fliers, and outbreaks can expand to less desirable habitats during intense outbreaks.

Tree-ring chronologies
LBM is detectable in the tree-rings by reduced ring width, reduced earlywood, reduced density of the latewood, and missing rings (Weber and Schweingruber, 1997; Esper *et al.*, 2007; Nola *et al.*, 2006; Rolland *et al.*, 2001). Often so many rings are missing during suppression periods that wood cannot be cross-dated and is discarded by investigators (large numbers of missing rings also occur with Jack pine sawfly (*Neodiprion swainei* Midd., Hymenoptera: Diprionidae)) (O'Neil 1963). Growth is suppressed 3.6±1.3 years (computed from numbers presented by Nola *et al.* 2006), somewhat longer than the 2.9±0.2 years documented in the shorter historical record (Baltensweiler and Fischlin, 1988). Growth recovery is rapid, and is complete within 3 years (Rolland *et al.*, 2001). Tree-ring reconstructions have confirmed a regular 8–9-year periodicity in the LBM population dynamics across the Alps that has persisted for 12 centuries (Table 7.3). LBM outbreaks are less frequent at lower elevations (Büntgen *et al.*, 2009) and often exhibit an easterly expansion pattern across the Alps (Bjørnstad *et al.*, 2002; Nola *et al.*, 2006).

Associations with temperature and twentieth-century patterns
Tree-ring isotope chronologies found that the LBM cycle has been strongly coupled with cyclicity of below-average late summer temperatures for 300 years (Kress *et al.*, 2009), and has persisted for 1200 years, through the Medieval Climate Anomaly and Little Ice Age, but ceased after 1981 (Esper *et al.*, 2007). Degradation of the stable periodic pattern corresponds to regional warming that is exceptional in the last 1000+ years (Büntgen *et al.*, 2009; Esper *et al.*, 2007). Populations still exhibit synchrony at the local scale (Büntgen *et al.*, 2009).

Periodicity and synchrony
LBM population dynamics are the best-synchronized oscillatory system observed in forest entomology. True periodicity in LBM cycles has been recognized for some time, and dendrochronology studies have established that epizootics have recurred with remarkable regularity every 8–9 years (Table 7.3). Rolland *et al.* (2001) and Nola *et al.* (2006) identified significant harmonic periods at 18 and 27 years, respectively. The most intense outbreaks in the time series are recorded in the years when short- and long-period cycles coincide (Nola *et al.*, 2006). Synchrony, spread, and intensity are not uniform over the full LBM range (Bjørnstad *et al.*, 2002; Nola *et al.*, 2006; Cooke *et al.*, 2007).

Synchrony is greater in the western Alps and in the valleys with continental climates (Nola *et al.*, 2006). Outbreaks in suboptimal areas depend largely on immigration of moths from more optimal areas (Baltensweiler and Fischlin, 1988; Rolland *et al.*, 2001).

7.3.7 Processionary caterpillars

Biology
Pine processionary caterpillars (*Thaumetopoea pityocampa* Dennis and Schiffermüller and *T. wilkinsoni* Tams) defoliate pines, cedars (*Cedrus* spp.), and larch in the circum-Mediterranean region and the Alps roughly corresponding to the distribution of *Pinus nigra*. They are univoltine, with all or part of populations exhibiting semivoltinism. Larval development and pupal diapause are variable, and adults emerge in summer. Pupal diapause may be extended for several years until weather conditions are suitable for adult emergence. Larvae feed on warm winter days and in the summer and autumn of alternate years, with maximum defoliation in February and March (Carus, 2004; Halperin, 1990; Masutti and Battisti, 1990), acting primarily as a dormant-season defoliator. Range expansion northward and to higher elevations in the last 3 years is associated with warmer winter temperature (Hódar *et al.*, 2003; Battisti *et al.*, 2005; Toffolo *et al.*, 2006). The cedar processionary moth, *T. ispartaensis* (Doğanlar and Avcí), is a univoltine early-season defoliator of *Pistacia* spp., *Cypressus sempervirens*, and *Schinus tere-binthifolius*, and a new threat on *Cedrus libani* in Turkey (Avcí and Carus, 2005).

Tree-ring chronologies
Dendrochronologic investigations have mostly been directed toward quantifying the impact of pine processionary caterpillars on timber impacts, but have also documented outbreak regimes. Tree-ring studies indicate a high frequency periodicity of 5–7 years in Turkey for pine processionary caterpillars (Table 7.3). Growth is usually suppressed for 4–6 years, reducing volume growth rate up to 45% over the long term. Trees barely recover from one outbreak before they incur another (Carus, 2004, 2009). Cedar processionary moth outbreaks exhibit a more variable return interval (Table 7.3) but outbreaks are of similar duration. Unusually long, large-scale outbreaks are followed by lengthy intervals before the subsequent outbreak. Outbreaks are associated with dry winter and spring weather prior to larval feeding (Avcí and Carus, 2005). Processionary moths on both pine and cedar in Turkey exhibited local- and regional-scale outbreaks (Carus, 2004, 2010; Avcí and Carus, 2005).

7.4 Conclusions

Using dendrochronology, researchers can date past forest insect disturbances with annual or near-annual precision and map their extent with considerable confidence. There is some loss of temporal precision because of delays between insect population increase, defoliation, and tree radial growth response, but the general timing of past outbreaks can usually be reconstructed with confidence.

Once developed, tree-ring chronologies of insect outbreaks provide proxy representations of population cycles. They are excellent for revealing the timing of events, quantifying the extent and strength of temporal and spatial variability in outbreak frequency and duration, determining the nature of population oscillations, and determining the presence and strength of spatial synchrony. With additional proxy or observational data, inferences are made about associations with other disturbance regimes and climate.

Tree-ring proxy records for many insects are several centuries long, and are composed of multiple outbreak events which occurred under variable forest and climatic conditions. Besides providing information about the ecology and population dynamics of individual species, these dendrochronology studies provide more broadly applicable insights into forest insect populations.

7.4.1 Insect outbreaks and forest health

Dendrochronology studies repeatedly show that host species and insects have coexisted for many centuries, surviving considerable variation in climate and many ecological upheavals. LBM/European larch, larch sawfly/eastern larch, Pandora moth/ponderosa pine, SBW/balsam fir, and WSBW/mixed-conifer systems have endured for centuries. This is evidence that insect outbreaks do not necessarily reflect unhealthy conditions or systems, and that alternating periods of outbreaks and non-outbreak conditions often represent stable systems. Ecosystems with natural cycles of perturbation and recovery may be highly resilient. It follows that many systems are self-regulating such that outbreaks modify stand or forest composition and structure such that reproduction of host species is favored in the long run, and natural development of forest structure and composition provides suitable habitat for the insect, as proposed by Mattson and Addy (1975). While tree-ring chronologies do not prove this hypothesis, they do provide evidence that it is so, since long chronologies identify repeated outbreaks, and recruitment dates make it clear that host species reestablish stands.

Forest ecosystems experience insect outbreaks much of the time. This fact seems self-evident, but the comprehensive nature of this observation can be startling. For some species, there are few years without an outbreak somewhere in the region. Other systems incur outbreaks 20–50% of the time. And in some systems, variation in tree growth is influenced more by insects than by climate, including two-year cycle budworm (Parish and Antos, 2002), larch sawfly (Girardin *et al.*, 2002; Busque and Arseneault, 2005), pine processionary caterpillars (Carus, 2004, 2009), and, though not explicitly documented, probably larch budmoth at higher elevations.

7.4.2 Outbreak periodicity and synchrony

The great length of tree-ring records provides an opportunity to study spectral properties of insect population oscillations. Esper *et al.* (2007) provide the best example of how informative, even profound, the results of such studies can be. All of the species for which multiple outbreak cycles have been reconstructed exhibit some degree of periodicity. Most of these systems are weakly periodic or quasi-periodic with local factors and exogenous Moran effects entraining persistent cycles, while other systems appear to be regulated by dispersal.

The dendrochronology record shows considerable evidence of large-scale synchrony in forest defoliator outbreaks, as well as indicating that synchrony is stronger is some places and periods than in others. There is consistent evidence that climate entrains the population dynamics of several species, especially for regional-scale events, acting as an exogenous Moran effect. SBW, WSBW, and LBM population fluctuations are closely associated with climate fluctuations. Evidence is particularly strong for WSBW and LBM. WSBW populations in geographically distant regions with different climate systems not only oscillate with precipitation (Ryerson *et al.*, 2003; Swetnam and Betancourt, 1998; Swetnam and Lynch, 1989, 1993; Swetnam *et al.*, 1995), but periodicity in the Oregon outbreak chronology changed frequency with a change in the precipitation cycle (Figure 7.1; Swetnam and Betancourt, 1998). LBM oscillated closely with late-summer temperatures for 1000+ years, and then abruptly lost regional synchrony when temperatures rose above those seen in the 1000-year record (Büntgen *et al.*, 2009; Esper *et al.*, 2007; Kress *et al.*, 2009), though populations are still locally periodic (Esper *et al.*, 2007). The chronologies also provide evidence that synchrony in some populations is associated with dispersal and natural enemy mechanisms.

Investigators speculate as to the mechanisms behind climatic influences on forest insect populations – food quality or quantity, larval development rates, overwintering survival, conditions promoting a good match between host and insect phenologic development, and so on – but causal relationships are difficult to discern through dendrochronology. The direct or indirect nature of climate-associated periodicity is not clear. Dendrochronologists tend to focus on information that can be extracted from the tree-ring record, and may overlook the many climate-related attributes that were not extracted in their analyses, such as flower events, phenology, and cold-season temperatures. Insect populations generally do respond to climate variability that can be quantified from tree-rings, but they also respond to climatic variability that cannot (so far) be reconstructed, such as bud phenology, and to nonclimatic variability. Identifying the coupled nature of insect and host-climate systems provides vital information to understanding ecosystem response to change, and other types of investigations can elucidate the nature of the relationship.

Cooke *et al.* (2007; see also Ranta *et al.*, 1999; Royama *et al.*, 2005) argue that periodicity and synchrony are inherent features of defoliator population dynamics that are determined by the ecological relationships among host plants, insect herbivores, and their natural enemies, and that variability in periodicity and synchrony is created through the influence of stochastic environmental variables. The dendrochronology record provides evidence that periodicity of is an inherent and strong characteristic of epizootic conifer defoliator population dynamics, and that exogenous factors often influence variability in the natural cycles. Climate is frequently implicated as an exogenous factor that influences cycle frequency and amplitude, including those of SBW, WSBW, and LBM. Dispersal is also implicated as a synchronizing force in SBW and LBM. In other cases, such as Pandora moth, periodicity persists with little change through long time frames that encompass considerable climate variability with little apparent change in outbreak periodicity. Dendrochronology studies also show repeatedly that synchrony is strongest in areas where habitat (including thermal habitat) is optimal for the insect, and that synchrony decays as habitat conditions become marginal. Relatively stable cycles with different characteristics amongst regions for the same insect species (Tables 7.1–7.3), and for the Oregon WSBW system, different cycle frequencies in different portions of the chronology (Figure 7.1) indicate that different environmental situations can create relatively stable, robust systems with different characteristics. Whether this means that periodicity in the insect cycle is inherently weak, that periodicity in the insect cycle is fairly strong but plastic, allowing adaption to various circumstances, or that strong periodicity in some other part of the system overwhelms the insect cycle, is not directly evidenced by the tree-ring record.

Some quasi-periodic populations exhibit multiple significant periodicities, with a high-amplitude long-period cycle harmonic with one or more shorter periods. It is not clear if harmonic cycles are inherent features of insect population dynamics or are due to exogenous factors such as climate or host population dynamics, but evidence indicates that such harmonics are not analytic artifacts, and that regional-scale outbreaks occur when multiple cycles coincide. There are several ways that harmonic periods could develop, including (1) the occurrence of fairly regular cycles, with a few outbreaks absent or of low intensity; (2) the occasional development of more extensive and/or severe outbreaks, resulting in greater numbers of trees or sites recording the outbreak (which subsequently might cause the next outbreak to be of low intensity if host-species survivors are immature or infrequent); or (3) the absence or diminution of entraining exogenous cycles (e.g., variability in climate cycles). High-amplitude events may result from the development of host forests to a mature condition over extensive areas (having once consisted of relatively immature stands over extensive areas because of a past large-scale disturbance), which may or may not be harmonic with the insect population cycles. Timing of the extraordinary event might be influenced by climate or other cyclic factors, but the size and severity of the event are influenced primarily by forest condition. The effects of tree pathogens, particularly those such as some root pathogens that have multi-decadal residence on sites and function at large spatial scales, on insect defoliator periodicity and synchrony have not received much attention in tree-ring or population dynamic modeling studies.

The tree-ring proxy data strongly indicate that Moran effects, especially climate-related mechanisms, synchronize some forest insect systems, and that other systems are regulated by dispersal and natural enemies. Proving synchrony or its causes requires the manipulation or elimination of one of the synchronizing mechanisms (Hudson and Cattadori, 1999), so arguments about Moran effects, random effects, and the mechanisms of synchrony will likely never be satisfied. Further insights in insect disturbance ecology will require combining information from long-term proxy data and short-term population studies in carefully designed model systems, and using those models for carefully structured comparisons (Cooke *et al.*, 2007; Royama *et al.*, 2005). By definition models are imperfect representations of real systems, but they provide considerable insight into system behavior. In order to have the best chance of the model system eluci-dating causal mechanisms, knowledge of the real system should encompass (1) the entire range of the insect species habitat, or at least large portions of it in a reasonably continuous fashion; and (2) endemic and epizootic population phases.

There is much to be gained through a better understanding of the relationships between insect population density, defoliation, and radial growth, especially dur-ing the initial outbreak stages. In this regard, I suggest that effort might be better spent studying eruptive defoliators such as LBM, DFTM, larch sawfly, and pro-cessionary caterpillars before trying to quantify these relationships with the SBW and WSBW systems, for which outbreaks often develop so slowly that variability in tree radial growth associated with other factors obscures the initial tree response to insect population densities. Also, studying differential effects of early- and late-season defoliators, and by comparing responses between defoli-ated and defoliator-excluded treatments (i.e., Morrow and LaMarche, 1978), will provide insight into lagged responses and negative feedback mechanisms. It should be noted that dendrochronologic investigations of the type reported here are restricted in large part to ecological systems in which tree radial growth var-ies temporally with one or more climatic variables.

7.4.3 Changes induced by altered climate and large-scale antecedent events

Dendrochronology has demonstrated that relatively small changes in climate can dramatically alter forest insect disturbance regimes. A relatively small change in atmospheric circulation and solar insulation at the end of the Little Ice Age, circa 1850, is associated with a change in SBW outbreak regimes in eastern Canada (Bergeron and Archambault, 1993; Bergeron and Leduc, 1998; Bergeron *et al.*, 2001; Carcaillet *et al.*, 2010). Similarly, a mid-nineteenth-century warming trend is associated with altered larch sawfly regimes in Québec (Arquillière *et al.*, 1990; Nishimura and Laroque, 2010). Millennium-long, near-continental scale syn-chrony of LBM outbreaks in the European Alps has disintegrated, perhaps ended entirely, with regional warming that is unprecedented in 1000+ years (Büntgen *et al.*, 2009; Esper *et al.*, 2007). In eastern Oregon, WSBW outbreak cycle fre-quency decreased when frequency of precipitation cycles decreased (Swetnam and Betancourt, 1998). In each case, change in climate profoundly altered insect disturbance regimes.

Dendrochronology studies demonstrate that landscape-scale effects can induce similarly scaled changes in forest vegetation succession and subsequent insect outbreak regimes. These effects include climate change in the systems described above, and anthropogenic fire exclusion in the western North American mixed-conifer and WSBW system. Perturbing precedent events have brought about subsequent disturbances at the same scale, and synchronized future disturbances. For example, an unusually long period of reduced WSBW activity in the southwestern United States, and subsequent extensive severe out-breaks, is associated with anthropogenic fire exclusion (Anderson *et al.*, 1987; Campbell *et al.*, 2006; Hadley and Veblen, 1993; McCune, 1983; Swetnam and Lynch, 1989, 1993).

7.4.4 *Choristoneura*

SBW, WSBW, and two-year cycle budworm chronologies exhibited many similarities, with periodicity and duration generally within 25–46 and 11–15 years, respectively (though both are highly variable), strong local periodicity, regional (and for SBW, almost subcontinental scale) synchronicity, and occasional high-amplitude synchronous events. Though not in any way identical, gross similarity in outbreak regimes across different species at such long geographic distances, with vastly different climates, and with different host species (from relatively short-lived balsam fir to long-lived Douglas-fir), suggests that periodicity and synchrony are inherent characteristics of the population dynamics of conifer-feeding *Choristoneura*, though they may not be strong. It also suggests that cold system–inhabiting *Choristoneura* may be replaced by relatively warmer system–inhabiting *Choristoneura* species if climate warms, except of course in the warmest systems. Analyses of the chronologies indicate that common regional Moran effects (with climate-related effects implicated frequently) help synchronize outbreaks of conifer-feeding *Choristoneura*, and that migration is a synchronizing feature in some regions.

Significant periodicities at circa 25, 34–38, and 43–46 are frequently found in SBW, WSBW, and two-year cycle budworm chronologies from different areas (Tables 7.1 and 7.2). Repeatedly finding similar frequencies not only indicates that the systems are inherently periodic, but also indicates that ecological relationships among host plants, insect herbivores, their natural enemies, and climate can produce stable systems at different frequencies. Several possible factors may affect the frequency that each system acquires – habitat quality for host, herbivore, or natural enemies; fire regime; tree pathogen activity; climate; or other agents. But it appears that *Choristoneura* populations can adapt to different cycle frequencies.

7.4.5 Other implications

Outbreak regimes are characterized by temporal and spatial variability at many scales. Periodicity and synchrony are strong enough in several systems to provide guidance in ecosystem management. The information found in the tree-ring chronologies provides a better indicator of the potential range of variability than do historical records for many systems. However, no matter how strongly regulated, periodicity and synchrony are affected by many external forces, and it is unreasonable to expect perfect mathematical description and prediction of these systems. The tree-ring chronologies also provide indication that long-established patterns can be modified by external factors, and provide an indication of what factors have been related to shifts in system dynamics in the past (e.g., climate change, changes in fire regimes, and land use history). Predicting the timing, size, and severity of future events is likely to be successful in general terms, providing managers with a sense of how often serious events may occur, what precedent conditions might moderate or intensify outbreaks, and whether outbreak regimes are likely to be altered with climate change. Predicting precisely just when and where outbreaks will occur is less likely.

Of the many insect species that occur in forests, tree-ring chronologies have been developed for only a few. Chronologies usually encompass single-species systems (sometimes two insect species simultaneously) of native conifer defoliators. Only a few chronologies have been developed for hardwood defoliators, sap suckers, bark beetles, or pathogens. Chronologies are insect species–oriented rather than pertaining to ecosystem disturbance history. Studying the disturbance regimes of individual insect species helps us anticipate future outbreaks, develop an understanding of population behavior, and understand which climatic patterns and forest conditions are associated with outbreak oscillations, but developing an understanding of how ecosystems respond to changing conditions requires studying tree population demographic changes coincident with disturbance regimes of multiple agents – defoliators, bark beetles, pathogens, fire, and climate. The attributes of parasitoid and predator population dynamics are poorly known, but undoubtedly influence defoliator population behavior.

By understanding the record of past forest insect outbreaks that is found in tree-ring chronologies, we can better interpret short-term records and experimental results, determine the extent to which current trends have departed from past trajectories, and predict ecosystem response to future disturbance events. And, above all, we may ultimately understand the mechanisms driving ecosystem response to disturbance.

Acknowledgments

I wish to thank Thomas Swetnam and two anonymous reviewers for their helpful comments on an earlier version of the manuscript, and David Meko and Barry Cooke for helpful discussions. I also thank René Alfaro, Paul Shepperd, Jim Speer, and Thomas Swetnam for providing photographs and the charts used in Plate 7.1 and Figure 7.1.

References

Alfaro, R. I., and MacDonald, R.N. 1988. Effects of defoliation by the western false hemlock looper on Douglas-fir tree-ring chronologies. *Tree-Ring Bull.* 48:3–11.

Alfaro, R. I., Thomson, A. J., and Van Sickle, G. A. 1985. Quantification of Douglas-fir growth losses caused by western spruce budworm defoliation using stem analysis. *Can. J. Forest Res.* 15:5–9.

Alfaro, R. I., Van Sickle, G. A., Thomson, A. J., and Wegwitz, E. 1982. Tree mortality and radial growth losses caused by the western spruce budworm in a Douglas-fir stand in British Columbia. *Can. J. Forest Res.* 12:780–787.

Anderson, L., Carlson, C. E., and Wakimoto, R. H. 1987. Forest fire frequency and western spruce budworm outbreaks in western Montana. *Forest Ecol. Manag.* 22:251–260.

Arquillière, S., Filion, L., Gajewski, K., and Cloutier, C. 1990. A dendrochronological analysis of eastern larch (*Larix laricina*) in subarctic Quebec. *Can. J. Forest Res.* 20:1312–1319.

Avcí, M., and Carus, S. 2005. The impact of cedar processionary moth [*Tramatocampa ispartaensis* Doğanlar & Avcí (Lepidoptera: Notodontidae)] outbreaks on radial growth of Lebanon cedar (*Cedrus libani* A. Rich.) trees in Turkey. *J. Pest Sci.* 78:91–98.

Baltensweiler, W., and Fischlin, A. 1988. The larch budmoth in the Alps. Pp. 331–351 in Berryman A. (editor), *Dynamics of Forest Insect Populations*. Plenum, New York.

Barrette, M., Bélanger, L., and De Grandpré, L. 2010. Preindustrial reconstruction of a perhumid midboreal landscape, Anticosti Island, Quebec. *Can. J. Forest Res.* 40:928–942.

Battisti, A., Stastny, M., Netherer, S., Robinet, C., Scopf, A., Roques A., and Larsson S. 2005. Expansion of geographic range in the pine processionary moth caused by increased winter temperatures. *Eco. Appl.* 15:2084–2096.

Belyea, R. M. 1952. Death and deterioration of balsam fir weakened by spruce budworm defoliation in Ontario. *J. Forest.* 50:729–338.

Berg, E. E., Henry, J. D., Fastie, C. L., De Volder, A. D., and Matsuoka, S. M. 2006. Spruce beetle outbreaks on the Kenai Peninsula, Alaska, and Kluane National Park and Reserve, Yukon Territory: relationship to summer temperatures and regional differences in disturbance regimes. *Forest Ecology and Management* 227:219–232.

Bergeron, Y., and Archambault, S. 1993. Decreasing frequency of forest fires in the southern boreal zone of Québec and its relation to global warming since the end of the "Little Ice Age." *The Holocene* 3:255–259

Bergeron, Y., Gauthier, S., Kafka, V., Lefort, P., and Lesieur, D. 2001. Natural fire frequency for the eastern Canadian boreal forest: consequences for sustainable forestry. *Can. J. Forest Res.* 31:384–391.

Bergeron, Y., and Leduc, A. 1998. Relationships between change in fire frequency and mortality due to spruce budworm outbreak in the southeastern Canadian boreal forest. *Journal of Vegetation Science* 9:492–500.

Bjørnstad, O. N., Peltonen, M., Liebhold, A. M., and Baltensweiler, W. 2002. Waves of larch budmoth outbreaks in the European Alps. *Science* 298:1020–1023.

Blais, J. R. 1954. The recurrence of spruce budworm infestations in the past century in the Lac Seul area in northwestern Ontario. *Ecology* 35:62–71.

Blais, J. R. 1961. Spruce budworm outbreaks in the lower St. Lawrence and Gaspe regions. *For. Chron.* 37:192–202.

Blais, J. R. 1981a. Mortality of balsam fir and white spruce following a spruce budworm outbreak in the Ottawa River watershed in Quebec. *Can. J. Forest Res.* 11:620–629.

Blais, J. R. 1981b. Recurrence of spruce budworm outbreaks for two hundred years in western Quebec. *For. Chron.* 57:273–275.

Blais, J. R. 1983. Trends in the frequency, extent, and severity of spruce budworm outbreaks in eastern Canada. *Can. J. Forest Res.* 13:539–547.

Bouchard, M., Kneeshaw, D., and Bergeron, Y. 2006. Forest dynamics after successive spruce budworm outbreaks in mixedwood forests. *Ecology* 87:2319–2329.

Bouchard, M., and Pothier, M. 2010. Spatiotemporal variability in tree and stand mortality caused by spruce budworm outbreaks in eastern Quebec. *Can. J. Forest Res.* 40:86–94.

Boulanger, Y., and Arseneault, D. 2004. Spruce budworm outbreaks in eastern Quebec over the last 450 years. *Can. J. Forest Res.* 34:1035–1043.

Brubaker, L. G. 1978. Effects of defoliation by Douglas-fir tussock moth on ring sequences of Douglas-fir and grand fir. *Tree-Ring Bull.* 38:49–60.

Brubaker, L. B., and Greene S. K. 1979. Differential effects of Douglas-fir tussock moth and western spruce budworm defoliation on radial growth of grand fir and Douglas-fir. *Can. J. Forest Res.* 9:95–105.

Büntgen, U., Frank, D., Liebhold, A., Johnson D., Carrer M., Urbinati C., Grabner M., Nicolussi K., Lavanic T., and Esper J. 2009. Three centuries of insect outbreaks across the European Alps. *New Phytologist* 182:929–941.

Burleigh, J. S., Alfaro, R. I., Borden, J. H., and Taylor S. 2002. Historical and spatial characteristics of spruce budworm *Choristoneura fumiferana* (Clem.) (Lepidoptera: Tortricidae) outbreaks in northeastern British Columbia. *Forest Ecol. Manag.* 168:301–309.

Busque, D., and Arseneault, D. 2005. Fire disturbance of larch woodlands in string fens in northern Québec. *Can. J. Botany* 83:599–609.

Camarero, J. J., Martin, E., and Gil-Pelegrín, E. 2003. The impact of a needleminer (*Epinotia subsequana*) outbreak on radial growth of silver fir (*Abies alba*) in the Aragón Pyrenees: a dendrochronological assessment. *Dendrochronologia* 21:3–12.

Campbell, R., Smith, D. J., and Arsenault, A. 2006. Multicentury history of western spruce budworm outbreaks in interior Douglas-fir forest near Kamloops, British Columbia. *Can. J. Forest Res.* 36:1758–1769.

Candau, J-N., and Fleming, R. A. 2005. Landscape-scale spatial distribution of spruce budworm defoliation in relation to bioclimatic conditions. *Can. J. Forest Res.* 35:2218–2232.

Carcaillet, C., Bergeron, Y., Ali, A. A., Gauthier, S., Girardin, M. P., and Hely C. 2010. Holocene fires in eastern Canada, towards a forest management perspective in light of future global change. *PAGES* 18:68–70.

Carus, S. 2004. Impact of defoliation by the pine processionary moth (*Thaumetopoea pityocampa*) on radial, height and volume growth of Calabria pine (*Pinus brutia*) trees in Turkey. *Phytoparasitica* 32:459–469.

Carus, S. 2009. Effects of defoliation caused by the processionary moth on growth of Crimean pines in western Turkey. *Phytoparasitica* 37:105–114.

Carus, S. 2010. Effect of defoliation by the pine processionary moth (PPM) on radial, height and volume growth of Crimean pine (*Pinus nigra*) trees in Turkey. *J. Environ. Biol.* 31:453–460.

Carus, S., and Avci, M. 2005. Growth loss of Lebanon cedar (*Cedrus libani*) stands as related to periodic outbreaks of the cedar shoot moth (*Dichelia cedricola*). *Phytoparasitica* 33:33–48.

Cook, E. R., and Kairiukstis, L. A. (editors). 1990. *Methods of Dendrochronology: Applications in Environmental Sciences*. Kluwer Academic Publishers, Dordrecht, the Netherlands.

Cooke, B. J., Nealis, V. G., and Régnière, J. 2007. Insect defoliators as periodic disturbances in northern forest ecosystems. Pp. 487–525 in E. A. Johnson E. A., Miyanishi K (editors), *Plant Disturbance Ecology: The Process and the Response*. Elsevier, Burlington, MA.

Drooz, A. T. 1971. Larch sawfly. US Forest Service, Forest Pest Leaflet 8. US Forest Service, Washington, DC.

Duff, G. H., and Nolan, J. N. 1953. Growth and morphogenesis in the Canadian forest species. I. The controls of cambial and apical activity in *Pinus resinosa* Ait. *Can. J. Bot.* 31:471–513.

Elliott, K. R. 1960. A history of recent infestations of the spruce budworm in northwestern Ontario, and an estimate of resultant timber losses. *For. Chron.* 36:61–82.

Esper, J., Büntgen, U., Frank, D. C., Nievergelt, D., and Liebhold A. 2007. 1200 years of regular outbreaks in alpine insects. *Proc. R. Soc. B* 274:671–679.

Evenden, J. D. 1940. Effects of defoliation by the pine butterfly upon ponderosa pine. *J. Forest.* 38:949–955.

Fellin, D. G., Shearer, R. C., and Carlson, C. E. 1983. Western spruce budworm in the northern Rocky Mountains: biology, ecology, and impacts. *Western Wildlands* 9:2–7.

Filion, L., Payette, S., Delwaide, A., and Bhiry, N. 1998. Insect defoliators as major disturbance factors in the thigh-altitude balsam fir forest of Mount Mégantic, southern Quebec. *Can. J. Forest Res.* 28:1832–1842.

Filion, L., and Cournoyer, L. 1995. Variation in wood structure of eastern larch defoliated by the larch sawfly in subarctic Quebec, Canada. *Can. J. Forest Res.* 25:1263–1268.

Fraver, S., Seymour, R. S., Speer, J. H., and White, A. S. 2007. Dendrochronological reconstruction of spruce budworm outbreaks in northern Maine, USA. *Can. J. Forest Res.* 37:523–529.

Fritts, H. C. 1966. Growth-rings of trees: their correlation with climate. *Science* 154:973–979.

Fritts, H. C. 1976. *Tree-rings and Climate*. Academic Press, New York. Reprinted 2001, Blackburn Press, Caldwell, NJ.

Ghassemi, M., Painter, P., Painter, P., Quinlivan, S., and Dellarco, M. 1983. *Bacillus thuringiensis*, nucleo-polyhedrosis virus, and pheromones: environmental considerations and uncertainties in large-scale insect control. *Environ. Int.* 9:39–49.

Girardin, M. P., Tardif, J., and Bergeron, Y. 2001. Radial growth analysis of *Larix laricina* from the Lake Duparquet area, Québec, in relation to climate and larch sawfly outbreaks. *Écoscience* 8:127–138.

Girardin, M-P., Berglund, E., Tardif, J. C., and Monson, K. 2005. Radial growth of tamarack (*Larix laricina*) in the Churchill Area, Manitoba, Canada, in relation to climate and larch sawfly (*Pristiphora erichsonii*) herbivory. *Arct. Antarct. Alp. Res.* 37:206–217.

Girardin, M-P., Tardif, J., and Bergeron, Y. 2002. Dynamics of eastern larch stands and its relationships with larch sawfly outbreaks in the northern Clay Belt of Quebec. *Can. J. Forest Res.* 32:206–216.

Gray, D. R., Régnière, J., and Boulet, B. 2000. Analysis and use of historical patterns of spruce budworm defoliation to forecast outbreak patterns in Quebec. *Forest Ecol. Manag.* 127:217–231.

Greenbank, D. O. 1956. The role of climate and dispersal in the initiation of outbreaks of the spruce budworm in New Brunswick. I. The role of climate. *Can. J. Zool.* 34:453–476.

Hadley, K. S., and Veblen, T. T. 1993. Stand response to western spruce budworm and Douglas-fir bark beetle outbreaks, Colorado Front Range. *Can. J. Forest Res.* 23:479–491.

Halperin, J. 1990. Life history of *Thaumetopoea* spp. (Lep., Thaumetopoeidae) in Israel. *J. Appl. Ent.* 110: 1–6.

Hódar, J. A., Castro, J., and Zamora, R. 2003. Pine processionary caterpillar *Thaumetopoea pityocampa* as a new threat for relict Mediterranean Scots pine forest under climatic warming. *Biological Conservation* 110:123–129.

Hudson, P. J., and Cattadori, I. M. 1999. The Moran effect: a cause of population synchrony. *Trends Ecol. Evol.* 14:1–2.

Jardon, Y., Filion, L., and Cloutier, C. 1994. Tree-ring evidence for endemicity of the larch sawfly in North America. *Can. J. Forest Res.* 24:742–747.

Jardon, Y., Morin, H., and Dutilleul, P. 2003. Périodicité et synchronisme des épidémies de la tordeus des borgions de l'épinette au Québec. *Can. J. Forest Res.* 33:1947–1961.

Keane, R. E., Ryan, K. C., Veblen, T. T., Allen, C. D., Logan, J., and Hawkes, B. 2002. Cascading effects of fire exclusion in Rocky Mountain ecosystems: a literature review. US Forest Service, Gen. Tech. Rept., RMRS-GTR-91. US Forest Service, Rocky Mountain Research Station, Fort Collins, CO.

Krause, C. 1997. The use of dendrochronological material from buildings to get information about past spruce budworm outbreaks. *Can. J. Forest Res.* 27:69–75.

Krause, C., Gionest, F., Morin, H., and MacLean, D. A. 2003. Temporal relations between defoliation caused by spruce budworm (*Choristoneura fumiferana* Clem.) and growth of balsam fir (*Abies balsamea* (L.) Mill.). *Dendrochronologia* 21:23–31.

Krause, C., and Morin H. 1995. Impact of spruce budworm defoliation on the number of latewood tracheids in balsam fir and black spruce. *Can. J. Forest Res.* 25:2029–2034.

Kress, A., Saurer, M., Büntgen, U., Treydte, K. S., Bugmann, H., and Siegwolf, R. T. W. 2009. Summer temperature dependency of larch budmoth outbreaks revealed by Alpine tree-ring isotope chronologies *Oecologia* 160:353–365.

Kulman, H. M. 1971. Effects of insect defoliation on growth and mortality of trees. *Annu. Rev. Entomol.* 16:289–324.

Kurkela, T., Aalto, T., Varama, M., and Jalkanen, R. 2005. Defoliation by the common pine sawfly (*Diprion pini*) and subsequent growth reduction in Scots pine: a retrospective approach. *Silva Fenn.* 39:467–480.

Laurent-Hervouët, N. 1986I. Mesure des pertes de croissance radiale sur quelques espèces de *Pinus* dues à deux défoliateurs forestiers. I. Cas de la processionnaire du pin en région méditerranéenne. *Annals of Forest Science* 43:239–262.

Laurent-Hervouët, N. 1986b. Mesure des pertes de croissance radiale sur quelques espèces de Pinus dues à défoliateurs forestiers. II. Cas du Lophyre du pin dans le Bassin parisien. *Annals of Forest Science* 43:419–440.

Lemieux, C., and Filion, L. 2004. Tree-ring evidence for a combined influence of defoliators and extreme climatic events in the dynamics of a high-altitude balsam fir forest, Mount Mégantic, southern Quebec. *Can. J. Forest Res.* 34:1436–1443.

Lynch, A. M., and Swetnam, T. W. 1992. Old-growth mixed-conifer and western spruce budworm in the southern Rocky Mountains. In: Kaufmann, Merrill R.; Moir, W. H.; Bassett, Richard L., technical coordinators. *Old-growth Forests in the Southwest and Rocky Mountain Regions: Proceedings of a Workshop*, pp. 66–80, March 9–13, Portal, AZ. Gen. Tech. Rep. RM-213. US Department of Agriculture, For. Serv., Rocky Mountain Forest and Range Experiment Station, Fort Collins, CO.

Mason, R. R., Wickman, B. E., and Paul, H. G. 1997. Radial growth response of Douglas-fir and grand fir to larval densities of the Douglas-fir tussock moth and the western spruce budworm. *Forest Sci.* 43:194–205.

Masutti, L., and Battisti, A. 1990. *Thametopoea pityocampa* (Den. And Schiff.) in Italy: bionomics and perspectives of integrated control. *Journal Applied Entomology* 110:229–234.

Mattson, W. J., and Addy, N. D. 1975. Phytophagous insects as regulators of forest primary productivity. *Science* 190:515–522.

Mayfield, A.E. III, Allen, D.C., and Briggs, R.D. 2005. Radial growth impact of pine false webworm defoliation on eastern white pine. *Can. J. Forest Res.* 35:1071–1086.

McCune, B. 1983. Fire frequency reduced two orders of magnitude in the Bitterroot Canyons, Montana. *Can. J. Forest Res.* 13:212–218.

Morin, H., Jardon, Y., and Gagnon, R. 2007. Relationship between spruce budworm outbreaks and forest dynamics in eastern North America. Pp. 555–577 in Johnson, E. A., and K. Miyanishi (editors), *Plant Disturbance Ecology: The Process and the Response*. Elsevier, Burlington, MA.

Morin, H., and Laprise, D. 1990. Histoire récente des épidémies de la Tordeuse des bourgeons de l'épinette au nord du lac Saint-Jean (Québec): une analyse dendrochronologique. *Can. J. Forest Res.* 20:1–8.

Morin, H., Laprise, D., and Bergeron, Y. 1993. Chronology of spruce budworm outbreaks near Lake Duparquet, Abitibi region, Quebec. *Can. J. Forest Res.* 23:1497–1506.

Morrow, P. A., and LaMarche, V. C. 1978. Tree-ring evidence for chronic insect suppression of productivity in subalpine Eucalyptus. *Science* 201:1244–1246.

Nishimura, P. H., and Laroque, C. P. 2010. Tree-ring evidence of larch sawfly outbreaks in western Labrador, Canada. *Can. J. Forest Res.* 40:1542–1549.

Nola, P., Morales, M., Motta, R., and Villalba, R. 2006. The role of larch budmoth (*Zeiraphera diniana* Gn.) on forest succession in a larch (*Larix deciduas* Mill.) and Swiss stone pine (*Pinus cembra* L.) stand in the Susa Valley (Piedmont, Italy). *Trends Ecol. Evol.* 20:371–382.

Olofsson, E. 1988. Environmental persistence of the nuclear polyhedrosis virus of the European pine sawfly in relation to epizootics in Swedish Scots pine forests. *J. Invertebr. Pathol.* 52:119–129.

O'Neil, L. C. 1963. The suppression of growth rings in jack pine in relation to defoliation by the Swaine jack-pine sawfly. *Can. J. Botany* 41:227–235.

Parish, R., and Antos, J. A. 2002. Dynamics of an old-growth, fire-initiated, subalpine forest in southern interior British Columbia: tree-ring reconstruction of 2 year cycle spruce budworm outbreaks. *Can. J. Forest Res.* 32:1947–1960.

Pilon, J. G., and Blais, J. R. 1961. Weather and outbreaks of the spruce budworm in the Province of Quebec from 1939 to 1956. *Can. Entomol.* 93:118–123.

Pretzsch, H. 2009. *Forest Dynamics, Growth and Yield.* Springer-Verlag, Berlin.

Ranta, E., Kaitala, V., and Lindstöm, J. 1999. Spatially autocorrelated disturbances and patterns in population synchrony. *Proc. R. Soc. Lond B* 266:1851–1856.

Rauchfuss, J., Ziegler, S.S., Speer, J. H., and Siegert, N. W. 2009. Dendroecological analysis of spruce budworm outbreaks and their relation to climate near the prairie–forest border in northwestern Minnesota. *Phys. Geogr.* 30:185–204.

Rolland, C., Baltensweiler, W., and Petitcolas, V. 2001. The potential for using *Larix deciduas* ring widths in reconstructions of larch budmoth (*Zeiraphera diniana*) outbreak history: dendrochronological estimates compared with insect surveys. *Trends Ecol. Evol.* 15:414–424.

Royama, T., MacKinnon, W. E., Kettela, E. G., Carter, N. E., and Hartlling, N. E. 2005. Analysis of spruce budworm outbreak cycles in New Brunswick, Canada, since 1952. *Ecology* 86:1212–1224.

Ryerson, D. E., Swetnam, T. W., and Lynch, A. M. 2003. A tree-ring reconstruction of western spruce budworm outbreaks in the San Juan Mountains, Colorado, U.S.A. *Can. J. Forest Res.* 33:1010–1028.

Schmid, J. M., and Bennett, D. D. 1986. The North Kaibab Pandora moth outbreak 1978–1984. US Forest Service, Gen. Tech. Rept. RM-153. US Forest Service, Rocky Mountain Forest and Range Experiment Station, Fort Collins, CO.

Schweingruber, F. H. 1996. *Tree-rings and Environment Dendrochronology.* Birmensdorf, Swiss Federal Institute for Forest, Snow and Landscape Research, Berne.

Simard, M., and Payette, S. 2001. Black spruce decline triggered by spruce budworm at the limit of lichen woodland in eastern Canada. *Can. J. For. Res.* 31:2160–2172.

Speer, J. H. 2010. *Fundamentals of Tree-ring Research.* University of Arizona Press, Tucson.

Speer, J. H., and Jensen, R.R. 2003. A hazards approach towards modeling pandora moth risk. *J. Biogeogr.* 30:1899–1906.

Speer, J. H., Swetnam, T. W., Wickman, B. E., and Youngblood, A. 2001. Changes in Pandora moth outbreak dynamics during the past 622 years. *Ecology* 82:670–697.

Swetnam, T. W. 1990. Fire history and climate in the Southwestern United States. In J. S. Krammes, Tech. Coord, *Proceedings of Symposium on Effects of Fire in Management of Southwestern US Natural Resources*, pp. 6–17, November 15–17, 1988, Tucson, AZ. US Forest Service, Gen. Tech. Rept. RM-191. US Forest Service, Washington, DC.

Swetnam, T. W., and Betancourt, J. L. 1998. Mesoscale disturbance and ecological response to decadal climatic variability in the American Southwest. *J. Climate* 11:3128–3147.

Swetnam, T. W., and Lynch, A. M. 1989. A tree-ring reconstruction of western spruce budworm history in the southern Rocky Mountains. *Forest Sci.* 35:962–986.

Swetnam, T. W., and Lynch, A. M. 1993. Multicentury, regional-scale patterns of western spruce budworm outbreaks. *Ecol. Monogr.* 63:399–424.

Swetnam, T. W., Thompson, M. A., and Sutherland, E. K. 1985. Using dendrochronology to measure radial growth of defoliated trees. US Forest Service, Coop. St. Res. Serv., Agric. Handb. 639. US Forest Service, Washington, DC.

Swetnam, T. W., Wickman, B. E., Paul, H. G., and Baisan, C. H. 1995. Historical patterns of western spruce budworm and Douglas-fir tussock moth outbreaks in the northern Blue Mountains, Oregon, since A.D.

1700. US Forest Service, Res. Paper PNW-RP-484. US Forest Service, Pacific Northwest Research Station, Portland, OR.

Tailleux, I., and Cloutier, C. 1993. Defoliation of tamarack by outbreak populations of larch sawfly in subarctic Quebec: measuring the impact on tree growth. *Can. J. Forest Res.* 23:1444–1452.

Telewski F. W., and Lynch A. M. 1990. Measurement of stem growth. In Lassoie J. P., Hinckley T. M. (editors), *Techniques and Approaches in Forest Tree Ecophysiology*. CRC Press, Boca Raton, FL.

Toffolo, E. P., Bernardinelli, I., Stergulc, F., and Battisti, A. 2006. Climate change and expansion of the pine processionary moth, *Thaumetopoea pityocampa*, in northern Italy. Pp. 331–340 in *Proceedings of the Workshop on Methodology of Forest Insect and Disease Survey in Central Europe*, 11–14 September 2006, Gmunden, Austria. IUFRO Working Party.

Veblen, T. T., Hadley, K. S., Reid, M. S., and Rebertus, A. 1991. Methods of detecting past spruce beetle outbreaks in Rocky Mountain subalpine forests. *Can. J. Forest Res.* 21:242–254.

Volney, W. J. A. 1994. Multi-century regional western spruce budworm outbreak patterns. *Trends Ecol. Evol.* 9:43–45.

Volney, W. J. A., and Fleming R. A. 2007. Spruce budworm (*Choristoneura* spp.) biotype reactions to forest and climate characteristics. *Global Change Biology* 13:1630–1643.

Weber, U. M. 1997. Dendrochronological reconstruction and interpretation of larch budmoth (*Zeiraphera diniana*) outbreaks in two central alpine valleys of Switzerland from 1470–1990. *Trends Ecol. Evol.* 11:277–290.

Weber, U. M., and Schweingruber, F. H. 1995. A dendrochronologic reconstruction of western spruce budworm outbreaks (*Choristoneura occidentalis*) in the Front Range, Colorado from 1720 to 1986. *Trends Ecol. Evol.* 9:204–213.

Wickman, B. E., Henshaw, D. L., and Gollob, S. K. 1980. Radial growth in grand fir and Douglas-fir related to defolation by the Douglas-fir tussock moth in the Blue Mountains outbreak. US Forest Service Gen. Tech. Rep. PNW-269. US Forest Service, Pacific Northwest Research Station, Portland, OR.

Williams, C. B. Jr. 1967. Spruce budworm damage symptoms related to radial growth of grand fir, Douglas-fir, and Engelmann spruce. *Forest Sci.* 13:274–285.

Williams, D. W., and Birdsey, R. A. 2003. Historical patterns of spruce budworm defoliation and bark beetle outbreaks in North American conifer forests: an atlas and description of digital maps. US Forest Service Gen. Tech. Rep. NE-308. US Forest Service, Northeastern Research Station, Newtown Square, PA.

Zhang, Q-B., and Alfaro, R. I. 2002. Periodicity of two-year cycle spruce budworm outbreaks in central British Columbia: a dendro-ecological analysis. *Forest Sci.* 48:722–731.

Zhang, Q-B., and Alfaro, R. I. 2003. Spatial synchrony of the two-year cycle budworm outbreaks in central British Columbia, Canada. *Oikos* 102:146–154.

8

Insect-Associated Microorganisms and Their Possible Role in Outbreaks

Yasmin J. Cardoza, Richard W. Hofstetter, and Fernando E. Vega

8.1 Introduction

The analysis of insect pest outbreaks in natural systems and development of effective pest management practices in managed systems should go hand in hand with an understanding of the biology, behavior, and ecology of the target pest. A common but often overlooked aspect of insect biology is the symbiotic relationships with microorganisms, which have direct impacts on insect performance. We adopt the definition of symbiosis proposed by de Bary (1879), meaning "to live together." Symbiotic associations can be beneficial to one partner to the detriment of the other (parasitic), beneficial to one partner without detriment to the other (commensalistic), beneficial to both partners (mutualistic), or detrimental to both partners (competitive) (Klepzig *et al.* 2009).

Symbionts can be housed inside (endosymbionts) or outside (ectosymbionts) the insect's body and associations can be considered primary (obligate) if neither organism can live alone or secondary (facultative) if either organism can survive

Insect Outbreaks Revisited, First Edition. Edited by Pedro Barbosa, Deborah K. Letourneau and Anurag A. Agrawal.
© 2012 Blackwell Publishing Ltd. Published 2012 by Blackwell Publishing Ltd.

and reproduce without the other (Breznak 2004). Many insects rely on symbiotic microorganisms to enhance food digestion, communication, nutrition, or defense (Buchner 1965, Brand *et al.* 1975, Dowd 1991, Douglas 1998, Vega and Dowd 2005). Insect–microbe associations are of a complex nature and potentially play important roles in the evolution of insect biotypes that can overcome common management practices such as natural host plant resistance, natural enemies, or chemical pesticides. As a consequence, these associations help the insect hosts adapt to ever-changing conditions. Facilitation of all these adaptive strategies potentially contributes to insect outbreaks.

Knowledge of mechanisms governing insect–symbiont associations can shed light on the population dynamics of important pests such as aphids, bark and ambrosia beetles, and leaf-cutting ants, among others. In this chapter, we present examples of how symbionts can influence insect behavior, physiology, and ecology, with consequences for insect fitness that ultimately drive insect population outbreaks. In addition, we discuss how implementation of modern molecular tools for the study of insect symbiotic interactions can lead to their exploitation in future population management programs for pestiferous insect species.

8.2   Microbial assemblages within Insects

Recent studies using molecular tools indicate that the microbiota associated with insects is extremely diverse. For example, Suh *et al.* (2005) and Suh and Blackwell (2005) discovered more than 150 new yeast species in the guts of 27 beetle families, increasing the number of known yeasts by 15%. Similarly, Warnecke *et al.* (2007) reported 216 bacterial phylotypes in the hindgut of *Nasutitermes* termites, and working with the glassy-winged sharpshooter (*Homalodisca vitripennis*). Hail *et al.* (2011) reported 38 genera of bacteria in 17 orders and 28 families. Considering the vast diversity of insect fauna, estimated at 5–15 million species with nearly 1 million described species (Scudder 2009), there is much to be discovered with regard to their complex microbial associates.

Insect symbionts contribute to their hosts' reproduction, digestion, nutrition, and communication (Buchner 1965, Brand *et al.* 1975, Campbell 1990, Dillon and Charnley, 2002, Dillon and Dillon, 2004, Kane and Breznak 1991, Kaufman and Klug 1991) and their removal have negative physiological consequences such as stunted growth, reduced fecundity, and premature death (Hill and Campbell 1973, Griffiths and Beck 1974, Schlein 1976, Nogge 1976, Prosser and Douglas 1991, Rada *et al.* 1997), all of which directly impact population dynamics. Furthermore, effects can be of an indirect nature as microbial fauna have been shown to be affected by the host plant, which can increase insect susceptibility to harmful organisms (Hwang *et al.* 1995, Broderick *et al.* 2003, 2004). To better understand the influence of symbionts on insect population dynamics it is essential to determine whether associations are obligate or facultative, and whether such associations can be manipulated to minimize the risk of population outbreaks.

8.2.1   Obligate and facultative associations

The relative effect of the symbiont on insect hosts strongly depends on whether the symbiont is obligate, facultative, or transient with the host insect. Obligate associations are those in which the symbiont and insect cannot exist and perpetuate themselves without the presence of the partner, while facultative associations are those in which both partners can exist with or without the presence of the other. Thus facultative symbionts can often exist as free-living organisms, while obligate symbionts cannot. Alternatively, other associations could be transient, based on the presence of other organisms in the habitat where the insect occurs, or on the hosts on which it feeds.

Theoretically, associations between insects and their symbiotic partners, including pathogens, parasites, commensals, and mutualists, primarily depend on their transmission modes. Obligate associations occur in all insect species, as all insects harbor gut microbes essential for digestion (Dillon and Dillon 2004). Additionally, many insects have specialized structures that house obligate symbionts. Many of these obligate associations involve transmission of bacteria or fungi that live in mycetocytes or bacteriocytes from parent to offspring (vertical transmission) (Douglas 1989, 2010) or fungi present in the mycangia (Paine *et al.* 1997). Many obligate symbionts provide their hosts with essential nutrients that are absent or lacking in their diet (Douglas 2009, 2010). Associations that involve obligate symbionts are believed to be ancient as the phylogenies are congruent over long periods of evolutionary time (Wernegreen 2002). Other obligate symbionts are acquired from the environment or neighboring insects, termed "horizontal transmission" (Kikuchi *et al.* 2007).

The fidelity of vertical or horizontal transmission often differs between the obligate symbionts and facultative ones (Kikuchi *et al.* 2007). Horizontal transmission across different insect hosts tends to facilitate the virulence (antagonism, parasitism) of the associates, whereas vertical transmission tends to reduce the virulence, leading to mutualisms (Ewald 1987). However, the evolution of mutualism without vertical transmission could occur if vertical transmission of the symbiont incurs some cost for the insect or if exploitation by the symbiont negatively affects the insect (Hoeksema and Bruna 2000). Alternatively if the insect drives the vertical transmission process, the insect utilizes byproducts of the symbiont, or the insect is able to discriminate mutualists from antagonists and parasites, then horizontal transmission could ensue (Genkai-Kato and Yamamura 1999). A key question is whether horizontal transmission of symbionts has more significant effects than vertical transmission on insect population outbreaks.

Facultative symbionts are found in a wide range of insects and are common in insects that also harbor obligate symbionts. Like many obligate symbionts, facultative symbionts can occur externally to the insect or within insect tissues, living both intra- and extracellularly (Dale *et al.* 2006, Kikuchi 2009). Whereas the roles of many facultative symbionts are not well understood, some are beneficial and harmful to their hosts under certain conditions. For instance, facultative symbionts of the pea aphid, *Acyrthosiphon pisum*, have a beneficial effect by increasing host resistance to attack by parasitoid wasps (Oliver *et al.* 2003) and fungal pathogens of the aphid (Scarborough *et al.* 2005). Furthermore, they protect their host against heat stress (Montllor *et al.* 2002). However, facultative symbionts can have negative consequences to the insect depending on environmental conditions and plant species (Montllor *et al.* 2002, Sakurai *et al.* 2005). A classic example of two facultative ectosymbionts involves ants and aphids (Stadler and Dixon 2005). Basically, aphids excrete energy rich honeydew that is fed upon by ants, and ants aggressively defend aphids from natural enemies (Nixon 1951). Ants that have evolved ways to better collect and process honeydew from aphids typically have larger foraging areas and are more abundant. However, there are many constraints and exceptions to this general pattern (Stadler and Dixon 2005). For instance, the mobility, feeding behavior, and presence of winged aphids reduce the likelihood of the aphid–ant symbiosis (Dixon 1998). Interestingly, aphid-attending ants are often successful invaders of new habitats because they have the ability to tend native aphid species (Davidson 1998). "Loose association" between species of nonnative ants and native aphids helps explain why ants are not only able to attend multiple aphid species but also able to occupy novel environments (Stadler and Dixon 2005).

8.2.2 Parasitism, commensalism, and mutualism

As previously discussed, there are three main types of symbioses: mutualism, parasitism (or antagonism), and commensalism. These interspecific associations can be an integral part of the host insects' population dynamics. Does the symbiont allow expansion of host plant range by the insect? Does the symbiont facilitate the invasion of novel ecological zones? Does the symbiont promote rapid population growth, creating unstable population dynamics? The three types of symbioses are not always clear, and there are frequent transitions between them. In many cases, individual relationships include both beneficial and detrimental effects to each partner during various phases of their life histories or as environmental conditions change. Of these three types of symbioses, parasitism is the most studied association and a likely mechanism to keep insect population in check. However, parasitism is often the precursor to mutualistic and commensalistic associations. Parasitic associations can eventually lead to beneficial associations that affect insect fitness and ultimately insect outbreak dynamics (Dale *et al.* 2002, Mueller and Gerardo 2002, Aanen *et al.* 2009).

With regard to insect population dynamics, parasites function similarly to predators but often have strategies to deal with insect internal defenses and have a long co-evolved history with their hosts. For instance, *Wolbachia* and *Cardinium* bacteria are commonly occurring sexual parasites in insects (Harris *et al.* 2010). These bacteria are maternally transmitted and inhabit testes and ovaries, causing cytoplasmic incompatibility, induction of parthenogenesis, and feminization of genetic males (O'Neill *et al.* 1997, Werren 1997). However, in the case of the coffee berry borer (*Hypothenemus hampei*; Scolytinae), the parasitic role of *Wolbachia* contributes to increases in female population levels due to a 10:1 skewed sex ratio favoring females (Vega *et al.* 2002). The net effect is the production of 10 times the number of females, each one of which, depending on temperature, can lay up to 300 eggs in galleries inside the coffee berry (Jaramillo 2008), thus greatly contributing to the pest status of this insect.

Parasites can indirectly cause insect outbreaks by negatively impacting natural enemies, competitors, or host plants, thus releasing the insect host from regulatory factors (Keane and Crawley 2002, Hardin *et al.* 1995). Direct effects of parasites on insect hosts are not likely to lead to outbreaks, unless the insect outbreak occurs because of the release of parasitic pressure on the insect population. Otherwise, parasites negatively affect host populations by reducing host density or driving local populations to extinction (Anderson 1982). For the most part, parasites regulate their host populations and play a key role in reducing or keeping insect densities low. A key question is whether processes at the individual level translate to effects at the population level.

Mathematical models predict different population dynamics for insects infected with parasites that reduce host fecundity versus those infected with parasites that reduce insect survival (Anderson 1982). Insect density is predicted to decrease with increasing negative effects of a parasite on host fecundity. In contrast, insect population density is predicted to first decrease and then increase as parasite-induced host mortality rises. This is because parasites that kill their hosts rapidly are less likely to be transmitted to other insect hosts and will, therefore, remain at low prevalence. Alternatively, parasites with little effect on insect mortality will have minimal effect on host demographics. Parasitic–insect associations are also likely to cause population fluctuations where insect host density decrease as a parasite exhibits an increasingly negative effect on insect host survival and fecundity (Anderson 1982). According to these models, density fluctuations increase the chance of extinction of small host populations because host density is more likely to drop to zero during population bottlenecks (May 1974, McCallum and Dobson 1995).

Commensalism is a form of symbiotic relationship where one organism gains benefit from the relationship while the other is unaffected. Upon examining the interaction between cohabiting immature of the pitcher-plant midge, *Metriocnemus knabi*, and the pitcher-plant mosquito, *Wyeomyia smithii* (Coquillet), Heard (1994) found that while both insects feed on organic material derived from captured invertebrates, the mosquito populations benefited from midge-mediated resource processing. This provides a clear example of how commensalistic associations have direct impacts on insect populations. Though the species examined in this study were not economically important, these types of associations, even if yet unknown, surely occur among cohabiting organisms of medically important mosquito vectors. Mutualistic interactions fall within a continuum from loose to tight mutualisms depending on the ecological context surrounding them, and commensalisitic interactions are not the exception (Klepzig and Six 2004). For example, *Enterobacter* are commensal bacteria in caterpillar hosts, but were shown to be the main cause of insect death following exposure to Bt toxin in gypsy moth caterpillars: *Enterobacter* are able to move through the gut wall pores formed by the Bt toxin, replicate in the hemolymph, and cause death by septicemia (Broderick *et al.* 2006). As described in the following section of this chapter, bark beetles share their galleries with a plethora of other organisms, such as mites and nematodes that may ultimately contribute to the success of these insects by keeping microbial communities in check (Cardoza *et al.* 2008), thereby facilitating beetle reproductive success.

Mutualistic associations between species are widespread and abundant (Buchner *et al.* 1965, Bronstein 2001, Gibson and Hunter 2010) but have received little attention with regard to insect outbreaks (Mattson and Haack 1987; Sabelis *et al.* 1999; Klepzig *et al.* 2009). There are difficulties in modeling insect populations that contain mutualisms because positive associations within models are destabilizing and thus unwieldy in models. However, it has been shown that mutualisms can be stabilized by a variety of mechanisms, such as strong negative density dependence, nonlinearities, frequency dependence, or even antagonism (Boucher *et al.* 1982). In any case, mutualistic symbionts play vital roles in insect nutrition, reducing plant defenses; protecting from natural enemies and pathogens; and improving development and reproduction, communication, and survival in variable or harsh environments (Klepzig *et al.* 2009, Gibson and Hunter 2010).

Examples of mutualisms among insect herbivores and plant pathogenic microbes are common. Plant defensive responses against plant pathogens, regardless of whether they are vectored by insects or not, can have a negative effect on the plant's ability to cope with herbivory, thereby facilitating an insect's attack upon the plant (Karban *et al.* 1987, Moran and Telang 1998, Bostock 1999, Felton *et al.* 1999, Fidantsef *et al.* 1999, Stout *et al.* 1999, Cardoza *et al.* 2002, 2003a, Cardoza and Tumlinson 2006). Moran (1998) found that cucumber leaves infected with *Cladosporium* fungi are preferably fed on by cucumber beetles, *Diabrotica undecimpunctata howardii*. In choice tests, beet armyworm (*Spodoptera exigua*) larvae feed preferentially on leaves from peanut plants previously infected by the white mold fungus, *Sclerotium rolfsii* (Cardoza *et al.* 2002) and larvae feeding on *S. rolfsii*–infected plants had significantly higher survival, produced significantly heavier pupae, and had shorter time to pupation than those on healthy plants (Cardoza *et al.* 2003b). *Spodoptera exigua* also fed and performed better on bell pepper (*Capsicum anuum*) plants infected with compatible and incompatible *Xanthomonas* strains (Cardoza and Tumlinson 2006). Indirect defenses (i.e., induced volatile compounds) of the plant were also reduced in response to disease progression in the plants. Therefore, exploitation of infected plants or transmission of plant pathogens can be adaptive for herbivorous insects. This hypothesis is reinforced by the fact that plant defense pathways against herbivorous insects and most plant pathogens conflict with one another, often benefiting the herbivore, except in cases where increased volatile emissions leads to greater recruitment of natural enemies (Cardoza *et al.* 2003b). Thus, symbiotic complexes can modulate or facilitate ecological interactions that result in insect pest outbreaks and associated plant mortality.

8.2.3 Multipartite symbiotic associations

Symbionts can also have multiple functions within the same host: for example, they can serve as a food source for larvae and provide protection from natural enemies (Weis 1982, Heath and Stireman 2010). Moreover, studies attempting to elucidate the nuances governing symbiotic interactions largely focused on bipartite systems, involving one host and one symbiont species. However, the existence and roles of symbiont complexes associated with a single host insect are common. For example, leaf-cutting ants in the genera *Acromyrmex* and

 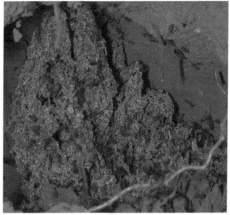

Figure 8.1   Leaf-cutting ants (left, YJC) have symbiotic associations with *Leucocoprinus/ Leucoagaricus gongylophorous*, which is grown in fungal gardens (right, RWH) and provides most of the colonies' nourishment.

Atta, some of the most prevalent species of herbivores in tropical and subtropical regions of the world, utilize plant foliage to cultivate fungi within their nests (Hölldobler and Wilson 1990, Schultz *et al.* 2005). Leaf-cutting ants (Figure 8.1) associate with basidiomycete fungal symbionts (e.g., *Leucocoprinus* and *Leucoagaricus*), which provide most of the colonies' nourishment, in addition to symbionts in the form of actinomycete bacteria that are active against fungi (Currie 2001, Sen *et al.* 2009). This multipartite association, together with the social complexity of these ant species, plays a significant role in the inefficacy of chemical and biological control measures against leaf cutter ant pests, as well as in their capacity for population growth and expansion. Another example involves the presence of an actinomycete bacterium in mycangia of the southern pine beetle (*Dendroctonus frontalis*) that protects the insect-vectored symbiotic fungus *Entomocorticium*, which serves as food for the beetles' larvae, from *Ophiostoma minus*, a fungal competitor (Scott *et al.* 2008). However, the specificity of this tripartite association remains to be explored. Likewise, bacteria present in oral secretions of other bark beetle species act as inhibitors for various fungal species that could invade the galleries in the host tree and outcompete the symbiotic fungus vectored by the insect (Cardoza *et al.* 2006a).

Microbial symbionts also impact their hosts' interactions with natural enemies. For example, *Amylostereum areolatum*, the main fungal symbiont of the wood wasp *Sirex noctilio* and a yeast, *Saccharomyces* sp., associated with its feces, produce volatile compounds that attract parasitoids (Madden 1975, Martinez *et al.* 2006). Attraction of bark beetle parasitoids and predators by volatiles produced by the beetles' symbiotic fungi has also been documented (Adams and Six 2008, Boone *et al.* 2008). Gut bacteria in the desert locust, *Schistocerca gregaria*, protect against invasion by other pathogenic bacteria (Dillon *et al.* 2005) and reduce viability of fungal entomopathogens (Dillon and Charnley 1986). Additionally, other fungal symbionts have been shown to build

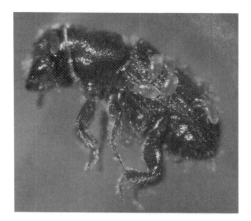 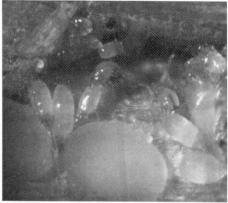

Figure 8.2 Mites are common associates of bark beetle species. Phoretic mite morphs on the southern pine beetle, *Dendroctonus frontalis* (left, RWH), and feeding morphs of the most prevalent mite species, *Histiogaster arborsignis*, associated with the spruce beetle, *D. rufipennis*, feeding within the beetles' oviposition galleries (right, YJC).

reinforced domatia (housing structures) to protect their hosts from parasitoid attack (Weis 1982). However, multipartite interactions are not limited to the microbial world alone.

Other animals can influence symbiotic relationships between insects and microorganisms. For instance mites, nematodes, or other community members promote, impair, or discourage symbiotic associations and thus influence insect population dynamics. Mites associated with *D. frontalis* often feed on and introduce microbes into beetle-infested trees (Lombardero *et al.* 2003, Hofstetter *et al.* 2006a). In the southeast United States, *Tarsonemus* mites associated with *D. frontalis* can introduce *Ophiostoma minus*, a bluestain fungus that negatively affects beetle development by disrupting the growth of beetle's symbiotic fungi (Bridges 1985). Increases of the fungus in trees, as a result of increasing mite densities, negatively affect beetle populations (Hofstetter *et al.* 2006b). Alternatively, the same mite species can promote the presence of beetle-symbiotic fungi by introducing spores of the fungus into beetle-infested trees (Moser *et al.* 2010), as well as feed on competing and harmful microbes or create conditions that promote beetle-symbiotic fungal growth and reproduction. Similarly, nematodes associated with the wood wasp, *Sirex noctilio*, can be parasitic of the wood wasp or affect the wood wasp's primary fungal symbiont (Bedding 1967, 1972).

Conversely, mite and nematode species associated with bark beetles can help maintain a balance within the micro-ecosystem of the insects' galleries. For example, *Dendroctonus rufipennis* is associated with eight mite and six nematode species (Cardoza *et al.* 2008) (Figures 8.2, 8.3, and Plate 8.1). Interestingly, the main ophiostomatoid fungus associated with *D. rufipennis*, *Leptographium abietinum*, plays an important role in these gallery faunal communities, by serving as the preferred food and oviposition substrate for at least one mite and nematode species housed within the insects' nematangia (Cardoza *et al.* 2008). These mite and nematode associates feed on fungi, bacteria, mites, or nematodes and help maintain a balance within the gallery.

Figure 8.3 *Ips pini* showing ventral blooms of an unidentified nematode (left, YJC) and *Dendroctonus rufipennis* nematangia containing female *Ektaphelenchus obtusus* (right, YJC).

The subtle complexity of multipartite symbiotic associations is influenced by changes in the biological and physical setting, a concept known as conditional mutualism or context-dependent symbiosis (Cushman and Whitman 1989, McCreadie *et al.* 2005). This context-dependent symbiosis concept can be applied to any symbiotic system, such as those of bark beetles, in which each species is associated with a specific cohort of symbiotic fungi (Plate 8.2), as well as numerous other organisms, yet the benefits of having multiple associations has not been completely discerned. The study of symbiotic associations in these systems would greatly benefit from a context dependent approach (Klepzig and Six 2004; Six and Klepzig 2004). For instance, Klepzig *et al.* (2001) and Hofstetter *et al.* (2006b) found that in *D. frontalis* populations, temperature altered the performance of the fungal species *O. minus*, normally associated with phoretic mites, relative to the beetle-symbiotic fungi. Moreover, the performance of two mutualistic fungi with the mountain pine beetle, *D. ponderosa* (Six and Bentz 2007) and *D. frontalis* (Hofstetter *et al.* 2007), differ based on geographical and climatic conditions. Thus, the multiple fungal associates of bark beetles can avoid direct competition by growing at different times, as dictated by varying temperatures. At the same time the host beetles exploit their multiple symbiotic associates' combined growth optima to facilitate their exploitation of a broad geographic range with highly variable environmental conditions. Ultimately, differential environmental effects on members of the beetles' symbiotic assemblage help explain some of the population fluctuations typical of these insect species, based on their direct impacts on the insects' performance (Hofstetter *et al.* 2006b).

8.3 Can microbial genetic contributions facilitate host insect outbreaks?

Although not widely explored, one possible consequence of insect–microbe symbioses is the transfer of genes from microorganisms to their insect hosts. Elucidation of such transfer would be aided if the genome for the insect under study is available. For example, Moran and Jarvik (2010) used the genome of the pea aphid (*Acyrthosiphon pisum*) to infer proteins and determine the origin of carotenoids in the aphid. Carotenoids are produced by fungi, bacteria, and plants, and animals must obtain them from food they consume. Bacterial endosymbionts in the pea aphid were ruled out because their elimination did not affect carotenoid presence and, furthermore, their genomes did not reveal genes homologous with carotenoid biosynthesis. Carotenoid production in the pea aphid was found to be the result of integration of fungal genes into the aphids' genome. Moran and Jarvik (2010) hypothesize that the gene donor could have been an aphid fungal symbiont or pathogen, or a fungus present on a host plant. This genetic integration provides the insect with a unique advantage: its own carotenoid factory without having to depend on a food or a symbiont source. Another example of lateral transfer of fungal genes involves polygalacturonase production by the rice weevil (*Sitophilus oryzae*; Shen *et al.* 2003). According to the authors, "acquisition of polygalacturonase could provide a selective advantage to phytophagous insects, as the ability to digest pectin should increase the availability of nutrients released from leaves or other plant tissues high in pectin." The critical tests needed to determine how acquisition of these fungal genes influence fitness is to engineer aphids without the ability to produce carotenoids or polygalacturonase and compare them to their producing counterparts. This information could reveal whether lateral gene transfer could contribute to insect outbreaks.

8.4 Symbiosis-facilitated insect outbreaks in new habitats

The introduction of an insect species to new habitats can result in significant and sudden increases in insect abundance as well as negative effects on host plant species. In many cases, this sudden explosion of insect populations can be attributed to microbial symbionts and their effects on the plant or insect hosts. The red bay ambrosia beetle, *Xyloborus glabratus* (Curculionidae), is an example of a species that is not a pest in its native range but has devastated coastal forests in new habitats. Since its first report in South Carolina (United States), *X. glabratus* has caused the death of more than 75% of red bay (*Persea borbonia*) due to the laurel wilt caused by the beetle's most common fungal symbiont, *Raffaelea lauricola* (Fraedrich *et al.* 2007). This provides an example of how a fungal symbiont can mediate an ecological switch resulting in great economic and ecological losses as well as an outbreak of an insect. In this case, the exotic fungal symbiont is extremely virulent in the new habitat due to decreased resistance or increased susceptibility to the symbiotic complex, which results in the outbreak of the insect.

Another example of symbiosis-facilitated outbreaks involves wood borers in the genus *Monochamus* (Cerambycidae), which introduce the nematode *Bursaphelenchus xylophilus* into trees resulting in tree death, thus providing substrate for *Monochamus* larvae to grow (Wingfield *et al.* 1984). This symbiosis results in extensive mortality to trees and insect population outbreaks when the complex is introduced into new habitats (Mota and Vieira 2008).

Other symbiotic complexes that resulted in sudden and extensive outbreaks and associated plant mortality involve elm bark beetles and woodwasps. These bark beetles and their phoretic mites vector the Ascomycete fungus *Ophiostoma novo-ulmi*, on which the insects feed, and which ends up killing the tree (Moser *et al.* 2010). The woodwasp, *Sirex noctilio* introduces the plant pathogenic fungus, *Amylostereum areolatum*, into pine trees resulting in tree death. In its native range of Europe, Asia, and northern Africa, *S. noctilio* is rarely a pest. However, in areas where it was accidentally introduced such as Australia, New Zealand, South Africa and parts of North and South America, extensive outbreaks of the insect occur (Hoebeke *et al.* 2005). Similarly, the red turpentine beetle, *Dendroctonus valens*, was introduced into China where it has become an aggressive pine-killing species (Yan *et al.* 2005) while in its native range in North America it is considered a secondary pest and rarely reaches outbreak status (Smith 1961). The differences could be attributed to its fungal symbionts; more virulent and competitive symbionts that are better adapted to kill host trees are associated with *D. valens* in China (Lu *et al.* 2010).

Facilitated outbreaks can be quite unexpected and result from associations between native and non-native species such as that involving the invasive Japanese beetle (*Popillia japonica*) and the native green June beetle (*Cotinis nitida*). Grape feeding by *P. japonica* results in wounds that allow *C. nitida* to use an otherwise unusable resource due to its blunt mandibles. Feeding by *P. japonica* results in deposition of at least eight yeast species in five genera that produce volatiles attractive to *C. nitida*, which also deposits yeasts that elicit volatiles that act as aggregation kairomones (Hammons *et al.* 2009). Thus, feeding by one species can result in an outbreak in a different species.

8.5 Microbial symbionts as modulators of pest population dynamics

Symbioses enrich the possibilities for feedback systems. Positive feedback, resulting from mutualistic interactions among symbionts, can promote insect outbreaks by accelerating processes that lead to population growth. For instance, microbial symbionts that help provide sudden increases in resources (e.g., *Raffaelea* sp. result in extensive mortality to Redbay trees and population growth of the ambrosia beetle, *X. glabratus*; Fraedrich *et al.* 2008), nutrients (e.g., amino acids synthesized by the intracellular symbiont *Buchnera aphidicola* for the pea aphid; Douglas *et al.* 2001), reproductive output (e.g., changes in sex ratio due to *Wolbachia* infections; Vega *et al.* 2002), or insect survival (e.g., defensive compounds produced by actinobacteria; Kaltenpoth 2009) would result in rapid insect population growth. Additionally, if the microbial symbiont(s) becomes more pathogenic or virulent towards the host plant due either to genotypic changes or exposure to susceptible host plants through introductions to new regions, as was discussed above for the *X. glabratus–Raffaelea* complex in the United States and the *D. valens* symbionts in China (Yan *et al.* 2005, Lu *et al.* 2010).

Insect population dynamics could shift as a result of symbiotic relationships. For instance, many ambrosia beetles build gallery systems in trees or plants where they cultivate fungi for food (Batra 1966, Francke-Grosmann 1967). The conventional view of ambrosia beetles' typical hosts is that of unhealthy, physiologically stressed, dying, or dead trees (Wood 1982). Many ambrosia beetles can also be classified as secondary pests, attacking a dying or dead host that is already exploited by more aggressive, tree-killing insects (Furniss and Carolin 1977). However, some species of ambrosia beetles, such as *Austroplatypus incompertus* (Kent and Simpson 1992) and *Platypus mutalus* (Costilla and Venditti 1992), are preferential or obligatory attackers of living trees. One principal difference in aggressiveness between bark and ambrosia beetles is the relative absence among ambrosia beetles of species that build up large populations in weakened hosts and then attack living trees. In recent years, however, there have been reports and observations of normally secondary ambrosia beetles attacking apparently healthy, standing trees, and causing tree death. Possible causes of this transition in behavior are the rapid evolution of introduced plant pathogens via hybridization (Brasier 2001), addition of a new microbial symbiont, or change in host plant resistance or host type (Kühnholz *et al*. 2001).

The presence of multiple microbial symbionts can also facilitate insect population growth and outbreaks by buffering environmental variability, diverting plant defenses, and/or interfering with natural enemies and competitors (all of which have been discussed in this chapter). The dependence of an insect on its symbionts could promote increased virulence and pathogenicity of the symbionts to the detriment of host plants. Increased virulence of plant pathogenic microbial symbionts can also drastically change the population dynamics of the insect. A classic example for rapid evolution of pathogenic fungi is the Dutch elm disease caused by *Ophiostoma ulmi*. The non-aggressive strain of this fungus was introduced into the United States in imported elm logs from Europe (Stipes and Campana 1981). In North America, an aggressive strain of the pathogen evolved (Manion 1981), which was subsequently transported back to Europe and caused extensive mortality of European elms (Brasier and Gibbs 1973).

Likewise, negative feedback, resulting from antagonistic interactions among symbionts or failed mutualistic interactions, can result in insect population decline. Temporal variability in the abundance or association of an antagonistic symbiont can result in insect population fluctuations and the prevention or decline of an insect outbreak. The *D. frontalis*, mites, and *O. minus* fungus system is a good example of how the delayed increase in an antagonistic fungus, promoted by mites, can result in beetle population decline (Hofstetter *et al*. 2006a).

The dependency of an insect species on its microbial symbionts can destabilize insect populations because it may amplify negative effects on individuals, mutual interactions, or distribution of symbionts and mutualists. If the mutualism is obligate, the likelihood of exogenous factors affecting each species increases. This has been predicted in mathematical models embedded with positive feedbacks (May and Oster 1976) which tend to amplify perturbations to the system resulting in greater fluctuations and instability of the population. Thus, one prediction is that insect species with obligate mutualists exhibit greater population fluctuations or growth potential than insect species without obligate mutualists. This partially contributes to the eruptive cycles of some forest pest species, including bark beetles. Only the future will tell if the same interactions hold the key to preventing such damaging pest population cycles.

8.6   Manipulating microbes to affect insect outbreaks

The development of molecular biology techniques that do not require the cultivation of microbial organisms (e.g., pyrosequencing) has greatly facilitated the identification and characterization of microbial organisms (Tyson *et al.* 2004, Handelsman 2004, Allen and Banfield 2005). Recently, Hail *et al.* (2011) used pyrosequencing to elucidate the microbiome associated with the glassy-winged sharpshooter (*Homoladisca vitripennis*) and, more specifically, with the hemolymph, alimentary canal, and whole insect. A total of 38 bacteria were identified to genus level, revealing the complex diversity within the insect. More importantly, this type of study raises several questions such as How does this diversity differ within populations collected in different areas and on different host plants? Are any members of the microbiome essential for insect survival (i.e., are they mutualists)? and Do they have a specific metabolic role? As molecular techniques such as pyrosequencing become more widely used in entomological research, it will become clear that insects are associated with a myriad of microbes and, hopefully, we'll be able to determine which ones might be contributing to the success of an insect in becoming a pest and whether they can be manipulated to reduce pest outbreaks.

Significant steps have already been taken in the manipulation of insect symbionts that may facilitate application of novel techniques such as paratransgenesis, population displacement through symbiont-mediated traits, and cytoplasmic incompatibility for the management of insect pest populations (Durvasula *et al.* 1997, Ben Beard *et al.* 2002, Dobson 2003, Zabalou *et al.* 2004, Sinkins and Gould 2006). For instance, the life span of the mosquito *Aedes aegypti*, a vector for the dengue fever virus, can be shortened by infection with *Wolbachia* (McMeniman *et al.* 2009). Since only older mosquitoes are carriers, this is a promising strategy to reduce the transmission of pathogens, without the ethically untenable eradication of a vector species.

Vector paratransgenesis is another novel strategy aimed at reducing insect vector efficacy to minimize and ultimately negate transmission of pathogens such as those responsible for dengue and malaria, but has applicability for other systems mentioned throughout this chapter. Paratransgenesis focuses on utilizing genetically modified insect symbionts to express molecules within the vector that negatively affect the vectored parasites. Population replacement and genetic drive mechanisms are receiving significant interest by the scientific community for their potential in genetic pest and disease management. Population replacement relies on massive releases of modified insects or insects containing modified symbionts to fill the niche left by the resident insect population knock-down through pesticide applications. Although this is a promising research tool, large-scale management is not feasible. Other transgenic strategies include the production of insects, especially mosquitoes, carrying a dominant lethal gene; however, reproductive incompatibilities among subspecies and migration of mosquitoes have proven to be major challenges for field application. An additional negative consequence of this approach is the production of empty ecological niches that could be filled by other pathogen-carrying insect species. For these reasons, researchers are now testing genes that yield insect species with compromised vector abilities (refractory). This technique can be used to engineer mosquitoes containing modified symbionts that inhibit or kill malaria during one of its developmental phases within the insect. Replacement of wild populations with transgenic mosquitoes carrying symbiotic organisms with refractory genes may provide a more practical alternative to methods resulting on severe population suppression or eradication. The main challenge to this approach lies in that it requires mass releases of biting insects, which is ethically unsound.

In spite of these advances in the field of host–symbiont genetics, and the enticing potential of these technologies to prevent pest outbreaks, many obstacles still need to be overcome before their applicability becomes feasible. There are many factors that should be considered such as environmental risk, cost-effectiveness, availability of comparably effective alternative management strategies, general accessibility, and long-term sustainability. Because of our appreciation for the complexity and uncertainty inherent in the biology of insect symbioses, we feel adamant that before field releases of any genetically modified organism, significant effort and resources need to be invested on studies to evaluate the ecological, economical, ethical, and political impacts of deploying such organisms.

References

Aanen, D. K., B Slippers, and M. J. Wingfield. 2009. Biological pest control in beetle agriculture. *Trends in Microbiology* 17:179–182.

Adams, A. S., and D. L. Six. 2008. Detection of host habitat by parasitoids using cues associated with mycangial fungi of the mountain pine beetle, *Dendroctonus ponderosae*. *Canadian Entomologist* 140:124–127.

Allen, E. E., and J. F. Banfield. 2005. Community genomics in microbial ecology and evolution. *Nature Reviews Microbiology* 3:489–498.

Anderson, R. M. 1982. Theoretical basis for the use of pathogens as biological control agents of pest species. *Parasitology* 84:3–33.

Batra, L. R. 1966. Ambrosia fungi: extent of specificity to ambrosia beetles. *Science* 173:193–195.

Bedding, R. A. 1967. Parasitic and free-living cycles in entomogenous nematodes of the genus *Deladenus*. *Nature* 214:174–175.

Bedding, R. A. 1972. Biology of *Deladenus siricidicola* (Neotylenchidae), an entomogenous nematode parasitic in siricid woodwasps. *Nematologica* 18:482–493.

Ben Beard C., C. Cordon-Rosales, and R. V. Durvasula. 2002. Bacterial symbionts of the triatominae and their potential use in control of Chagas disease transmission. *Annual Review of Entomology* 47:123–141.

Boone, C. K., D. L. Six, Y. Zheng, and K. F. Raffa. 2008. Parasitoids and dipteran predators exploit volatiles from microbial symbionts to locate bark beetles. *Environmental Entomology* 37:150–161.

Bostock, R. M. 1999. Signal conflicts and synergies in induced resistance to multiple attackers. *Physiological and Molecular Plant Pathology* 55:99–109.

Boucher, D. H., S. James, and K. H. Keeler. 1982. The ecology of mutualism. *Annual Review of Ecology and Systematics* 13:315–347.

Brand, J. M., J. W. Bracke, A. J. Markovetz, D. L. Wood, and L. E. Browne. 1975. Production of verbenol pheromone by a bacterium isolated from bark beetles. *Nature* 254:136–137.

Brasier, C. M. 2001. Rapid evolution of introduced plant pathogens via interspecific hybridization. *Bioscience* 51:123–133.

Brasier, C. M., and J. N. Gibbs. 1973. Origin of the Dutch elm disease epidemic in Britain. *Nature* 242:607–609.

Breznak, J. A. 2004. Invertebrates–insects. In A. T. Bull (editor), *Microbial Diversity and Bioprospecting*, pp. 191–203. ASM Press, Washington, DC.

Bridges, J. R. 1985. Relationship of symbiotic fungi to southern pine beetle population trends. In S. J. Branhan and R. C. Thatcher (editors), *Integrated Pest Management Research Symposium: The Proceedings*, pp. 127–135. US Forest Service, Southern Forest Experiment Station, General Technical Report SO-56. US Forest Service, Asheville, NC.

Broderick, N. A., R. M. Goodman, J. Handelsman, and K. F. Raffa. 2003. Effect of host diet and insect source on synergy of gypsy moth (Lepidoptera: Lymantriidae) mortality to *Bacillus thuringiensis* subsp. *kurstaki* by zwittermicin A. *Environmental Entomology* 32:387–391.

Broderick, N. A., K. F. Raffa, R. M. Goodman, and J. Handelsman. 2004. Census of the bacterial community of the gypsy moth larval midgut using culturing and culture-independent methods. *Applied and Environmental Microbiology* 70:293–300.

Broderick, N. A., K. F. Raffa, and J. Handelsman. 2006. Midgut bacteria required for *Bacillus thuringiensis* insecticidal activity. *Proceedings of the National Academy of Sciences USA* 103:15196–15199.

Bronstein, J. L. 2001. The exploitation of mutualisms. *Ecology Letters* 4:277–287.

Buchner, P. 1965. *Endosymbionts of Animals with Plant Microorganisms*. Interscience Publishers, New York.

Campbell, B. C. 1990. On the role of microbial symbiotes in herbivorous insects. In E. A. Bernays (editor), *Insect–Plant Interactions*, vol. 2, pp. 1–44. CRC Press, Boca Raton, FL.

Cardoza, Y. J., and J. H. Tumlinson. 2006. Compatible and incompatible *Xanthomonas* infections differentially affect herbivore-induced volatile emission by pepper plants. *Journal of Chemical Ecology* 32:1755–1768.

Cardoza, Y. J., H. T. Alborn, and J. H. Tumlinson. 2002. *In vivo* volatile emissions of peanut plants induced by fungal infection and insect damage. *Journal of Chemical Ecology* 28:161–174.

Cardoza, Y. J., C. G. Lait, E. A. Schmelz, J. Huang, and J. H. Tumlinson. 2003a. Fungus induced biochemical changes in peanut plants and their effect on development of beet armyworm, *Spodoptera exigua* Hübner (Lepidoptera: Noctuidae) larvae. *Environmental Entomology* 32:220–228.

Cardoza, Y. J., P. E. A. Teal, and J. H. Tumlinson. 2003b. Effect of peanut plant fungal infection on oviposition preference by *Spodoptera exigua* and on host-searching behavior by *Cotesia marginiventris*. *Environmental Entomology* 32:970–976.

Cardoza, Y. J., K. D. Klepzig, and K. F. Raffa. 2006. Bacteria in oral secretions of an endophytic insect inhibit antagonistic fungi. *Ecological Entomology* 31:636–645.

Cardoza, Y. J., J. C. Moser, K. D. Klepzig, and K. F. Raffa. 2008. A multipartite symbiosis between fungi, mites, nematodes and the spruce beetle, *Dendroctonus rufipennis*. *Environmental Entomology* 37:956–953.

Costilla, M. A., and M. E. Venditti. 1992. Importancia y control del taladrillo *Platypus sulcatus*. *Revista Industrial y Agrícola de Tucumán* 69:63–166.

Currie, C. R. 2001. A community of ants, fungi, and bacteria: a multilateral approach to studying symbiosis. *Annual Review of Microbiology* 55:357–380.

Cushman, J. H., and T. G. Whitham. 1989. Conditional mutualism in a membracid ant association: temporal, age-specific, and density-dependent effects. *Ecology* 70:1040–1047.

Dale, C., M. Beeton, C. Harbison, T. Jones, and M. Pontes. 2006. Isolation, pure culture and characterization of *Candidatus Arsenophonus arthropodicus*, an intracellular secondary endosymbiont from the hippoboscid louse fly, *Pseudolynchia canariensis*. *Applied Environmental Microbiology* 72:2997–3004.

Dale, C., G. R. Plague, B. Wang, H. Ochman, and N. A. Moran. 2002. Type III secretion systems and the evolution of mutualistic endosymbiosis. *PNAS* 99:12397–12402.

Davidson, D. W. 1998. Resource discovery versus resource domination in ants: a functional mechanism for breaking the trade-off. *Ecological Entomology* 23:484–490.

de Bary, A. 1897. *Die erscheinung der symbiose*. Karl J. Trubner, Strasbourg.

Dillon, R. J., and A. K. Charnley. 1986. Inhibition *of Metarhizium anisopliae* by the gut bacterial flora of the desert locust, *Schistocerca gregaria*: evidence for an antifungal toxin. *Journal of Invertebrate Pathology* 47:350–360.

Dillon, R., and K. Charnley. 2002. Mutualism between the desert locust *Schistocerca gregaria* and its gut microbiota. *Research in Microbiology* 153:503–509.

Dillon, R. J., and V. M. Dillon. 2004. The gut bacteria of insects: nonpathogenic interactions. *Annual Review of Entomology* 49:71–92.

Dillon, R. J., C. T. Vennard, A. Buckling, and A. K. Charnley. 2005. Diversity of locust gut bacteria protects against pathogen invasion. *Ecology Letters* 8:1291–1298.

Dixon, A. F. G. 1998. *Aphid Ecology*. Chapman & Hall, London.

Dobson, S. L. 2003. Reversing *Wolbachia*-based population replacement. *Trends in Parasitology* 19:128–133.

Douglas, A. E. 1989. Mycetocyte symbiosis in insects. *Biological Reviews* 64:409–434.

Douglas, A. E. 1998. Nutritional interactions in insect–microbial symbioses: aphids and their symbiotic bacteria *Buchnera*. *Annual Review of Entomology* 43:17–37.

Douglas, A. E. 2009. The microbial dimension in insect nutritional ecology. *Functional Ecology* 23:38–47.

Douglas, A. E. 2010. *The Symbiotic Habit*. Princeton University Press, Princeton.

Douglas, A. E., L. B. Minto, and T. L. Wilkinson. 2001. Quantifying nutrient production by the microbial symbionts in an aphid. *Journal of Experimental Biology* 204:349–368.

Dowd, P. F. 1991. Symbiont-mediated detoxification in insect herbivores. In P. Barbosa, V. A. Krischik, and C. G. Jones (editors) *Microbial Mediation of Plant–Herbivore Interactions*, pp. 422–440. John Wiley & Sons, Inc., New York.

Durvasula R. V., A. Gumbs, A. Panackal, O. Kruglov, S. Aksoy, R. B. Merrifield, F. F. Richards, and C. B. Beard. 1997. Prevention of insect-borne disease: an approach using transgenic symbiotic bacteria. *Proceedings of the National Academy of Sciences USA* 94:3274–3278.

Ewald, P. W. 1987. Transmission modes and evolution of the parasite–mutualism continuum. *Annals of the New York Academy of Sciences* 503:295–305.

Felton, G. W., K. L. Korth, J. L. Bi, S. V. Wesley, D. V. Huhman, M. C. Mathews, J. B. Murphy, C. Lam, and R. A. Nixon. 1999. Inverse relationship between systemic resistance of plants to microorganisms and to insect herbivory. *Current Biology* 9:317–320.

Fidantsef, A. L., M. J. Stout, J. S. Thaler, S. S. Duffey, and R. M. Bostock. 1999. Signal interactions in pathogen and insect attack: Expression of lipoxygenase, proteinase inhibitor II, and pathogenesis-related protein P4 in the tomato, *Lycopersicon esculentum*. *Physiological and Molecular Plant Pathology* 54:97–114.

Fraedrich, S. W., T. C. Harrington, and R. J. Rabaglia. 2007. Laurel wilt: a new and devastating disease of redbay caused by a fungal symbiont of the exotic redbay ambrosia beetle. *Newsletter of the Michigan Entomological Society* 52:15–16.

Fraedrich, S. W., T. C. Harrington, R. J. Rabaglia, M. D. Ulyshen, A. E. Mayfield, J. L. Hanula, J. M. Eickwort, and D. R. Miller. 2008. A fungal symbiont of the redbay ambrosia beetle causes a lethal wilt in redbay and other *Lauraceae* in the southeastern United States. *Plant Disease* 92:215–224.

Francke-Grosmann, H. 1967. Ectosymbiosis in wood-inhabiting insects. In S. M. Henry (editor), *Symbiosis*, pp. 141–205. Academic Press, New York.

Furniss, R. L., and V. M. Carolin. 1977. *Western Forest Insects*. US Forest Service, Washington, DC.

Genkai-Kato, M., and N. Yamamura. 1999. Evolution of mutualistic symbiosis without vertical transmission. *Theoretical Population Biology* 55:309–323.

Gibson, C. M., and M. S. Hunter. 2010. Extraordinarily widespread and fantastically complex: comparative biology of endosymbiotic bacterial and fungal mutualists of insects. *Ecology Letters* 13:223–234.

Griffiths, G. W., and S. D. Beck. 1974. Effects of antibiotics on intracellular symbionts in the pea aphid, *Acrythosiphon pisum*. *Cell Tissue Research* 148:287–300.

Hail, D., I. Lauzìere, S. E. Dowd, and B. Bextine. 2011. Culture independent survey of the microbiota of the glassy-winged sharpshooter (*Homalodisca vitripennis*) using 454 pyrosequencing. *Environmental Entomology* 40:23–29.

Hammons, D. L., S. Kaan Kurtural, M. C. Newman, and D. A. Potter. 2009. Invasive Japanese beetle facilitate aggregation and injury by native scarab pest of ripening fruits. *Proceedings of the National Academy of Sciences USA* 106:3686–3691.

Handelsman, J. 2004. Metagenomics: application of genomics to uncultured microorganisms. *Microbiology and Molecular Biology Reviews* 68:669–685.

Hardin M. R., B. Benrey, M. Coll, W. O. Lamp, G. K. Roderick, and P. Barbosa. 1995. Arthropod pest resurgence: an overview of potential mechanisms. *Crop Protection* 2:3–18.

Harris, H. L., L. J. Brennan, B. A. Keddie, and H. R. Braig. 2010. Bacterial symbionts in insects: balancing life and death. *Symbiosis* 51:37–53.

Heard S. B. 1994. Pitcher-plant midges and mosquitoes: a processing chain commensalism. *Ecology* 75:1647–1660.

Heath, J. J., and J. O. Stireman III. 2010. Dissecting the association between a gall midge, *Asteromyia carbonifera*, and its symbiotic fungus, *Botryosphaeria dothidea*. *Entomologia Experimentalis et Applicata* 137:36–49.

Hill, P. D. S., and J. A. Campbell. 1973. The production of symbiont-free *Glossina morsitans* and an associate of female fertility. *Transactions of the Royal Society of Tropical Medicine and Hygiene* 67:727–728.

Hoebeke, E. R., D. A. Haugen, and R. A. Haack. 2005. *Sirex noctilio*: discovery of a Palearctic siricid woodwasp in New York. *Newsletter of the Michigan Entomological Society* 50:24–25.

Hoeksema, J. D., and E. M. Bruna. 2000. Pursuing the big questions about interspecific mutualism: a review of theoretical approaches. *Oecologia* 125:321–330.

Hofstetter, R. W., J. T. Cronin, K. D. Klepzig, J. C. Moser, and M. P. Ayres. 2006a. Antagonisms, mutualisms, and commensalisms affect outbreak dynamics of the southern pine beetle. *Oecologia* 147:679–691.

Hofstetter, R. W., K. D. Klepzig, J. C. Moser, and M. P. Ayres. 2006b. Seasonal dynamics of mites and fungi and their interaction with southern pine beetle. *Environmental Entomology* 35:22–30.

Hofstetter, R. W., T. D. Dempsey, K. D. Klepzig, and M. P. Ayres. 2007. Temperature-dependent effects on mutualistic and phoretic associations. *Community Ecology* 8:47–56.

Hölldobler, B., and E. O. Wilson. 1990. *The Ants*. Springer Verlag, Berlin.

Hwang, S. H., R. L. Lindroth, M. E. Montgomery, and K. S. Shields. 1995. Aspen leaf quality affects gypsy moth (Lepidoptera: Lymantriidae) susceptibility to *Bacillus thuringiensis*. *Journal of Economic Entomology* 88:278–282.

Jaramillo, J. 2008. Biology, ecology and biological control of the coffee berry borer, *Hypothenemus hampei* (Ferrari) (Coleoptera: Curculionidae: Scolytinae). Ph.D. thesis, Faculty of Natural Sciences, Gottfried Wilhelm Leibniz Universität Hannover, Germany.

Kaltenpoth, M. 2009. Actinobacteria as mutualists: general healthcare for insects? *Trends in Microbiology* 17:529–535.

Kane, M. D., and J. A. Breznak. 1991. Effect of host diet on production of organic acids and methane by cockroach gut bacteria. *Applied and Environmental Microbiology* 57:2628–2634.

Karban, R., R. Adamchack, and W. C. Schnathorst. 1987. Induced resistance and interspecific competition between spider mites and vascular wilt fungus. *Science* 235:678–679.

Kaufman, M. G., and M. J. Klug. 1991. The contribution of hindgut bacteria to dietary carbohydrate utilization by crickets (Orthoptera: Gryllidae). *Comparative Biochemistry and Physiology* 98:117–123.

Keane R. M., and M. J. Crawley. 2002. Exotic plant invasions and the enemy release hypothesis. *Trends in Ecology and Evolution* 17:164–170.

Kent, D. S., and J. A. Simpson. 1992. Eusociality in the beetle *Austroplaypus incompertus*. *Naturwissenschaften* 79:86–87.

Kikuchi, Y. 2009. Endosymbiotic bacteria in insects: their diversity and culturability. *Microbes and Environment* 24:195–204.

Kikuchi, Y., T. Hosokawa, and T. Fukatsu. 2007. Insect–microbial mutualism without vertical transmission: a stinkbug acquires a beneficial gut symbionts from the environment every generation. *Applied Environmental Microbiology* 73:4308–4316.

Klepzig, K. D., J. C. Moser, F. J. Lombardero, R. W. Hofstetter, and M. P. Ayres. 2001. Symbiosis and competition: complex interactions among beetles, fungi and mites. *Symbiosis* 30:83–96.

Klepzig, K. D., and D. L. Six. 2004. Bark beetle–fungal symbiosis: context dependency in complex associations. *Symbiosis* 37:189–205.

Klepzig, K. D., A. S. Adams, J. Handelsman, and K. F. Raffa. 2009. Symbioses: a key driver of insect physiological processes, ecological interactions, evolutionary diversification, and impacts on humans. *Environmental Entomology* 38:67–77.

Kühnholz, S., J. H. Borden, and A. Uzunovic. 2001. Secondary ambrosia beetles in apparently healthy trees: adaptations, potential causes and suggested research. *Integrated Pest Management Reviews* 6: 209–219.

Lombardero, M. J., M. P. Ayres, R. W. Hofstetter, J. C. Moser, and K. D. Klepzig. 2003. Strong indirect interactions of *Tarsonemus* mites (Acarina: Tarsonemidae) and *Dendroctonus frontalis* (Coleoptera: Scolytidae). *Oikos* 102:243–252.

Lu, M. M. J. Wingfield, N. E. Gillette, S. R. Mori, and J-H. Sun. 2010. Complex interactions among host pines and fungi vectored by an invasive bark beetle. *New Phytologist* 187:859–866.

Madden, J. L. 1975. Bacteria and yeasts associated with *Sirex noctilio. Journal of Invertebrate Pathology* 25:283–287.

Manion, P. D. 1981. *Tree Disease Concepts.* Prentice-Hall, Inc., Englewood Cliffs, NJ.

Martinez, A. S., V. Fernandez-Arhex, and J. C. Corley. 2006. Chemical information from the fungus *Amylostereum areolatum* and host-foraging behaviour in the parasitoid *Ibalia leucospoides. Physiological Entomology* 31:336–340.

Mattson, W. J., and R. A. Haack. 1987. The role of drought in outbreaks of plant-eating insects. *Bioscience* 37:110–118.

May R. M. 1974. *Stability and Complexity in Model Ecosystems.* Princeton University Press, Princeton.

May, R. M., and G. F. Oster. 1976. Bifurcations and dynamic complexity of simple ecological models. *American Naturalist* 110:573–599.

McCallum, H., and Dobson, A.P. 1995. Detecting disease and parasite threats to endangered species and ecosystems. *Trends in Ecology and Evolution* 10:190–194.

McCreadie, J. W., C. E. Beard, and P. H. Adler. 2005. Context-dependent symbiosis between black flies (Diptera: Simuliidae) and trichomycete fungi (Harpellales: Legeriomycetaceae). *Oikos* 108:362–370.

McMeniman, C. J., R. V. Lane, B. N. Cass, A. W. C. Fong, M. Sidhu, Y-F. Wang, and S. L. O'Neill. 2009. Stable introduction of a life-shortening *Wolbachia* infection into the mosquito *Aedes aegypti. Science* 323:141–144.

Montllor, C. B., A. Maxmen, and A. H. Purcell. 2002. Facultative bacterial endosymbionts benefit pea aphids *Acyrthosiphon pisum* under heat stress. *Ecological Entomology* 27:189–195.

Moran, P. 1998. Plant-mediated interactions between insects and a fungal plant pathogen and the role of plant chemical responses to infection. *Oecologia* 115:523–530.

Moran, N. A., and A. Telang. 1998. Bacteriocyte-associated symbionts of insects: a variety of insect groups harbor ancient prokaryotic endosymbionts. *Bioscience* 48:295–304.

Moran, N. A., and T. Jarvik. 2010. Lateral transfer of genes from fungi underlies carotenoid production in aphids. *Science* 328:624–627.

Moser, J. C., H. Konrad, S. R. Blomquist, and T. Kirisits. 2010. Do mites phoretic on elm bark beetles contribute to the transmission of Dutch elm disease. *Naturwissenschaften* 97:219–227.

Mota, M. M., and P. R. Vieira. 2008. *Pine Wilt Disease: A World-Wide Threat to Forest Ecosystems.* Springer Science, Berlin.

Mueller, U. G., and N. Gerardo. 2002. Fungus-farming insects: multiple origins and diverse evolutionary histories. *Proceedings of the National Academy of Sciences USA* 99:15247–15249.

Nixon, G. E. J. 1951. *The Association of Ants with Aphids and Coccids.* Commonwealth Institute of Entomology, London.

Nogge, G. 1976. Sterility in tsetse flies (*Glossina morsitans* Westwood) caused by loss of endosymbionts. *Experientia* 32:995–996.

Oliver K. M, J. A. Russell, N. A. Moran, and M. S. Hunter. 2003. Facultative bacterial symbionts in aphids confer resistance to parasitic wasps. *Proceedings of the National Academy of Sciences USA* 100:1803–1807.

O'Neill, S. L., A. A. Hoffmann, and J. H. Werren (editors). 1997. *Influential Passengers.* Oxford University Press, New York.

Paine, T. D., K. F. Raffa, and T. C. Harrington. 1997. Interactions among scolytid bark beetles, their associated fungi, and live host conifers. *Annual Review of Entomology* 42:179–206.

Prosser, W. A., and A. E. Douglas. 1991. The aposymbiotic aphid: an analysis of chlortetracycline-treated pea aphid, *Acrythsiphon pisum. Journal of Insect Physiology* 37:713–719.

Rada, V., M. Machova, J. Hulk, M. Marounek, and D. Duskova. 1997. Microflora in the honeybee digestive tract-counts, characteristics and sensitivity to veterinary drugs. *Apodologie* 26:357–365.

Sabelis, M. W., M. van Baalen, F. M. Bakker, J. Bruin, B. Drukker, C. J. M. Egas, A. Janssen, I. Lesna, B. Pels, P. C. J. van Rijn, and P. Scutareanu. 1999. The evolution of direct and indirect plant defence against herbivorous arthropods. In H. Olff, V. A. Brown, and R. H. Drent (editors), *Herbivores: Between Plants and Predators*, pp. 109–166. Blackwell Science Ltd, Oxford.

Sakurai, M., R. Koga, T. Tsuchida, X-Y. Meng, and T. Fukatsu. 2005. Rickettsia symbiont in the pea aphid *Acyrtosiphon pisum*: novel cellular tropism, effect on host fitness, and interaction with the essential symbiont *Buchnera*. *Applied and Environmental Microbiology* 71:4069–4075.

Scarborough, C. L., J. Ferrari, and H. C. J. Godfray. 2005. Aphid protected from pathogen by endosymbiont. *Science* 310:1781.

Schlein, Y. 1976. Lethal effect of tetracycline on tsetse following damage to bacterial symbionts. *Experientia* 33:450–451.

Schultz, T. R., U. G. Mueller, C. R. Currie, and S. A. Rehner. 2005. Reciprocal illumination: a comparison of agriculture of humans and ants. In F. E. Vega and M. Blackwell (editors), *Insect–Fungal Associations: Ecology and Evolution*, pp. 149–190. Oxford University Press, New York.

Scott, J. D., D. C. Oh, M. C. Yuceer, K. D. Klepzig, J. Clardy, and C. R. Currie. 2008. Bacterial protection of beetle–fungus mutualism. *Science* 322:63.

Scudder, G. G. E. 2009. The importance of insects. In R. Foottot and P. Adler (editors), *Insect Biodiversity: Science and Society*, pp. 7–32. Blackwell Publishing, Chichester.

Sen, R., H. D. Ishak, D. Estrada, S. E. Dowd, E. Hong, and U. G. Mueller. 2009. Generalized antifungal activity and 454-screening of *Pseudonocardia* and *Amycolatopsis* bacteria in nests of fungus-growing ants. *Proceedings of the National Academy of Sciences USA* 106:17805–17810.

Shen, Z., M. Denton, N. Mutti, K. Pappan, M. R. Kanost, J. C. Reese, and G. R. Reeck. 2003. Polygalacturonase from *Sitophilus oryzae*: possible horizontal transfer of a pectinase gene from fungi to weevils. *Journal of Insect Science* 3:24–33.

Sinkins, S. P., and F. Gould. 2006. Gene drive systems for insect disease vectors. *Nature Reviews Genetics* 7:427–435.

Six, D. L., and K. D. Klepzig. 2004. *Dendroctonus* bark beetles as model systems for studies on symbiosis. *Symbiosis* 37:207–232.

Six, D. L., and B. J. Bentz. 2007. Temperature determines symbiont abundance in a multipartite bark beetle–fungus ectosymbiosis. *Microbial Ecology* 54:112–118.

Smith, R. H. 1961. *Red Turpentine Beetle*. US Forest Service, Washington, DC.

Stadler, B., and A. F. G. Dixon. 2005. Ecology and evolution of aphid–ant interactions. *Annual Review of Ecology, Evolution, and Systematics* 36:345–372.

Stipes, R. J., and R. S. Campana. 1981. *Compendium of Elm Diseases*. American Phytopathological Society, St. Paul, MN.

Stout, M. J., A. L. Fidantsef, S. S. Duffey, and R. M. Bostock. 1999. Signal interactions in pathogen and insect attack: systemic plant-mediated interactions between pathogen and herbivores of the tomato, *Lycopersicon esculentum*. *Physiological and Molecular Plant Pathology* 54:115–130.

Suh, S-O., and M. Blackwell. 2005. The beetle gut as a habitat for new species of yeasts. In F. E. Vega and M. Blackwell (editors), *Insect–Fungal Associations: Ecology and Evolution*, pp. 244–256. Oxford University Press, New York.

Suh, S-O., J. V. McHugh, D. D. Pollock, and M. Blackwell. 2005. The beetle gut: a hyperdiverse source of novel yeasts. *Mycological Research* 109:261–265.

Tyson, G. W., J. Chapman, P. Hugenholtz, E. E. Allen, R. J. Ram, P.M. Richardson, V. V. Solovyev, E. M. Rubin, D. S. Rokhsar, and J. F. Banfield. 2004. Community structure and metabolism through reconstruction of microbial genomes from the environment. *Nature* 428:37–43.

Vega, F. E., P. Benavides, J. Stuart, and S. L. O'Neill. 2002. Wolbachia infection in the coffee berry borer (Coleoptera: Scolytidae). *Annals of the Entomological Society of America* 95:374–378.

Vega, F. E., and P. F. Dowd. 2005. The role of yeasts as insect endosymbionts. In F. E. Vega, and M. Blackwell (editors), *Insect–Fungal Associations: Ecology and Evolution*, pp. 211–243. Oxford University Press, New York.

Warnecke, F., P. Luginbühl, N. Ivanova, M. Ghassemian, T. H. Richardson, J. T. Stege, M. Cayouette, A. C. McHardy, G. Djordjevic, N. Aboushadi, R. Sorek, S. G. Tringe, M. Podar, H. Garcia Martin, V. Kunin, D. Dalevi, J. Madejska, E. Kirton, D. Platt, E. Szeto, A. Salamov, K. Barry, N. Mikhailova, N. C. Kyrpides, E. G. Matson, E. A. Ottesen, X. Zhang, M. Hernández, C. Murillo, L. G. Acosta, I. Rigoutsos, G. Tamayo, B. D. Green, C. Chang, E. M. Rubin, E. J. Mathur, D. E. Robertson, P. Hugenholtz, and J. R. Leadbetter. 2007. Metagenomic and functional analysis of hindgut microbiota of a wood-feeding higher termite. *Nature* 450:560–565.

Weis, A. E. 1982. Use of a symbiotic fungus by the gall maker *Asteromyia carbonifera* to inhibit attack by the parasitoid *Torymus capite*. *Ecology* 63:1602–1605.

Wernegreen, J.J. 2002. Genome evolution in bacterial endosymbionts of insects. *Nature Reviews Genetics* 3:850–861.

Werren, J. H. 1997. Biology of *Wolbachia*. *Annual Review of Entomology* 42:587–609.

Wingfield, M. J., R. A. Blanchette, and T. H. Nicholls. 1984. Is the pine wood nematode an important pathogen in the United States? *Journal of Forestry* 82:232–235.

Wood, S. L. 1982. The bark and ambrosia beetles of North and Central America (Coleoptera: Scolytidae), a taxonomic monograph. *Great Basin Naturalist Memoirs* 6:129–203.

Yan, Z., J. Sun, O. Don, and Z. Zhang. 2005. The red turpentine beetle, *Dendroctonus valens* LeConte (Scolytidae): an exotic invasive pest of pine in China. *Biodiversity and Conservation* 14:1735–1760.

Zabalou S., M. Riegler, M. Theodorakopoulou, C. Stauffer, C. Savakis, and K. Bourtzis. 2004. *Wolbachia*-induced cytoplasmic incompatibility as a means for insect pest population control. *Proceedings of the National Academy of Sciences USA* 101:15042–15045.

Part III

Population, Community, and Ecosystem Ecology

9

Life History Traits and Host Plant Use in Defoliators and Bark Beetles: Implications for Population Dynamics

Julia Koricheva, Maartje J. Klapwijk, and Christer Björkman

9.1   Introduction

Life history traits of a species determine the survival, reproduction and fitness of its offspring and hence may have consequences for species population dynamics (Cole 1954). If we can predict population dynamics of insects based on their life history characteristics, it would be useful for pest forecasting and predicting population dynamics of poorly studied, rare or pest species (Nylin 2001, Veldtman *et al.* 2007). Therefore, a number of studies have been undertaken to identify life history traits that predict insect population dynamics. Ideally, the role of life history characteristics in insect population dynamics should be explored via phylogenetically controlled comparisons of various traits between large number of outbreaking and non-outbreaking species. In the past such comparisons have been precluded by the paucity of data available on biology of non-outbreaking species. For instance, in the first edition of "Insect outbreaks"

Insect Outbreaks Revisited, First Edition. Edited by Pedro Barbosa, Deborah K. Letourneau and Anurag A. Agrawal.

Nothnagle and Schultz (1987) noted that "although much is known about some irruptive pest species, the majority of the nonirruptive species remains virtually uncharacterized in any meaningful way. Hence, comparing the biological traits of pest and nonpest species is very difficult." Instead, Nothnagle and Schultz (1987) resorted to analysis of common traits between outbreaking species arguing that "if the ability to undergo population outbreaks derives from having certain biological traits, we should see convergence of parallelism among outbreak species." The problem with this approach is that it is not clear whether the detected suites of traits indeed distinguish outbreaking species from non-outbreaking ones. Moreover, "outbreak species constitute a minor proportion of all phytophagous insects and may represent the exception rather than the rule, and ... broad principles should not be drawn from this behavior alone" (Morris 1964).

Another common approach to assessment of population dynamics consequences of life history traits includes pairwise comparisons of species which are closely related or share the same host plant but display contrasting population dynamics (Dodge and Price 1991, Auerbach *et al.* 1995, Eber *et al.* 2001, Veldtman *et al.* 2007). Although such comparisons allow very detailed (and often phylogenetically controlled) comparison of life histories of the species, such studies in isolation have little statistical power. It is only by studying large number of species or by bringing together large number of studies that we can detect general differences between life histories of outbreaking and non-outbreaking species.

Large data sets on densities of local insect populations covering many species and years can be collected by means of continuous sampling by using light or suction trapping. Several such data sets have been used to analyze the relationship between long term temporal variability of insect species and their life history traits (Spitzer *et al.* 1984, Redfearn and Pimm 1988, Miller 1996, Leps *et al.* 1998, Kozlov *et al.* 2010). The results of such analyses, however, may be difficult to interpret because of the use of very different statistical approaches, with at least 23 published sampling estimators of population variability (Gaston and McArdle 1994). Moreover, species shown to exhibit large fluctuations in abundance may not necessarily reach outbreaking densities; for instance a change in abundance from 0.01 to 1 larvae per tree is an equivalent level of fluctuation to a change from 1000 to 100 000 larvae per tree, but only the latter case is likely to be classified or reported as an outbreak (Kozlov *et al.* 2010).

During the last 20 years several extensive data sets have been assembled, most notably on Macrolepidoptera (Hunter 1991, 1995b) and diprionid sawflies (Hanski 1987, Larsson *et al.* 1993), allowing detailed phylogenetically controlled comparisons of life history traits between outbreaking and non-outbreaking species within broader insect taxa. However, no previous studies have compared the degree of similarity/difference in the life history traits and host plant characteristics that determine insect population dynamics *across* different herbivore taxa. In this chapter we attempt such a comparison in order to find out whether there is a suite of common traits associated with outbreaking species across different taxa or whether those traits are phylogenetically constrained and differ between taxa. This is important in order to increase our ability to predict insect population dynamics, e.g. for rare or non-native species.

We have complemented the existing data sets on forest Macrolepidoptera (Hunter 1991, 1995b) and diprionid sawflies (Hanski 1987, Larsson *et al.* 1993) by collecting comparable data set on bark beetles. We have deliberately chosen bark beetles as a group which is ecologically very different from two defoliator groups analyzed previously. Comparing three groups with very different ecology will allow a stronger test of the hypothesis that life history traits are associated with population dynamics. We a priori expect that Macrolepidoptera and sawflies will share more life history traits with each other than with bark beetles, but the question is whether the traits that predispose species to outbreak are idiosyncratic and different in each of the three groups or whether some common life history traits can be identified across different taxa. This analysis will also shed the light on the relative importance of intrinsic factors, bottom-up effects and phylogenetic constraints in determining insect population dynamics.

In this chapter, we (1) describe theoretical advances that took place since the first edition of the "Insect outbreaks" with respect to implications of life history and host plant characteristics for insect herbivore population dynamics, and (2) compare life history characteristics of outbreaking and non-outbreaking species across three taxonomic groups: Macrolepidoptera, diprionid sawflies (Hymenoptera) and bark beetles (Coleoptera).

9.2 Theoretical advances

9.2.1 Effects of insect life history characteristics

The concept of r- and K-selection introduced by MacArthur and Wilson (1967) has been responsible for stimulating much of the research into life history patterns in 1970s and 1980s. Pianka (1970) proposed a set of life history characteristics expected to be associated with the r- and K-selection: r-selected organisms are supposed to have rapid development, high intrinsic rate of increase (r), early reproduction, small body size and single reproductive episode, whereas K- selected species have the opposite characteristics. Pianka (1970) suggested that on the whole the majority of insects are r-selected and the majority of vertebrates are K-selected. Southwood *et al.* (1974) argued that selection strategies may vary within these groups and provided examples of species at different end of r–K selection spectrum within birds and Lepidoptera. Wallner (1987) adapted Pianka's trait list for outbreaking insect species and suggested that they may exhibit the r-selected life history characteristics such as high fecundity, egg clumping, larval aggregation, short adult life span, no adult feeding, and high adult vagility whereas non-outbreaking species display characteristics of K strategists.

Peter Price (1994, 2003; Price *et al.* 1990) has further developed the argument that the evolutionary biology of a species predetermines its current ecology and population dynamics. His phylogenetic constraints hypothesis predicts that insect herbivore population dynamics is a product of interplay between phylogenetic constraints (key ancient life history traits that are conserved in the phylogeny of a species or a group of species, which limit the range of life history patterns that can evolve) and adaptive syndromes (evolutionary responses to the phylogenetic constraints that minimize the limitations and maximize larval performance, often involving female oviposition behavior). The population dynamic consequences of phylogenetic constraints and adaptive syndromes are called "emergent properties." Using the phylogenetic constraints hypothesis, Price (1994) predicted that life histories in which the female does not select the place of larval feeding in relation to resource quality provide potential for eruptive population dynamics, whereas "latent" species are likely to have high preference–performance linkage. This is because species with high preference-performance linkage evolve specificity for high-quality plant modules, which are relatively rare, and population densities of these species will be constrained by availability of these modules. In contrast, species with no preference–performance linkage evolve with a capacity to feed on a wide range of foliage quality and can, at least occasionally, reach outbreak population densities because they are not constrained by availability of specific food type.

The poor link between oviposition preference and larval performance occurs when females oviposit away from larval feeding sites in space or in time. Several life history characteristics, particularly those associated with female dispersal ability, feeding and oviposition behaviour, may result in absence of linkage between oviposition preference and larval performance (Price 1994). For instance, if females lay eggs in autumn, those eggs overwinter and larvae feed on new foliage the following spring, and there is little capacity for females to choose the optimal leaves or shoots for the larvae. Also, if females are weak fliers or wingless, they have limited ability to choose the best feeding sites for their larvae and usually lay eggs in clusters on the bark of the tree trunks, which in turn may lead to gregarious behaviour of larvae (Barbosa *et al.* 1989, Hunter 1995a). Therefore, insects which overwinter as eggs, have low flying ability, lay eggs in clusters and have gregarious larvae and/or larvae feeding in spring are likely to have no link between oviposition preference and larval performance and hence might be expected to be more prone to outbreaks. Moreover, Miller (1996) and Tammaru and Haukioja (1996) have provided evidence that absence of adult feeding is associated with eruptive population dynamics in Lepidoptera. Among species with non-feeding adults (capital breeders) strong natural selection must exist for a large female abdomen, which in turn impedes flight, making females indiscriminate in relation to oviposition substrate (Price 1994, Tammaru and Haukioja 1996). However, given that the absence of female feeding is fixed within Lepidoptera at either the family or tribe level (Tammaru and Haukioja 1996), it cannot explain variation in outbreaking status of species within the same family, tribe or genus.

Another important conceptual advance that took place since the first edition of the "Insect outbreaks" is the growing appreciation of phenotypic plasticity in life history traits (Nylin and Gotthard 1998). The majority of earlier work on life history evolution, including the original model of *r*- and *K*-selection and research on implications of life history traits for insect population dynamics, tried to ignore this variation and attempted to identify the set of possible combinations of life history traits that maximize fitness and the trade-offs that limit this set (Roff 1992). However, many life history traits show considerable phenotypic variation which is particularly high in fluctuating environments. For instance, the maternal effects hypothesis of herbivore outbreaks (Rossiter 1994) predicts that paternal environments cause alterations in offspring life history traits which may predispose offspring populations to outbreaks. Parental crowding has been recently shown, for instance, to reduce female oviposition times and to increase offspring size in *Locusta migratoria* (Chapuis *et al.* 2010).

New evidence for plasticity of life history traits in insect herbivores has also been provided by the research on effects of climate change on insects. It has been shown that climatic warming may affect not only distributional patterns and physiological responses of insect herbivores, but also induce changes in their life cycle such as shifts in overwintering stages, feeding phenology, diapause patterns and voltinism (Altermatt 2010, Bale and Hayward 2010, Robinet and Roques 2010, Chapter 20 this volume). Climate change can lead to an increased likelihood for outbreaks through microevolution if there is variation among individuals (genotypes) in their response to increased temperature in traits such as oviposition rate, in turn, linked to realized fecundity (Björkman *et al.* 2011).

In some insect species phenotypic plasticity in life history traits is so extreme that it leads to presence of several discrete phenotypes (morphs) within the species. The switch behind the morphs could be environmental (e.g. overcrowding) or genetic. For instance, phase polyphenism occurs in locusts in response to changes in local population density, resulting in distinct solitary and gregarious phases which differ in the expression of numerous physiological, morphological and behavioural traits (Pener and Simpson 2009). Similarly, wing polymorphism is well known in some species of aphids and has been proposed to evolve as a strategy to avoid the trade-off between flight ability and fecundity (Brisson 2010).

This body of research indicates that life history traits may vary considerably not only between species, but also within species. However, different life history traits vary in degree of plasticity (reaction norm) that they display. In general, life history traits like fecundity, body size and developmental time which are closely associated with fitness have lower heritabilities and display more variation within species than behavioral and physiological traits like diapause patterns, overwintering stages, feeding phenology and egg laying behaviour (Mousseau and Roff 1987). We hypothesize therefore that population dynamics of a given species is more closely associated with life history traits that exhibit narrower reaction norms because more plastic traits are less likely to impose constraints on species population dynamics. Moreover, we hypothesize that the degree of phenotypic plasticity in individual life history traits will be higher in outbreaking species. This is because phenotypic plasticity is favoured by selection when the environment is changing and such environmental changes can be generated by fluctuating population densities (Svanbäck *et al.* 2009). We tested the above hypotheses using the data on life history traits of Macrolepidoptera, diprionid sawflies and bark beetles.

9.2.2 Effects of host plant characteristics

Host plant characteristics may also be important determinants of herbivore population dynamics. For instance, Price *et al.* (2005) have compared the number of outbreak species of tenthredinid sawflies on herbaceous versus woody plants and on conifer versus angiosperm hosts in three different geographic regions

(North America, Europe, and Japan). In all geographic regions more outbreaking tenthredinid species were associated with woody plants than with herbaceous plants and with conifers than with angiosperms. Price *et al.* (2005) interpreted higher frequency of outbreaking species on woody plants as the result of higher predictability of woody plants in space and time as compared to herbaceous plants, which have more patchy distribution and hence are less likely to support high densities of sawflies. To account for difference in dynamics of sawflies on angiosperms and conifers, Price *et al.* (2005) suggested the determinate growth of conifers hypothesis. Conifer trees have determinate shoot growth with new tissues already present in the overwintering buds (Kozlowski 1971); this allows for rapid growth in early season and provides only short window of opportunity for ovipositing sawfly females to lay eggs into young needles. As a result, there may be no link between oviposition preference and larval performance, which in turn leads to eruptive population dynamics (Price 1994). In contrast, angiosperms have indeterminate growth with a longer growing period, providing more time for oviposition. To test the above hypothesis, we have compared the proportions of different types of hosts among outbreaking and non-outbreaking species of Macrolepidoptera and bark beetles.

9.3   Case study 1: Macrolepidoptera

Hunter (1991) compared life history traits of outbreaking and non-outbreaking hardwood-feeding Macrolepidoptera of northern North America. She later extended that work to include conifer-feeding and European species (Hunter 1995b). The complete database included 858 Macrolepidoptera species from 17 different families. Species were assigned to outbreaking category if there were recorded episodes of greater than 50% defoliation of their hosts (Hunter 1995b).

Analyses by Hunter (Hunter 1991, 1995a, b) revealed that outbreaking species are more likely to be spring feeding; overwinter as eggs or larvae; cluster their eggs and feed gregariously; have poor female flying ability; be well defended against predators with bright coloration, hairs, tents, or leaf ties; and have greater host breadth than non-outbreaking species (Figure 9.1). Hunter also conducted phylogenetically controlled analyses (e.g., paired sister-group comparisons, the distribution test, and the contingent states test) which revealed that associations between population dynamics and the above traits have arisen independently in several lineages. It should be noted, however, that these analyses were based on incomplete phylogenies, especially for the moths, and hence the results should be considered as preliminary. The outbreaking species are also on average larger and more fecund than non-outbreaking species, but this appears to be due to the phylogenetic conservatism of size and the tendency of families with large species to have more outbreaking species. The proportion of univoltine and multivoltine species does not differ between outbreaking and non-outbreaking species.

In order to test the hypothesis that feeding on conifers rather than angiosperms promotes the probability of insect outbreaks (Price *et al.* 2005), we have used

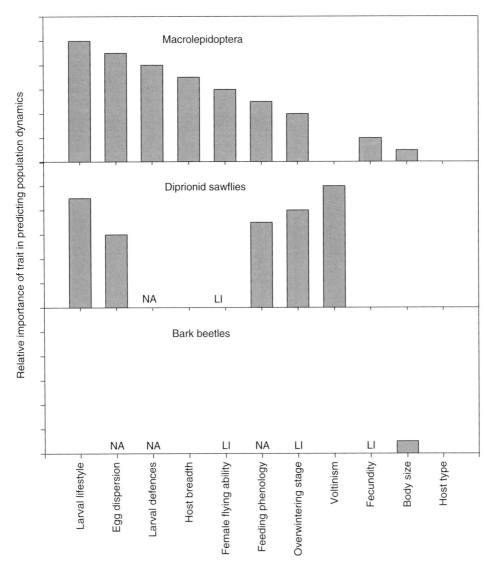

Figure 9.1 Comparison of life history traits distinguishing outbreaking and non-outbreaking species across three insect taxa. NA=not applicable; LI=lack of information (trait could not be included in the analysis). Ranking of relative importance of traits is made on the basis of test statistics comparing the given traits between outbreaking and non-outbreaking categories.

Hunter's data to compare the proportion of outbreaking and non-outbreaking species among Macrolepidoptera feeding on conifers and angiosperms. There was no significant association between the propensity to outbreak and the host plant type (χ^2=1.55, df=1, P=0.213). The result remained the same when analysis was restricted to Geometridae, the largest family with sufficient number numbers of observations in both groups (χ^2=0.43, df=1, P=0.514).

9.4 Case study 2: Diprionid sawflies (Hymenoptera: Diprionidae)

The life history traits of outbreaking and non-outbreaking diprionid sawflies have previously been compared by Larsson *et al.* (1993) using multivariate analyses (CDA), which allow to detect combination of traits that differentiate outbreaking and non-outbreaking species. However, species were then divided into three classes; outbreaking, intermediate, and non-outbreaking. The intermediate group was defined as "species that less frequently, and often only locally, occur at outbreak densities." Here we have reclassified the diprionids into the two groups, outbreaking and non-outbreaking, to facilitate the comparison with the Macrolepidoptera data set. Species were assigned to outbreaking category if there were recorded episodes of greater than 50% defoliation of their hosts at least at local spatial scale. Almost all species in the intermediate group were transferred to the outbreaking group. The data set included 56 sawfly species, 20 of which are native to Europe and 36 to North America. European and North American species were analyzed together to get high enough expected numbers for each life history trait. Due to missing information on life history characteristics for some species, number of outbreak species used in the analysis ranged between 11 and 29 for different traits, and the number of non-outbreak species ranged between 11 and 26.

The main conclusions from the previous multivariate analysis remain valid when we have re-analyzed the diprionid data (Figure 9.1). The largest difference between the groups was found for number of generations per year with the outbreaking species having one or fewer generations per year whereas almost all non-outbreaking species had one or more generations per year ($\chi^2 = 12.50$, df$=1$, P<0.001). The second largest difference was that outbreaking species to a larger extent than non-outbreaking species had larvae that feed gregariously ($\chi^2 = 10.46$, df$=1$, P<0.001). Almost all non-outbreaking species overwinter as cocoons, whereas outbreaking species overwinter as commonly in the egg stage as in the cocoon stage ($\chi^2 = 4.32$, df$=1$, P$=0.038$). Needle age preference and egg dispersion both showed trends for a difference between groups ($\chi^2 = 3.78$, df$=1$, P$=0.052$ and $\chi^2 = 3.16$, df$=1$, P$=0.076$, respectively). Outbreaking species tended to feed more on older foliage ($\chi^2 = 3.78$, df$=1$, P$=0.052$) and lay more eggs per needle than non-outbreaking species ($\chi^2 = 3.16$, df$=1$, P$=0.076$). Preference for older foliage is associated with spring feeding because only needles from previous-year age classes are available in spring.

No other traits showed significant differences between the two groups (Figure 9.1), including the traits such as sex ratio (t$=0.05$, df$=6$, P$=0.96$) and cocoon location (Fisher exact test, P$=1$). Larval defenses could not be analyzed since all diprionids have some level of defense during the larval stage, although the defensive capacity may vary between host plant species or genotypes (Björkman and Larsson 1991, Codella and Raffa 1995). Female flying ability was another trait not analyzed here because of lack of information. Association between the host plant type and population dynamics could not be analyzed because all species of diprionids have coniferous hosts.

9.5   Case study 3: Bark beetles (Curculionidae: Scolytinae)

To complement the existing data on Macrolepidoptera and diprionid sawflies, we have collected data on body size, host plant type, and diet breadth for 210 European and North American species of bark beetles belonging to three tribes (Ipini, Hylesinini, and Scolytini); however, only Ipini and Hylesinini are used in the analyses as the data for the tribe Scolytini were restricted to a single genus, *Scolytus*. In addition, more detailed information on life history traits was collected for 69 species from the genera *Scolytus*, *Dendroctonus*, and *Ips*. Species were classified as outbreaking if we found at least one report of an outbreak involving the species independent of the extent of forest area damaged. While *Dendroctonus* and *Ips* contained significant proportion of both outbreaking and non-outbreaking species, a vast majority of species in the genus *Scolytus* are non-outbreaking (Table 9.1). Moreover, most life history characteristics varied more between genera than within genera. Therefore, we conducted separate analyses for each genus, where possible, and compared the results. We follow the bark beetle taxonomy as outlined by Alonso-Zarazaga and Lyal (2009). Most of the information on life history traits of individual species was obtained from the online database http://www.barkbeetles.org maintained by the University of Georgia Bugwood Network and from the "Forest Insect and Diseases" leaflets of the US

Table 9.1 Summary data for the bark beetle traits included in the analysis. Missing values resulted in differences from the total number of species included in the χ^2 test for each trait.

Tribes	*Ipini*		*Hylesinini*		*Scolytini*		*Total*	
	OS	NOS	OS	NOS	OS	NOS	OS	NOS
Number of species	23	98	9	48	2	30	34	176
Monophagous	8	41	4	21	1	17	13	79
Polyphagous	15	55	4	26	1	11	19	92
Coniferous host	17	68	8	34	1	13	26	115
Deciduous host	5	28	0	13	1	15	6	56
Body size:								
1.00–1.95 mm	1	46	0	12	0	0	1	58
2.00–2.95 mm	6	22	0	19	0	15	6	56
>3.00 mm	15	16	7	15	2	14	24	45

Genera	*Ips*		*Dendroctonus*		*Scolytus*		*Total*	
	OS	NOS	OS	NOS	OS	NOS	OS	NOS
Number of species	11	16	7	4	2	29	20	49
Female attack	0	0	7	3	0	4	7	7
Male attack	11	16	0	0	0	1	11	17
1 female/gallery	0	0	7	4	0	9	7	13
>1 female/gallery	7	9	0	0	0	2	7	11
Facultative multivoltine	4	5	2	2	2	9	8	16
Strictly multivoltine	6	3	2	–	–	4	8	7

Forest Service, Washington, DC. A full list of references used is provided in the appendix at the end of this chapter.

The degree of host specialization did not vary between the outbreaking and non-outbreaking species (monophagous vs. polyphagous, $\chi^2 = 0.08$, df=1, P=0.89, n=173); separate analysis for the tribes produced similar results (Hylesinini: $\chi^2 = 0.011$, df=1, P=0.92, n=57; Ipini: $\chi^2 = 0.093$, df=1, P=0.76, n=121). At the genus level, the degree of host specialization varied only within the genus *Ips*, but the difference between outbreaking and non-outbreaking species was not significant ($\chi^2 = 0.12$, df=1, P=0.073, n=34).

There was no association between the type of host (coniferous vs. angiosperm) and population dynamics at the tribe level ($\chi^2 = 1.27$, df=1, P=0.26, n=173). The majority of bark beetle species are associated with coniferous hosts (Table 9.1). At the genera level, *Scolytus* is the only genus which contains species that feed on both angiosperms and conifers; *Dendroctonus* and *Ips* species feed solely on conifers. The low number of outbreak species in the genus *Scolytus* precluded separate analysis of association between type of host plant and population dynamics within this genus.

To overcome tree defenses, bark beetles use mass attack which can be initiated either by males or by females. However, there was no difference in mass attack strategy (male vs. female) between the outbreaking and non-outbreaking species ($\chi^2 = 0.48$, df=1, P=0.49). Instead, attack strategy varies more between bark beetle genera. For instance, in the genus *Ips* only males execute the attack and in *Dendroctonus* it is mostly females, whereas in *Scolytus* both sexes may be involved.

Neither was there a difference between outbreaking and non-outbreaking species in the number of females in the mating chamber when both *Ips* and *Dendroctonus* were included in the analysis (one vs. more than one; $\chi^2 = 0.02$, df=1, P=0.90) or when *Dendroctonus* was removed from the analysis ($\chi^2 = 1.81$, df=1, P=0.18). Variation in the number of females in the mating chamber could only be compared for species within the genus *Ips* and no difference between outbreaking and non-outbreaking species was found (t=0.568, p=0.58, 95% CI –2.062253 to 3.490824). There were enough observations to examine gallery length only within genus *Ips*, and differences between outbreaking and non-outbreaking species were found neither in the absolute gallery length (t=0.79, p =0.49, n=8, 95% CI –43.38 to 73.18) nor in variation in gallery length (max–min) within species (t=–0.2705, p=0.80, n=6, 95% CI –45.24 to 37.74).

Within the two tribes, outbreaking species had a larger body size than non-outbreaking species ($\chi^2 = 28.47$, df=2, p<0.001, n=190). However, body size varied more between genera than between outbreaking and non-outbreaking species, and there was no difference in body size between the above species groups within the genera *Ips* and *Dendroctonus* ($\chi^2 = 1.20$, df=1, p=0.549). Analyzing the plasticity in body size between outbreaking and non-outbreaking species by using the difference between maximum and minimum body size reported for each species showed larger variation in outbreaking species in both tribes ($F_{1,146} = 47.34$, p<0.001; Figure 9.2).

Finally, there was no difference in voltinism (facultative multivoltine vs. strictly multivoltine) between the outbreaking and non-outbreaking species regardless of whether genus *Scolytus* was included in the analysis (including *Scolytus*: $\chi^2 = 1.53$, df=1, p=0.22; excluding *Scolytus*: $\chi^2 = 1.25$, df=1, p=0.27).

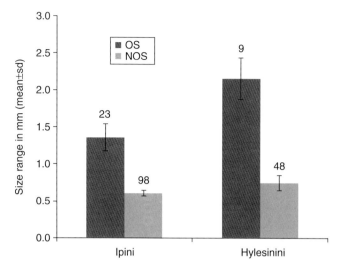

Figure 9.2 Differences in body size variation between outbreaking (OS) and non-out-breaking (NOS) species of bark beetles in tribes Ipini and Hylesinini. Numbers next to error bars indicate number of species in each group.

9.6 Discussion and conclusions

9.6.1 Life history characteristics

Comparisons of life history traits associated with type of population dynamics in Macrolepidoptera, diprionid sawflies, and bark beetles revealed that none of the life history traits examined was associated with propensity for outbreaks in all three taxa. However, in two more ecologically similar groups, Macrolepidoptera and diprionid sawflies, outbreaking species shared several common life history traits such as higher likelihood of overwintering at egg stage, egg clustering, gregarious larvae, and spring feeding (Figure 9.1). All of the above traits are characteristic of species in which females oviposit away from larval feeding site in space or in time resulting in poor linkage between oviposition preference and larval performance. Price (1994) suggested that eruptive population dynamics is more likely in such species, and our results provide some indirect support for his prediction. However, Price predicted that combinations of life history traits leading to weak preference–performance linkage and, ultimately, eruptive population dynamics are phylogenetically constrained (Price *et al.* 1990, Price 1994). In contrast, in both Macrolepidoptera and diprionid sawflies traits like overwintering stage, egg clustering, gregariousness of larvae, and feeding phenology varied among closely related species within the same genus or family. Ideally, the phylogenetic constraints hypothesis should be tested directly by examining the relative strength of the link between oviposition preference and larval performance in outbreaking and non-outbreaking species. However, recent meta-analysis on the preference–performance relationship by Gripenberg *et al.* (2010) concluded that available data are not sufficient to allow testing the above hypothesis. We suggest therefore that future studies are specifically designed to compare the strength of linkage between oviposition preference and larval performance for a variety of insect taxa, including outbreaking and non-outbreaking species.

We hypothesized that life history traits exhibiting a narrower reaction norm would be better predictors of species population dynamics and that the degree of phenotypic plasticity in individual life history traits will be higher in outbreaking species. We found some evidence for both of the hypotheses. Life history traits that were most important in predicting population dynamics of species (clustering of offspring, feeding phenology, and overwintering stage) are less plastic and hence may impose more constraints on population dynamics as compared to traits like fecundity, body size, and developmental time. The second hypothesis could be tested only for the bark beetles and we found that, as predicted, variation in body size was significantly larger in outbreaking species than in non-outbreaking species in both tribes examined (although there were no differences in variation in gallery length and number of females in mating chambers). Higher variation in body size in outbreaking bark beetles is not necessarily adaptive and may be a consequence of increased intraspecific competition at high population densities.

Some of the life history traits associated with outbreaking population dynamics in Macrolepidoptera and diprionid sawflies might help to reduce the risk of predation. Cocoon predation by small mammals is an important mortality factor for many insect species, and Hanski (1987) suggested that overwintering as eggs instead of pupae may have evolved as a life history strategy to reduce cocoon predation. He also proposed that gregarious species are more likely to escape from the control of predators than solitary species. Larvae feeding in groups suffer less predation and have higher survival than solitary larvae (Lawrence 1990, Hunter 2000) and tend to have better defenses than solitary species (Hunter 1991, 1995b, 2000), although they may be more prone to pathogen infections (Hochberg 1991). Therefore, combination of traits that distinguish outbreaking and non-outbreaking species in Macrolepidoptera and diprionids are associated not only with poor linkage between oviposition preference and larval performance, but also with predation risk avoidance.

In both Macrolepidoptera and diprionid sawflies, outbreaking species tend to feed in spring, which might be a direct consequence of overwintering as eggs. For Macrolepidoptera associated with broadleaves, spring feeding allows larvae to exploit young leaves which are richer in water and nutrients and less tough than mature leaves (Feeny 1970). Since leaf quality changes very rapidly during leaf maturation (Koricheva and Barton in press), larval development should be very well synchronized with leaf development and any deviations from normal pheno-logical patterns may have disastrous consequences for herbivore population dynamics (Visser and Holleman 2001), making spring-feeding species more prone to extreme density fluctuations. Indeed, Forkner *et al.* (2008) have shown that population variability in oak-feeding Lepidoptera was considerably higher in spring-feeding species than in species feeding in summer months and attributed this result to higher unpredictability of timing of availability of ephemeral resources such as new leaves.

For Macrolepidoptera and diprionid sawflies associated with coniferous hosts, only older needle age classes are available in spring because young needles in conifers appear later in the season as compared to leaves of angiosperms (Larsson *et al.* 1993). Feeding on older needles may be beneficial because they contain lower levels of certain resin acids which are repellent to diprionids (Ikeda *et al.* 1977) and may have negative effects on larval development (Larsson *et al.* 1986). However, high resin acid concentrations may also be beneficial because sawfly larvae sequester resin acids and use them in their defense against predators. Larsson *et al.* (2000) have shown that risk of outbreak in sawflies depends on both resin acid concentrations and larval predation pressure, resulting in unstable population dynamics.

Although it might be expected that the rates of population increase should be higher for multivoltine species and hence these species may be more likely to reach outbreaking densities, we found no evidence that species with more generations per year are more likely to outbreak. In fact, in forest Lepidoptera univoltinism tends to be associated with life history traits like flightlessness in females, overwintering at egg or larval stage, and polyphagy (Barbosa *et al.* 1989), which predispose to eruptive population dynamics. In diprionid sawflies, outbreaking species had one or fewer generations per year whereas almost all non-outbreaking species had one or more generations per year. Most of the outbreaking diprionid species have been reported to have a prolonged diapause, resulting in partial semivoltinism. The duration of diapause often varies between populations within the same sawfly species and may reduce intraspecific competition as well as allow to escape parasitism and/or predation (Knerer 1983, Larsson *et al.* 1993).

For Macrolepidoptera and bark beetles, there was a tendency for outbreaking species to have larger body size than non-outbreaking species. This is contrary to the prediction that *r*-selected organisms should have small body size (Pianka 1970). However, *r*-selected organisms should also have higher fecundity than *K*-selected organisms (Wallner 1987) and in insects adult body size is usually positively correlated with female fecundity (Honek 1993). Indeed, in the case of Macrolepidotera maximum fecundity of outbreaking species was significantly higher than in non-outbreaking species (Hunter 1995b). Positive relationships between fecundity and abundance of noctuid moths have also been earlier reported by Spitzer *et al.* (1984). However, most of the differences in body size and fecundity in Macrolepidoptera were explained by phylogeny, and within a family outbreaking species are not larger or more fecund (Hunter 1995b). Similarly, in bark beetles most of the variation in body size was between genera rather than between the outbreaking and non-outbreaking species.

Effects of herbivore diet breadth on population dynamics have been much disputed in the literature. MacArthur (1955) suggested that restricted diet lowers population stability because specialists are more susceptible to fluctuations in the abundance of their particular host species. Redfearn and Pimm (1988) provided some support for this prediction by analyzing long-term data on population variability of British aphids, British moths, and Canadian Macrolepidoptera: populations of highly polyphagous species tended to be less variable than populations of specialists, but the relationship was weak. In contrast, Watt (1964) suggested that populations with a wide variety of host plants may be more variable because during favorable conditions they would be able to exploit a larger proportion of the total habitat and spend less time searching as compared to more specialized herbivores. Another argument for positive relationship between diet breadth and population variability could be formulated based on the strength of linkage between oviposition preference and larval performance. Polyphagous insects have weaker coupling between female preference and offspring performance than oligophagous insects (Gripenberg *et al.* 2010), and insects with weak preference–performance linkage are more likely to exhibit eruptive population dynamics (Price 1994).

Among the three data sets examined in this chapter, the diet breadth was associated with outbreak propensity only in Macrolepidoptera with outbreaking species being more polyphagous than non-outbreaking species. Sister-group comparisons showed that outbreaking Macrolepidoptera species have on average seven more genera in their diets than non-outbreaking sister species (Hunter 1995b). No association between diet breadth and outbreak propensity was found for bark beetles and diprionids. In diprionid sawflies the majority of species have relatively narrow feeding specialization, and therefore the lack of difference between the outbreaking and non-outbreaking species may be due to the low variation in diet breadth in this taxon. In bark beetles, diet breadth varied considerably within tribes and among species of genus *Ips*, but no clear relationship between degree of feeding specialization and population dynamics of bark beetles was found.

Several other life history traits may be associated with outbreaking behavior only in specific taxa. For instance, Miller (1996) and Tammaru and Haukioja (1996) provided the evidence that absence of adult feeding is associated with eruptive population dynamics in Lepidoptera. In diprionid sawflies all species are capital breeders *sensu* Tammaru and Haukioja (1996) and rely on energy derived from larval feeding for reproduction, although they may occasionally feed as adults, and hence absence of obligatory adult feeding cannot separate outbreaking and non-outbreaking diprionids. In bark beetles, young adults display maturation feeding either by extending galleries near the pupation place or by feeding on other trees of the same species (Rudinsky 1962). Since most bark beetles display this behavior, this trait cannot be used to separate outbreaking and non-outbreaking species. Adult feeding is also common in other beetle families (e.g. Chrysomelidae) and other insect taxa such as aphids.

Another life history trait that appears to be associated with eruptive population dynamics in Macrolepidoptera, but not in other groups is poor female flying ability (Hunter 1995a). Not much information is available on flying ability of diprionid sawflies, but it is often claimed that diprionid females are poor fliers in general (e.g. Griffiths 1959) although *Neodiprion sertifer* females showed a decent flight capacity in one of the few published experimental studies (Björkman *et al.* 1997). In bark beetles, all species display the dispersal flight to locate the host tree after completing their development and the majority of species are capable of long flights (Rudinsky 1962).

Contrary to Macrolepidoptera and diprionid sawflies, we found no life history traits which were associated with propensity to outbreak in bark beetles. This is not surprising given that the majority of bark beetle life history traits examined varied largely between genera and hence could not explain different population dynamics between species within the same genera. In contrast, majority of life history traits in Macrolepidoptera and diprionid sawflies varied considerably within genera and families, suggesting that phylogenetic signals are low. This was particularly true for behavioral traits associated with oviposition and larval development, which proved to be most important in distinguishing between outbreaking and non-outbreaking species (Larsson *et al.* 1993, Hunter 1995b, Kozlov *et al.* 2010). This suggests that the same combinations of life history traits have evolved repeatedly within different taxa. Although our bark beetle data set covered large number of species, only a few genera were represented, which restricted our ability to conduct phylogenetically controlled analyses. Interestingly, type of population dynamics varied only in some of the genera whereas in the genus *Scolytus* the majority of species were non-outbreaking. It would be interesting to extend the bark beetles data set to include more genera, which would allow testing of the phylogenetic constraints theory.

Overall, while combination of several life history traits may predispose herbivores to outbreaks, life history alone does not determine population dynamics of a species. This is clearly demonstrated by the fact that in many eruptive species outbreaks occur only in populations at some parts of species distribution. For instance, outbreaks of the autumnal moth regularly occur in northern Fennoscandia, but not in southern Finland (Ruohomäki *et al.* 2000); outbreaks of the larch budmoth occur only at high elevations (Baltensweiler *et al.* 1977); and outbreaks of *Diprion pini* usually occur in lowland rather than highland populations (Larsson *et al.* 1993). In these cases, abiotic factors and/or natural enemies prevent populations from reaching outbreaking densities in the rest of their distribution range.

9.6.2   Host plant characteristics

With respect to host plant characteristics, Price *et al.* (2005) predicted that outbreaking species are more common on conifer hosts than on angiosperms and found support for this hypothesis for tenthredinid sawflies. We tested this hypothesis for Macrolepidoptera and bark beetles. There was no association between the propensity to outbreak and the host plant type in either Macrolepidoptera or bark beetles. Price *et al.* (2005) explained increased likelihood of outbreaking species of tenthredinid sawflies to be associated with conifer hosts by determinate growth of conifers which results in rapid shoot growth, short time available for oviposition, and, hence, poor linkage between oviposition preference and larval performance. However, determinate shoot growth in which winter buds contain primordia of all the leaves is not restricted to conifers and occurs in many angiosperm trees such as beech, green ash, and some maples (Kozlowski 1971, Pallardy 2008). Moreover, not all conifers have preformed shoots in the winter bud, for instance larches, eastern hemlock, and some southern pine species also produce leaves from leaf primordia that formed during the current year (Pallardy 2008). As a result, seasonal duration of shoot growth varies greatly both among gymnosperms and among angiosperms, often among different species of the same genus (Pallardy 2008). Therefore, it might be more useful in the future to examine the link between population dynamics of herbivores and the seasonal duration of the shoot growth of their host plants rather than crudely categorize hosts into conifers and angiosperms.

We were unable to test the importance of host plant characteristics for diprionid sawflies since all of them are associated with coniferous hosts. However, outbreaking species of diprionids showed preference for older needles, which might be an adaptation to the determinate shoot growth of conifers. Proportion of species overwintering as eggs was also higher in outbreaking diprionid species, which means that females ovipositing in autumn would have only older needles available for oviposition and larvae feeding the following spring would initially also have only older needles available to them. When new needles become available, older larvae of some sawfly species switch from older to newer needles (Larsson *et al.* 1993). However, given that many outbreaking sawfly species oviposit in older needles, the argument of Price *et al.* (2005) that the saw-like ovipositor acts as a phylogenetic constraint because oviposition is limited to young tissues is not supported in the case of diprionid sawflies.

Differences in types of shoot growth between conifers and angiosperms are unlikely to influence wood-boring taxa like bark beetles. Instead, differences in induced defenses between the above host types may be more relevant. For instance, formation of traumatic resin ducts in xylem and induced accumulation of terpene resins in response to bark beetle attack is specific to conifers (Schmidt *et al.* 2005). However, the intensity of the above induced responses varies considerably among different species of conifers (Lewinsohn *et al.* 1991) and bark beetles are able to exhaust these defenses by mass attack (Raffa and Berryman 1983). We found no evidence that bark beetles associated with conifers are more prone to outbreaks than those associated with angiosperms. In summary, while host plant characteristics are undoubtedly important in determining population dynamics of herbivores, broad categorizations of host plants like conifers versus angiosperms are unlikely to be able to predict behavior of individual insect species. We suggest that in future studies, proportions of outbreaking bark beetle species are compared between host species displaying variable induced responses.

9.6.3 Conclusions

To summarize, we did find evidence of convergence or parallelism among outbreak species in Macrolepidoptera and diprionid sawflies in several interrelated life history traits such as overwintering at egg stage, egg clustering, gregarious larvae, and spring feeding. The above traits vary within taxonomic lines, suggesting that there are no phylogenetic constraints associated with them. On the other hand, several life history traits such as polyphagy, lack of feeding in adults, and reduced flight ability in females were associated with eruptive population dynamics only in Lepidoptera, and none of the examined life history traits differed between outbreaking and non-outbreaking species of bark beetles. It appears that the occurrence of bark beetle outbreaks depends largely upon availability of susceptible host plant material rather than on combination of specific life history traits (Rudinsky 1962). Finally, we did not find support for the hypothesis that insects are more likely to exhibit outbreaks on conifer hosts than on angiosperms.

Given considerable phenotypic plasticity in majority of insect life history traits and in host plant characteristics, we suggest that future studies examine associations between population dynamics of insect herbivores with the degree of phenotypic variation (reaction norm) of specific life history traits or host plant characteristics rather than treating the above traits as fixed and taxa-specific. The importance of being able to predict population dynamics of a species is increasing due to problems with introduced alien pests and changes in herbivore population dynamics in response to climate change, resulting in appearance of new pests (Tenow *et al.* 1999; Chapter 20, this volume). We believe that more research on phenotypic plasticity of life history traits and studies on species currently displaying latent population dynamics would improve our ability to predict future population dynamics of insect herbivores. Furthermore, several other insect groups have well-known examples of outbreaking species, including Coleoptera (e.g. Chrysomelidae), Hemiptera (aphids, scale insects), and Orthoptera (locusts). When complete phylogenies for these groups become available, it would be interesting to expand the current analysis to the above groups.

Acknowledgments

MJK and CB were funded through Future Forests (a multidisciplinary program supported by Mistra, the Swedish Forestry Industry, SLU, Umeå University, and Skogforsk). We would like to thank Milos Knizek, Åke Lindelöw, and Martin Schroeder for their help with specific bark beetle enquiries and Dragos Cocos for his help with compiling the bark beetle data set.

Appendix: Literature used to collect information on life history traits of bark beetles

CABI and EPPO for the European Union (prep.). 1997. Data sheets on quarantine pests for the European Union and for the European and Mediterranean Plant Protection Organization. CABI, London.

DeMars, C. J. Jr., and B. H. Roettgering. 1982. *Western Pine Beetle*. Forest Insect & Disease Leaflet 1 (revised). US Forest Service, Washington, DC.

Ferrell, G. T. 1986. *Fir Engraver*. Forest Insect & Disease Leaflet 13 (revised). US Forest Service, Washington, DC.

Gibson, K., S. Kegley, and B. Bentz. 2009. *Mountain Pine Beetle*. Forest Insect & Disease Leaflet 2 (revised); FS-R6-RO-FIDL#2/002-2009. US Forest Service, Portland, OR.

Holsten, E. H., R. W. Their, A. S. Munson, and K. E. Gibson. 1999. *The Spruce Beetle*. Forest Insect & Disease Leaflet 127 (revised). US Forest Service, Washington, DC.

Kegley, S. J., R. L. Livingston, and K. E. Gibson. 1997. *Pine Engraver, Ips pini (Say), in the Western United States*. Forest Insect & Disease Leaflet 122 (revised). US Forest Service, Washington, DC.

Kruse, J. 2002. *The Northern Spruce Engraver:* Ips perturbatus. US Forest Service, Forest Health Protection, Alaska Region, Fairbanks.

McMillin, J., and T. E. DeGomez. 2008. *Arizona Fivespined Ips,* Ips lecontei Swaine, in the Southwestern United States. Forest Insect & Disease Leaflet 116 (revised). US Forest Service, Washington, DC.

Massey, C. L., D. D. Lucht, and J. M. Schmid. 1977. *Roundheaded Pine Beetle*. Forest Insect & Disease Leaflet 155. US Forest Service, Washington, DC.

Payne, T. L. 2002. Life history and habits. In: *The Southern Pine Beetle*. R. C. Thatcher; J. L. Searcy; J. E. Coster, and G. D. Hertel (eds). 2002, University of Georgia, Athens. http://www.bugwood.org

Rudinsky, J. A. 1962. Ecology of Scolytidae. *Annual Review of Entomology* 7:327–348.

Schmitz, R. F., and K. E. Gibson. 1996. *Douglas-Fir Beetle*. Forest Insect & Disease Leaflet 5 (revised). US Forest Service, Washington, DC.

Seybold, S. J., Albers, M. A., and Katovich, S. A. 2002. *Eastern Larch Beetle*. Forest Insect & Disease Leaflet 175. US Forest Service, Washington, DC.

Smith, R. H. 1971. *Red Turpentine Beetle*. Forest Pest Leaflet 55 (revised). US Forest Service, Washington, DC.

Smith, R. H. and R. E. Lee III. 1972. *Black Turpentine Beetle*. Forest Pest Leaflet 12 (revised). US Forest Service, Washington, DC.

Smith, S. L., Borys, R. R. and P. J. Shea. 2009. *Jeffrey Pine Beetle*. Forest Pest Leaflet 11 (revised). FS-R6-RO-FIDL#11/003-2009. US Forest Service, Portland, OR.

Struble, G. R. 1970. *Monterey Pine Ips*. Forest Pest Leaflet 56 (revised). US Forest Service, Washington, DC. http://www.barkbeetles.org

References

Alonso-Zarazaga, M. A., and C. H. C. Lyal. 2009. A catalogue of family and genus group names in Scolytinae and Platypodinae with nomenclatural remarks (Coleoptera: Curculionidae). *Zootaxa* 2258:1–134.

Altermatt, F. 2010. Climatic warming increases voltinism in European butterflies and moths. *Proceedings of the Royal Society of London B* 277:1281–1287.

Auerbach, M. J., E. F. Connor, and S. Mopper. 1995. Minor miners and major miners: population dynamics of leaf-mining insects. Pages 83–110 *in* N. Cappuccino and P. W. Price, editors. *Population Dynamics*. Academic Press, San Diego.

Bale, J. S., and S. A. L. Hayward. 2010. Insect overwintering in a changing climate. *Journal of Experimental Biology* 213:980–994.

Baltensweiler, W., G. Benz, P. Bovey, and V. DeLucchi. 1977. Dynamics of larch budmoth populations. *Annual Review of Entomology* 22:79–100.

Barbosa, P., V. A. Krischik, and D. Lance. 1989. Life history traits of forest-inhabiting flightless Lepidoptera. *American Midland Naturalist* 122:262–274.

Björkman, C., S. Kindvall, S. Höglund, A. Lilja, L. Bärring, and K. Eklund. 2011. High temperature triggers latent variation among individuals: oviposition rate and probability for outbreaks. *PLoS ONE* 6:e16590.

Björkman, C., and S. Larsson. 1991. Pine sawfly defense and variation in host plant resin acids – a trade-off with growth. *Ecological Entomology* 16:283–289.

Björkman, C., S. Larsson, and R. Bommarco. 1997. Oviposition preferences in pine sawflies: a trade-off between larval growth and defense against natural enemies. *Oikos* 79:45–52.

Brisson, J. A. 2010. Aphid wing dimorphisms: linking environmental and genetic control of trait variation. *Philosophical Transactions of the Royal Society B* 365:605–616.

Chapuis, M. P., L. Crespin, A. Estoup, A. Auge-Sabatier, A. Foucart, M. Lecoq, and Y. Michalakis. 2010. Parental crowding influences life history traits in *Locusta migratoria* females. *Bulletin of Entomological Research* 100:9–17.

Codella, S. G., and K. F. Raffa. 1995. Host-plant influence on chemical defense in conifer sawflies (Hymenoptera, Diprionidae). *Oecologia* 104:1–11.

Cole, L. C. 1954. The population consequences of life history phenomena. *Quarterly Review of Biology* 29:103–137.

Dodge, K. L., and P. W. Price. 1991. Eruptive versus noneruptive species: a comparative study of host plant use by a sawfly, {IEuura exiguae} (Hymenoptera: Tenthredinidae) and a leaf beetle, {IDisonycha pluriligata} (Coleoptera: Chrysomelidae). *Environmental Entomology* 20:1129–1133.

Eber, S., H. P. Smith, R. K. Didham, and H. V. Cornell. 2001. Holly leaf-miners on two continents: what makes an outbreak species? *Ecological Entomology* 26:124–132.

Feeny, P. P. 1970. Seasonal changes in oak leaf tannins and nutrients as a cause of spring feeding by winter moth caterpillars. *Ecology* 51:565–581.

Forkner, R. E., R. J. Marquis, J. T. Lill, and J. Le Corff. 2008. Timing is everything? Phenological synchrony and population variability in leaf-chewing herbivores of *Quercus*. *Ecological Entomology* 33:276–285.

Gaston, K. J., and B. H. McArdle. 1994. The temporal variability in animal abundances: measures, methods and patterns. *Philosophical Transactions of the Royal Society of London B* 345:335–358.

Griffiths, K. J. 1959. Observations on the European pine sawfly, *Neodiprion sertifer* (Geoff.) and its parasites in southern Ontario. *Canadian Entomologist* 91:501–512.

Gripenberg, S., P. J. Mayhew, M. Parnell, and T. Roslin. 2010. A meta-analysis of preference–performance relationships in phytophagous insects. *Ecology Letters* 13:383–393.

Hanski, I. 1987. Pine sawfly population dynamics: patterns, processes, problems. *Oikos* 50:327–335.

Hochberg, M. E. 1991. Viruses as costs to gregarious feeding behaviour in the Lepidoptera. *Oikos* 61:291–296.

Honek, A. 1993. Intraspecific variation in body size and fecundity in insects – a general relationship. *Oikos* 66:483–492.

Hunter, A. F. 1991. Traits that distinguish outbreaking and non-outbreaking Macrolepidoptera feeding on northern hardwood trees. *Oikos* 60:275–282.

Hunter, A. F. 1995a. The ecology and evolution of reduced wings in forest macrolepidoptera. *Evolutionary Ecology* 9:275–287.

Hunter, A. F. 1995b. Ecology, life history, and phylogeny of outbreak and non-outbreak species. Pages 42–64 *in* N. Cappuccino and P. W. Price, editors. *Population Dynamics: New Approaches and Synthesis.* Academic Press, San Diego.

Hunter, A. F. 2000. Gregariousness and repellent defenses in the survival of phytophagous insects. *Oikos* 91:213–224.

Ikeda, T., F. Matsumura, and D. M. Benjamin. 1977. Mechanism of feeding discrimination between matured and juvenile foliage by 2 species of pine sawflies. *Journal of Chemical Ecology* 3:677–694.

Knerer, G. 1983. Diapause strategies in diprionid sawflies. *Naturwissenschaften* 70:203–205.

Koricheva, J., and K. E. Barton. In press. Temporal changes in plant secondary metabolite production: patterns, causes and consequences. *In* G. R. Iason, M. Dicke, and S. E. Hartley, editors. *Ecology of Plant Secondary Metabolites: Genes to Landscapes.* Cambridge University Press, Cambridge.

Kozlov, M. V., M. D. Hunter, S. Koponen, J. Kouki, P. Niemelä, and P. W. Price. 2010. Diverse population trajectories among coexisting species of subarctic forest moths. *Population Ecology* 52:295–305.

Kozlowski, T. T. 1971. *Growth and Development of Trees.* Academic Press, New York.

Larsson, S., C. Björkman, and N. A. C. Kidd. 1993. Outbreaks in diprionid sawflies: why some species and not others? Pages 453–483 *in* M. R. Wagner and K. F. Raffa, editors. *Sawfly Life History Adaptations to Woody Plants.* Academic Press, San Diego.

Larsson, S., C. Björkman, and R. Gref. 1986. Responses of *Neodiprion sertifer* (Hym, Diprionidae) larvae to variation in needle resin acid concnetration in Scots pine. *Oecologia* 70:77–84.

Larsson, S., B. Ekbom, and C. Björkman. 2000. Influence of plant quality on pine sawfly population dynamics. *Oikos* 89:440–450.

Lawrence, W. S. 1990. The effects of group size and host species on development and surivorship of a gregarious caterpillar *Halisidota caryae* (Lepidoptera, Arctiidae). *Ecological Entomology* 15: 53–62.

Leps, J., K. Spitzer, and J. Jaros. 1998. Food plants, species composition and variability of the moth community in undisturbed forest. *Oikos* 81:538–548.

Lewinsohn, E., M. Gijzen, and R. Croteau. 1991. Defense mechanisms of conifers: differences in constitutive and wound-induced monoterpene biosynthesis among species. Plant Physiology 96:44–49.

MacArthur, R. H. 1955. Fluctuations of animal populations and a measure of community stability. *Ecology* 36:533–536.

MacArthur, R. H., and E. O. Wilson. 1967. *The Theory of Island Biogeography.* Princeton University Press, Princeton.

Miller, W. E. 1996. Population behavior and adult feeding capability in Lepidoptera. *Environmental Entomology* 25:213–226.

Morris, R. F. 1964. The value of historical data in population research, with particular reference to *Hyphantria cunea* Drury. *Canadian Entomologist* 96:356–368.

Mousseau, T. A., and D. A. Roff. 1987. Natural selection and heritability of fitness components. *Heredity* 59:181–198.

Nothnagle, P. J., and J. C. Schultz. 1987. What is a forest pest? Pages 59–79 *in* P. Barbosa and J. C. Schultz, editors. *Insect Outbreaks.* Academic Press, San Diego.

Nylin, S. 2001. Life history perspectives on pest insects: what's the use? *Austral Ecology* 26:507–517.

Nylin, S., and K. Gotthard. 1998. Plasticity in life history traits. *Annual Review of Entomology* 43:63–83.

Pallardy, S. G. 2008. *Physiology of Woody Plants*, 3rd edition. Academic Press, Amsterdam.

Pener, M. P., and S. J. Simpson. 2009. Locust phase polyphenism: an update. *Advances in Insect Physiology* 36:1–272.

Pianka, E. R. 1970. On r- and K-selection. *American Naturalist* 104:592–597.

Price, P. W. 1994. Phylogenetic constraints, adaptive syndromes, and emergent properties: from individuals to population dynamics. *Researches on Population Ecology* 36:3–14.

Price, P. W. 2003. *Macroevolutionary Theory on Macroecological Patterns*. Cambridge University Press, Cambridge.

Price, P. W., N. Cobb, T. P. Craig, G. W. Fernandes, J. K. Itami, S. Mopper, and R. W. Preszler. 1990. Insect herbivore popultion dynamics on trees and shrubs: new approaches relevant to latent and eruptive species and life table development. Pages 1–38 *in* E. A. Bernays, editor. *Insect–Plant Interactions*. CRC Press, Boca Raton, FL.

Price, P. W., H. Roininen, and T. Ohgushi. 2005. Adaptive radiation into ecological niches with eruptive dynamics: a comparison of tenthredinid and diprionid sawflies. *Journal of Animal Ecology* 74:397–408.

Raffa, K. F., and A. A. Berryman. 1983. The role of host plant resistance in the colonization behavior and ecology of bark beetles (Coleoptera: Scolytidae). *Ecological Monographs* 53:27–49.

Redfearn, A., and S. L. Pimm. 1988. Population variability and polyphagy in herbivorous insect communities. *Ecological Monographs* 58:39–55.

Robinet, C., and A. Roques. 2010. Direct impacts of recent climate warming on insect populations. *Integrative Zoology* 5:132–142.

Roff, D. A. 1992. *The Evolution of Life Histories: Theory and Analysis*. Chapman & Hall, New York.

Rossiter, M. C. 1994. Maternal Effects Hypothesis of Herbivore Outbreak. *Bioscience* 44:752–763.

Rudinsky, J. A. 1962. Ecology of Scolytidae. *Annual Review of Entomology* 7:327–348.

Ruohomäki, K., M. Tanhuanpää, M. P. Ayres, P. Kaitaniemi, T. Tammaru, and E. Haukioja. 2000. Causes of cyclicity of *Epirrita autumnata* (Lepidoptera, Geometridae): grandiose theory and tedious practice. *Population Ecology* 42:211–223.

Schmidt, A., G. Zeneli, A. Hietala, C. G. Fossdal, P. Krokene, E. Christiansen, and J. Gershenzon. 2005. Induced chemical defenses in conifers: biochemical and molecular approaches to studying their function. Pages 1–28 *in* J. T. Romeo, editor. *Chemical Ecology and Phytochemistry of Forest Ecosystems*. Elsevier, Amsterdam.

Southwood, T. R. E., R. M. May, M. P. Hassell, and G. R. Conway. 1974. Ecological strategies and population parameters. *American Naturalist* 108:791–804.

Spitzer, K., M. Rejmanek, and T. Soldan. 1984. The fecundity and long-term variability in adundance of noctuid moths (Lepidoptera, Noctuidae). *Oecologia* 62:91–93.

Svanbäck, R., M. Pineda-Krch, and M. Doebeli. 2009. Fluctuating population dynamics promotes the evolution of phenotypic plasticity. *American Naturalist* 174:176–189.

Tammaru, T., and E. Haukioja. 1996. Capital breeders and income breeders among Lepidoptera – consequences to population dynamics. *Oikos* 77:561–564.

Tenow, O., A. C. Nilssen, B. Holmgren, and F. Elverum. 1999. An insect (*Argyresthia retinella*, Lep., Yponomeutidae) outbreak in northern birch forests, released by climatic changes? *Journal of Applied Ecology* 36:111–122.

Veldtman, R., M. A. McGeoch, and C. H. Scholtz. 2007. Can life history and defense traits predict the population dynamics and natural enemy responses of insect herbivores? *Ecological Entomology* 32:662–673.

Visser, M., and L. Holleman. 2001. Warmer springs disrupt the synchrony of oak and winter moth phenology. *Proceedings of the Royal Society of London B* 268:289–294.

Wallner, W. E. 1987. Factors affecting insect population dynamics: differences between outbreak and non-outbreak species. *Annual Review of Entomology* 32:317–340.

Watt, K. E. F. 1964. Comments on fluctuations of animal populations and measures of community stability. *Canadian Entomologist* 96:1434–1442.

10

The Ecological Consequences
of Insect Outbreaks

Louie H. Yang

10.1 Introduction

While ecologists have long studied the *causes* of insect outbreaks, less attention
has been focused on understanding the ecological *consequences* of these events.
For example, the previous edition of this book was largely focused on

Insect Outbreaks Revisited, First Edition. Edited by Pedro Barbosa, Deborah K. Letourneau
and Anurag A. Agrawal.
© 2012 Blackwell Publishing Ltd. Published 2012 by Blackwell Publishing Ltd.

understanding why and when insect outbreaks occur (Barbosa and Schultz 1987). Some chapters examined key explanatory factors such as the weather and climate, the surrounding context of community interactions, and characteristics of the outbreaking species, while other chapters developed conceptual and mathematical models to better understand and predict outbreak dynamics. This initial emphasis on the causes of insect outbreaks makes sense from both basic and applied perspectives. From ancient times to the modern day, efforts to understand and explain insect "outbreaks," "plagues," and "irruptions" have provided an important foundation for efforts to predict, prevent, and control these potentially destructive events. Examining the causes of the events has also led to a richer understanding of the intrinsic and extrinsic factors that influence population dynamics more generally.

By comparison, past efforts to understand the community consequences of insect outbreaks have often emphasized impacts of economic, rather than ecological, importance. This may be changing. In recent years, increasing attention has been focused on understanding the ecological impacts of insect outbreaks. In part, this trend may reflect greater research into the role of insects in structuring communities and regulating ecosystem functions more generally (e.g., Weisser and Siemann 2004). In addition, growing interest in the ecological consequences of insect outbreaks could reflect a greater recognition of non-equilibrium or transient processes in ecology (Chen and Cohen 2001, Hastings 2001, 2004), and the importance of event-driven dynamics in our more variable climatic future (Meehl *et al.* 2000, Parmesan *et al.* 2000, Bell *et al.* 2004, Jentsch *et al.* 2009). Together, these emerging trends have focused greater attention on understanding how insect outbreaks affect natural systems.

Carson *et al.* (2004) suggested three reasons why understanding the consequences of insect outbreaks can lend general insights into the structure and dynamics of communities and ecosystems. First, insect outbreaks occur in a wide range of systems. Second, although they may appear infrequent from a human perspective, they commonly occur with considerable regularity on relevant ecological time scales. Finally, insect outbreaks can have important impacts on species interactions and alter ecosystem processes. As potentially strong episodic perturbations, these events provide useful opportunities to examine the propagation of indirect effects and the factors that influence community responses and resilience. By investigating how communities and ecosystems respond to outbreak perturbations, we may better understand the structure and dynamics of communities, and how these complex systems will respond to future variability.

Outbreaking insects can affect communities via multiple and interacting mechanisms, including important effects as consumers, ecosystem engineers, competitors, and resources. This chapter seeks to revisit fundamental questions related to each of these four ecological mechanisms: (1) Does herbivory during insect outbreaks reduce plant growth? (2) How do insect outbreaks affect community succession? (3) Do insect outbreaks increase competition for other members of the community? (4) Do insect outbreaks increase resource availability? Examining these four ecological mechanisms may also provide an entry to understand the role of insect outbreaks in regulating the productivity and diversity of ecosystems.

10.2 Outbreaking insects as consumers: Does outbreak herbivory reduce plant growth?

It seems intuitive that outbreaks of insect herbivores should reduce the growth of the plants they consume. However, while many systems provide evidence that insect outbreaks can have direct, negative effects on plant growth, the consumptive effects of insect outbreaks can also have other, more complex consequences.

10.2.1 Direct negative effects of consumption

While herbivores generally consume a relatively small fraction of primary productivity under non-outbreak conditions (Landsberg and Ohmart 1989, Coupe and Cahill 2003), herbivory during outbreak events can entail much more severe biomass losses (Schowalter *et al.* 1986, Brown 1994, Carson and Root 2000). Severe defoliation events are commonly associated with reduced growth in many species of trees (e.g., Kulman 1971, Piene 1980, Jardon *et al.* 1994), and the negative effects of herbivory in a variety of species are generally much larger during outbreak events than during non-outbreak intervals (Schowalter *et al.* 1986, Hoogesteger and Karlsson 1992, Carson and Root 2000, Coupe and Cahill 2003). For example, a dendrochronological analysis of eastern larch (*Larix laricina*) in a boreal forest of Québec suggested that at least eight multiyear periods of reduced growth and increased mortality have occurred in this forest since 1782; these patterns were consistent with patterns of reduced growth observed during a recent outbreak of larch sawfly (*Pristiphora erichsonii*; Jardon *et al.* 1994). Similarly, an 1173-year reconstruction of tree ring growth increments in the subalpine forests of the European Alps was used to infer a consistent pattern of insect outbreaks every 8–9 years based on outbreak-associated patterns of reduced growth (Esper *et al.* 2007). Mounting evidence suggests that the negative effects of outbreak herbivory on plant growth are probably quite general. For example, in a meta-analysis of 24 insect suppression experiments conducted between 1952 and 2001, Coupe and Cahill (2003) found that insect herbivores generally had a negative effect on primary production, and these effects were significantly larger under outbreak conditions.

In severe cases, the negative effects of outbreak herbivory can be associated with widespread host plant mortality. Examples of outbreak-mediated plant mortality include widespread die-offs of bush lupine shrubs (*Lupinus arboreus*) due to the root herbivory of ghost moth (*Hepialus californicus*) caterpillars in a California coastal grassland (Strong *et al.* 1995), high rates of lodgepole pine (*Pinus contorta*) death caused by native mountain pine beetles (*Dendroctonus ponderosae*) in Wyoming forests (Romme *et al.* 1986), and the loss of eastern hemlock stands (*Tsuga canadensis*) in the wake of exotic hemlock wooly adelgid (*Adelgies tsugae*) outbreaks (Cobb 2010).

10.2.2 Other effects of consumption

However, not all insect outbreaks have strong negative effects on plant growth. In many cases, the effects of outbreaking insects as consumers of plant growth are only weakly negative, transient or highly specific. For example, while high-density migratory bands of gregarious Mormon crickets (*Anabrus simplex*) in the American West are capable of extensively defoliating preferred food plants, including sagebrush and crop plants (Criddle 1926), these outbreak events seem to have minimal effects on most rangeland grasses and forbs, possibly due to the combined effects of their itinerant behavior and their preference for consuming sagebrush (Redak *et al.* 1992). In other outbreak systems, host plants show a high tolerance of herbivory, and are capable of resilient regrowth despite extensive biomass loss. For example, outbreaks of tussock moths (*Orgyia vestusa*) in California coastal grasslands cause heavy and rapid defoliation of lupine bushes, but their effects on lupine growth appear to be transient (Harrison and Maron 1995). These outbreaks often have relatively stable spatial distributions for more than a decade, with locally severe defoliation occurring each season. Defoliation reduces leaf quality and seed output during the first 1 or 2 years under outbreak conditions, but defoliated bushes are able to produce leaves and seeds that are similar to nondefoliated bushes thereafter, despite continued and repeated herbivory.

These strong regrowth responses do not necessarily indicate weak effects of outbreak herbivory on overall plant growth, however. For example, Hoogesteger and Karlsson (1992) experimentally defoliated mountain birch (*Betula pubescens*) in order to mimic 50% and 100% leaf area loss to outbreaks of autumnal moth (*Epirrita autumnata*) in northern Sweden. These trees responded to even 100% experimental defoliation with a substantial regrowth of leaves about one month after the experimental defoliation, with total leaf areas returning to pre-defoliation levels within two years. This rapid compensatory response indicated considerable resilience in leaf production. However, the effects of this one-time defoliation event severely reduced mean tree-ring growth increments to less than 25% of control trees for at least three years, suggesting a large and persistent overall cost, perhaps due to the reallocation of stored resources to foliage regrowth responses. Importantly, these costs were not readily apparent from the outward appearance of the affected trees.

In many cases, however, plant communities respond to outbreak herbivory with increases in primary productivity. These overcompensatory responses could occur via multiple co-occurring mechanisms, with simultaneous explanations at multiple levels of causation. At a proximate level, overcompensatory regrowth responses may reflect the transient re-allocation of stored reserves, or more persistent changes in nutrient or light availability which allow greater growth (e.g.,Wickman 1980, Carson and Root 2000). At an ultimate level, rapid regrowth responses following severe defoliation may have evolved due to past correlations between outbreak consumption and changes in resource availability, if outbreak conditions have historically been associated with future opportunities for increased growth. The key insight that emerges here may be that the direct consumptive effects of herbivory by outbreaking insects frequently co-occur with other impacts on species composition, resource availability, nutrient cycling, or changes in habitat structure (Lovett *et al.* 2002), and separating the combined effects of these multiple drivers may be difficult.

10.3 Outbreaking insects as ecosystem engineers: How do insect outbreaks affect succession?

In many systems, insect outbreaks can profoundly change the physical structure of habitats as a secondary consequence of plant consumption. Through their effects on the environment, insect outbreaks can create or maintain habitats for other species, and thereby act as ecosystem engineers (Jones *et al*. 1994, Wright and Jones 2006).

10.3.1 Effects on habitat heterogeneity

In many systems, outbreak herbivory can have strong effects on community structure and succession through the formation of gaps in canopy cover. This gap formation causes underlying changes in the composition and diversity of plant communities, resulting in greater spatial heterogeneity and successional patchiness. For example, spruce budworm outbreaks in northern Québec are known to create early successional gaps in mature forest stands that allow greater species diversities to persist over a landscape scale (de Grandpre and Bergeron 1997). These gaps allow for the re-establishment of diverse understory communities which tend to reduce the likelihood of species losses during subsequent disturbance events, such as those caused by future insect outbreaks, fire, windthrow, or logging.

The formation of successional gaps during insect outbreaks does not promote greater diversity in all systems, however. In a California coastal grassland and shrub system, episodic outbreaks of root-feeding ghost moth caterpillars (*Hepialis californicus*) cause widespread mortality of bush lupine (*Lupinus arboreus*); these die-off events clear the lupine canopy layer, resulting in greater light penetration to nitrogen-rich soils (Strong *et al*. 1995, Maron and Jefferies 1999). These changes to the soil environment allow the rapid invasion of non-native grasses at the expense of native plants, and these effects persist until new bush lupines re-establish a dense canopy after several years (Maron and Jefferies 1999). While both bush lupines and the ghost moths are native species, their outbreak interaction creates structural habitat gaps that are most rapidly exploited by a relatively small number of non-native annual grass species, to the detriment of native plant diversity (Maron and Jefferies 1999).

Despite their different effects on species diversity, these studies and others do suggest a potentially general pattern in which insect outbreaks reduce the canopy cover of mature and competitively dominant species, effectively resetting successional processes. According to this hypothesis, early successional plant species are able to establish within gap communities with reduced asymmetric competition for light and space following outbreak events, with potential effects on the diversity of other trophic levels. For example, herbivory by an outbreaking leaf beetle (*Trirhabda canadensis*) in an old-field community in Minnesota reduced the cover of a dominant goldenrod species (*Solidago missouriensis*), increasing light penetration, soil water content, and N mineralization rate. These changes to the habitat structure promoted the establishment of disturbance-adapted understory plant species which are otherwise characteristic of newly abandoned fields and gopher tillage at this site

(Brown 1994). The observed shift in plant species compositions was consistent with a resetting of the successional trajectory in these communities.

   However, even this pattern may not be entirely general. A long-term experiment that suppressed insect herbivores in a similar New York old-field community showed that a multiyear outbreak of leaf beetles (*Microrhopala vittata*) reduced the dominance of tall goldenrod (*Solidago altissima*), increasing light penetration and understory diversity (Carson and Root 2000). In this case, however, the outbreak effects facilitated the establishment of invading trees, accelerating successional encroachment of woody species into the old-field community. Similar to the old-field study by Brown (1994), the mechanisms of these effects involve changes to the physical and abiotic environment consistent with broad-sense definitions of ecosystem engineering (Jones *et al.* 1994, 1997). However, in contrast to the study by Brown (1994), this study showed how outbreak-mediated gap formation can accelerate successional processes, instead of resetting them.

10.3.2   The effects of heterogeneity at higher trophic levels

The effects of outbreak-mediated gap formation also extend to higher trophic levels. For example, the forest gaps and resulting habitat heterogeneity created by European bark beetle (*Ips typographus*) outbreaks contributed to the diversity of other arthropod species by promoting a successional mosaic that included forests, forest edges, and meadows (Müller *et al.* 2008). Similarly, extensive tree mortality caused by a particularly severe gypsy moth outbreak created early successional gaps in a West Virginia deciduous forest which favored eastern towhees (*Pipilo erythrophthalmus*); these birds are known to favor early successional habitats, but avoid consuming gypsy moths (Bell and Whitmore 1997).

   Insect outbreaks also affect other disturbance drivers that influence successional processes. Several authors have compared the effects of outbreak defoliation with the effects of low-intensity fires, emphasizing their potential to regulate ecosystem function in forest systems (e.g., Mattson and Addy 1975, McCullough *et al.* 2010). While insect outbreaks and fire can have similar effects on forest ecosystem dynamics, these impacts may also differ in key ways. In the boreal forests of Québec, the spatial scale of spruce budworm gap formation tends to be much smaller than the landscape scale of fire effects (de Grandpre and Bergeron 1997). As a result, outbreak-mediated microsite heterogeneity in this system may help to maintain the coexistence of diverse species at local scales (de Grandpre and Bergeron 1997). These two disturbance drivers also interact in complex and reciprocal ways. In some cases, severe insect outbreaks can increase the likelihood and intensity of future forest fires by contributing to the accumulation and quality of fuel loads (McCullough *et al.* 2010). However, less intense outbreak events can also reduce fuel accumulation by resetting successional processes and increasing rates of nutrient cycling (McCullough *et al.* 2010). Reciprocally, the immediate effects of fire can make trees more susceptible to attacks by many outbreaking insects, including several species of bark beetles and wood-boring beetles; however, long-term fire suppression has also increased the spatial scale and intensity of many insect outbreaks, due to effects on forest species composition, density, and demography (Bergeron and Leduc 1998, McCullough *et al.* 2010).

These examples suggest that some of the strongest and most persistent community effects of insect outbreaks are due to their ability to change the physical environment as ecosystem engineers, but predicting the mechanism and direction of these effects in specific systems may be difficult. Thus, even transient outbreaks can have persistent effects on habitat structure with variable long-term consequences for community composition and succession.

10.4 Outbreaking insects as competitors: Do insect outbreaks increase competition?

Examples of simple, symmetrical, and strong resource competition among herbivores are generally thought to be uncommon in nature (Lawton and Strong 1981, Lawton and Hassell 1981, Denno *et al.* 1995, Kaplan and Denno 2007). While competitive interactions of some kind do appear to be widespread forces structuring interactions among herbivores in ecological communities, these interactions are often asymmetrical, relatively weak or variable, not clearly linked with specific resources, and difficult to separate from co-occurring facilitation or indirect effects (Kaplan and Denno 2007). However, ecologists have long suspected that competition may be most apparent and intense during periods of resource scarcity (i.e., "ecological crunches"; Wiens 1977, Schoener 1982, Work and McCullough 2000), such as those that could occur for herbivores following outbreak events. Examining the ways in which insect outbreaks could affect competition may provide an entry to begin integrating these two perspectives.

10.4.1 Effects mediated by changes in plant quantity and quality

There is mixed support for the idea that outbreaks of insect herbivores increase competitive effects on other herbivores. For example, a multiyear analysis of insect community dynamics in several New York old-field communities suggested little evidence for competitive interactions among co-occurring insects, even for insects in the same guild (Root and Cappuccino 1992). These data suggest that insect herbivores in this system rarely reduce plant biomass sufficiently to cause strong interspecific competition, even during outbreak conditions. In contrast, outbreaks of forest tent caterpillars (*Malacosoma disstria*) reduced the densities of several species of galling and leaf-mining herbivores on quaking aspen (*Populus tremuloides*) in a Canadian forest (Roslin and Roland 2005). However, the responses of individual species were variable, with 3 of 10 species showing no significant responses, and one species that responded positively to prior outbreak defoliation. Conversely, gypsy moth outbreaks in a northern Michigan forest appeared to reduce the densities of other oak-feeding caterpillars, but did not affect caterpillar species that feed on other plants (Work and McCullough 2000). These results suggest potential competition mediated by changes in the quality or quantity of host plant resources available, but cannot exclude the possibility of apparent competition, or other indirect effects.

In many systems, competition among herbivores is influenced by induced changes in the quality of host plants as much as – or more than – changes in the quantity of host plant biomass (Williams and Myers 1984, Faeth 1986, Bryant *et al.* 1991, Work and McCullough 2000, Nykanen and Koricheva 2004, Kaplan and Denno 2007). In particular, the widespread induction of plant defensive traits following outbreak defoliation can have consequences for the persistence of the outbreak itself, as well as effects on other herbivores in the system. However, specific host plant responses to outbreak herbivory appear to be widely variable. In some cases, herbivory increases plant quality, while in other cases, herbivory decreases plant quality (e.g.,Williams and Myers 1984, Bryant *et al.* 1991, Kaplan and Denno 2007). For example, the outbreak defoliation of mountain birch (*Betula pubescens*) by autumnal moth (*Epirrita autumnata*) reduced leaf quality in the following year, probably due to delayed induced resistance traits (Kaitaniemi *et al.* 1999). While the effects of delayed induced resistance were probably insufficient by themselves to terminate outbreak dynamics in this system, they show the potential for self-limiting intraspecific competition. Similarly, an experimental study mimicking the outbreak-associated defoliation of red pine (*Pinus resinosa*) by pine sawflies (*Neodiprion lecontei*) in a Wisconsin plantation forest showed evidence for induced changes in foliar quality in the year following defoliation, but these effects were highly variable and dependent on specific characteristics of the herbivory and nutrient-specific measures of leaf quality (Raffa *et al.* 1998). Overall, the nonlinearity and contingency of these outbreak-mediated effects suggest that the consequences of induced responses to outbreaking herbivores may prove to be complex and specific.

10.4.2   Temporally asymmetric competition

In many systems, the competitive effects of outbreak consumption may occur most intensely after the outbreak itself has waned. These cases represent instances of temporally asymmetric competition (Lawton and Hassell 1981), mediated either through changes in plant quantity or quality. For example, outbreak densities of early-season grasshoppers in a Montana mixed-grass prairie site significantly reduced the survival of a later season grasshopper species (*Ageneotettix deorum*) by dramatically reducing the biomass of grasses (Branson 2010). Similarly, a combination of field and laboratory experiments suggest that high densities of woolly bear caterpillars (*Platyprepia virginalis*) on bush lupines (*Lupinus arboreous*) in the early spring reduced the subsequent growth rates of outbreaking tussock moths (*Orgyia vetusta*) later in the season due to delayed effects on plant quality (Harrison and Karban 1986). Such temporally asymmetric competitive interactions may be especially common in systems with short-duration or early-season outbreaks, though some degree of temporal asynchrony may be the rule rather than the exception (Lawton and Hassell 1981).

10.4.3   Two cautionary conclusions

While some studies have attempted to identify broad, predictive patterns among host plant responses to herbivory (e.g., Bryant *et al.* 1991), the range and variability of these responses are striking (Nykanen and Koricheva 2004, Kaplan and Denno 2007). Two conclusions from recent meta-analyses suggest caution in attempting to identify general patterns in the ecology of insect outbreaks. First, in an extensive meta-analysis of induced responses among woody plants, Nykänen and Koricheva (2004) show a disconcertingly strong temporal pattern by publication year for the reported effect sizes describing induced responses to herbivory; the authors suggest that these observed shifts in reported effects may correspond with paradigm shifts in the history of the field. If so, this result suggests that our ability to detect real generalities in nature may be blurred by the strength of our collective preconceptions. Second, Kaplan and Denno's (2007) meta-analysis of 333 herbivore–herbivore interactions in 145 published studies showed an unexpectedly weak and flat relationship between defoliation intensity and the effect sizes of interspecific competition. This result questions the fundamental assumption that competition should be most intense and most apparent following episodes of outbreak herbivory. While these two results should not discourage future attempts to identify outbreak-mediated effects on competition, they do suggest that even fundamental assumptions about competition should be considered with caution.

10.5 Outbreaking insects as resources: Do insect outbreaks increase resource availability?

In addition to their effects as consumers, ecosystem engineers, and competitors, outbreaking insects may also affect communities by increasing the availability of resources in the environment. This general pattern could occur through different mechanisms in both the above-ground and below-ground components of ecosystems.

10.5.1 Aboveground effects

Aboveground, insect outbreaks often represent superabundant resource pulses for insectivorous consumers (Yang *et al.* 2010). For example, synchronously emerging 17-year periodical cicadas (*Magicicada* spp.) in North American deciduous forests are consumed by many vertebrate and invertebrate species, including birds (Howard 1937, Steward *et al.* 1988, Kellner *et al.* 1990, Stephen *et al.* 1990, Koenig and Liebhold 2005), small mammals (Hahus and Smith 1990, Krohne *et al.* 1991), reptiles (Williams and Simon 1995), spiders (Smith *et al.* 1987), and insects (Williams and Simon 1995). Similar examples are numerous: California gulls (*Larus californicus*) opportunistically consume Mormon crickets (*Anabrus simplex*) during outbreak events in the North American Great Basin (Wakeland 1959), and insectivorous bats rapidly shift their diets to consume more western spruce budworm caterpillars (*Choristoneura occidentalis*) during outbreaks in British Columbia (Wilson and Barclay 2006). In a riparian habitat of Japan, aggregated outbreaks of leaf beetles are consumed by ladybird beetles (*Aiolocaria hexapilota*) and web spiders (*Agelena opulenta*, Nakamura *et al.* 2005).

While insect outbreaks seem to present a superabundance of resources, many outbreaking insects have defensive strategies to avoid predation. Caterpillars of the pine processionary moth (*Thaumetopoea pityocampa*) are well-defended with urticating hairs (Battisti *et al*. 2010) which strongly limit the predators that are able to consume them (Pimentel and Nilsson 2009, Barbaro and Battisti 2011). Although periodical cicadas are often described as "predator food-hardy" (Lloyd and Dybas 1966), the fly–squawk response of male cicadas substantially reduces the success of red-winged blackbird attacks. More generally, predator satiation may commonly limit predation rates during pulsed insect outbreaks. For example, gregarious-phase Mormon crickets (*Anabrus simplex*) form vast migrating bands that can extend for 10 km and contain millions of individuals (Simpson *et al*. 2006); these high-density traveling outbreaks effectively reduce per capita predation risk (Sword *et al*. 2005). Predator satiation also reduces the ability of consumers to control populations of periodical cicadas (Karban 1982, Williams *et al*. 1993) and may contribute to the stable spatial distribution of western tussock moth (*Orgyia vetusta*) outbreaks by reducing predation risk within established outbreak areas (Harrison and Wilcox 1995, Maron and Harrison 1997). In general, effective predator satiation during outbreak events could redirect pulsed resources to scavenging consumers and detrital pathways at the interface between above-ground and below-ground food webs (Yang 2004, Yang 2006, Nowlin *et al*. 2007).

While many outbreaking insects have strategies to avoid predation, many insectivorous consumers also employ strategies to capitalize on these ephemeral pulses of resource availability, including opportunistic diet shifts, aggregative and reproductive numerical response mechanisms, and the ability to anticipate or forecast future resource availability. Ovenbirds consumed spruce budworm during an outbreak event in Ontario, and showed reduced territory sizes and larger clutches during that period (Zach and Falls 1975). Similarly, both house sparrows (*Passer domesticus*) and Eurasian tree sparrows (*Passer montanus*) readily consumed periodical cicadas during an emergence event in Missouri, and showed significant increases in reproductive success compared to non-emergence years (Anderson 1977). Many consumers also show numeric responses to pulsed resources, including increased aggregation in areas of locally increased resource availability, increased reproduction via offspring production or survival, or a combination of these two consumer response mechanisms (Yang *et al*. 2010). In addition, some consumers of resource pulses may anticipate future outbreak events with increasing reproductive investments. Two recent studies suggest that white-footed mice used pre-emergence cues to forecast and anticipate cicada emergence events with greater investments of early-season reproductive effort (Marcello *et al*. 2008, Vandegrift and Hudson 2009).

In addition to their effects on consumer abundance, outbreak events may also affect the diversity of consumer guilds. For example, an 11-year data set documenting a large outbreak of mountain pine beetle (*Dendroctonus pondero-sae*) in British Columbia showed increases in both the abundance and local diversity of bark-feeding insectivore birds (Drever *et al*. 2009). Similarly, observational and experimental studies in a New Brunswick balsam fir insect community showed cascading increases in the abundance and diversity of mobile generalist consumers at higher trophic levels during spruce budworm (*Choristoneura fumiferana*) outbreaks (Eveleigh *et al*. 2007). These studies suggest insect outbreaks could allow a mechanistic approach to investigate how temporal variation in resource availability affects the species richness of consumer guilds.

Across the diversity of insect outbreaks in nature, the range and variety of outbreak–consumer interactions seem to include a confusing array of species-specific responses and exceptional cases. The effects of outbreak-mediated resource pulses on specific communities can be highly variable, and seem to be strongly influenced by specific characteristics of the resource, the consumer and the broader community context (Ostfeld and Keesing 2000, Yang *et al.* 2008, 2010). For example, a 30-year study of a Norwegian bird community that included multiple outbreaks of two geometrid moth species (*Epirrita autumnata* and *Operophtera brumata*) showed consistent, positive numerical responses of brambling (*Fringilla montifringilla*) populations to *E. autumnata* outbreaks, but weak population responses for seven other species of birds (Hogstad 2005). In this system, the unusual long-lived and nomadic life history of bramblings may make them particularly adept at responding to patchy insect outbreaks aggregatively and reproductively. Similarly, an analysis of North American Breeding Bird Surveys from 1966 to 2002 showed that 63% of bird species that were likely cicada consumers showed numerical responses that could be correlated with the emergence periodical cicadas, but these responses included both population increases and decreases, with effects of varying size, timing, and duration (Koenig and Liebhold 2005). Consumer responses to outbreaks are not always positive. For example, outbreaks of cockchafers (*Melolontha melolontha*) in a Slovakian agricultural landscape were readily consumed by lesser gray shrikes (*Lanius minor*), and cockchafer outbreak years were associated with earlier clutch initiation, larger clutch size, and larger nestlings than non-outbreak years. However, outbreak years were also associated with higher hatch failure rates, resulting in fledging success rates that were unchanged overall (Hoi *et al.* 2004).

Despite the apparent complexity and variety of outbreak–consumer interactions in nature, some general patterns may be emerging. Drawing on a meta-analysis of 189 pulsed resource–consumer interactions from 68 published sources, Yang *et al.* (2010) suggested that characteristics of the resource pulse (including its magnitude, trophic level, and duration), the consumer (including its body size, and whether its numeric responses are reproductive, aggregative, or both), and the surrounding community (including broad categories of habitat type) explained variation in the magnitude, timing, and duration of consumer responses to resource pulses observed across a wide range of systems. One of the particularly strong patterns to emerge from this analysis suggested that understanding whether consumer numerical responses are primarily reproductive, aggregative, or both provides a robust basis for predicting several aspects of the consumer response magnitude, duration, and lag. In the context of insect outbreaks, this analysis suggests the encouraging possibility that a relatively small number of key factors may allow insight into complex and species-specific responses to outbreaks.

10.5.2 Belowground effects

Insect outbreaks can also have strong effects on resource availability in the below-ground components of communities and ecosystems. Hunter (2001) outlined seven mechanisms by which insect herbivory can influence nutrient cycling: (1) direct inputs of frass, (2) direct inputs of dead bodies, (3) nutrient inputs dissolved in the precipitation falling through damaged vegetation (i.e., throughfall), (4) changes in litter quality, (5) changes in plant species composition, (6) changes in root exudates and below-ground symbioses, and (7) changes in habitat structure and associated edaphic factors. This list highlights the diversity of interacting mechanisms by which herbivory could affect nutrient cycling, but does not attempt to determine the direction and size of their combined effects on nutrient availability. In the case of insect outbreaks, many of the same processes that can increase nutrient cycling and availability in some systems could also contribute to nutrient losses and immobilization in others. Moreover, multiple interacting mechanisms could occur with opposing simultaneous effects, or sequential effects on different time scales.

Outbreak-mediated increases in nutrient availability have been observed in a wide range of natural systems. In a prairie grassland system, high densities of grasshoppers increased plant abundance, apparently by increasing nitrogen availability though accelerated litter cycling (Belovsky and Slade 2000). A similar pattern of increased nitrogen cycling with increasing herbivory was found in a Puerto Rican tropical forest; the authors suggested that herbivory could have potentially large effects on ecosystem dynamics if these observed effects scale to outbreak levels of defoliation (Fonte and Schowalter 2005). However, while high levels of herbivory in eucalyptus forests could promote greater productivity under certain circumstances, there doesn't appear to be a general pattern of outbreak-associated effects on nutrient cycling in these systems (Ohmart *et al.* 1983, Lamb 1985). At high densities, direct inputs of dead insect bodies can also increase nutrient availability in the soil, with effects on microbes and plants (e.g., Yang 2004, Gratton *et al.* 2008). Similar increases in nutrient availability have also been observed in aquatic systems. For example, Carlton and Goldman (1984) observed marked increases in nitrogen availability and primary productivity in a California oligotrophic subalpine lake following a massive deposition of alate ants (*Lasius alienus*) during a mating swarm. Similar increases in nutrient availability and phytoplankton have been observed following the allochthonous input of periodical cicada bodies in woodland ponds and streams (Nowlin *et al.* 2007).

These examples suggest that outbreaks generally tend to accelerate nutrient cycling and availability in the short term, with positive effects on primary productivity. However, insect outbreaks can also cause net system losses due to the rapid export and loss of dissolved nutrients. For example, an outbreak of fall cankerworms (*Alsophila pometaria*) in a mixed hardwood forest of North Carolina increased nitrogen losses during periods of heavy defoliation, probably due to the export of labile nutrients from frass and litter inputs (Swank *et al.* 1981). In a California oak–grassland, an outbreak of California oak moth (*Phryganidia californica*) on two native oaks (*Quercus agrifolia* and *Q. lobata*) dramatically increased the movement of nitrogen and phosphorus into the soil, primarily as accumulated insect frass and remains (Hollinger 1986). However, these nutrients were rapidly mobilized during the first seasonal rains, probably resulting in a brief pulse of dissolved nutrient availability in the short term, followed by a net loss of nutrients from the system (Hollinger 1986). In this system, the nutrients mobilized as frass and insect remains during outbreak events represented a substantial fraction of the total nitrogen and phosphorus inputs in this system, suggesting that even occasional outbreak events could strongly influence nutrient dynamics.

Instead of increasing nutrient losses, insect outbreaks may also increase rates of net immobilization as available nutrients are assimilated into soil microbial pools (Bardgett 2005). In particular, a number of studies suggest that frass-derived nitrogen is commonly immobilized in the soil, limiting nitrogen availability for primary productivity, but also limiting leaching losses from the system (Lovett and Ruesink 1995, Christenson *et al*. 2002). Lovett *et al*. (2002) suggest that while outbreaks can create system-wide nutrient losses under certain conditions (e.g., tree mortality, or thin soils; Lovett *et al*. 2002) or in particular systems (e.g., the combination of rapid leaching, strongly seasonal rainfall observed in the California oak moth system; Hollinger 1986), most outbreak events probably lead to the redistribution of, rather than the loss of, nitrogen from ecosystems.

Outbreaking insects may also have strong and lasting indirect effects on nutrient cycling mediated though changes in plant species composition. If the direct negative effects of outbreak herbivory are generally focused on palatable plant species, insect outbreaks can shift plant community composition to species with lower quality or more highly defended tissues that tend to decompose more slowly (Hunter 2001, Weisser and Siemann 2004). This suggests one potentially general pattern: if selective herbivory on higher quality plants increases the proportion of slow-decomposing plant species in the community, this herbivory could reduce nutrient availability over the long term. However, this generalization relies on assumptions about the nature of selective herbivory and the associated decomposition characteristics of the resulting plant community. Not all outbreaking herbivores follow these assumptions, and the resulting changes in plant community structure can have profound (and less predictable) effects. For example, outbreaks of invasive hemlock wooly adelgids (*Adelgies tsugae*) on eastern hemlocks (*Tsuga canadensis*) in North American forests can increase nutrient cycling in the short term though direct increases the quality and quantity of hemlock litter inputs, but larger changes in nutrient cycling emerge in the long term through shifts in forest structure and composition (Jenkins *et al*. 1999, Cobb 2010). In a Massachusetts forest, adelgid outbreaks cause slow-cycling hemlock stands to be replaced by black birch (*Betula lenta*), whose more labile litter results in faster decomposition and accelerated nitrogen cycling (Cobb 2010). This pattern differs from the prediction of decreased long-term nutrient cycling based on simple assumptions of selective herbivory, suggesting that our ability to generalize about the system-specific effects of selective herbivory may be limited.

Recognizing the multitude of ways by which outbreak herbivory can increase or decrease nutrient availability seems to suggest that few general patterns exist. However, such specific exceptions could serve to refine our generalizations, rather than erode them; the observed range of possibilities could reflect meaningful differences in time scale (i.e., separating the immediate effects of outbreak-associated detrital inputs from the longer time scale effects of species composition shifts), or perhaps the key differences between the effects of selective herbivory in specialists and generalist herbivores.

10.6   Key themes and future directions

Ecologists have long suggested that outbreak herbivory may play an important role in regulating the productivity and diversity of natural systems (e.g., see Mattson and Addy 1975, Romme *et al.* 1986, Schowalter *et al.* 1986, Haack and Byler 1993, Carson and Root 2000, Hunter 2001, McCullough *et al.* 2010). The mechanisms of these effects are potentially wide-ranging, including virtually all of the various ways in which outbreaking insects affect communities as consumers, ecosystem engineers, competitors, and resources. However, insect outbreaks do not always regulate primary productivity or maintain species coexistence, and often seem to have destabilizing effects. Given the wide range of mechanisms and consequences that are linked with insect outbreaks, how should we understand the effects of insect outbreaks on the productivity and diversity of communities? Have any general patterns emerged?

10.6.1   Multiple interacting mechanisms

One readily evident theme is that insect outbreaks affect communities through multiple, complex, and sometimes counter-acting mechanisms. These multiple and interacting mechanisms probably explain some of the specificity in the cumulative effects of outbreak events. For example, an analysis of Breeding Bird Census data from six US sites showed variable but predictable responses to gypsy moth outbreaks (Gale *et al.* 2001). While cuckoos (*Coccyzus* spp.) readily consumed gypsy moth caterpillars and showed consistent population increases during outbreak years, other species that favored more open habitats also tended to increase during outbreak years, while species that favored closed-canopy forests were negatively affected during gypsy moth outbreaks. This study suggests that gypsy moth outbreaks affect bird communities via multiple mechanisms, including bottom-up effects as resource pulses, and both positive and negative effects via habitat modification. Similarly, forest bird communities in mixed coniferous–deciduous forests of British Columbia respond to multiple, interacting resource pulses resulting from mountain pine beetle (*Dendroctonus ponderosae*) outbreaks (Norris and Martin 2010). Mountain pine beetle outbreaks provide superabundant food for several insectivorous bird species, including red-breasted nuthatches (*Sitta canadensis*), mountain chickadees (*Poecile gambeli*), black-capped chickadees (*Poecile atricapillus*), and downy woodpeckers (*Picoides pubescens*). However, these outbreaks can also increase the number of dead trees available for birds that actively excavate cavity nests, with secondary effects for the bird species that typically reuse abandoned cavity nests in subsequent years (Norris and Martin 2010). These "dual resource pulses" (i.e., insect prey and cavity sites) may affect habitat selection and species interactions among cavity-nesting birds in complex ways, as the outbreak alters the availability of two key resources in the system (Norris and Martin 2008).

The emerging picture of outbreak-mediated effects on plant communities and primary productivity also balances multiple and often conflicting process. In order to investigate some of the factors which determine the cumulative effects of herbivory on primary productivity, de Mazancourt and Loreau (2000) developed an equilibrium-based model that considered important herbivore-mediated effects, such as direct consumption and altered plant species composition. This model suggested that herbivory can increase primary productivity at the community level under a wide range of circumstances, even with simultaneous (and potentially counteracting) changes in plant species composition. However, these conclusions are predicated on an equilibrium model that does not consider transient dynamics, such as the herbivory-mediated effects on nutrient cycling that commonly occur in outbreaking systems (de Mazancourt *et al.* 1998). While integrative models such as these have great promise to clarify our understanding of the multiple processes associated with outbreak events, many questions about the multiple and interacting effects of insect outbreaks remain unanswered.

10.6.2   Outbreaks are not created equal

Insect outbreaks are exceptional events whose specific characteristics (such as the species involved and climatic conditions) can determine their ecological consequences. Even within specific systems, variation in the magnitude, frequency, duration, and spatial extent of individual outbreak events may be key factors modulating the community effects of an outbreak. This event specificity suggests that it may be difficult to arrive at general patterns that apply to all outbreak events. However, this complexity and variability among individual events could also provide a powerful means of separating the key factors that influence the ecological consequences of outbreak events through efforts at broader synthesis (e.g., Carson *et al.* 2004, Yang *et al.* 2010), possibility allowing insight into the diversity of community responses. For example, Carson *et al.* (2004) suggested a "host concentration hypothesis" whereby insect outbreaks generally have larger per capita effects on plant communities when their host plants form large, persistent, and dense stands, while Yang *et al.* (2010) used a meta-analysis approach to identify factors that explain variation in community responses to diverse resource pulses, including insect outbreaks. While these studies use very different approaches, both draw upon the diversity of outbreak phenomena to identify key factors that could explain community responses to these seemingly rare or extreme events.

  While it is useful to look for general patterns among the diversity of outbreak phenomena in nature, many insect outbreaks and their consequences do not fit easily into semantic boundaries. For example, outbreaks of the 17-year periodical cicada do not correspond to the usual paradigm of multigenerational population dynamics. Instead, these outbreaks represent the conspicuous adult stage of a single massive generation that is otherwise feeding on plant roots below-ground (Williams and Simon 1995). Although driven by a complex life history rather than multigenerational population dynamics, these events are meaningfully understood as outbreaks in the sense that above-ground communities experience a rapid increase in the abundance of insects, with consumers experiencing many of the same general mechanisms that occur in more conventional insect outbreaks. Other systems stretch the conventional outbreak concept even further: while this chapter has focused on outbreaks of herbivores, outbreaks may also occur at other trophic levels. For example, outbreaks of detritivorous periodical train millipedes occur at 8-year intervals in Japan, causing increased litter decomposition, bioturbation, and carbon sequestration (Niijima 1998, Hashimoto *et al.* 2004, Toyota *et al.* 2006). Similarly, outbreaks of prey species commonly cause indirect increases in the densities of their predators (Yang *et al.* 2008, 2010); in these cases, we may see "outbreaks" of increased predator densities following herbivore outbreaks. Given the "exceptional nature of nature" (R.B. Root, personal communication; Bartholomew 1982), such system-specific idiosyncrasies are likely to be more the rule than the exception. Any attempt to develop a general understanding of the consequences of outbreak phenomena will undoubtedly be challenged by, and perhaps benefit from, this diversity.

10.6.3   Spatial and temporal scale

The ultimate effects of insect outbreaks also vary with spatial and temporal scales. For example, outbreak defoliation events that reduce species richness and productivity over the short-term or at local spatial scales can increase species richness and primary productivity over the long term or at larger spatial scales. While the immediate effects of outbreaking insects as consumers may be most apparent in the short term, they may have more persistent effects via other, less conspicuous mechanisms. Several authors have emphasized the importance of considering outbreak effects at multiple spatial and temporal scales (e.g., Schowalter *et al.* 1986, de Grandpre and Bergeron 1997, Carson and Root 2000, Uriarte 2000, Hunter 2001). Often, some of the most profound effects of insect outbreaks may be manifest as increased heterogeneity (e.g., a successional mosaic) observed over long time scales and large spatial scales.

The possible evolutionary consequences of episodic outbreak phenomena remain little known, but outbreak events could represent brief episodes of strong selection for both host plants and the insectivorous consumers of outbreaking insects. For example, the apparent evolution of anticipatory reproductive responses among consumers of insect outbreaks (Marcello *et al.* 2008, Vandegrift and Hudson 2009) suggests that such episodic selection events could have long-term consequences. Changes in habitat structure could also have lasting effects on the selective environment following outbreak events. Future studies will be necessary to better understand how these periods of potentially strong selection could affect the combined ecology and evolution of populations (Schoener 2011).

10.6.4   When do we see resilience?

In future studies, it may be useful to consider whether specific characteristics of the outbreaking species or the surrounding community are associated with greater or lesser community resilience to outbreak perturbations. This question provides a counterpoint to long-standing questions about the ability of outbreak phenomena to regulate natural systems; instead of focusing on the ability of the outbreaking insects to drive community dynamics, it asks about how communities return to a pre-outbreak state following strong perturbations.

Some examples from the literature suggest that communities may show greater resilience to outbreaks of co-evolved species than to outbreaks of exotic species. For example, many plants are able to tolerate very high levels of outbreak defoliation from native herbivores with limited or transient effects on growth or survival (e.g., Hoogesteger and Karlsson 1992, Harrison and Maron 1995), while outbreaks of hemlock wooly adelgids commonly lead to high rates of host mortality, with large and persistent community consequences (e.g., Jenkins *et al.* 1999, Cobb 2010). If this pattern is general, it would suggest an insight that is simultaneously profound and trivial: the communities and ecosystems that we observe today show resilience to co-evolved perturbations – even severe outbreak events – because past community configurations that lacked sufficient resilience were not able to persist. Thus, our current co-evolved communities have been "pre-filtered" by their ecological history; their past resilience allowed them to persist to the present day. By comparison, the introduction of exotic species has resulted in communities and ecosystems that lack a shared co-evolutionary history, and thus have not been pre-filtered for their resilience characteristics. This hypothesis of co-evolved community resilience suggests a somewhat discouraging conclusion, namely, that a co-evolved history may be one of the only characteristics can reliably predict the resilience characteristics of a complex system. Given the complexity of species interactions and the "indeterminacy" of indirect effects (Yodzis 1988), if resilience is a prerequisite for persistence, then past persistence may be our best predictor of future resilience.

10.6.5 Non-equilibrium dynamics and the effects of climate change

Ecologists have shifted from a primarily equilibrium view of communities to increasingly emphasize the importance of transient, event-driven dynamics (Hastings 2001, 2004). The study of insect outbreaks has reflected these broader shifts; while early studies sought to understand the causes of outbreak events as exogenous deviations from equilibrium or as stable limit cycles, more recent work is increasingly investigating the complex transient consequences of these events. Continuing this trend, future studies will likely place greater emphasis on understanding the consequences of different event regimes, building on our greater understanding of single-event dynamics to incorporate multi-event series, including aspects of outbreak frequency, intensity, duration, and variation in inter-outbreak intervals.

While it would be a mistake to understand outbreaks as solely as destructive deviations from an equilibrium state, ongoing climate change suggests that we may be entering an unfamiliar era in which the dynamics and consequences of outbreak regimes may be more variable and less predictable (Raffa *et al.* 2008). Climatic models now commonly predict extreme weather events of greater intensity and frequency in the future (Easterling *et al.* 2000, Parmesan *et al.* 2000, Pachauri and Intergovernmental Panel on Climate Change 2008). The ecological consequences of these changes for the resilience of ecosystems are not yet clear, but we should consider the possibility that we are entering a realm of uncharted consequences, where even past co-evolved history may no longer provide a reliable indication for future resilience and persistence. For the ecology of insect outbreaks, the uncertainty of our shared climatic future could be changing the rules in natural systems even as we seek to understand and identify them.

10.7 Conclusions

The community consequences of insect outbreaks seem unlikely to be predicted by simple and general rules because insect outbreaks commonly have multiple and interacting mechanisms of effects. The challenge, then, is to understand how multiple kinds of outbreak effects combine. Considering the multiple and interacting ways that outbreaking insects affect communities as consumers, ecosystem engineers, mediators of resource availability, and competitors may help ecologists identify general patterns in the community consequences of insect outbreaks. This perspective encourages both mechanistic studies that attempt to evaluate the relative importance of multiple mechanisms, and efforts at broad synthesis that consider the multiple interacting effects of insect outbreaks on communities.

Acknowledgments

This chapter greatly benefited from the comments and suggestions kindly provided by Deborah Letourneau, Dan Gruner, Rick Karban, and Chelse Prather. Additional thanks to Evan Preisser, Lee Dyer, and Angela Smilanich for their valuable suggestions on references. Finally, thanks to Pedro Barbosa for the encouragement to write this chapter.

References

Anderson, T. R. 1977. Reproductive responses of sparrows to a superabundant food supply. *The Condor* 79:205–208.

Barbaro, L., and A. Battisti. 2011. Birds as predators of the pine processionary moth (Lepidoptera: Notodontidae). *Biological Control* 56:107–114.

Barbosa, P., and J. C. Schultz. 1987. *Insect Outbreaks*. Academic Press.

Bardgett, R. D. 2005. *The Biology of Soil: A Community and Ecosystem Approach*. Oxford University Press, New York.

Bartholomew, G. A. 1982. Scientific innovation and creativity: a zoologist's point of view. *American Zoologist* 22:227–235.

Battisti, A., G. Holm, B. Fagrell, and S. Larsson. 2010. Urticating hairs in arthropods: their nature and medical significance. *Annual Review of Entomology* 56:203–220.

Bell, J. L., L. C. Sloan, and M. A. Snyder. 2004. Regional changes in extreme climatic events: a future climate scenario. *Journal of Climate* 17:81–87.

Bell, J. L., and R. C. Whitmore. 1997. Eastern Towhee numbers increase following defoliation by gypsy moths. *The Auk* 114:708–716.

Belovsky, G. E., and J. B. Slade. 2000. Insect herbivory accelerates nutrient cycling and increases plant production. *Proceedings of the National Academy of Sciences of the USA* 97:14412–14417.

Bergeron, Y., and A. Leduc. 1998. Relationships between change in fire frequency and mortality due to spruce budworm outbreak in the southeastern Canadian boreal forest. *Journal of Vegetation Science* 9:493–500.

Branson, D. 2010. Density-dependent effects of an early season insect herbivore on a later developing insect herbivore. *Environmental Entomology* 39:346–350.

Brown, D. G. 1994. Beetle folivory increases resource availability and alters plant invasion in monocultures of goldenrod. *Ecology* 75:1673–1683.

Bryant, J. P., I. Heitkonig, P. Kuropat, and N. Owen-Smith. 1991. Effects of severe defoliation on the long-term resistance to insect attack and on leaf chemistry in six woody species of the southern African savanna. *American Naturalist* 137:50–63.

Carlton, R. G., and C. R. Goldman. 1984. Effects of a massive swarm of ants on ammonium concentrations in a subalpine lake. *Hydrobiologia* 111:113–117.

Carson, W. P., J. P. Cronin, and Z. T. Long. 2004. A general rule for predicting when insects will have strong top-down effects on plant communities: on the relationship between insect communities and host concentration. Pages 27–52 *Insects and Ecosystem Function*. Springer-Verlag, Berlin.

Carson, W. P., and R. B. Root. 2000. Herbivory and plant species coexistence: community regulation by an outbreaking phytophagous insect. *Ecological Monographs* 70:73–99.

Chen, X., and J. E. Cohen. 2001. Transient dynamics and food-web complexity in the Lotka–Volterra cascade model. *Proceedings of the Royal Society of London Series B-Biological Sciences* 268:869–877.

Christenson, L. M., G. M. Lovett, M. J. Mitchell, and P. M. Groffman. 2002. The fate of nitrogen in gypsy moth frass deposited to an oak forest floor. *Oecologia* 131:444–452.

Cobb, R. 2010. Species shift drives decomposition rates following invasion by hemlock woolly adelgid. *Oikos* 119:1291–1298.

Coupe, M. D., and J. F. Cahill. 2003. Effects of insects on primary production in temperate herbaceous communities: a meta-analysis. *Ecological Entomology* 28:511–521.

Criddle, N. 1926. The life-history and habits of *Anabrus longipes* Caudell (Orthop). *Canadian Entomologist* 58:261–265.

de Grandpre, L., and Y. Bergeron. 1997. Diversity and stability of understorey communities following disturbance in the southern boreal forest. *Journal of Ecology* 85:777–784.

de Mazancourt, C., and M. Loreau. 2000. Effect of herbivory and plant species replacement on primary production. *American Naturalist* 155:735–754.

de Mazancourt, C., M. Loreau, and L. Abbadie. 1998. Grazing optimization and nutrient cycling: when do herbivores enhance plant production? *Ecology* 79:2242–2252.

Denno, R. F., M. S. McClure, and J. R. Ott. 1995. Interspecific interactions in phytophagous insects: competition reexamined and resurrected. *Annual Review of Entomology* 40:297–331.

Drever, M., J. Goheen, and K. Martin. 2009. Species-energy theory, pulsed resources, and regulation of avian richness during a mountain pine beetle outbreak. *Ecology* 90:1095–1105.

Easterling, D. R., G. A. Meehl, C. Parmesan, S. A. Changnon, T. R. Karl, and L. O. Mearns. 2000. Climate extremes: observations, modeling, and impacts. *Science* 289:2068–2074.

Esper, J., U. Büntgen, D. C. Frank, D. Nievergelt, and A. Liebhold. 2007. 1200 years of regular outbreaks in alpine insects. *Proceedings of the Royal Society B: Biological Sciences* 274:671–679.

Eveleigh, E. S., K. S. McCann, P. C. McCarthy, S. J. Pollock, C. J. Lucarotti, B. Morin, G. A. McDougall, D. B. Strongman, J. T. Huber, J. Umbanhowar, and L. D. B. Faria. 2007. Fluctuations in density of an outbreak species drive diversity cascades in food webs. *Proceedings of the National Academy of Sciences USA* 104:16976–16981.

Faeth, S. H. 1986. Indirect interactions between temporally separated herbivores mediated by the host plant. *Ecology* 67:479–494.

Fonte, S. J., and T. D. Schowalter. 2005. The influence of a neotropical herbivore (*Lamponius portoricensis*) on nutrient cycling and soil processes. *Oecologia* 146:423–431.

Gale, G. A., J. A. DeCecco, M. R. Marshall, W. R. McClain, and R. J. Cooper. 2001. Effects of gypsy moth defoliation on forest birds: an assessment using breeding bird census data. *Journal of Field Ornithology* 291–304.

Gratton, C., J. Donaldson, and M. J. V. Zanden. 2008. Ecosystem linkages between lakes and the surrounding terrestrial landscape in northeast Iceland. *Ecosystems* 11:764–774.

Haack, R. A., and J. W. Byler. 1993. Insects and pathogens: regulators of forest ecosystems. *Journal of Forestry* 91.

Hahus, S. C., and K. G. Smith. 1990. Food habits of *Blarina*, *Peromyscus* and *Microtus* in relation to an emergence of periodical cicadas *Magicicada*. *Journal of Mammalogy* 71:249–252.

Harrison, S., and R. Karban. 1986. Effects of an early-season folivorous moth on the success of a later-season species, mediated by a change in the quality of the shared host, *Lupinus arboreus* Sims. *Oecologia* 69:354–359.

Harrison, S., and J. L. Maron. 1995. Impacts of defoliation by tussock moths (*Orgyia vetusta*) on the growth and reproduction of bush lupine (*Lupinus arboreus*). *Ecological Entomology* 20:223–229.

Harrison, S., and C. Wilcox. 1995. Evidence that predator satiation may restrict the spatial spread of a tussock moth (*Orgyia vetusta*) outbreak. *Oecologia* 101:309–316.

Hashimoto, M., N. Kaneko, M. T. Ito, and A. Toyota. 2004. Exploitation of litter and soil by the train millipede *Parafontaria laminata* (Diplopoda: Xystodesmidae) in larch plantation forests in Japan. *Pedobiologia* 48:71–81.

Hastings, A. 2001. Transient dynamics and persistence of ecological systems. *Ecology Letters* 4:215–220.

Hastings, A. 2004. Transients: the key to long-term ecological understanding? *Trends in Ecology & Evolution* 19:39–45.

Hogstad, O. 2005. Numerical and functional responses of breeding passerine species to mass occurrence of Geometrid caterpillars in a subalpine birch forest: a 30-year study. *Ibis* 147:77–91.

Hoi, H., A. Kristin, F. Valera, and C. Hoi. 2004. Clutch enlargement in Lesser Gray Shrikes (*Lanius minor*) in Slovakia when food is superabundant: a maladaptive response? *Auk* 121:557–564.

Hollinger, D. Y. 1986. Herbivory and the cycling of nitrogen and phosphorus in isolated California oak trees. *Oecologia* 70:291–297.

Hoogesteger, J., and P. S. Karlsson. 1992. Effects of defoliation on radial stem growth and photosynthesis in the mountain birch (*Betula pubescens* ssp. *tortuosa*). *Functional Ecology* 6:317–323.

Howard, W. 1937. Bird behavior as a result of emergence of seventeen year locusts. *The Wilson Bulletin* 43–44.

Hunter, M. D. 2001. Insect population dynamics meets ecosystem ecology: effects of herbivory on soil nutrient dynamics. *Agricultural and Forest Entomology* 3:77–84.

Jardon, Y., L. Filion, and C. Cloutier. 1994. Long-term impact of insect defoliation on growth and mortality of eastern larch in boreal Québec. *Ecoscience Sainte-Foy* 1:231–238.

Jenkins, J. C., J. D. Aber, and C. D. Canham. 1999. Hemlock woolly adelgid impacts on community structure and N cycling rates in eastern hemlock forests. *Canadian Journal of Forest Research* 29:630–645.

Jentsch, A., J. Kreyling, J. Boettcher-Treschkow, and C. Beierkuhnlein. 2009. Beyond gradual warming: extreme weather events alter flower phenology of European grassland and heath species. *Global Change Biology* 15:837–849.

Jones, C. G., J. H. Lawton, and M. Shachak. 1994. Organisms as ecosystem engineers. *Oikos* 69:373–386.

Jones, C. G., J. H. Lawton, and M. Shachak. 1997. Positive and negative effects of organisms as physical ecosystem engineers. *Ecology* 78:1946–1957.

Kaitaniemi, P., K. RuohomaKI, T. Tammaru, and E. Haukioja. 1999. Induced resistance of host tree foliage during and after a natural insect outbreak. *Journal of Animal Ecology* 68:382–389.

Kaplan, I., and R. Denno. 2007. Interspecific interactions in phytophagous insects revisited: a quantitative assessment of competition theory. *Ecology Letters* 10:977–994.

Karban, R. 1982. Increased reproductive success at high densities and predator satiation for periodical cicadas. *Ecology* 63:321–328.

Kellner, C. J., K. G. Smith, N. C. Wilkinson, and D. A. James. 1990. Influence of periodical cicadas on foraging behavior of insectivorous birds in an Ozark forest. *Studies of Avian Biology* 13:375–380.

Koenig, W. D., and A. M. Liebhold. 2005. Effects of periodical cicada emergences on abundance and synchrony of avian populations. *Ecology* 86:1873–1882.

Krohne, D. T., T. J. Couillard, and J. C. Riddle. 1991. Population responses of *Peromyscus leucopus* and *Blarina brevicauda* to emergence of periodical cicadas. *American Midland Naturalist* 126:317–321.

Kulman, H. M. 1971. Effects of insect defoliation on growth and mortality of trees. *Annual Review of Entomology* 16:289–324.

Lamb, D. 1985. The influence of insects on nutrient cycling in eucalypt forests: a beneficial role? *Australian Journal of Ecology* 10:1–5.

Landsberg, J., and C. Ohmart. 1989. Levels of insect defoliation in forests: patterns and concepts. *Trends in Ecology & Evolution* 4:96–100.

Lawton, J. H., and D. R. Strong. 1981. Community patterns and competition in folivorous insects. *American Naturalist* 118:317–338.

Lawton, J. H., and M. P. Hassell. 1981. Asymmetrical competition in insects. *Nature* 289:793–795.

Lloyd, M., and H. S. Dybas. 1966. The periodical cicada problem. II. Evolution. *Evolution* 466–505.

Lovett, G. M., and A. E. Ruesink. 1995. Carbon and nitrogen mineralization from decomposing gypsy-moth frass. *Oecologia* 104:133–138.

Lovett, G. M., L. M. Christenson, P. M. Groffman, C. G. Jones, J. E. Hart, and M. J. Mitchell. 2002. Insect defoliation and nitrogen cycling in forests. *Bioscience* 52:335–341.

Marcello, G., S. Wilder, and D. Meikle. 2008. Population dynamics of a generalist rodent in relation to variability in pulsed food resources in a fragmented landscape. *Journal of Animal Ecology* 77:41–46.

Maron, J. L., and S. Harrison. 1997. Spatial pattern formation in an insect host–parasitoid system. *Science* 278:1619–1621.

Maron, J. L., and R. L. Jefferies. 1999. Bush lupine mortality, altered resource availability, and alternative vegetation states. *Ecology* 80:443–454.

Mattson, W. J., and N. D. Addy. 1975. Phytophagous insects as regulators of forest primary production. *Science* 190:515–522.

McCullough, D. G., R. A. Werner, and D. Neumann. 2010. Fire and insects in northern and boreal forest ecosystems of North America. *Annual Review of Entomology* 43:107–127.

Meehl, G. A., T. Karl, D. R. Easterling, S. Changnon, R. Pielke, D. Changnon, J. Evans, P. Y. Groisman, T. R. Knutson, K. E. Kunkel, L. O. Mearns, C. Parmesan, R. Pulwarty, T. Root, R. T. Sylves, P. Whetton, and F. Zwiers. 2000. An introduction to trends in extreme weather and climate events: observations, socio-economic impacts, terrestrial ecological impacts, and model projections. *Bulletin of the American Meteorological Society* 81:413–416.

Müller, J., H. Bußler, M. Goßner, T. Rettelbach, and P. Duelli. 2008. The European spruce bark beetle *Ips typographus* in a national park: from pest to keystone species. *Biodiversity and Conservation* 17:2979–3001.

Nakamura, M., S. Utsumi, T. Miki, and T. Ohgushi. 2005. Flood initiates bottom-up cascades in a tri-trophic system: host plant regrowth increases densities of a leaf beetle and its predators. *Journal of Animal Ecology* 74:683–691.

Niijima, K. 1998. Effects of outbreak of the train millipede *Parafontaria laminata armigera* Verhoeff (Diplopoda: Xystodesmidae) on litter decomposition in a natural beech forest in central Japan. 1. Density and biomass of soil invertebrates. *Ecological Research* 13:41–53.

Norris, A. R., and K. Martin. 2008. Mountain pine beetle presence affects nest patch choice of red-breasted nuthatches. *Journal of Wildlife Management* 72:733–737.

Norris, A., and K. Martin. 2010. The perils of plasticity: dual resource pulses increase facilitation but destabilize populations of small-bodied cavity-nesters. *Oikos* 119:1126–1135.

Nowlin, W. H., M. J. González, M. J. Vanni, M. H. Stevens, M. W. Fields, and J. J. Valente. 2007. Allochthonous subsidy of periodical cicadas affects the dynamics and stabilty of pond communities. *Ecology* 88:2174–2186.

Nykanen, H., and J. Koricheva. 2004. Damage-induced changes in woody plants and their effects on insect herbivore performance: a meta-analysis. *Oikos* 104:247–268.

Ohmart, C. P., L. G. Stewart, and J. R. Thomas. 1983. Leaf consumption by insects in 3 eucalyptus forest types in southeastern Australia and their role in short-term nutrient cycling. *Oecologia* 59:322–330.

Ostfeld, R. S., and F. Keesing. 2000. Pulsed resources and community dynamics of consumers in terrestrial ecosystems. *Trends in Ecology & Evolution* 15:232–237.

Pachauri, R., and Intergovernmental Panel on Climate Change. 2008. *Climate Change 2007 Synthesis Report*. IPCC, Geneva, Switzerland.

Parmesan, C., T. L. Root, and M. R. Willig. 2000. Impacts of extreme weather and climate on terrestrial biota. *Bulletin of the American Meteorological Society* 81:443–450.

Piene, H. 1980. Effects of insect defoliation on growth and foliar nutrients of young balsam fir. *Forest Science* 26:665–673.

Pimentel, C., and J. Nilsson. 2009. Response of passerine birds to an irruption of a pine processionary moth *Thaumetopoea pityocampa* population with a shifted phenology. *Ardeola* 56:189–203.

Raffa, K. F., B. H. Aukema, B. J. Bentz, A. L. Carroll, J. A. Hicke, M. G. Turner, and W. H. Romme. 2008. Cross-scale drivers of natural disturbances prone to anthropogenic amplification: the dynamics of bark beetle eruptions. *Bioscience* 58:501–517.

Raffa, K. F., S. C. Krause, and P. B. Reich. 1998. Long-term effects of defoliation on red pine suitability to insects feeding on diverse plant tissues. *Ecology* 79:2352–2364.

Redak, R. A., J. L. Capinera, and C. D. Bonham. 1992. Effects of sagebrush removal and herbivory by Mormon crickets (Orthoptera: Tettigoniidae) on understory plant biomass and cover. *Environmental Entomology* 21:94–102.

Romme, W. H., D. H. Knight, and J. B. Yavitt. 1986. Mountain pine beetle outbreaks in the Rocky Mountains: regulators of primary productivity? *American Naturalist* 484–494.

Root, R. B., and N. Cappuccino. 1992. Patterns in population change and the organization of the insect community associated with goldenrod. *Ecological Monographs* 62:393–420.

Roslin, T., and J. Roland. 2005. Competitive effects of the forest tent caterpillar on the gallers and leaf-miners of trembling aspen. *Ecoscience* 12:172–182.

Schoener, T. W. 1982. The controversy over interspecific competition. *American Scientist* 70:586–595.

Schoener, T. W. 2011. The newest synthesis: understanding the interplay of evolutionary and ecological dynamics. *Science* 331:426–429.

Schowalter, T. D., W. W. Hargrove, and D. A. Crossley. 1986. Herbivory in forested ecosystems. *Annual Review of Entomology* 31:177–196.

Simpson, S. J., G. A. Sword, P. D. Lorch, and I. D. Couzin. 2006. Cannibal crickets on a forced march for protein and salt. *Proceedings of the National Academy of Sciences of the USA* 103:4152–4156.

Smith, K. G., N. C. Wilkinson, K. S. Williams, and V. B. Steward. 1987. Predation by spiders on periodical cicadas (Homoptera, *Magicicada*). *Journal of Arachnology* 15:277–279.

Stephen, F. M., G. W. Wallis, and K. G. Smith. 1990. Bird predation on periodical cicadas in Ozark forests: ecological release for other canopy arthropods. *Studies in Avian Biology* 13:369–374.

Steward, V. B., K. G. Smith, and F. M. Stephen. 1988. Red-winged blackbird predation on periodical cicadas (Cicadidae: *Magicicada* spp.): bird behavior and cicada responses. *Oecologia* 76:348–352.

Strong, D. R., J. L. Maron, P. G. Connors, A. Whipple, S. Harrison, and R. L. Jefferies. 1995. High mortality, fluctuation in numbers, and heavy subterranean insect herbivory in bush lupine, *Lupinus arboreus. Oecologia* 104:85–92.

Swank, W. T., J. B. Waide, D. A. Crossley, and R. L. Todd. 1981. Insect defoliation enhances nitrate export from forest ecosystems. *Oecologia* 51:297–299.

Sword, G. A., P. D. Lorch, and D. T. Gwynne. 2005. Migratory bands give crickets protection. *Nature* 433:703.

Toyota, A., N. Kaneko, and M. T. Ito. 2006. Soil ecosystem engineering by the train millipede *Parafontaria laminata* in a Japanese larch forest. *Soil Biology and Biochemistry* 38:1840–1850.

Uriarte, M. 2000. Interactions between goldenrod (*Solidago altissima* L.) and its insect herbivore (*Trirhabda virgata*) over the course of succession. *Oecologia* 122: 521–528.

Vandegrift, K., and P. Hudson. 2009. Response to enrichment, type and timing: small mammals vary in their response to a springtime cicada but not a carbohydrate pulse. *Journal of Animal Ecology* 78:202–209.

Wakeland, C. 1959. *Mormon Crickets in North America*. USDA Agricultural Research Service Technical Bulletin 1202. USDA, Washington, DC.

Weisser, W. W., and E. Siemann. 2004. The various effects of insects on ecosystem functioning. Pages 3–24 *Insects and Ecosystem Function*. Springer-Verlag.

Wickman, B. E. 1980. Increased growth of white fir after a Douglas-fir tussock moth outbreak. *Journal of Forestry* 78:31–33.

Wiens, J. A. 1977. Competition and variable environments. *American Scientist* 65:590–597.

Williams, K. S., and J. H. Myers. 1984. Previous herbivore attack of red alder may improve food quality for fall webworm larvae. *Oecologia* 63:166–170.

Williams, K. S., and C. Simon. 1995. The ecology, behavior, and evolution of periodical cicadas. *Annual Review of Entomology* 40:269–295.

Williams, K. S., K. G. Smith, and F. M. Stephen. 1993. Emergence of 13-yr periodical cicadas (Cicadidae, *Magicicada*) – phenology, mortality, and predator satiation. *Ecology* 74:1143–1152.

Wilson, J. M., and R. M. R. Barclay. 2006. Consumption of caterpillars by bats during an outbreak of western spruce budworm. *American Midland Naturalist* 155:244–249.

Work, T. T., and D. G. McCullough. 2000. Lepidopteran communities in two forest ecosystems during the first gypsy moth outbreaks in northern Michigan. *Environmental Entomology* 29:884–900.

Wright, J. P., and C. G. Jones. 2006. The concept of organisms as ecosystem engineers ten years on: progress, limitations, and challenges. *Bioscience* 56:203.

Yang, L. H. 2004. Periodical cicadas as resource pulses in North American forests. *Science* 306:1565–1567.

Yang, L. H. 2006. Interactions between a detrital resource pulse and a detritivore community. *Oecologia* 147:522–532.

Yang, L. H., J. L. Bastow, K. O. Spence, and A. N. Wright. 2008. What can we learn from resource pulses? *Ecology* 89:621–634.

Yang, L. H., K. Edwards, J. E. Byrnes, J. L. Bastow, A. N. Wright, and K. O. Spence. 2010. A meta-analysis of resource pulse–consumer interactions. *Ecological Monographs* 80:125–151.

Yodzis, P. 1988. The indeterminacy of ecological interactions as perceived through perturbation experiments. *Ecology* 69:508–515.

Zach, R., and J. B. Falls. 1975. Response of ovenbird (Aves-Parulidae) to an outbreak of spruce budworm. *Canadian Journal of Zoology–Revue Canadienne De Zoologie* 53:1669–1672.

11

Insect Outbreaks in Tropical Forests: Patterns, Mechanisms, and Consequences

Lee A. Dyer, Walter P. Carson, and Egbert G. Leigh Jr.

11.1 Introduction

"First of all, the statement that there are no outbreaks in tropical forests (Elton 1958) is wrong." Wolda (1983) issued this verdict when he demonstrated that over six-year periods, tropical insect populations fluctuated as strongly as temperate ones. Nevertheless, the paradigm that tropical forests are free of outbreaks (Elton 1958) has persisted in the face of strong evidence to the contrary (Nair 2007). In fact, the term "tropical outbreaks" is usually associated with

Insect Outbreaks Revisited, First Edition. Edited by Pedro Barbosa, Deborah K. Letourneau and Anurag A. Agrawal.

disease outbreak literature or, to a lesser extent, with insect outbreaks in agricultural ecosystems or tree plantations (Nair 2007). However, as studies of insect natural history and population dynamics in the tropics accumulate, it is clear that insect outbreaks occur in most natural tropical terrestrial ecosystems and that these outbreaks may occur moderately frequently (Table 11.1). Our goal is to review what is currently known about the frequency, causes, and consequences of outbreaks both within and among tropical forests worldwide.

In this chapter, we do the following: (1) define an outbreak, identify different general types of outbreaks, and outline challenges associated with quantifying and detecting outbreaks in tropical forests; (2) point out potential biotic linkages between outbreaks in managed landscapes and those in more natural forests; (3) identify taxa that are known to outbreak and traits that make species prone to outbreak; (4) predict where outbreaks are likely to occur both within a stand and across gradients from dry to wet forests; (5) discuss the consequences of outbreaks for plant communities and species coexistence; (6) make predictions about how global climate change will alter the frequency and severity of outbreaks; and, finally, (7) propose a series of testable hypotheses concerning outbreaks in the tropics to serve as a guide for future research. We focus mostly on natural systems and native herbivores. Nair (2007) provides a thorough review of outbreaks in tropical agriculture, and the Food and Agriculture Organization (FAO 2011) of the United Nations provides an extensive review and database of introduced and indigenous pest outbreaks in natural and planted forests throughout the tropics (http://www.fao.org/forestry/pests/en/).

11.2   Defining, categorizing, and detecting tropical insect outbreaks

11.2.1   Definitions and types of insect outbreaks

We define a tropical insect outbreak the same as a temperate outbreak, specifically a high-population density relative to long-term mean densities or ecosystem-wide densities (Berryman 1999, Nair 2007, Singh and Satyanarayana 2009). Here we focus on differences in the patterns and scales of tropical versus temperate outbreaks and examine mechanisms that cause population changes leading to outbreaks. We utilize the definitions provided by Berryman (1999) for two major categories of tropical outbreaks: (1) eruptive outbreaks are population explosions that re-occur in time and space because they are regulated by positive feedback or density-dependent processes (these outbreaks often spread to adjacent patches or ecosystems), and (2) gradient outbreaks are driven by local environmental conditions (e.g., temperature, precipitation, host plant availability, and quality) and track changes in these conditions in time or space. These outbreaks are typically confined to patches within an ecosystem.

Density-dependent processes with time lags are typically responsible for predictable cycles between herbivore outbreaks and low population densities, thus outbreaks can be characterized either by the presence of cycles or by pulses, which are short-lived increases in insect abundances. The eruptive-gradient and

Table 11.1 Examples of tropical insects for which outbreaks of native herbivores in natural ecosystems have been documented. The herbivores in these examples are assumed to be native, but natural history data for most tropical herbivores are so poor that it is difficult to determine the recent origin of most herbivores within a tropical forest. Additional references are provided for outbreaks of other species within the same broad taxonomic feeding guild. Diet breadth for guilds is categorized based on host family – generalists consume greater than one family of host plant, and the facultative generalist increases its diet breadth when outbreaking. Species categorized as "solitary" may be gregarious in earlier instars, or adults may oviposit large clutches. "Long-term" observational or experimental studies utilized at least 5 years of data; all others were short-term data sets. The Van Bael et al. (2004) and Torres (1992) references document one large outbreak in Panama and Puerto Rico, respectively. Only outbreaks recorded in tropical forests and measured or described as outbreaks or eruptions are included here. Other sources summarize outbreaks in tropical agriculture or of invasive insect species in the tropics (e.g., Nair 2007; Food and Agricultural Organization (FAO) 2011).

Order	Family	Example	Guild	Documentation	References	Outbreak damage
Coleoptera	Buprestidae	Agrilus kalshoveni in Indonesia	Specialist; wood borer	Long-term observational; tree mortality	Kalshoven 1953	Large-scale tree mortality
Coleoptera	Buprestidae	Sphenoptera aterrima in India	Specialist; wood borer	Observational; tree mortality	Singh et al. 2001	Large-scale tree mortality
Coleoptera	Cerambycidae	Hoplocerambyx spinicornis in India	Specialist; wood borer	Long-term observational; tree mortality	Roonwal 1978	3–7 million trees attacked
Coleoptera	Curculionidae	Ambates spp. in Costa Rica	Specialist; stem borer	Long-term observational; experimental; stem damage	Letourneau and Dyer 1998; Letourneau 1998	0.58±0.1% shrub mortality
Coleoptera	Curculionidae	Cryptorhynchus rufescens in India	Specialist; wood borer	Observational; tree mortality	Singh et al. 2001	Large-scale tree mortality
Coleoptera	Curculionidae	Dendroctonus frontalis in Central America	Specialist; wood borer	Long-term observational, experimental; tree mortality	Billings et al. 2004	1700–2 million ha
Coleoptera	Curculionidae	Platypus biformis in India	Specialist; wood borer	Observational; tree mortality	Singh et al. 2001	Large-scale tree mortality
Coleoptera	Curculionidae	Polygraphus longifolia in India	Specialist; wood borer	Observational; tree mortality	Singh et al. 2001	Large-scale tree mortality
Coleoptera	Various	Various species in Costa Rica	Generalist and specialist leaf chewers	Observational, leaf damage	Janzen 1981	Localized defoliation, >90% herbivory

(Continued)

Table 11.1 (Cont'd).

Order	Family	Example	Guild	Documentation	References	Outbreak damage
Hemiptera	Pentatomidae	*Udonga montana* in India	Generalist, seed predator, leaf chewer	Observational; adult numbers	Reviewed by Nair 2007	Millions of individuals
Hemiptera	Psyllidae	*Phytolyma* spp. In Africa	Specialist gall	Observational; gall numbers	Wagner *et al.* 1991	Large-scale mortality
Hymenoptera	Agridae	*Shizocera* spp. In Vietnam	Generalist; gregarious leaf chewer	Observational; leaf damage	Tin 1990	Severe leaf damage
Hymenoptera	Formicidae	*Atta* and *Acromyrmex* in the Neotropics	Generalist; leaf harvester	Long-term observational, experimental, leaf damage, nest density	Holldobler and Wilson 1990; Terborgh *et al.* 2001; Feeley and Terborgh 2008; Meyer *et al.* 2009	Doubled rates of colony growth; 5–11 times increases in density
Lepidoptera	Arctiidae	*Cosmosoma myrodora* in Puerto Rico	Specialist; gregarious leaf chewer	Observational; rough estimates of larval densities and leaf damage	Torres 1992	Herbivory per tree up to 99%
Lepidoptera	Arctiidae	*Eucereon tesselatum* in Panama	Generalist ; solitary leaf chewer	Long-term observational; larval densities; leaf damage	Van Bael *et al.* 2004	0.32 larvae per young leaf
Lepidoptera	Arctiidae	*Munona iridescens* in Panama	Generalist; solitary leaf chewer	Long-term observational; larval densities; leaf damage	Van Bael *et al.* 2004	0.4 larvae per young leaf
Lepidoptera	Epiplemidae	*Syngria druidaria* in Panama	Specialist; solitary leaf chewer	Long-term observational; larval densities; leaf damage	Van Bael *et al.* 2004	1.09 larvae per young leaf
Lepidoptera	Geometridae	*Cleora injectaria* in Thailand	Specialist; solitary leaf chewer	Observational; larval densities; leaf damage	Piyakarnchana 1981	Up to 100% defoliation on trees in a large area

Order	Family	Species	Feeding type	Study type	Reference	Herbivory
Lepidoptera	Geometridae	*Eois* spp. in Costa Rica	Specialist; solitary leaf chewer	Observational and experimental; larval densities and leaf damage	Letourneau and Dyer 1998; Letourneau 1998; Dyer and Gentry 2010	47±10% herbivory
Lepidoptera	Geometridae	*Miliona basalis* in Indonesia	Specialist; solitary leaf chewer	Observational	Reviewed by Nair 2007	Repeated outbreaks
Lepidoptera	Geometridae	*Scotorythra paludicola* in Hawaii	Specialist; gregarious leaf chewer	Long-term observational; larval and adult densities; leaf damage	Haines *et al.* 2009	Repeated outbreaks; up to 16km² full defoliation
Lepidoptera	Hesperiidae	*Perichares philetes* in Puerto Rico	Specialist; solitary shelter builder	Observational; rough estimates of larval densities and leaf damage	Torres 1992	Herbivory per plant up to 75%
Lepidoptera	Hyblaeidae	*Hyblaea puera* across southern Asia	Specialist; shelter-building leaf chewer		Reviewed by Nair 2007	100% leaf herbivory, up to 100% herbivory per tree or shrub, large scale out breaks of up to 350 million larvae
Lepidoptera	Lasiocampidae	*Euglyphis* spp. in Panama	Generalist; gregarious leaf chewer	Long-term observational; larval densities; leaf damage	Van Bael *et al.* 2004	1.54 larvae per young leaf
Lepidoptera	Lasiocampidae	*Voracia casuariniphaga* in East Java.	Generalist; gregarious shelter builder	Observational; leaf damage	Kalshoven 1953	Up to 100% defoliation in 800ha of forest
Lepidoptera	Limacodidae	*Acharia hyperoche* in Costa Rica	Generalist; gregarious leaf chewer	Observational; larval densities, leaf damage	Dyer and Gentry 2010	90±12% herbivory per shrub
Lepidoptera	Lymantriidae	*Lymantria galinaria* in Indonesia	Generalist; gregarious leaf chewer	Observational; larval densities	Kalshoven 1953	100% of host trees attacked

(Continued)

Table 11.1 (Cont'd).

Order	Family	Example	Guild	Documentation	References	Outbreak damage
Lepidoptera	Noctuidae	*Condica cupentia* in Puerto Rico	Specialist; gregarious leaf chewer	Observational; rough estimates of larval densities and leaf damage	Torres 1992	Herbivory per vine up to 95%
Lepidoptera	Noctuidae	*Dyops dotata* in Panama	Specialist; solitary leaf chewer	Long-term observational; larval densities; leaf damage	Van Bael et al. 2004	5.07 larvae per young leaf
Lepidoptera	Noctuidae	*Eulepidotis* spp. in Panama	Generalist; gregarious leaf chewer	Long-term observational; larval densities; leaf damage	Wong et al. 1990; Pogue and Aiello 1999; Nascimento and Proctor 1994	95% trees with herbivory; herbivory per tree up to 100%
Lepidoptera	Noctuidae	*Ophiusa* spp. in Indonesia	Generalist; leaf chewer	Observational, leaf damage	Kalshoven 1953; Whitten and Damanik 1986	500–1000 ha
Lepidoptera	Noctuidae	*Spodoptera eridania* in Puerto Rico	Generalist; solitary leaf chewer	Observational; rough estimates of larval densities and leaf damage	Torres 1992	Herbivory per tree up to 100%
Lepidoptera	Notodontidae	*Anaphe venata* in Ghana	Specialist; gregarious leaf chewer	Long-term observational; leaf damage	Wagner et al. 1991	100% herbivory per host tree
Lepidoptera	Notodontidae	*Malocampa* spp. in Panama	Specialist; solitary leaf chewer	Long-term observational; larval densities; leaf damage	Van Bael et al. 2004	1.27 larvae per young leaf
Lepidoptera	Notodontidae	*Rifargia distinguenda* in Panama	Specialist; solitary leaf chewer	Long-term observational; larval densities; leaf damage	Van Bael et al. 2004	0.33 larvae per leaf

Lepidoptera	Notodontidae	*Zunacetha annulata* in Panama	Facultative generalist; gregarious leaf chewer	Long-term observational; adult and larval densities; leaf damage	Wolda and Foster 1978; Richards and Coley 2008	Up to 95% herbivory on 95% of host shrubs; over 10 000 larvae on 50 shrubs
Lepidoptera	Nymphalidae	*Actinote* spp. in Ecuador	Specialist; gregarious leaf chewer	Observational; larval densities, leaf damage	Dyer et al. 2010	43 ± 8% herbivory per shrub
Lepidoptera	Nymphalidae	*Antillea pelops* in Puerto Rico	Specialist; solitary leaf chewer	Observational; rough estimates of larval densities and leaf damage	Torres 1992	50% herbivory per plant
Lepidoptera	Nymphalidae	*Hypanartia paullus* in Puerto Rico	Specialist; solitary leaf chewer	Observational; rough estimates of larval densities and leaf damage	Torres 1992	100% herbivory on most host saplings
Lepidoptera	Psychidae	*Eumeta variegata* in Indonesia	Generalist; gregarious shelter builder	Observational	Reviewed by Nair 2007	Infrequent outbreaks
Lepidoptera	Psychidae	*Pteroma* spp. in Indonesia and Malaysia	Generalist; gregarious shelter builder	Long-term observational; shelter densities; leaf damage	Kalshoven 1953	Severe outbreaks
Lepidoptera	Pyralidae	*Deuterollyta nigripunctata* in Panama	Specialist; solitary shelter builder	Long-term observational; larval densities; leaf damage	Van Bael et al. 2004	0.05 larvae per leaf
Lepidoptera	Pyralidae	*Eutectona machaeralis* in India	Generalist; solitary shelter builder	Observational; larval densities; leaf damage	Reviewed by Nair 2007	100% of host trees with skeletonized leaves in a 32 mile forest transect

(Continued)

Table 11.1 (Cont'd).

Order	Family	Example	Guild	Documentation	References	Outbreak damage
Lepidoptera	Pyralidae	*Pantographa expansalis* in Costa Rica	Generalist; solitary shelter builder	Observational; larval densities, leaf damage	Dyer and Gentry 2010	72% of trees attacked
Lepidoptera	Pyralidae	*Syllepte silicalis* in Puerto Rico	Generalist; solitary shelter builder	Observational; rough estimates of larval densities and leaf damage	Torres 1992	More than 50% damage on most host saplings
Lepidoptera	Saturniidae	*Hylesia* spp. throughout the Neotropics	Generalist; gregarious leaf chewer	Observational; larval and adult densities	Janzen 1984; Pescador 1993, 1995; Carrillo-Sanchez 2002	Up to 100% herbivory on multiple shrubs
Lepidoptera	Sphingidae	*Perigonia lusca* in Puerto Rico	Specialist; solitary leaf chewer	Observational; rough estimates of larval densities and leaf damage	Torres 1992	100% herbivory on most host shrubs
Orthoptera	Acrididae	*Anacridium malanorhodon* in Sudan	Generalist, gregarious leaf chewer	Long-term observational; adult densities, degree of defoliation	Food and Agricultural Organization (FAO) 2011	Up to 150 insects per tree

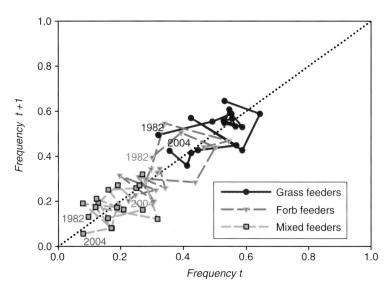

Plate 1.1 Among-year dynamics of grasshopper densities of three feeding guilds from Konza Prairie, Kansas, over 25 years. Reprinted with kind permission from Springer Science+Business Media: *Oecologia*, Grasshopper (Orthoptera: Acrididae) communities respond to fire, bison grazing and weather in North American tallgrass prairie: a long-term study, volume 153, 2007, pages 699–711, Jonas, J.L. and A. Joern.

Insect Outbreaks Revisited, First Edition. Edited by Pedro Barbosa, Deborah K. Letourneau and Anurag A. Agrawal.
© 2012 Blackwell Publishing Ltd. Published 2012 by Blackwell Publishing Ltd.

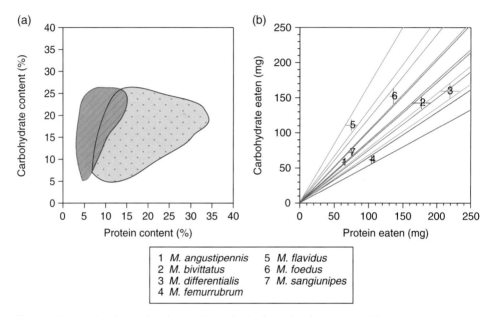

Plate 1.2 A mixed grassland protein–carbohydrate landscape, and the protein–carbohydrate intake targets (means ± SEM) for seven coexisting *Melanoplus* grasshopper species from the sandhills of western Nebraska. Panel (a) shows a hypothetical representation of protein–carbohydrate nutrient space for grasses (hatched dark green area) and forbs (stippled light green area). In panel (b), each number in the figure corresponds to a particular grasshopper species. All species differ from one, expect for *M. angustipennis* (1) and *M. sanguinipes* (7). The protein–carbohydrate intake targets for each species were identified experimentally, and the colored lines bracketing each protein–carbohydrate intake target define the wedge of nutrient space occupied by that particular species (as determined by the SEM of the intake targets). Overlap in nutrient space is obvious only for species 1 and 7. Panel (b) is a figure that originally appeared in Behmer and Joern (2008), copyright (2008) National Academy of Sciences, USA.

Plate 5.1 Herbivore-induced responses affecting arthropod community composition. (a) The mirid bug *Tupicoris notatus* induces changes in primary and secondary metabolism of (b) the wild tobacco plant *Nicotiana attenuata* in the Great Basin Desert, United States. (c) Herbivory-induced changes in plant volatile emission attract the predatory bug *Geocoris pallens* that selectively feeds on (d) tomato hawkmoth (*Manduca quinquemaculata*) eggs and larvae. Induced direct resistance responses cause *Manduca* larvae to grow slower on mirid-attacked plants. Therefore the synergistic function of mirid-induced direct and indirect resistance causes reduced survival and slowed population growth of the more damaging herbivore, *Manduca* (Kessler and Baldwin 2004).

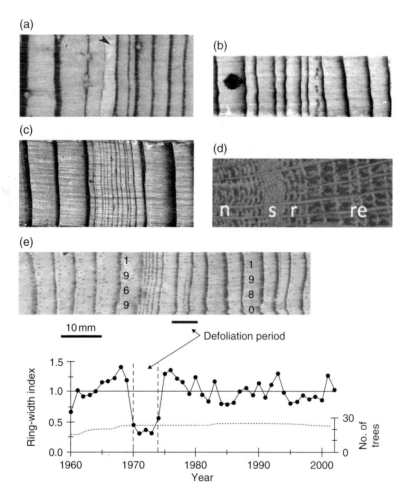

Plate 7.1 Examples of tree-ring response to disturbance. (a) *Pinus ponderosa* defoliated by Pandora moth (J. Speer). Note that the ring formed the first year of defoliation is half the size of normal rings, followed by several small rings. The initial "white ring" and reduced latewood are also characteristic of Pandora moth defoliation: (b) *Abies lasiocarpa* defoliated by two-year cycle budworm (R. Alfaro). Note the alternating narrow and wide growth rings typical of this insect: (c) *Abies concolor* defoliated by western spruce budworm (T.W. Swetnam). Note that the 12 years of reduced growth are the width of 2 years of normal growth, (d) in *Quercus gambelii*; normal radial growth (n), probably in a closed-canopy situation, suppressed growth (s), followed by circa 4 years of recovery (r) and growth release (re) after defoliation and mortality associated with an unknown defoliator (R. Adams). (e) *Pinus leiophylla* defoliated by *Zadiprion falsus* Smith (Hymenoptera: Diprionidae) (photo) and standardized but uncorrected ring width series (chart) (P. Sheppard).

Plate 8.1 The spruce beetle, *D. rufipennis* (Top left, YJC), has a complex assemblage of microbial associates that include oral bacteria (top right), ophiostomatoid fungi (bottom left, YJC), and antagonistic green spore fungi such as *Trichoderma harzianum* (bottom right, YJC).

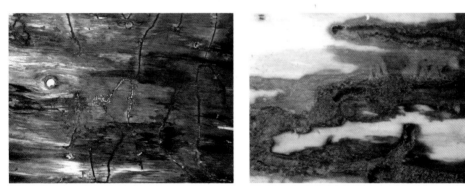

Plate 8.2 Fungi are the best documented associates of bark beetles. Oviposition galleries and symbiotic blue stain fungi, *Ophiostoma* spp., in wood infested by the southern pine beetle, *D. frontalis* (left, RWH). Spruce beetle, *D. rufipennis*, oviposition gallery with associated microbial activity (right, YJC).

Plate 12.1 Series of photos with manipulated variation in extent of pine crown mortality shown randomly to survey participants to evaluate visual preference for forests affected by insects or diseases. Note the progressively greater crown defoliation of the central tree from (a) through (c). Courtesy of F. Baker.

Plate 12.2 Herbivore modification of succession in central Sierran mixed-conifer forest during 1998. White fir, *Abies concolor*, increases in abundance in the absence of fire but becomes increasingly stressed by competition for water in this arid forest type. An outbreak of Douglas-fir tussock moth, *Orgyia pseudotsugata*, completely defoliated white fir (brown trees), reducing the abundance of this tree species to historic levels and restoring the forest to a more stable condition dominated by drought- and fire-tolerant pines and sequoia (green trees). Reprinted from Schowalter (2006, 459, fig. 15.8), with permission from Elsevier.

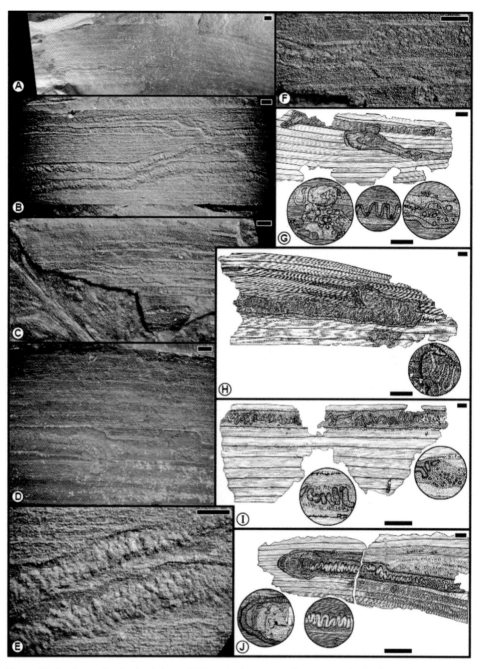

Plate 13.1 Extensive leaf mining of DT71 at the Aasvoëlberg 311 site in the early Late Triassic (Carnian Stage, about 228 Ma) of the Molteno Formation, Karoo Basin, Western Cape Province, South Africa. This is the earliest example of extensive leaf mining in the fossil record, probably produced by an unknown polyphagan beetle, which has extensively targeted the broad-leafed voltzialean conifer, *Heidiphyllum elongatum*. Photo images in (a–f) are of heavily herbivorized leaves, one with a complete mine (a), and others indicating multiple mines (b, c, e) and characteristic enlargements (d, f). In (g–j) are camera lucida drawings of leaf mines showing sinusoidal frass trail variability, the processing of mesophyll and other parenchymatic tissues by a mandibulate insect, and larval mine widths. Scale bars indicate 1 millimeter. This material is housed at the Bernard Price Institute of Palaeontology at the University of the Witwatersrand, Johannesburg, South Africa. See Anderson and Anderson (1989), Scott *et al.* (2004), and Labandeira (2006b) for additional details.

Plate 13.2 A panorama of highly herbivorized leaves of morphotype HC81 (Urticaceae) leaves at the Somebody's Garden locality (DMNH loc. 2203) from the latest Cretaceous (Maastrichtian Stage, 65.5 Ma) of the Hell Creek Formation, Williston Basin, southwestern North Dakota, United States. The predominant damage type is DT57, distinctive hole feeding at the axils of primary and secondary veins, seen in (a–i), and enlarged in (l–n). Accompanying DT57 in (a–i) are circular, elliptical, and polylobate holes, magnified in (k), (m), and (o), attributed principally to DT3–DT5. A thin, delicate, leaf mine occurs in (j), attributable to a nepticulid moth. Note in (k–o) the presence of antiherbivore resiniferous or mucilage-laden epidermal glands, appearing as amber dots. This damage was probably produced by a single insect species, most likely a chrysomelid beetle. Scale bars are in millimeter increments. This material is housed at the Denver Museum of Nature and Science, in Denver, CO. Consult Labandeira *et al.* (2002a, 2002b) and Johnson (2002) for additional information.

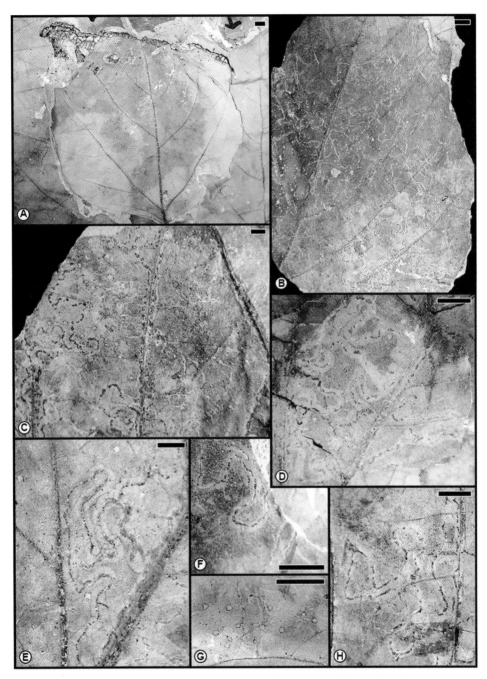

Plate 13.3 Intensively mined *Platanus raynoldsi* (Platanaceae) leaves at the Mexican Hat locality (USNM loc. 42090), from the early Eocene (Danian Stage, about 64.4 Ma), of the Fort Union Formation, Powder River Basin, eastern Montana, United States. The leaf miner is an agromyzid dipteran, the earliest occurrence in the fossil record for this clade. An uncommonly unmined leaf is in (A), but typical are leaves riddled with mines in B–F and H. Leaf mining commonly occurs where primary veins subtend secondary veins (E), at leaf edges (F), or pervasively throughout the entire leaf. Ovipositional probes prior to egg deposition by the leaf miner fly are indicated in (G). Scale bars indicate 1 millimeter. This material is housed at the National Museum of Natural History in Washington, DC. Further details are available from Wilf *et al.* (2006) and Winkler *et al.* (2010).

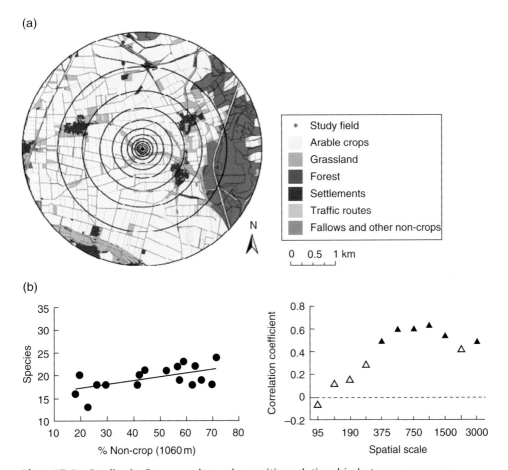

Plate 17.1 Studies in Germany showed a positive relationship between noncrop habitat and spider species richness. (a) An example landscape showing the composition of land cover and spatial scales of analysis. (b) Response of spider species richness to noncrop percentage within 1060 m of fields, and coefficient of correlation for multiple spatial scales showing peak responsiveness at the 1060 m scale. Reprinted with permission from Schmidt *et al.* (2008).

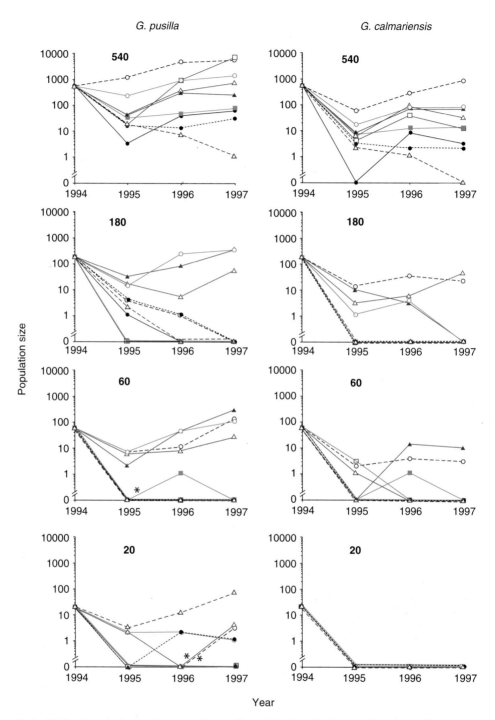

Plate 19.1 Population trajectories for replicated isolated releases of *Galerucella* spp. leaf beetles against purple loosestrife in New York State using four release sizes (20, 60, 180, and 540). Asterisks indicate populations for which eggs were found but not adults. The probabilities of establishment and population growth rates were significantly affected by release size (see Grevstad 1999a).

the pulse-cycle dichotomies are often combined to characterize different types of outbreaks throughout the outbreak literature (reviewed by Singh and Satyanarayana 2009, Nair 2007), and yield these main classes of outbreaks: pulse eruptive, cyclical eruptive, pulse gradient, and cyclical gradient. Other classifications consider the length of the cycle or the persistence of the outbreak (Berryman 1999, Singh and Satyanarayana 2009), thus terms such as "sustained gradient outbreak" refer to sites within an ecosystem that experience persistently high levels of herbivory compared to other sites. There is a perception that most outbreaks in tropical forests are pulse gradient. This suggests that disturbances or clear changes in resource availability are putatively responsible for tropical outbreaks that are ephemeral, unpredictable, and occurring unseen during a short-term nutrient flux or high up in forest canopies (Wolda 1978, Wolda and Foster 1978, Lowman 1997, Singh and Satyanarayana 2009).

We argue that when considering outbreaks, it is critical to consider both the temporal and spatial scales of outbreaks and perhaps most importantly to take a phytocentric view. As Strong *et al.* (1984) made clear, rare events such as outbreaks cannot be ignored because they may greatly influence plant diversity and relative abundance. Indeed, an outbreak that occurs once every 40–50 years and causes major defoliation might seem infrequent or rare, yet it would occur several times in the life span of an individual tree that has managed to reach the canopy (Carson *et al.* 2004). Outbreaking insects that cause major defoliation or mortality on abundant species can be keystone species in plant communities (e.g., Carson and Root 2000) and occur in nearly all biomes (Carson *et al.* 2004). Additionally outbreaks in tropical forests may be as frequent as in temperate regions but go unnoticed because they are highly localized on a subset of patches or individuals of a fairly common species (Janzen 1981), they occur on uncommon and sparsely distributed species (Nair 2007), or they are out of sight, high in the canopy (Lowman 1997). We note that some pulse gradient outbreaks can occur at very small spatial and temporal scales, and even eruptive outbreaks can involve rare insect species that increase substantially in abundance. However, these outbreaks may never cause major defoliation or seriously injure host populations. Thus, our focus below is primarily on outbreaks that lead to major defoliation, host mortality, or both (Table 11.1).

11.2.2 Quantifying and detecting insect outbreaks

Some authors have suggested that cyclical gradient and cyclical eruptive outbreaks do not exist in tropical ecosystems (Wolda 1978, Myers 1988, but see Bigger 1976). Long-term cyclical outbreaks are well documented in temperate ecosystems (e.g., the larch bud moth experienced dramatic outbreak cycles from 800 to 1980 CE; Johnson *et al.* 2010), thus it is premature to dismiss their importance. They are difficult to detect in tropical ecosystems for three reasons. First, detection of most natural outbreaks, cyclical or otherwise, is very difficult because on average tropical herbivores are more specialized (Dyer *et al.* 2007), resulting in more localized outbreaks on one or a few species (Janzen 1981). Consequently, dismissing cyclical outbreaks is not possible without consistent (e.g., weekly) monitoring of entire populations of host plant genera over a number of years.

Second, detection of cyclical outbreaks in particular requires very detailed time series of population data (e.g., weekly) to reveal cycles (Haines *et al.* 2009) because most tropical insects are multivoltine. This means that the standard cycle periodicity common in temperate systems should be longer than in the tropics because both population changes and time lags take longer for univoltine populations that go through a long seasonal quiescent phase. Third, outbreak detection methods that utilize tree rings are less useful because only a few tropical trees have annual growth rings; this diminishes the possibility of recording cyclical histories of defoliation (e.g., Buntgen *et al.* 2009). Thus, quantifying cyclical population dynamics will require years of data (e.g., Roland 1993) on focal host genera or species, long time series that may require weekly censuses of marked plants, and selection of plant–herbivore study systems that might be more susceptible to outbreaks.

While documenting and verifying outbreaks in the tropics in some regards are more difficult than in temperate regions, they may also be easier under certain conditions. The diversity of plants and insects, and the increase in the number of insect generations per year, may improve a tropical ecologist's ability to detect outbreaks over shorter periods of time and for many species simultaneously. Indeed, Wolda (1983) was able to monitor population fluctuations of numerous species of Auchenorrhyncha (formerly known as Homoptera) using a single light trap. Thus the task, although challenging, is not daunting.

11.2.3　Monitoring insect population dynamics versus defoliation

Overall, an outbreak can be quantified by monitoring either the population dynamics of the insects, or defoliation as an indicator of an outbreak, including percentage of leaf removal or percentage of tree or shrub mortality (reviewed by Nair 2007). However, while high levels of damage often occur during outbreaks, it is problematic to use the degree of damage as an indicator of outbreaks when testing hypotheses generated by mathematical models or other theory and it is problematic when attempting any comparison of tropical and temperate outbreaks. This is because there are many factors that decouple measures of herbivory from herbivore population dynamics, including difficulties in measuring herbivory at multiple scales, inability to measure defoliation by specific herbivores (particularly when entire leaves are removed), error variance introduced by compensatory growth, changes in damage associated with herbivore density, and rapid leaf growth which can obscure damage. All of these problems can prevent accurate parameter estimates in models and can make it difficult to compare tropical outbreaks with quantitative studies of temperate outbreaks that focus on insect populations rather than herbivory (e.g., larval counts in Johnson *et al.* 2010). Nevertheless, both metrics of quantifying outbreaks have value, and typically the most common measure is the degree of defoliation or host mortality associated with an outbreak (e.g., Janzen 1981). Indeed, the degree of defoliation and its timing with other stressful abiotic factors (e.g., drought) may be the best metric for evaluating the impact of the outbreak on plant population and community dynamics.

Useful models have been developed for defining and characterizing outbreaks, and have yielded relevant hypotheses about the patterns and causes of outbreaks.

For example, Johnson *et al.* (2010) utilized a tritrophic discrete-space, discrete-time model to examine population densities of larch bud moths, host plant quality, and parasitoid population densities. Their modeling approach revealed that the elevation of outbreak epicenters for larch bud moths has been increasing with mean winter temperatures since 1870 CE, yielding a clear hypothesis that the disruption of long-term cyclical eruptive outbreaks is caused by limits to how far the moth distribution can shift upslope in response to warming temperatures. However, most traditional models for insect outbreaks focus on such population cycles and delayed density dependence (Myers 1988, Logan 2003, Turchin 2003). These models may be useful for generating testable hypotheses for tropical insect herbivores that exhibit cyclical outbreaks linked with density dependence (Godfray and Hassell 1989, Ray and Hastings 1996, reviewed by Wallner 1987), but there are still very few documented examples of cyclical outbreaks in the tropics (Wolda 1983, 1992, reviewed by Wallner 1987). Outbreaks are characterized by large departures from equilibrium with considerable spatial variation, thus models that assume fluctuations around equilibrium (i.e., cyclic outbreaks) without spatial structure may not provide relevant insight into the causes and consequences of outbreaks (Hastings 1999). Several spatial-modeling techniques explicitly incorporate the role of spatial structure, including metapopulation models, reaction–diffusion models, and cellular automata approaches. These models address important questions, such as the role of dispersal or resource patchiness in the spread of outbreaks (Hastings 1999). In a tropical forest, where many host plants are rare and resources can be very patchy, metapopulation models, which examine subpopulations at scales as small as a single tree, can be useful for examining determinants of tropical outbreaks. Thus, if local outbreaks are common in tropical forests, the metapopulation approach (e.g., Kondoh 2003) is best suited for examining the causes and consequences of dynamics such as an outbreak on a single tree or in one patch. Such outbreaks can spread to other patches, or, alternatively, dynamics across the entire ecosystem can remain stable (Hastings 1999).

11.2.4   Knowledge of outbreaks in the tropics is often anecdotal

Many examples of tropical forest outbreaks are only known through anecdote or personal observations (reviewed by Nair 2007; D. Janzen, D. Wagner, V. Novotny, H. Morais, H. Greeney, and G. Gentry, personal communication). For example, Grogan (2001) described his encounter in Brazilian Amazonia with an unidentified Thysanoptera that "appeared by the tens of millions during the 1997 dry season in a 500-m^2 patch of forest, which was completely defoliated." Torres (1992) qualitatively described an eruption of larvae of *Spodoptera eridania* (Lepidoptera, Noctuidae) in Puerto Rico, following Hurricane Hugo in 1989. The larvae erupted and invaded nearby human habitats, whereby they "crossed forest roads in the Luquillo Mountains when their food was exhausted. USDA Forest Service personnel assigned to the Luquillo Mountains observed larvae feeding on tomatoes, potato skins, and lettuce inside trashcans. Larvae entered houses in the forest and even fed on dog food." Beyond these types of descriptions, few studies of tropical outbreaks were designed to parameterize mathematical models or test hypotheses that underlie outbreaks. For example, outbreaking *Hylesia* spp. (Saturniidae) caterpillars are a common and noticeable feature of forests and managed landscapes across the Neotropics (Janzen 1981, 1984, Pescador 1995, Specht *et al.* 2006, Hernandez *et al.* 2009). These caterpillars have urticating spines as larvae and urticating scales as adults. Studies on this genus therefore have focused on the human impacts resulting from these outbreaks, particularly the widespread cases of painful dermatitis they cause (e.g., Iserhard *et al.* 2007, Moreira *et al.* 2007). We searched for *Hylesia* on ISI Web of Science in August 2010 and found that only 6 of 34 studies of this genus were ecological and none used quantitative data to examine causes and consequences of the outbreaks. Nevertheless, there are a growing number of observational and experimental studies that document outbreaks of tropical insect herbivores though the bulk of these come from managed ecosystems (reviewed by Nair 2007; Table 11.1). Consequently, while future research on tropical outbreaks should be hypothesis driven (see Section 11.8), we argue that the simple documentation of the occurrence of outbreaks remains important. Indeed, these occurrences need to make their way into the peer-reviewed literature as short communications where the identity of the herbivore, its hosts, its location, and other salient aspects of the outbreak are reported.

11.3   Outbreaks in managed systems have biotic linkages to intact forests and vice versa

The best documented cases of outbreaks are in agricultural landscapes (e.g., Azerefegne *et al.* 2001), but these outbreaks are not isolated from natural ecosystems because the arthropod communities often have host plants in both places (e.g., arthropods on Heliconiaceae in forests near banana plantations, Matlock and de la Cruz 2002; see also Azerefegne and Solbreck 2009). Tropical forests are likely to affect outbreaks in nearby agroecosystems (and vice versa) by serving as sources of outbreaking insects or natural enemies, or by more complex pathways (Azerefegne and Solbreck 2009, Vandermeer *et al.* 2010). In fact, tropical forests act as reservoirs for many pests in tropical agriculture and vice versa, and it can sometimes be difficult to distinguish native forest herbivores from exotic pests that may have immigrated from adjacent agroecosystems (Nair 2007, Azerefegne and Solbreck 2009). For example, in Costa Rican banana plantations that were formerly lowland rainforest, frequent outbreaks by 53 native species of lepidopteran defoliators occurred for over twenty years (1950–1973). These outbreaks followed aerial applications of the insecticides dieldrin and carbaryl to the plantations (Lara 1970, Ostmark 1974, Stephens 1984). In 1973, these applications were stopped and lepidopteran larvae have rarely exceeded their economic thresholds since (Stephens 1984, Thrupp 1990), presumably due to control by parasitoids that were susceptible to insecticide treatments (Dyer *et al.* 2005). These caterpillars are all species that occur in nearby wet forest fragments (Stireman *et al.* 2004, Dyer *et al.* 2005, Dyer and Gentry 2010), but they have not been documented to outbreak in natural forests. Although speculative, it is likely that natural forests are the original source of these eruptions and that when they occur in plantations they probably spill back into natural forests.

Can outbreaks within managed landscapes cause smaller-scale outbreaks in nearby intact forests? Trophic relationships suggest this possibility. Specifically, experimental and observational studies (Stireman *et al*. 2004, Dyer *et al*. 2005, Dyer unpublished data) indicate that most interactions between these caterpillars and their host plants and enemies are similar in their natural and managed settings. For example, parasitism and predation rates in low-pesticide banana plantations and in wet tropical forest at the La Selva Biological Station were identical for two of the most common outbreak Lepidoptera, *Caligo memnon* and *Opsiphanes tam-arindi* (both Nymphalidae) (Stireman *et al*. 2004). Thus, indirect evidence suggests that large-scale outbreaks of these banana pests may spill back and cause outbreaks at small scales, perhaps on relatively dense patches of host plants or under conditions when natural enemy populations are depressed by climatic events or disturbances (e.g., in response to flooding, droughts, strong El Niño weather, or other disturbances; van Bael *et al*. 2004, Stireman *et al*. 2005). If spill-back of insect outbreaks from agriculture to forest isn't a rare phenomenon, then outbreaks may increase in intact forests as fragmentation becomes more and more widespread.

11.4 What taxa are likely to outbreak, and which traits predispose species to outbreak?

Lepidoptera accounted for 73% and Coleoptera for 19% of the outbreaks listed in Table 11.1; together they accounted for 92% of these outbreaks. Similarly, Carson *et al*. (2004) found that 94% of outbreaks of native insects in natural habitats worldwide were caused by species in these two orders (Lepidoptera 58% and Coleoptera 36%). This may simply represent the high diversity and abundance of herbivores within these two orders. Specialists (herbivores feeding on plants within one family) comprised 56% of these taxa whereas 44% were generalists. Leaf chewers, wood borers, and shelter builders were the most important guilds (58%, 17%, and 12.5% respectively). Outbreaks occurred in all major tropical forest areas including Asia (21%), India (15%), Africa (3%), and the mainland Neotropics (42%). Only two outbreaks were documented in South America (Ecuador and Brazil) whereas 17% of all outbreaks were recorded from Puerto Rico. Thirty one percent of all outbreaks were documented in Panama and Costa Rica. These results must be interpreted with a big grain of salt because of the small sample size and the various means used to documenting outbreaks. Still, they do show that outbreaks occur in nearly all tropical forest regions, are most often documented in areas where numerous researchers have been active for years (Panama, Costa Rica, and Puerto Rico) and that specialists cause rather more outbreaks than generalists, especially as generalist outbreaks are more likely to be noticed.

A number of authors have compiled key life history traits that may make certain temperate taxa prone to outbreak, including high mean densities, efficient dispersal, large size, high intrinsic rates of increase, and polyphagous diets (see citations in Table 11.2). These generalizations are generated primarily from temperate studies, and it is worth considering whether or not outbreaking tropical herbivores possess these traits. Among these attributes, high mean density is a trait that is the most consistent characteristic of outbreak species (Table 11.2) because insect populations that fluctuate around high mean abundances are also

Table 11.2 Attributes of outbreaking insect populations and relevance to tropical ecosystems. Selected references have proposed these attributes as conducive to outbreaks or have provided empirical evidence demonstrating the roles of these attributes in creating outbreaks.

Life history trait of pest or ecosystem parameters	Relevance to tropical ecosystems
High density (Hunter 1991, but see Root and Cuppuccino 1992)	More common in dry forests (Janzen 1993)
Gregarious larvae (Nothnagle and Schultz 1987, Hunter 1991)	Common for many tropical insects (Dyer and Gentry 2010)
Large clutch size (Hunter 1991)	Present in many tropical herbivores (Nair 2007)
Single large egg mass (Hunter 1991)	Present in many tropical herbivores (Nair 2007)
High intrinsic rate of increase (Nothnagle and Schultz 1987, Root and Cappuccino 1992)	True for all tropical ecosystems
Large size (Hunter 1991)	Common for many tropical insects
High dispersal (Greenbank 1956)	Unknown for most outbreaking herbivores in tropical forests
Polyphagous (Nothnagle and Schultz 1987, Hunter 1991, but see Redfearn and Pimm 1988)	Tropical herbivores are more specialized, but highly polyphagous species are present (Dyer et al. 2007)
Well defended (Hunter 1991, Veldtman et al. 2007)	Tropical herbivores may be better defended (Gauld et al. 1992)
Overwinter as eggs (Hunter 1991)	Relevant to dry forests
Reduced wings (Hunter 1995)	Not true for most outbreaking herbivores in tropical forests (Nair 2007)
Capital breeders (Tammaru and Haukioja 1996)	The most damaging tropical herbivores feed as immatures (Nair 2007)
Disturbance (Schowalter and Lowman 1999, Torres 1992)	Common in gaps and fragmented tropical forests
Favorable climate for population growth (Greenbank 1956, Roland 1998, McCloskey et al. 2009)	True for all tropical ecosystems
Low diversity at all trophic levels (Pimentel 1961, Root 1973, Letourneau 2009)	Only true in tropical agroecosystems or in highly disturbed areas

more likely to exhibit wide fluctuations. One might expect abundance and variability to be linked in some cases (Root and Capuccino 1992) because it is not unusual to observe data where the variance increases with the mean. Abundant species, however, can be stable whereas rare species can hardly be given to outbreaks because wide population fluctuation could cause their extinction (Root and Capuccino 1992). Common species, whether they have narrow or broad diet breadths, have the potential to reach outbreak levels whenever ecosystem parameters or microhabitats are favorable (Table 11.2).

Other putative outbreak traits are not as well supported by empirical data on tropical outbreaks and research here could help clarify the effects of these traits on propensity to outbreak. For example, Redfearn and Pimm (1988) conclude that diet breadth is not tightly linked to propensity to outbreak. Thus, just because temperate herbivores have higher mean diet breadth than tropical ones (Scriber 1973, Dyer *et al.* 2007), does not necessarily indicate that temperate herbivores are likely to outbreak. In fact, the outbreaks reported by Janzen (1981) were mostly specialists and Janzen argues that most outbreaking tropical herbivores are either specialists or only consume a small proportion of available plant species. While this assertion is not supported by the documented cases of generalist outbreaks in Table 11.1, it is clear from those empirical examples that numerous tropical specialists do not abide by the temperate paradigm that polyphagous herbivores are more likely to outbreak.

11.5 Likelihood of outbreaks within a stand and across transitions from dry to wet forests

11.5.1 Patchiness and scale of insect outbreaks within tropical forests

A number of studies have examined determinants of high levels of herbivory or high arthropod abundances within a forest and most have concluded that high density of new leaves and high light availability are two key factors predicting higher herbivore abundance (Basset 1992, Itioka and Yamauti 2004, Richards and Coley 2008). Furthermore, Eveleigh *et al.* (2007) argue that diversity across trophic levels should be a potent stabilizing mechanism, preventing outbreaks, and they demonstrated that outbreaks were more likely in less diverse forest plots, which is likely due to resource concentration and other mechanisms that are responsible for higher herbivory in less diverse communities (Pimentel 1961, Root 1973, Letourneau *et al.* 2009, Long *et al.* 2003, Carson *et al.* 2004). There is compelling evidence that outbreaks, whether in temperate or tropical systems, are more likely when host species become aggregated or dense at either neighborhood or regional spatial scales (reviewed by Carson *et al.* 2004, Nair 2007). Nair (2001) found that in Indonesian forests outbreaks were generally restricted to tree species that occurred "gregariously" and that a high host density could precipitate outbreaks even in natural assemblages (see also Nair 2000). For example, an outbreak by a Noctuid moth (*Eulepidotis phrygionia*) in the Amazonian rainforest caused severe defoliation of a leguminous tree (*Peltogyne gracilipes*) within monodominant stands but not in those in nearby mixed stands where this tree species was less frequent (Nascimento and Proctor 1994). Furthermore, the famous outbreak of a lepidopteran in Panama occurred on the most abundant woody species on Barro Colorado Island (Wolda and Foster 1978). In addition, high insect herbivore densities and outbreaks might be more frequent in gaps or edges, where new growth and light availability are greatest (Richards and Coley 2008) and where hosts are aggregated or relatively abundant. Localized outbreaks within the intact understory have been documented as well (Table 11.1), but it is likely that they are smaller in scale and shorter in duration, and only outbreaks on the commonest understory species attract attention. The relationship of most outbreaks to host concentration typically remains unknown but a recent meta-analysis found a significant reduction in herbivory in more diverse forests (Jactel and Brockerhoff 2007).

Local outbreaks could also occur if the impact of enemies on herbivores is interrupted. Experimental removal of ants from *Piper* ant-plants cause local outbreaks of both specialist and generalist herbivores on these plants; these outbreaks decrease *Piper* density in the understory and spill over onto plants of other species (Letourneau 2004).

11.5.2 Predicting outbreaks across broad gradients: the dry-to-wet forest transition

We predict that outbreaks will be more likely to occur in more seasonal tropical forests than in wet forests, with an increase in frequency as precipitation decreases and seasonality increases. As total annual rainfall increases and the length of the dry season decreases, tropical forests generally have higher plant species richness (Hall and Swaine 1976, Huston 1980, Gentry 1982, 1988, Leigh *et al.* 2004), greater primary productivity (Philips *et al.* 1994), less seasonal production of new vegetation (Opler *et al.* 1980, van Schaik *et al.* 1993), higher levels of anti-herbivore defense (Coley and Aide 1991), higher levels of predation and parasitism (Stireman *et al.* 2005, Janzen, Coley, and Dyer, unpublished data), and lower herbivore densities. These changes will alter the nature of tritrophic interactions across these gradients for at least three reasons. First, plant investments in qualitative anti-herbivore defenses, such as alkaloids, often increase with nutrient availability, whereas quantitative defenses, such as polyphenols and fiber increase with leaf lifetime (reviewed by Massad *et al.* 2011); thus, plant defenses should increase with rainfall in tropical forests (Coley and Aide 1991). Second, enhanced plant defenses with increasing rainfall will lead to significantly lower leaf damage and herbivore densities in wet tropical forests (Janzen 1993, Coley and Barone 1996). Third, rates of parasitism tend to increase with rainfall and variance in precipitation (Stireman *et al.* 2005; Dyer, Coley, Janzen, unpublished data). Parasitoids may be less abundant in dry forests because the hotter, drier conditions often may make it difficult for small-bodied insects to survive, and increased seasonality of herbivore abundance in drier forests makes it harder for parasitoids to track them (e.g., Janzen 1983, 1993, Parry *et al.* 2003, Stireman *et al.* 2005). Overall, the top-down impact of natural enemies and the bottom-up effect of plant defenses increase with greater rainfall, both of which make for lower annual herbivore densities in wetter tropical forests. As higher densities are correlated with greater incidence of outbreak (Hunter 1991), outbreaks should be most common in dry forests. Though it remains untested, the frequency and severity of herbivore outbreaks should decrease from dry to wet tropical forests. In fact, Janzen (1981) found that insect herbivores caused greater than 90% damage on numerous individuals of 25 tree species during just two growing seasons in a dry tropical forest in Costa Rica. We are not aware of the same extent of defoliation occurring in any wet forest. Probably no one has bothered to look, because outbreaks may be so hard to detect in wetter forests!

11.6   The consequences of outbreaks for plant communities and species coexistence

The long-term effects of outbreaks are more intense than gradient or eruptive increases in mean levels of herbivory (Carson and Root 2000). The effects of outbreaks can be comparable to abiotic disturbances such as storms, fires, floods, droughts, or other disturbances (Schowalter and Lowman 1999). Because outbreaks rarely occur within the lifetime of a single research grant, the large number of studies examining tropical community structure, such as tests of the Janzen-Connell hypothesis (Carson *et al.* 2008), tests of trophic cascades hypotheses (Knight *et al.* 2005), or tests of niche hypotheses (Wiens *et al.* 2010), have not included the impact of outbreaks as part of theoretical considerations, observational methods, or experimental design. As Carson and Root (2000) point out, most hypotheses in community ecology assume that outbreaks are rare or exceptional. An outbreak, however, can have long-lasting effects with profound impacts on its community. We suggest that the roles of outbreaks in tropical ecosystems could be particularly important for the maintenance of diversity. Outbreaks of insect herbivores in many systems worldwide commonly occur on dominant plant species that are widespread and reach high abundance both locally and regionally (Carson and Root 2000, Carson *et al.* 2004). These outbreaks tend to substantially reduce the abundance and performance of these species, thereby opening up forest canopies, promoting the growth of other species and plant species richness, making these insect herbivores classic keystone species (Carson and Root 2000, Carson *et al.* 2004, Nair 2007). For example, in India, outbreaks of the wood borer *Hoplocerambyx spinicornis* (Cerambycidae) have been documented for more than a century in tropical deciduous Sal forests of India (Nair 2007, Ghimre and Dongre 2001). These wood borers prefer to attack the largest sal trees (*Shorea robusta*) and can kill millions of individuals over vast areas (500,000 ha); outbreaks are thought to be triggered at least in part by high host concentrations (Ghimre and Dongre 2001, Nair 2007). Overall, if outbreaks are more likely to occur when tree species in tropical forest become locally or regionally abundant or aggregated, then this will decrease the performance and competitive ability of the common trees and promote local species richness. This possibility deserves serious consideration at both small and large spatial scales.

11.7   Global change, disturbance, and outbreaks

Empirical studies and models predict intensified insect outbreaks in response to climate change, invasive species, habitat fragmentation, and other global changes (Roland 1993, Coley 1998, Logan *et al.* 2003, Stireman *et al.* 2005, Moreau *et al.* 2006). Insect populations respond rapidly to a wide variety of disturbances and environmental changes, and outbreaks are one common response to such changes (Schowalter and Lowman 1999). Massad and Dyer (2010) utilized meta-analysis to quantify increases in herbivory in response to major categories of global change and concluded that increases in temperature, CO_2, nutrient inputs, and decreases in diversity all create favorable conditions for herbivores, particularly for generalist insects, which can be important for outbreaks (Table 11.1). Fossil evidence suggests that global warming increases herbivory: might it also enhance the frequency or severity of outbreaks? For example, Currano *et al.* (2008) examined fossil records of herbivory and found that abrupt increases in the partial pressure of CO_2 and in temperature, comparable to current rates and levels of increase, were associated with distinct increases in herbivory.

For over a decade, models of global climate change have consistently predicted that temperature increases over the next century in the tropics will be relatively moderate, but extreme weather events, such as droughts and floods, will become more frequent (e.g., Hulme and Viner 1998, Easterling *et al.* 2000). These extreme weather events could trigger outbreaks in tropical forests via disrupting natural enemy control of herbivores and altering host plant abundance and quality. Stireman *et al.* (2005) found that parasitic wasps were particularly vulnerable to extreme weather events; they concluded that increases in climatic variation could lead to higher frequencies and intensities of outbreaks via diminished control by parasitoids, possibly due to phenological asynchrony between the parasitoids and herbivores. Empirical studies in the tropics have demonstrated that outbreaks do increase with climatic changes associated with El Niño Southern Oscillation (ENSO) events (Van Bael *et al.* 2004, Srygley *et al.* 2010). For dry or moist tropical forests that are strongly seasonal, early onset of rainfall and increased precipitation in El Niño years can cause outbreaks via increased leaf flushing early in the season (Jaksic 2001). Srygley *et al.* (2010) documented such outbreaks for *Aphrissa statira* (Pieridae) butterflies in Panama in response to an increase in dry season rainfall and greater leaf flushing due to ENSO.

In addition to these global changes, increases in fragmentation are likely to enhance conditions for outbreaks via creating patches of high herbivore density, limiting movement of natural enemies (Schowalter and Lowman 1999) and increasing the abundance of well-lit edge habitats that favor more edible pioneers. Tropical ecosystems are subject to all of these changes and are likely to experience more outbreaks. Furthermore, in the tropics, the projected increases in extreme weather events due to climate change are potentially more important for plant insect interactions because outbreaks are also predicted as a result of disruption of parasitism (Stireman *et al.* 2005).

11.8 Critical hypotheses need to be tested: A guide for future research on outbreaks

Tropical insect outbreak research has been limited mainly to applied research in agricultural (Nair 2007) or reforested (Massad 2009) ecosystems and observational studies in natural forests. Future research in natural systems should incorporate careful models, a strong experimental approach, a focus on multiple mechanisms that cause outbreaks, and details of how key herbivore populations change throughout a forest (Myers 1988). The modeling approach used for tropical outbreaks should not always mirror approaches used for outbreaks in temperate forests (Turchin 2003). In particular, a focus on metapopulation models or spatially explicit models may be more useful for examining the potential dynamics and consequences of outbreaks at small spatial and temporal scales across a heterogeneous forest landscape. It is likely that low-density specialists on patches of host plants within a forest that have the appropriate population fluctuations necessary for local outbreaks will also experience local extinctions (Hunter 1991). Such dynamics, combined with dispersal from patch to patch, are perfectly suited for metapopulation models. On the other hand, for generalist herbivores, whose dynamics may mimic temperate outbreaks but with shorter periodicity, traditional models may be most appropriate.

While mathematical models will be useful for exploring theoretical possibilities and posing additional hypotheses, the existing literature, reviewed here, is also rich enough to provide hypotheses to drive observational and experimental research approaches to tropical forest outbreaks. Here we provide a series of hypotheses that can be addressed via quantitative observational or careful experimental approaches. The first four hypotheses focus on where outbreaks are hypothesized to occur and assume that plant chemistry and natural enemies are important factors in preventing outbreaks. These are followed by several hypotheses about the impacts of outbreaks on tropical ecosystems.

Hypothesis 1. Outbreaks that are more localized and shorter in duration will occur in tropical wet forests (compared to dry forests and temperate forests) due to greater levels of plant defense.

Leaf quality declines with leaf lifetime, which is longer in wet forests. There is considerable preliminary evidence that a gradient of increasing chemical defense from dry to wet forests exists, which could lower frequency and amplitude of outbreaks. Testing this hypothesis will require appropriate methods for quantifying particular defensive compounds and their efficacy, long-term field experiments that manipulate leaf age and levels of chemical defenses, and assessment of effects on herbivore densities or incidence of outbreaks.

Hypothesis 2. Lower incidence of outbreaks in wet forests is the result of higher rates of parasitism or predation, which may interact with plant defenses.

If parasitism is higher in wet forests than in dry forests, this may lower the density of herbivores. The following approaches could help assess how predation or parasitism rates vary among forest types: (1) analyze life table data derived from following cohorts of insect herbivores over their larval lifetimes at multiple sites; (2) manipulate leaf age, defenses, and exposure to natural enemies and evaluate herbivore responses or outbreak incidence; and (3) model the dynamics of the focal outbreak system with parameters from rearing data and field experiments.

Hypothesis 3. Lower incidence of outbreaks in wet forests is the result of reduced tree species aggregation and higher tree species diversity.

Long-term monitoring studies of herbivore population abundance would be a first step in testing this hypothesis. It would be especially valuable to examine incidences of defoliation of individuals within a species that vary in abundance at multiple spatial scales both within a region and across rainfall gradients. Multivariate techniques that can distinguish the effects of the degree of host concentration from other explanations (enemies and defenses) would be required. We note here that theories that predict outbreaks based upon host concentration are more parsimonious than those based upon enemies or plant defenses because only aggregation predisposes a population to an herbivore outbreak whether it be on a local or regional scale.

Hypothesis 4. The absence of a distinct and forest-wide pulse in new leaf production in wet forests reduces food quantity and quality, thus limiting herbivore density and outbreaks.

In dry forests new leaf production is concentrated forest-wide at the beginning of the wet season, whereas in wet forests new leaves are produced at much lower rates throughout the year. This minimizes large, high-quality food pulses and hence outbreaks. Seasonality of new leaf production is intermediate in moist forests. Leaf quality parallels patterns in food quantity. Both the seasonal abundance and high quality of food in dry forests should work together to generate seasonal peaks in herbivore population densities, some of which may reach outbreak levels. This hypothesis can be tested via documenting the seasonality of new leaf production for selected plant species and examining correlations and partial correlations between the number of new leaves, leaf quality, the relative density of herbivores per host plant, and the incidence of outbreaks in sites that vary in precipitation and seasonality.

Hypothesis 5. In dry tropical forests, the short, sharp peaks in larval densities at the beginning of the wet season are difficult for parasitoids to track and regulate, resulting in lower parasitism rates in dry forests than wet forests.

This is best addressed via observational data (long-term rearing and population studies on focal plants), simulation models, and analytical models followed by careful long-term experiments designed to determine the population-level consequences of leaf quality and abundance, and parasitoid density on herbivore population dynamics.

Hypothesis 6: Outbreaks in tropical forests represent significant nutrient inputs, affecting plant physiology and growth and enemy dynamics.

Resource pulses are not commonly reported for tropical forests (Yang *et al.* 2010), presumably for the same reasons that outbreaks are not commonly reported – small spatial and temporal scales. Nevertheless, the resource pulse created by outbreaks, including greater nutrient availability from insect feces and mortality as well as greater prey availability for predators, is likely to have measurable effects on plant and natural enemy communities (reviewed by Yang *et al.* 2010).

Hypothesis 7. Outbreaks in managed habitats (e.g., plantations) will spill over into intact forests, cause localized outbreaks on shared hosts, and increase with fragmentation.

As deforestation and fragmentation continue to increase, outbreaks in these managed landscapes are also likely to increase and these herbivores may colonize shared hosts within intact forests. Thus the frequency of outbreaks in natural forest stands nearby may also increase, particularly at the scale of individual and patches of host plants.

Hypothesis 8. Outbreaking insects can function as keystone species and thereby increase plant diversity at many scales.

We need to go beyond studies that only documented the occurrence of outbreaks (though this remains important) and document the consequences of outbreaks for forest regeneration and the maintenance of diversity (Carson *et al.* 2008). As Schowalter (1996) concluded, outbreaks of insect herbivores that attack abundant trees in many temperate systems reduce dominance, thereby enhancing diversity. They likely do the same thing in tropical forests though at much smaller spatial scales of local tree species aggregation. Nonetheless, the relevant studies remain to be done. A first and basic step would be to evaluate whether the host plants attacked are relatively common forest-wide and then evaluate patterns of regeneration in the understory following these events.

11.9   Conclusions

Outbreaks of herbivorous insects do occur in tropical forests and may be far more frequent than previously thought. Nonetheless, documentation of outbreaks in the tropics remains scant and it remains important that such events find their way into the peer-reviewed literature. Outbreaks usually occur in restricted areas on sets of a few related plant species, but sometimes spread to other species. Outbreaks often follow disturbances. These disturbances typically increase light ability, abundance of new leaves, or the abundance of pioneers with tastier leaves, all of which may precipitate outbreaks, particularly at smaller spatial scales.

Outbreaks appear to attack the commonest species or uncommonly large aggregations of individual species. Whatever the scale and duration of such outbreaks, they may act as keystone species by influencing the diversity and relative abundance of species by opening space for rarer, or better defended, less competitive species. A variety of global changes, including global warming, increases in extreme weather events, and habitat fragmentation will likely enhance the frequency and magnitude of outbreaks in natural forests by opening abundant opportunities for pioneers.

We know little about the relationship between tropical climate and insect outbreaks. Wetter forests, however, are more diverse, and have longer lived, better defended leaves and parasitoids that are better able to track their herbivore prey. This suggests that insect outbreaks will be more frequent in dry forest, where new leaves flush in overwhelming abundance early in the rainy season. We ended this chapter by outlining a series of testable hypotheses to address the large number of unknowns that currently surround the prevalence and impact of insect outbreaks in tropical forests.

Acknowledgments

We thank an anonymous reviewer, D. Letourneau, and especially V. Novotny for helpful comments on earlier drafts. We thank P. Barbosa and D. Letourneau for inviting us to submit this chapter.

References

Aiello, A. 1992. Dry season strategies of two Panamanian butterfly species, *Anartia fatima* (Nymphalinae) and *Pierella luna luna* (Satyrinae) (Lepidoptera: Nymphalidae). Pages 573–575 *in* D. Quintero and A. Aiello (editors), *Insects of Panama and Mesoamerica: Selected Studies*. Oxford University Press, New York.

Azerefegne, F., C. Solbreck, and A. R. Ives. 2001. Environmental forcing and high amplitude fluctuations in the population dynamics of the tropical butterfly *Acraea acerata* (Lepidoptera: Nymphalidae). *Journal of Animal Ecology* 70:1032–1045.

Azerefegne, F., and C. Solbreck. 2009. Oviposition preference and larval performance of the sweet potato butterfly *Acraea acerata* on *Ipomoea* species in Ethiopia. *Agricultural and Forest Entomology* 12:161–168.

Basset, Y. 1992. Influence of leaf traits on the spatial-distribution of arboreal arthropods within an overstory rain-forest tree. *Ecological Entomology* 17:8–16.

Berryman A. A. 1999. *Principles of Population Dynamics and Their Application*. Stanley Thornes Publishers, Glasgow.

Bigger, M. 1976. Oscillations of tropical insect populations. *Nature* 259:207–209.

Billings, R. F., S. R. Clarke, V. Espino Mendoza, P. Cordon Cabrera, B. Melendez Figueroa, J. Ramon Campos, and G. Baeza. 2004. Bark beetle outbreaks and fire: a devastating combination for Central America's pine forests. *Unasylva* 55:15–21.

Buntgen, U., D. Frank, A. Liebhold, D. Johnson, M. Carrer, C. Urbinati, M. Grabner, K. Nicolussi, T. Levanic, and J. Esper. 2009. Three centuries of insect outbreaks across the European Alps. *New Phytologist* 182:929–941.

Carson, W. P., J. T. Anderson, E. G. Leigh Jr., and S. A. Schnitzer. 2008. Challenges associated with testing and falsifying the Janzen–Connell hypothesis: a review and critique. Pages 210–241 *in* W. P. Carson, and S. A. Schnitzer (editors), *Tropical Forest Community Ecology*. Wiley-Blackwell, Oxford.

Carson, W. P., and R. B. Root. 2000. Herbivory and plant species coexistence: community regulation by an outbreaking phytophagous insect. *Ecological Monographs* 70:73–99.

Carson, W. P., J. P. Cronin, and Z. T. Long. 2004. A general rule for predicting when insects will have strong top-down effects on plant communities: on the relationship between insect outbreaks and host concentration. Pages 193–212 *in* W. W. Weisser and E. Siemann (editors), *Insects and Ecosystem Functions*. Springer-Verlag, Berlin.

Coley, P. D. 1998. The effects of climate change on plant–herbivore interactions in moist tropical rainforests. *Climate Change* 39:455–472.

Coley, P. D., and T. M. Aide. 1991. Comparison of herbivory and plant defenses in temperate and tropical broad-leaved forests. Pages 25–49 *in* P. W. Price, T. M. Lewinsohn, G. W. Fernandes, and W. W. Benson (editors), *Plant–Animal Interactions: Evolutionary Ecology in Tropical and Temperate Regions*. John Wiley & Sons, Inc., New York.

Coley, P. D., and J. A. Barone. 1996. Herbivory and plant defenses in tropical forests. *Annual Review of Ecology and Systematics* 27:305–335.

Currano, E. D., P. Wilf, S. L. Wing, C. C. Labandeira, E. C. Lovelock, and D. L. Royer. 2008. Sharply increased insect herbivory during the Paleocene-Eocene thermal maximum. *Proceedings of the National Academy of Sciences of the USA* 105:1960–1964.

Dyer, L. A., and Gentry G. L. 2010. Caterpillars and parasitoids of a Costa Rican tropical wet forest. http://www.caterpillars.org

Dyer, L. A., R. B. Matlock, D. Chehrezad, and R. O'Malley. 2005. Predicting caterpillar parasitism in banana plantations. *Environmental Entomology* 34:403–409.

Dyer, L. A., M. S. Singer, J. T. Lill, J. O. Stireman, G. L. Gentry, R. J. Marquis, R. E. Ricklefs, H. F. Greeney, D. L. Wagner, H. C. Morais, I. R. Diniz, T. A. Kursar, and P. D. Coley. 2007. Host specificity of Lepidoptera in tropical and temperate forests. *Nature* 448:696–699.

Easterling, D. R., G. A. Meehl, C. Parmesan, S. A. Changnon, T. R. Karl, and L. O. Mearns. 2000. Climate extremes: observations, modeling, and impacts. *Science* 289:2068–2074.

Elton, C. S. 1958. *The Ecology of Invasions by Animals and Plants*. Chapman & Hall, London.

Eveleigh, E. S., K. S. McCann, P. C. McCarthy, S. J. Pollock, C. J. Lucarotti, B. Morin, G. A. McDougall, D. B. Strongman, J. T. Huber, J. Umbanhowar, and L. D. B. Faria. 2007. Fluctuations in density of an outbreak species drive diversity cascades in food webs. *Proceedings of the National Academy of Sciences of the USA* 104:16976–16981.

Feeley, K. J., and J. W. Terborgh. 2008. Direct versus indirect effects of habitat reduction on the loss of avian species from tropical forest fragments. *Animal Conservation* 11:353–360.

Food and Agricultural Organization (FAO). 2011. Forest health. http://www.fao.org/forestry/pests/en/

Gauld, I. D., K. J. Gaston, and D. H. Janzen. 1992. Plant allelochemicals, tritrophic interactions and the anomalous diversity of tropical parasitoids: the "nasty" host hypothesis. *Oikos* 65:353–357.

Gentry, A. H. 1982. Patterns of neotropical plant-species diversity. *Evolutionary Biology* 15:1–85.

Gentry, A. H. 1988. Changes in plant community diversity and floristic composition on environmental and geographical gradients. *Annals of the Missouri Botanical Garden* 75:1–34.

Gerhardt, K. 1998. Leaf defoliation of tropical dry forest tree seedlings – implications for survival and growth. *Trees – Structure and Function* 13:88–95.

Ghimre, K., and A. Dongre. 2001. Sal borer problem in Indian Sal Forests: a serious threat to *Shorea robusta* forests ecosystems of India. http://www.iifm.ac.in/databank/problems/salborer.html

Gilman, S. E., M. C. Urban, J. Tewksbury, G. W. Gilchrist, and R. D. Holt. 2010. A framework for community interactions under climate change. *Trends in Ecology and Evolution* 25:325–331.

Godfray, H. C. J., and M. P. Hassell. 1989. Discrete and continuous insect populations in tropical environments. *Journal of Animal Ecology* 58:153–174.

Gray, B. 1972. Economic tropical forest entomology. *Annual Review of Entomology* 17:313–354.

Greenbank, D. O. 1956. The role of climate and dispersal in the initiation of outbreaks of the spruce budworm in New Brunswick. *Canadian Journal of Zoology* 34:453–476.

Grogan, J. 2001. Bigleaf mahogany (*Swietenia macrophylla* King) in southeast Pará, Brazil: a life history study with management guidelines for sustained production from natural forests. PhD dissertation, Yale University.

Haines, W. P., M. L. Heddle, P. Welton, and D. Rubinoff. 2009. A recent outbreak of the hawaiian koa moth, *Scotorythra paludicola* (Lepidoptera: Geometridae), and a review of outbreaks between 1892 and 2003. *Pacific Science* 63:349–369.

Hall, J. B., and M. D. Swaine. 1976. Classification and ecology of closed-canopy forest in Ghana. *Journal of Ecology* 64:913–951.

Hastings, A. 1999. Outbreaks of insects: a dynamic approach. Pages 206–215 *in* B.A. Hawkins and H.V. Cornell (editors), *Theoretical Approaches to Biological Control*. Cambridge University Press, Cambridge.

Hernandez, J. V., F. Osborn, B. Herrera, C. V. Liendo-Barandiaran, J. Perozo, and D. Velasquez. 2009. Larvae–pupae parasitoids of *Hylesia metabus* Cramer (Lepidoptera: Saturniidae) in northeastern Venezuela: a case of natural biological control. *Neotropical Entomology* 38:243–250.

Hölldobler B., and E. O. Wilson. 1990. *The Ants*. Harvard University Press, Cambridge, MA.

Hulme, M., and D. Viner. 1998. A climate change scenario for the tropics. *Climatic Change* 39:145–177.

Hunter, A. F. 1991. Traits that distinguish outbreaking and nonoutbreaking Macrolepidoptera feeding on northern hardwood trees. *Oikos* 60:275–282.

Hunter, A. F. 1995. Ecology, life history, and phylogeny of outbreak and nonoutbreak species. Pages 41–64 *in* N. Cappuccino and P. W. Price (editors), *Population Dynamics: New Approaches and Synthesis*. Oxford University Press, New York.

Hunter, M. D. 2001. Multiple approaches to estimating the relative importance of top-down and bottom-up forces on insect populations: experiments, life tables, and time-series analysis. *Basic and Applied Ecology* 2:295–309.

Huston, M. 1980. Soil nutrients and tree species richness in Costa Rican forests. *Journal of Biogeography* 7:147–157.

Iserhard, C. A., L. A. Kaminski, M. O. Marchiori, E. C. Teixeira, and H. P. Romanowski. 2007. Occurrence of lepidopterism caused by the moth *Hylesia nigricans* (Berg) (Lepidoptera: Saturniidae) in Rio Grande do Sul State, Brazil. *Neotropical Entomology* 36:612–615.

Itkao, T., and M. Yamauti. 2004. Severe drought, leafing phenology, leaf damage and lepidopteran abundance in the canopy of a Bornean aseasonal tropical rain forest. *Journal of Tropical Ecology* 20:479–482.

Jactel, H., and E. G. Brockerhoff. 2007. Tree diversity reduces herbivory by forest insects. *Ecology Letters* 10:835–848.

Jaksic, F. M. 2001. Ecological effects of El Niño in terrestrial ecosystems of western South America. *Ecography* 24:241–250.

Janzen, D. H. 1981. Patterns of herbivory in a tropical deciduous forest. *Biotropica* 13:271–282.

Janzen D. H. 1983. *Costa Rican Natural History*. University of Chicago Press, Chicago.

Janzen, D. H. 1984. Natural history of *Hylesia lineata* (Saturniidae, Hemileucinae) in Santa Rosa National Park, Costa Rica. *Journal of the Kansas Entomological Society* 57:490–514.

Janzen, D. H. 1993. Caterpillar seasonality in a Costa Rican dry forest. Pages 448–477 *in* N. E. Stamp, and T. M. Casey (editors), *Caterpillars: Ecological and Evolutionary Constraints on Foraging*. Chapman & Hall, New York.

Johnson, D. M., U. Buntgen, D. C. Frank, K. Kausrud, K. J. Haynes, A. M. Liebhold, J. Esper, and N. C. Stenseth. 2010. Climatic warming disrupts recurrent alpine insect outbreaks. *Proceedings of the National Academy of Sciences of the USA* 107:20576–20581.

Kalshoven, L. G. E. 1953. Important outbreaks of insect pests in the forests of Indonesia. *Transactions of the 9th International Congress of Entomology* 2:229–234.

Knight, T. M., M. W. McCoy, J. M. Chase, K. A. McCoy, and R. D. Holt. 2005. Trophic cascades across ecosystems. *Nature* 437:880–883.

Kondoh, M. 2003. Habitat fragmentation resulting in overgrazing by herbivores. *Journal of Theoretical Biology* 225:453–460.

Lara E. F. 1970. *Problemas y procedimientos bananeros en la zona Atlántica de Costa Rica*. Imprenta Trejos Anos, San Jose, Costa Rica.

Leigh, E. G., Jr., P. Davidar, C. W. Dick, J-P. Puyravaud, J. Terborgh, H. ter Steege, and S. J. Wright. 2004. Why do some tropical forests have so many species of trees? *Biotropica* 36:447–473.

Leigh, E. G., Jr., G. J. Vermeij, and M. Wikelski. 2009. What do human economies, large islands and forest fragments reveal about the factors limiting ecosystem evolution? *Journal of Evolutionary Biology* 22:1–12.

Letourneau, D. K. 1998. Ants, stem-borers, and fungal pathogens: experimental tests of a fitness advantage in *Piper* ant-plants. *Ecology* 79:593–603.

Letourneau, D. K. 2004. Mutualism, antiherbivore defense, and trophic cascades: *Piper* ant-plants as a mesocosm for experimentation. Pages 5–33 *in* L. A. Dyer and A. P. N. Palmer (editors), *Piper: A Model Genus for Studies of Phytochemistry, Ecology, and Evolution*. Kluwer Academic, New York.

Letourneau, D. K., and L. A. Dyer. 1998. Density patterns of *Piper* ant-plants and associated arthropods: top predator cascades in a terrestrial system? *Biotropica* 30:162–169.

Letourneau, D. K., J. A. Jedlicka, S. G. Bothwell, and C. R. Moreno. 2009. Effects of natural enemy biodiversity on the suppression of arthropod herbivores in terrestrial ecosystems. *Annual Review of Ecology Evolution and Systematics* 40:573–592.

Logan, J. A., J. Regniere, and J. A. Powell. 2003. Assessing the impacts of global warming on forest pest dynamics. *Frontiers in Ecology and the Environment* 1:130–137.

Long, Z., C. Mohler, and W. P. Carson. 2003. Herbivory, litter accumulation, and plant diversity: applying the resource concentration hypothesis to plant communities. *Ecology* 84:652–665.

Lowman, M. D. 1997. Herbivory in forests: from centimeters to megameters. Pages 135–149 *in* A. D. Watt, N. E. Stork, and M. D. Hunter (editors), *Forests and Insects*. Chapman & Hall, New York.

Maron, J. L., and S. Harrison. 1997. Spatial pattern formation in an insect host–parasitoid system. *Science* 278:1619–1621.

Massad, T. J. 2009. The efficacy and environmental controls of plant defenses and their application to tropical reforestation. PhD dissertation, Tulane University.

Massad, T. J., and L. A. Dyer. 2010. A meta-analysis of the effects of global environmental change on plant–herbivore interactions. *Arthropod–Plant Interactions* 4:181–188.

Massad, T. J., Fincher, R. M., Smilanich, A. M., and L. A. Dyer. 2011. A quantitative evaluation of major plant defense hypotheses, nature versus nurture, and chemistry versus ants. *Arthropod–Plant Interactions* 5:125–139.

Matlock, R. B., and R. de la Cruz. 2002. An inventory of parasitic Hymenoptera in banana plantations under two pesticide regimes. *Agriculture Ecosystems & Environment* 93:147–164.

McCloskey, S. P. J., L. D. Daniels, and J. A. Mclean. 2009. Potential impacts of climate change on western hemlock looper outbreaks. *Northwest Science* 83:225–238.

Meyer, S. T., I. R. Leal, and R. Wirth. 2009. Persisting hyper-abundance of leaf-cutting ants (*Atta* spp.) at the edge of an old Atlantic forest fragment. *Biotropica* 41:711–716.

Moreau, G., E. S. Eveleigh, C. J. Lucarotti, and D. T. Quiring. 2006. Ecosystem alteration modifies the relative strengths of bottom-up and top-down forces in a herbivore population. *Journal of Animal Ecology* 75:853–861.

Moreira, S. C., J. C. de Lima, L. Silva, and V. Haddad. 2007. Description of an outbreak of lepidopterism (dermatitis associated with contact with moths) among sailors in Salvador, State of Bahia. *Revista da Sociedade Brasileira de Medicina Tropical* 40:591–593.

Myers, J. H. 1988. Can a general hypothesis explain population-cycles of forest Lepidoptera? *Advances in Ecological Research* 18:179–242.

Nair, K. S. S. 2001. *Pest Outbreaks in Tropical Forest Plantations: Is There a Greater Risk for Exotic Tree Species?* Center for International Forestry Research, Bogor, Indonesia.

Nair, K. S. S. 2000. *Insect Pests and Diseases in Indonesian Forests: An Assessment of the Major Threats, Research Efforts and Literature.* Center for International Forestry Research, Bogor, Indonesia.

Nair, K. S. S. 2007. *Tropical Forest Insect Pests.* Cambridge University Press, Cambridge.

Nascimento, M. T., and J. Proctor. 1994. Insect defoliation of a monodominant Amazonian rainforest. *Journal of Tropical Ecology* 10:633–636.

Nothnagle, P. J., and J. C. Schultz. 1987. What is a forest pest? Pages 59–80 *in* P. Barbosa and J. C. Schultz (editors), *Insect Outbreaks.* Academic Press, San Diego.

Novotny, V., and Y. Basset. 2000. Rare species in communities of tropical insect herbivores: pondering the mystery of singletons. *Oikos* 89:564–572.

Opler, P. A., G. W. Frankie, and H. G. Baker. 1980. Comparative phenological studies of treelet and shrub species in tropical wet and dry forests in the lowlands of Costa Rica. *Journal of Ecology* 68:167–188.

Ostmark, E. H. 1974. Economical insect pests of bananas. *Annual Review of Entomology* 19:161–175.

Parry, D., D. A. Herms, and W. J. Mattson. 2003. Responses of an insect folivore and its parasitoids to multiyear experimental defoliation of aspen. *Ecology* 84:1768–1783.

Perfecto, I., J. H. Vandermeer, G. L. Bautista, G. I. Nunez, R. Greenberg, P. Bichier, and S. Langridge. 2004. Greater predation in shaded coffee farms: the role of resident neotropical birds. *Ecology* 85:2677–2681.

Pescador, A. R. 1993. The effects of a multispecies sequential diet on the growth and survival of a tropical polyphagous caterpillar. *Entomologia Experimentalis et Applicata* 67:15–24.

Pescador, A. R. 1995. Distribution and abundance of *Hylesia ineata* egg masses in a tropical dry forest in western Mexico. *Southwestern Entomologist* 20:367–375.

Philips, O. L., P. Hall, A. H. Gentry, S. A. Sawyer, and R. Vasquez. 1994. Dynamics and species richness of tropical rain forests. *Proceedings of the National Academy of Sciences of the USA* 91:2805–2809.

Piyakarnchana, T. 1981. Severe defoliation of *Avicennia alba* BL. by larvae of *Cleora injectaria* Walker. *Journal of the Science Society of Thailand* 7:33–36.

Pogue, M. G., and A. Aiello. 1999. Description of the immature stages of three species of *Eulepidotis guenee* (Lepidoptera: Noctuidae) with notes on their natural history. *Proceedings of the Entomological Society of Washington* 101:300–311.

Ray, C., and A. Hastings. 1996. Density dependence: are we searching at the wrong spatial scale? *Journal of Animal Ecology* 65:556–566.

Redfearn, A., and S. L. Pimm. 1988. Population variability and polyphagy in herbivorous insect communities. *Ecological Monographs* 58:39–55.

Richards, L. A., and P. D. Coley. 2008. Combined effects of host plant quality and predation on a tropical lepidopteran: a comparison between treefall gaps and the understory in Panama. *Biotropica* 40:736–741.

Roland, J. 1993. Large-scale forest fragmentation increases the duration of tent caterpillar outbreak. *Oecologia* 93:25–30.

Roonwal, M. L. 1978. The biology, ecology and control of the sal heartwood borer, *Hoplocerambyx spinicornis*: a review of recent work. *Indian Journal of Entomology* 1:107–120.

Root, R. B. 1973. Organization of a plant–arthropod association in simple and diverse habitats: the fauna of collards (*Brassica oleracea*). *Ecological Monographs* 43:95–124.

Root, R. B., and N. Cappuccino. 1992. Patterns in population-change and the organization of the insect community associated with goldenrod. *Ecological Monographs* 62:393–420.

Schowalter, T. D. 1996. *Insect Ecology: An Ecosystem Approach*. Academic Press, San Diego.

Schowalter, T. D., and M. D. Lowman. 1999. Forest herbivory: insects. Pages 253–270 *in* L. R. Walker (editor), *Ecosystems of Disturbed Ground*. Elsevier, Amsterdam.

Scriber, J. M. 1973. Latitudinal gradients in larval feeding specialization of the world Papilionidae (Lepidoptera). *Psyche* 73:355–373.

Singh, R., G. S. Goraya, C. Singh, H. Kumar, and S. Kumar. 2001. Mortality of chir pine trees by insect borers in Morni Hills, Haryana: a case study. *Indian Forester* 127:1279–1286.

Singh, T. V. K., and J. Satyanarayana. 2009. Insect outbreaks and their management. Pages 331–350 *in* R. Peshin, and A. K. Dhawan (editors), *Integrated Pest Management: Innovation-Development Process*. Springer, New York.

Specht, A., A. C. Formentini, and E. Corseuil. 2006. Biology of *Hylesia nigricans* (Berg) (Lepidoptera, Saturniidae, Hemileucinae). *Revista Brasileira de Zoologia* 23:248–255.

Srygley, R. B., R. Dudley, E. G. Oliveira, R. Aizprua, N. Z. Pelaez, and A. J. Riveros. 2010. El Nino and dry season rainfall influence hostplant phenology and an annual butterfly migration from neotropical wet to dry forests. *Global Change Biology* 16:936–945.

Stephens, C. S. 1984. Ecological upset and recuperation of natural control of insect pests in some Costa Rican banana plantations. *Turrialba* 34:101–105.

Stireman, J. O., L. A. Dyer, D. H. Janzen, M. S. Singer, J. T. Li, R. J. Marquis, R. E. Ricklefs, G. L. Gentry, W. Hallwachs, P. D. Coley, J. A. Barone, H. F. Greeney, H. Connahs, P. Barbosa, H. C. Morais, and I. R. Diniz. 2005. Climatic unpredictability and parasitism of caterpillars: implications of global warming. *Proceedings of the National Academy of Sciences of the USA* 102:17384–17387.

Stireman, J. O., L. A. Dyer, and R. B. Matlock. 2004. Top-down forces in managed versus unmanaged habitats. Pages 303–323 *in* P. Barbosa, and I. Castellanos (editors), *Ecology of Predator–Prey Interactions*. Oxford University Press, Oxford.

Strong D. R., J. H. Lawton, and T. R. E. Southwood. 1984. *Insects on Plants: Community Patterns and Mechanisms*. Harvard University Press, Cambridge, MA.

Terborgh, J., L. Lopez, P. Nunez, M. Rao, G. Shahabuddin, G. Orihuela, M. Riveros, R. Ascanio, G. H. Adler, T. D. Lambert, and L. Balbas. 2001. Ecological meltdown in predator-free forest fragments. *Science* 294:1923–1926.

Thrupp, L. A. 1990. Entrapment and escape from fruitless insecticide use: lessons from the banana sector of Costa Rica. *International Journal of Environmental Studies* 36:173–189.

Tin, N. T. 1990. Biology of the sawfly *Schizocera* sp. (Hymenoptera: Agridae) in Northern Vietnam. Pages 194–197 *in* C. Hutacharern, K. G. MacDicken, M. H. Ivory, and K. S. S. Nair (editors). FAO, Regional Office for Asia and the Pacific, Bangkok.

Torres, J. A. 1992. Lepidoptera outbreaks in response to successional changes after the passage of Hurricane Hugo in Puerto Rico. *Journal of Tropical Ecology* 8:285–298.

Turchin, P. 2003. *Complex Population Dynmaics: A Theoretical/Empirical Synthesis*. Princeton University Press, Princeton.

Van Bael, S. A., A. Aiello, A. Valderrama, E. Medianero, M. Samaniego, and S. J. Wright. 2004. General herbivore outbreak following an El Niño–related drought in a lowland Panamanian forest. *Journal of Tropical Ecology* 20:625–633.

Vandermeer, J., I. Perfecto, and S. Philpott. 2010. Ecological complexity and pest control in organic coffee production: uncovering an autonomous ecosystem service. *Bioscience* 60:527–537.

van Schaik, C. P., J. W. Terborgh, and S. J. Wright. 1993. Phenology of tropical forests: adaptive signficance and consequences for primary consumers. *Annual Review of Ecology and Systematics* 24:353–377.

Veldtman, R., M. A. McGeoch, and C. H. Scholtz. 2007. Can life-history and defence traits predict the population dynamics and natural enemy responses of insect herbivores? *Ecological Entomology* 32:662–673.

Vilela, E. F. 1986. Status of leaf-cutting ant control in forest plantations in Brazil. Pages 399–408 *in* C. S. Lofgren and R. K. Vander Meer (editors), *Fire Ants and Leaf-Cutting Ants: Biology and Management*. Westview Press, Boulder, CO.

Wagner M. R., S. K. N. Atuahene, and J. R. Cobbinah. 1991. *Forest Entomology in West Tropical Africa: Forest Insects of Ghana*. Kluwer Academic, Dordrecht.

Wallner, W. E. 1987. Factors affecting insect population dynamics: differences between outbreak and non-outbreak species. *Annual Review of Entomology* 32:317–340.

Whitten, A. J., and S. J. Damanik. 1986. Mass defoliation of mangroves in Sumatra, Indonesia. *Biotropica* 18:176.

Wiens, J. J., D. D. Ackerly, A. P. Allen, B. L. Anacker, L. B. Buckley, H. V. Cornell, E. I. Damschen, T. J. Davies, J. A. Grytnes, S. P. Harrison, B. A. Hawkins, R. D. Holt, C. M. Mccain, and P. R. Stephens. 2010. Niche conservatism as an emerging principle in ecology and conservation biology. *Ecology Letters* 13:1310–1324.

Wolda, H. 1978. Fluctuations in abundance of tropical insects. *American Naturalist* 112:1017–1045.

Wolda, H. 1983. Long-term stability of tropical insect populations. *Researches on Population Ecology* 112–126.

Wolda, H., and R. Foster. 1978. *Zunacetha annulata* (Lepidoptera: Dioptidae): an outbreak insect in a neotropical forest. *Geo-Eco-Trop* 2:443–454.

Wolda, H. 1992. Trends in abundance of tropical forest insects. *Oecologia* 89:47–52.

Wong, M., S. J. Wright, S. P. Hubbell, and R. B. Foster. 1990. The spatial pattern and reproductive consequences of outbreak defoliation in *Quararibea asterolepis*, a tropical tree. *Journal of Ecology* 78:579–588.

Yang, L. H., K. F. Edwards, J. E. Byrnes, J. L. Bastow, A. N. Wright, and K. O. Spence. 2010. A meta-analysis of resource pulse–consumer interactions. *Ecological Monographs* 80:125–151.

12

Outbreaks and Ecosystem Services

Timothy D. Schowalter

12.1 Introduction

Ecosystems provide a variety of services on which humans, and other organisms, depend for survival. Ecosystem services can be categorized as provisioning (harvestable production of food, fiber, water and other resources), cultural (spiritual and recreational values), supporting (primary production, pollination, and soil formation), and regulating (maintenance of consistent supply of other ecosystem services through density-dependent feedback) (Millenium Ecosystem Assessment 2005). Ascribing economic values to ecosystem services is difficult, because only provisioning services, some recreational services that provide user fees, and pollination services for crop production have established market values (Dasgupta *et al.* 2000).

Herbivorous insects, especially during outbreaks, affect a variety of ecosystem services through changes in primary production, vegetation cover, and fluxes of energy and nutrients (Klock and Wickman 1978, Leuschner 1980, White and Schneeberger 1981). Short-term losses in food and fiber production are obvious, and their economic value is easily measured and used to develop economic thresholds for pest control (Pedigo *et al.* 1986, Torrell *et al.* 1989). When the cost of pest control is less than the anticipated value of losses to insects, pest management appears to be warranted. However, outbreaks benefit some ecosystem products, and complex long-term effects on supporting and regulating

Insect Outbreaks Revisited, First Edition. Edited by Pedro Barbosa, Deborah K. Letourneau
and Anurag A. Agrawal.

services can compensate for short-term losses in provisioning services. Few attempts have been made to evaluate the relative benefits or costs of many of these outbreak effects for noncommodity resources (Klock and Wickman 1978, Leuschner 1980, White and Schneeberger 1981), but management decisions should be based on trade-offs among short- and long-term effects on multiple ecosystem services, as described in this chapter.

12.2   Effects on provisioning services

Ecosystems are sources of a variety of food, water, wood, and pharmaceutical and industrial products that have been the basis for agricultural or silvicultural production. Fruits, seeds, tubers, and other plant parts are widely used foods for humans or livestock. Woody materials are used in construction of homes, furnishings, and fencing, as well as for firewood. Plant defensive chemicals are the original sources for important pharmaceutical compounds (such as, salicylic acid, morphine, quinine, epinephrin, and camphor) (e.g., Zenk and Juenger 2007). Exploration continues for new pharmaceutical compounds (Helson *et al.* 2009), and many plant and animal materials are widely used in traditional remedies. Plant-derived tannins, resins, and other compounds are used as dyes, adhesives, lacquer, and other industrial products.

Insect outbreaks sometimes reduce harvest of such resources and may warrant substantial expenditures for pest control. For example, locust outbreaks have destroyed entire crop and rangeland production over large areas and caused massive human migration from devastated areas or required extraordinary economic aid to sustain farmers or ranchers on their land (Riley 1878, Smith 1954, Pfadt and Hardy 1987). Even during average years, 21–23% loss of available range vegetation due to grasshopper feeding represents a loss of about $393 million (Pfadt and Hardy 1987). Spread of boll weevil, *Anthonomus grandis* Boheman, from native cotton, *Gossypium* spp., throughout the cotton belt of the southern United States during the late 1800s and early 1900s devastated the southern economy, ended the reign of cotton as the dominant crop in the South, and led to a massive demographic shift from bankrupt farms and communities (Smith 2007). Tree mortality caused by gypsy moth, *Lymantria dispar* (L.), defoliation was >$104 million over a three-year period in Pennsylvania (Ticehurst and Finley 1988). Johnson *et al.* (2006) reported that more than $194 million was spent on monitoring and control of gypsy moth in the United States during 1985–2004. However, marginal benefits of control in forests and grasslands may warrant control for only targeted sites with very high outbreak populations or very high resource values (Shewchuk and Kerr 1993, Zimmerman *et al.* 2004).

Acceptance of control costs by producers and the general public varies with the degree of visual damage or food shortage (Torrell *et al.* 1989, Sheppard and Picard 2006). When control costs are substantially subsidized by the government, range or forest managers are inclined to control insects at lower densities than would be acceptable to taxpayers. For example, an individual rancher's 50% share of the $6.18 ha^{-1} cost for control at an economic injury level of 18 grasshoppers ha^{-1} is equivalent to a rancher's economic injury level of <3 grasshoppers ha^{-1} (Torrell *et al.* 1989). Gatto *et al.* (2009) conducted an economic analysis of pest management for processionary moth, *Thaumetopoea pityocampa* Denis and Schiffermüller, in Portugal and concluded that pest management costs outweighed market revenues for maritime pine, *Pinus pinaster* Aitonx plantations, at least in the short term, making control undesirable for private landowners. Taxpayer support also would be unwise, based on provisioning service values alone, but could be justified by potential benefits to the public through other types of ecosystem services, such as improved carbon sequestration, recreation, and public health (i.e., non-provisioning ecosystem services).

However, in other cases, pest management might be more detrimental to other ecosystem services than justified by its value for the provisioning service. For example, insecticides have documented toxicity for many nontarget species, especially fish and pollinators, threatening sustainability of provisioning and supporting services derived from these species (Smith *et al.* 1983, Claudianos *et al.* 2006, Baldwin *et al.* 2009). Furthermore, nontarget effects could undermine important ecosystem functions that contribute to the long-term sustainability of all ecosystem services (Downing and Leibold 2002, Hättenschwiler and Gasser 2005).

Wildlife and fish are important food sources worldwide, and maintenance of their populations often is a primary management goal. Many of these animals feed primarily or exclusively on insects, and their abundances may increase during insect outbreaks (Koenig and Liebhold 2005). Insects falling into streams comprise 30–80% of the diets of young salmon, with herbivores composing at least 20–25% (Allan *et al.* 2003, Baxter *et al.* 2005). Menninger *et al.* (2008) found that large numbers of emergent periodical cicadas falling into aquatic ecosystems provided sufficient pulses of carbon and nitrogen to stimulate aquatic productivity. Carbohydrate-rich honeydew from aphids, scale insects, and other Hemiptera is a rich food source for hummingbirds, many ant species, and honey bees, *Apis mellifera* L. (Edwards 1982), illustrating how insect outbreaks can benefit other provisioning and supporting services (see below). In addition, canopy opening during outbreaks increases light availability for understory plants (Collins 1961), many of which provide important floral resources for bees and other pollinators (Regal 1982, Kudo *et al.* 2008). Flowering and pollinator visitation are limited by light availability for many understory plants (Kudo *et al.* 2008). Collins (1961) reported that light intensity increased 50% under an oak canopy defoliated by gypsy moth, permitting continued growth of understory red maple, *Acer rubrum* L., for two weeks longer than seen for maples under intact canopies. Pollination by insects is necessary for 35% of global crop production, but in many areas depends on the proximity of crops requiring pollination to refuges in natural habitats that have a sufficient abundance of flowering plants to support wild bees or other pollinators (Kremen *et al.* 2002). Therefore, insect outbreaks may be largely beneficial for wildlife, fish, and pollinator management.

Natural ecosystems with intact mechanisms for filtering water and minimizing runoff are valued sources of fresh water, which often is the primary management goal for municipal watersheds. Insect outbreaks increase water yields (the amount of water leaving a watershed), as a result of reduced water uptake and evapotranspiration by plants (Klock and Wickman 1978, Leuschner 1980, White and

Schneeberger 1981), but may temporarily increase stream water concentrations of nitrogen and other elements leached from plant tissues and soil (Swank *et al.* 1981, Eshleman *et al.* 1998, Lovett *et al.* 2002, Hunter *et al.* 2003). Increased light availability to streams draining outbreak areas promotes the growth of algae and supported aquatic invertebrates (Kiffney *et al.* 2003). Large numbers of insects falling into lakes and streams also add substantial amounts of carbon, nitrogen, and phosphorus to aquatic ecosystems, contributing to the resource base of aquatic food webs (Menninger *et al.* 2008, Pray *et al.* 2009). Insect-induced increases in nutrient export generally are negligible and should have little effect on the quality of municipal water supplies (Lovett *et al.* 2002). The extent to which changes in water yield and quality are positive or negative depends on the needs of downstream communities. For example, increased water yield during a drought, a typical trigger for outbreaks (Mattson and Haack 1987, Schowalter *et al.* 1999, Van Bael *et al.* 2004), would maintain a higher supply of water to municipalities (than would occur without the outbreak) and be seen as a benefit, whereas excess yields in some cases could flood downstream communities.

Herbivorous insects represent valuable food resources in many parts of the world, and outbreaks increase availability of these resources. Although this service is not valued currently in Europe or North America, grasshoppers, cicadas, crickets, caterpillars, beetles, and other herbivorous insects make up 5–10% of dietary protein in some cultures (Ramos-Elorduy 2009, Yen 2009), including Native Americans in historic times. For example, pandora moth, *Coloradia pandora* Blake, larvae, and pupae were harvested by Native Americans in pine forests of Oregon and northern California (Furniss and Carolin 1977). Increased abundance of edible caterpillars (primarily two saturniids, *Gynanisa maja* Strand and *Gonimbrasia zambesina* Walker) in Zambian forests is of great food value to indigenous cultures, and caterpillar harvest is ritually regulated (Mbata *et al.* 2002), demonstrating the importance of this provisioning service. In a unique study of the costs and benefits of insect consumption versus control, Cerritos and Cano-Santana (2008) calculated that harvest of grasshoppers for sale during an outbreak in Mexico substantially reduced grasshopper damage and provided US$3000 in revenue per family, compared to a cost of US$150 per family for insecticide treatment.

Silkworms, *Bombyx mori* (L.), are the primary source of commercial silk. The economic value of silk supported the historic Silk Roads that connected Europe, the Middle East, and China for at least 500 years, and was responsible for the introduction of gypsy moth and other silk-producing species into North America as unsuccessful efforts to establish silk production in the West (Andrews 1868, Forbush and Fernald 1896, Anelli and Prischmann-Voldseth 2009). Silkworms and other insects provide a variety of medically useful compounds (Singh and Jayasomu 2002). Scale insects remain an important source of commercial shellac. Furthermore, insects can be used to identify plants with pharmaceutically active compounds (Helson *et al.* 2009). Outbreaks increase the availability of these insect-derived resources. Increased abundance of insects as a provisioning service, where recognized, requires consideration of trade-offs among various provisioning services in deciding whether or not to control an outbreak.

12.3 Effects on cultural services

Ecosystems provide various spiritual, recreational, and other cultural services, including hiking, backpacking, hunting and fishing, and educational and scientific activities. The global value of recreational services alone (which often can be calculated from usage fees) has been estimated at US $815 billion by Costanza *et al.* (1997). Effects of insect outbreaks on cultural services can be positive or negative depending on the extent of outbreak and public perceptions. Some insects have been important cultural symbols, and their increased abundance is sometimes viewed as a positive event. In China cicadas are symbols of rebirth, and crickets are symbols of good fortune; both insects often are caged and kept as pets for their songs. However, both are considered nuisances in many other cultures (Clausen 1954).

Outbreaks also can have negative impacts on cultural services. Trees killed by insects can create a safety hazard in camping or hiking areas. Defoliation reduces shade and may be perceived as unsightly. Insect frass and tissues falling on people or eating surfaces are considered a nuisance. Furthermore, some caterpillars are venomous or allergenic (Perlman *et al.* 1976, Schowalter 2011), exacerbating the nuisance. These detrimental effects can reduce visitation to recreational sites experiencing outbreaks.

Few studies have evaluated effects of insect outbreaks on cultural values. Downing and Williams (1978) reported that a Douglas-fir tussock moth, *Orgyia pseudotsugata* (McDunnough), outbreak in Oregon did not significantly affect recreational land use but that recreational use appeared to increase as a result of curiosity. Although 75% of visitors were aware of the outbreak, few chose to avoid the area. In fact, the only negative effect mentioned in their study was avoidance of salvage logging operations that were considered unappealing or hazardous. On the other hand, extensive defoliation or plant mortality may be viewed as unattractive or hazardous (Michalson 1975). Sheppard and Picard (2006) compiled a number of studies in which subjects were shown pairs of photos, one with insect damage and the other without (Plate 12.1). In general, visual preference depended on the type of outbreak (e.g., concentrated tree mortality caused by bark beetles vs. widespread defoliation). In some studies, overall visual preference increased for low levels of outbreak effect, but most studies showed decline in visual preference with increasing defoliation or plant mortality, sometimes showing a threshold of about 10% of visible landscape, above which additional defoliation or mortality had less effect. Some studies showed that visual preference was affected by the subject's awareness of the cause.

12.4 Effects on supporting services

Supporting services include primary production and soil formation. Primary production is the energy and matter accumulated by plants and determines the production of provisioning services. Soil formation provides the resources for primary production. Pollination also provides critical support for plant reproduction. Disruption of these processes interferes with provisioning and cultural services, as well as with carbon sequestration and climate modification services that depend on primary production and vegetation cover. Although insect outbreaks traditionally have been viewed as affecting these services negatively, fluxes of organic matter and nutrients during outbreaks contribute to soil formation, and plant compensatory growth following outbreaks may largely replace short-term reductions in primary production, depending on plant condition, resource availability, and timing and severity of the outbreak. These long-term compensatory effects require consideration of trade-offs in evaluating effects of outbreaks on supporting services.

12.4.1   Primary production

Plants have considerable capacity to compensate for herbivory, depending on plant condition, timing, and intensity of herbivory and on the availability of water and nutrients (Trumble *et al.* 1993, Feeley and Terborgh 2005, Schowalter 2011, Dungan *et al.* 2007). Healthy plants can compensate better than stressed plants. Larger plants typically have greater carbohydrate storage for reallocation to compensatory growth. Seedlings are particularly vulnerable to herbivores because of their limited resource storage capacity and limited ability to replace tissues lost to herbivores. Survival of tropical tree seedlings was highly correlated with the percentage of original leaf area present 1 month after germination and with the number of leaves present at 7 months of age (Clark and Clark 1985).

Herbivory typically is focused on less efficient and/or less defended foliage, often resulting from superfluous foliage production, allowing plants to reallocate resources to more productive foliage (Knapp and Seastedt 1986, Gutschick and Wiegel 1988, Trumble *et al.* 1993). Plants are best able to compensate for foliage loss in the spring, when environmental conditions favor continued growth to replace lost foliage, but become less able to compensate later in the season. Grasshoppers, *Aulocara elliotti* (Thomas), did not significantly reduce blue grama grass, *Bouteloua gracilis* (H.B.K.) Lag., biomass when feeding occurred early in the growing season but significantly reduced grass biomass when feeding occurred late in the growing season (Thompson and Gardner 1996). In fact, some plants, especially grasses, require pruning or low-to-moderate grazing to maintain production (Knapp and Seastedt 1986, Williamson *et al.* 1989). Lovett and Tobiessen (1993) found that experimental defoliation (80%) of red oak, *Quercus rubra* L., seedlings resulted in a significant 50% increase in photosynthetic rates; seedlings provided with elevated nitrogen were able to maintain high photosynthetic rates for a longer time than were seedlings at lower nitrogen levels (Figure 12.1). Short-term growth losses in defoliated conifers can be followed by several years, or decades, of growth rates that exceed pre-defoliation rates (Wickman 1980, Alfaro and Shepherd 1991), replacing at least some of the short-term losses. Similarly, annual wood production in at least some pine forests reached or exceeded pre-attack levels within 10–15 years following outbreaks of mountain pine beetle, *Dendroctonus ponderosae* Hopkins (Romme *et al.* 1986). Compensatory growth reflects selective removal of less efficient and less defended plant parts, permitting plant allocation of carbon and nutrients to new, more productive plant tissues (e.g., Gutschick 1999). Plant ability to replace lost production is affected by the following factors.

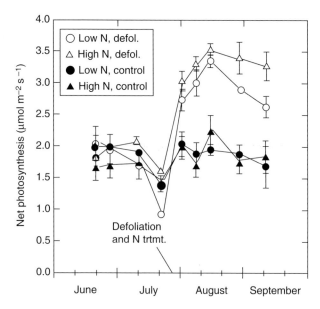

Figure 12.1   Mean net photosynthetic rate in old leaves from plants subjected to four combinations of nitrogen fertilization and defoliation intensity. Defoliation and fertilization treatments began on July 26. From Lovett and Tobiessen (1993, fig. 2) by permission from Oxford University Press.

Outbreak effects on above-ground plant parts can affect below-ground resources and plant ability to compensate. Defoliation of Douglas-fir, *Pseudotsuga menziesii* (Mirb.) Franco, seedlings decreased biomass but reduced water stress and increased photosynthesis, compared to undefoliated seedlings, improving seedling survival under drought conditions in greenhouse experiments (Kolb *et al.* 1999). Starch concentrations in roots were related inversely to the level of mechanical damage to shoots of a tropical tree, *Cedrela odorata* L. (Rodgers *et al.* 1995). Folivory on pinyon pines, *Pinus edulis* Engelm., adversely affected mycorrhizal fungi, perhaps through reduced carbohydrate supply to roots (Gehring and Whitham 1995).

Severe outbreaks by one species can weaken plants and increase vulnerability to other insect species. The Douglas-fir beetle, *Dendroctonus pseudotsugae* Hopkins, and fir engraver beetle, *Scolytus ventralis* LeConte, preferentially colonized Douglas-fir trees that had lost >90% of foliage to Douglas-fir tussock moth, although larval survival was greater in nondefoliated than in defoliated trees (Wright *et al.* 1986). However, Kolb *et al.* (1999) demonstrated that intense defoliation also can reduce moisture stress which limits survival of tree seedlings during dry periods (see above).

Differential herbivory among host and nonhost plants and plant species influences the rate and direction of change in community structure and explains distributions of many plant species (Louda *et al.* 1990). Low-to-moderate intensities of herbivory that prevail most of the time ensure a slow turnover of plant parts or individual plants, whereas high intensity of herbivory during outbreaks can dramatically reduce the abundance of preferred host species and favor replacement by nonhosts (Mattson and Addy 1975, McNaughton 1979, Knapp and Seastedt 1986, Williamson *et al.* 1989, Trlica and Rittenhouse 1993, Belovsky and Slade 2000, Feeley and Terborgh 2005, Dungan *et al.* 2007). Davidson (1993) compiled data indicating that herbivores retard or reverse succession during early seres but advance succession during later seres.

In at least some cases, alteration of vegetation composition by insects may tailor overall biotic demand for water and nutrients to prevailing conditions at a site (e.g., replacement of N-rich species by low-N species) (Ritchie *et al.* 1998, Belovsky and Slade 2000). Succession from pioneer pine forest to late successional fir forest in western North America can be retarded or advanced by insects, depending primarily on moisture availability and condition of the dominant vegetation (Schowalter 2008). When moisture is adequate (e.g., in riparian corridors and at high elevations), mountain pine beetle advances succession by facilitating the replacement of host pines by more shade tolerant, fire-intolerant, understory firs. However, limited moisture and short fire return intervals at lower elevations favor pine dominance. In the absence of fire during drought periods, western spruce budworm, *Choristoneura occidentalis* Freeman, Douglas-fir tussock moth, and bark beetles concentrate on the understory firs, truncating (or reversing) succession (Plate 12.2). Fire fueled by fir mortality also leads to eventual regeneration of pine forest. Canopy openings resulting from spruce budworm, *Choristoneura fumiferana* (Clemens), outbreaks had a greater diversity of saplings and trees and larger perimeter–area ratios than did canopy openings resulting from clear-cut harvest with protection of advance regeneration practices, suggesting that stand recovery and contributions by the surrounding forest should be greater in budworm-generated openings than in harvest-generated openings (Belle-Isle and Kneeshaw 2007).

Therefore, insect outbreaks often have less negative effects on primary production (and provisioning and other ecosystem services that it supports) over long time periods than is generally perceived. Biomass reduction by insects may not always be desirable, especially in intensively managed forest or rangeland, but can increase primary production in the same manner as prescribed mowing, pruning, and/or thinning. Improved survival of defoliated plants during drought would mitigate effects of drought, which is a common trigger for outbreaks (Mattson and Haack 1987, Van Bael *et al.* 2004). Furthermore, outbreaks may tailor vegetation structure to balance overall water and nutrient demands with prevailing conditions at a site. If harvest of ecosystem products (e.g., timber) can be delayed, the cost of unscheduled salvage harvest may be offset by long-term replacement of lost products (McNaughton 1979, Wickman 1980, Knapp and Seastedt 1986, Romme *et al.* 1986, Alfaro and Shepherd 1991). In other words, although the costs of plant growth loss, mortality, and unscheduled salvage harvest are easy to calculate, these costs might have been avoided by pruning or thinning, recommended management practices for preventing outbreaks.

12.4.2   Soil formation

Insect outbreaks move considerable organic matter from vegetation to the soil surface, affecting the rate and timing of litterfall, decomposition, and soil conditions. Insect feces, in particular, increase the rate and amount of nutrient fluxes to soil and support plant compensatory growth (Chapman *et al.* 2003, Frost and Hunter 2004, 2007, 2008, Classen *et al.* 2005, Fonte and Schowalter 2005). Organic matter accumulation, especially where outbreaks increase the amount of coarse woody debris, also affects soil texture, water-holding capacity, and fertility. Bark beetle excavation and inoculation of galleries with decay organisms advance the process of decomposition (Progar *et al.* 2000). As organic matter decomposes, its porosity and water retention capacity increase, facilitating aerobic decomposition, carbon loss, and nitrogen fixation (Harmon *et al.* 1986, Progar *et al.* 2000, Coleman *et al.* 2004). These processes contribute to soil water content and fertility that are necessary for primary production and for those provisioning and cultural services that it supports (as discussed in this chapter). Wood *et al.* (2009) demonstrated that experimental addition of litter on a plot scale increased foliage production and nitrogen and phosphorus content. These results indicate that biomass and nutrients transferred to the forest floor during outbreaks could be incorporated quickly into new plant tissues, as compensatory growth. The fertilization effects of herbivores potentially mitigate nutrient limitation and reduce the need for exogenous fertilizers in managed ecosystems.

12.5   Effects on regulating services

In contrast to supporting services, regulating services provide feedback that maintains more consistent supply of other services. For example, feedback control of climate and biogeochemical cycling rates through interactions among organisms maintains a more consistent supply of water and nutrients, and thereby maintains more consistent primary production and the services it supports, than would occur in the absence of such regulation (e.g., Foley *et al.* 2003, Schowalter 2011). Relatively few studies have addressed effects of insect outbreaks on regulating services, despite obvious outbreak effects on primary production, which controls fluxes of energy and nutrients among atmosphere, biosphere, and geosphere and modifies local and regional climate (Foley *et al.* 2003). Insects themselves may serve as regulators of primary production (Mattson and Addy 1975), which supports other ecosystem services (see Section 12.5.3). Therefore, effects of insect outbreaks on regulating services warrant special consideration.

12.5.1   Biogeochemical cycling

Crossley and Howden (1961) were the first to demonstrate that insect herbivores accelerate nutrient fluxes via consumption of foliage. Subsequent research has demonstrated that insect herbivores affect biogeochemical cycling through changes in vegetation structure and composition and altered rate, seasonal pattern, and quality of throughfall (net precipitation reaching the ground) and litterfall (see above).

Outbreaks affect ecosystem sequestration of carbon and nutrients. For example, widespread pine mortality during outbreaks of mountain pine beetle reduced carbon uptake and increased carbon emission from decaying trees (Kurz *et al.* 2008). A similar outcome was found as a result of defoliation by gypsy moth

(Clark *et al.* 2010). In both studies, the change in net carbon flux converted the forest from a carbon sink to a carbon source. However, forests recovering from mortality caused by mountain pine beetle can remain growing-season carbon sinks as a result of increased photosynthesis by surviving trees and understory vegetation, whereas nearby harvested stands may remain carbon sources 10 years after harvest (Brown *et al.* 2010). Brown *et al.* (2010) recommended deferral of salvage harvest of outbreak sites with substantial surviving trees and understory vegetation, to prevent such sites from being converted from carbon sinks to sources for extended periods. Ritchie *et al.* (1998) reported that insect herbivory reduced the abundance of N-rich plant species, leading to replacement by plant species with lower N concentrations in an oak savanna. Invasive insect species may have long-term effects on carbon sequestration through alteration of species composition, primary productivity, and nutrient fluxes (Peltzer *et al.* 2010).

Outbreaks affect uptake and use of water and nutrients by vegetation, affecting water quality and yield for downstream uses. Removal of foliage or other plant tissues reduces rates of precipitation interception and evapotranspiration (e.g., Foley *et al.* 2003). As a result, more precipitation reaches the ground, temporarily increasing soil water content. Leaching of excess water exports nutrients from the system (Swank *et al.* 1981, Eshleman *et al.* 1998, Lovett *et al.* 2002). However, reduced albedo resulting from canopy opening increases evaporation from the soil and reduces cloud formation and local precipitation (Foley *et al.* 2003).

Outbreaks increase the flux of nutrients from vegetation to soil in several ways. Herbivory increases nutrient flux in throughfall, precipitation enriched with nutrients leached from damaged foliage (Seastedt *et al.* 1983, Stachurski and Zimka 1984, Schowalter *et al.* 1991, Hunter *et al.* 2003). However, these nutrients may not contribute immediately to primary production. In ecosystems with high annual precipitation, herbivore-induced nutrient fluxes may be masked by greater inputs to soil via precipitation (Fonte and Schowalter 2005, Schowalter *et al.* 1991), and in ecosystems with high background levels of N, herbivore-induced N flux may be immobilized quickly by soil microorganisms (Lovett and Ruesink 1995, Stadler and Müller 1996, Stadler *et al.* 1998, Treseder 2008).

Herbivory increases the amount and alters seasonal pattern and form of nutrients in litterfall. In the absence of herbivory, litterfall is highly seasonal (i.e., concentrated at the onset of cold or dry conditions) and has low nutrient concentrations, especially of nitrogen or other nutrients that are reabsorbed from senescing foliage (Marschner 1995, Gutschick 1999, Fonte and Schowalter 2004). Herbivory increases litterfall during the growing season (as fragmented plant material, insect tissues, and feces), but the nutritional quality of litter is affected by herbivore-induced defenses (see Chapter 5, this volume), which may retard decomposition.

Insect tissues and feces have particularly high concentrations of nutrients, especially nitrogen, compared to plant material (Schowalter and Crossley 1983) and increase the rate of nutrient flux to soil (Frost and Hunter 2004). Hollinger (1986) reported that an outbreak of the California oak moth, *Phryganidia californica* Packard, increased fluxes of nitrogen and phosphorus from trees to litter by twofold, and feces and insect remains accounted for 60–70% of the total fluxes. Deposition of folivore feces can explain 62% of the variation in soil nitrate availability (Hunter *et al.* 2003). Christenson *et al.* (2002) and Frost and Hunter (2007) demonstrated, using ^{15}N, that early-season herbivore feces were rapidly decomposed, whereas leaf litter-N remained in litter, and some feces-N was incorporated into foliage and, subsequently, into late-season defoliators during the same growing season. To the extent that nutrients, especially nitrogen, often are immobilized in plant tissues and limited in availability for plant use (Gutschick 1999), such herbivore-induced turnover may be an important mechanism for nutrient incorporation into new plant tissues.

Outbreaks affect litter decomposition and mineralization through alteration of the soil/litter environment. Decomposition is strongly affected by litter moisture (Meentemeyer 1978, Whitford *et al.* 1981), a factor affected by canopy opening (Classen *et al.* 2005, Foley *et al.* 2003). Schowalter and Sabin (1991) reported that experimental defoliation increased abundance of three litter arthropod taxa. Similarly, experimental addition of herbivore feces or throughfall increased abundances of Collembola and fungal- and bacterial-feeding nematodes (Reynolds *et al.* 2003). Although long-term canopy opening may result in evaporation of soil moisture (Foley *et al.* 2003), outbreak-induced canopy opening and litter deposition increase soil moisture and decomposition (Classen *et al.* 2005).

12.5.2 Modification of climate and disturbance

Vegetation has considerable capacity to modify climate and mitigate disturbances, depending on vegetation height and density. Several studies have demonstrated the importance of vegetation to shading and protecting the soil surface, abating wind speed, and controlling water fluxes (Foley *et al.* 2003, Classen *et al.* 2005). Vegetation cover reduces albedo and diurnal soil surface temperatures (Foley *et al.* 2003). Evapotranspiration contributes to canopy cooling and to convection-generated condensation above the canopy, thereby increasing local precipitation (Meher-Homji 1991, Foley *et al.* 2003, Juang *et al.* 2007). Vegetation removal results in evaporation of soil moisture and loss of control of soil temperature. Exposed soil surfaces can reach midday temperatures lethal to most organisms (Seastedt and Crossley 1981). Unimpeded wind and precipitation erode and degrade soils.

Although effects of outbreaks on climate have not been studied directly, reduced vegetation cover over large areas during outbreaks could have similar effects on regional climate. However, unlike anthropogenic vegetation removal, defoliation or tree mortality due to insect outbreaks retains some shade and adds water-retaining litter to the soil surface. Schowalter *et al.* (1991) reported that 20% defoliation of experimental Douglas-fir saplings doubled the amount of water and litterfall at the soil surface, compared to undefoliated saplings. Classen *et al.* (2005) reported that canopy opening by manipulated abundances of scale insects, *Matsucoccus acalyptus* Herbert, increased soil temperature and moisture by 26% and 35%, respectively, similar to global change scenarios and sufficient to alter ecosystem processes (see Foley *et al.* 2003).

Outbreaks affect the probability or severity of future disturbances, especially fire or storms. Increased fuel accumulation generally has been considered to increase the likelihood and severity of fire (McCullough *et al.* 1998), but this is not necessarily the case. Bebi *et al.* (2003) concluded that spruce, *Picea engelmannii* Parry ex. Engelm., mortality to spruce beetle, *Dendroctonus rufipennis* (Kirby), did not increase the occurrence of subsequent fires. The probability of fire resulting from outbreaks depends on amount and decomposition rate of increased litter. Grasshopper outbreaks that reduce grass biomass should reduce the severity of subsequent grassland fire. Outbreaks that increase only fine litter material (e.g., foliage fragments) may increase the probability and spread of low-intensity fire, whereas outbreaks that cause tree mortality (and increase abundance of ladder fuels) are more likely to increase the risk of catastrophic fire (Jenkins *et al.* 2008; see also Plate 12.2). Insect outbreaks that open the canopy increase penetration of high wind speeds and the probability of tree fall, but also reduce wind resistance of defoliated trees. Pruning at least 80% of the canopy can reduce wind stress significantly (Moore and Maguire 2005). Wind-related tree mortality following spruce budworm defoliation in eastern Canada was related to outbreak severity (Taylor and MacLean 2009). Tree mortality during storms peaked 11–15 years after outbreak, due to greater exposure of surviving trees to wind. Obviously, insect-induced disturbances can interfere with harvest of provisioning services and likely reduce cultural values, but also may prevent undesirable changes in ecosystem structure or composition (e.g., succession from grassland to forest or from pine forest to fir forest in the absence of fire).

12.5.3   Outbreaks as a regulating service

Accumulating data suggest that outbreaks of native insect species (but not exotic species) may regulate primary production in a density- or stress-dependent manner, much as predators regulate prey populations (Mattson and Addy 1975, Schowalter 2011). Native insect populations, and herbivory, are regulated by a combination of bottom-up, top-down, and lateral factors (Schowalter 2011). Bottom-up regulation is provided by host plant density and defensive chemistry, including production of volatile elicitors that can induce defenses in neighboring, even unrelated, plants in advance of herbivory (e.g., Karban and Baldwin 1997, Dolch and Tscharntke 2000; see also Chapter 5, this volume). Top-down regulation is conferred by predation and parasitism (e.g., Marquis and Whelan 1994, Turlings *et al.* 1995, Letourneau and Dyer 1998, Mooney 2007, Letourneau *et al.* 2009). Lateral regulation is by intra- and interspecific competition. Increased competition, drought, or other environmental changes that stress host plants inhibit production of defensive chemicals and increase vulnerability to herbivores (Mattson and Haack 1987, Schowalter and Turchin 1993, Schowalter *et al.* 1999, Van Bael *et al.* 2004, Schowalter 2008). Disturbances also can reduce predator abundances and facilitate herbivore population growth (Schowalter *et al.* 1999).

Low to moderate levels of herbivory often stimulate primary production (compensatory growth), whereas high levels of herbivory reduce primary production (Figure 12.2; Mattson and Addy 1975, McNaughton 1979, Knapp and Seastedt 1986, Williamson *et al.* 1989, Trlica and Rittenhouse 1993, Belovsky and Slade 2000, Feeley and Terborgh 2005, Dungan *et al.* 2007; and see

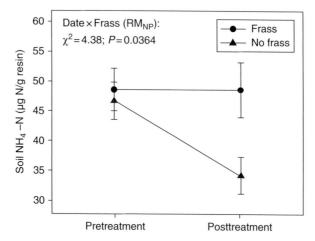

Figure 12.2 Regression of relationship between grasshopper, *Melanoplus sanguinipes* (Fabricius), density and primary production in experimental field microcosms at the National Bison Range in Montana. Peak natural densities at the site were 4–36 adults m². From Belovsky and Slade (2000) with permission from *Nature*, © Macmillan Magazine, Ltd.

Section 12.4.1). These effects, combined with the tendency of outbreaks to reduce the abundance of the most stressed plant species, should stabilize primary production at intermediate levels that may be most sustainable under prevailing environmental conditions. Outbreaks may, in fact, alleviate drought or competitive stresses that often trigger outbreaks (e.g., Kolb *et al.* 1999).

Tests of this hypothesis are difficult and depend on one's perspective. For example, the Douglas-fir tussock moth outbreak depicted in Plate 12.2 caused devastating losses in timber supply and increased the risk of wildfire in the short term. However, over longer time periods, reduced fir density returned forest structure to historic conditions that prevailed prior to fire suppression in Sierran forests and that are the recommended management goal for these forests (North *et al.* 2007). From this perspective, the outbreak improved forest stability and sustainability of ecosystem services from this forest. To the extent that outbreaks function to regulate primary production, suppression may be counterproductive to the sustainability of at least some ecosystem services, including timber production.

12.6 Conclusions

Insect outbreaks affect ecosystem services positively or negatively, depending on management goals and perspectives. Outbreaks have largely negative effects on provisioning of plant products and on cultural services, at least in the short term, but may benefit water yield, wildlife and fish, and even human ethnic groups who use insects as food. When harvest of ecosystem products can be delayed, long-term compensatory production of plant resources following outbreaks may largely replace short-term losses. Insects have complex effects on primary production and soil formation that support other ecosystem services over the long term. Ecosystem managers should consider potential increases in fire frequency, as a result of herbivore-induced fuel accumulation, but increased abundance of fine fuels may reduce fire severity and improve long-term primary production. To the extent that native herbivorous insects in natural ecosystems function as regulators of primary production (and thereby maintain more constant levels of ecosystem services than would occur in their absence), suppression of outbreaks would not be cost-effective and could be counterproductive (i.e., would hinder regulatory function or negatively affect ecosystem services). Furthermore, use of insecticides in agroecosystems often interferes with natural enemies of target "pests" and with pollinators, and the target pest populations recover or develop tolerance to pesticides more quickly than do pollinators or natural enemies (Kremen *et al.* 2002, Letourneau *et al.* 2009). Therefore, pest and ecosystem management decisions should consider trade-offs among short- and long-term effects on multiple ecosystem services.

Research clearly is needed to evaluate long-term effects and costs of outbreaks on primary production, soil formation, and sustainability of ecosystem functions and services. In particular, very limited data are available to evaluate costs of outbreaks versus costs of suppression for various ecosystem services. Clearly, more data are needed to make sound pest management decisions based on long-term, as well as short-term, benefits and costs for multiple ecosystem services. Identifying factors underlying compensatory growth would clarify necessary conditions and contributions to long-term primary production, as this mitigates costs of outbreaks. Research described in this chapter has demonstrated that defoliation can improve plant water balance and survival under drought conditions in the greenhouse, but the extent to which this contributes to sustainability of primary production and other ecosystem services in the field warrants evaluation. Finally, the extent to which outbreaks modify local climate and disturbance dynamics requires clarification. A number of long-term ecosystem research platforms are available for this research, including Long-Term Ecological Research (LTER) sites and National Environmental Observatory Network (NEON) sites in North America and International Long-Term Ecological Research (ILTER) sites worldwide. Insect ecologists are encouraged to use these sites to address the research priorities identified here.

References

Alfaro, R. I., and R. F. Shepherd. 1991. Tree-ring growth of interior Douglas-fir after one year's defoliation by Douglas-fir tussock moth. *Forest Science* 37:959–964.

Allan, J. D., M. S. Wipfli, J. P. Caouette, A. Prussian, and J. Rodgers. 2003. Influence of streamside vegetation on inputs of terrestrial invertebrates to salmonid food webs. *Canadian Journal of Fisheries and Aquatic Sciences* 60:309–320.

Andrews, W. V. 1868. The Cynthia silk-worm. *American Naturalist* 2:311–320.

Anelli, C. M., and D. A. Prischmann-Voldseth. 2009. Silk batik using beeswax and cochineal dye: an interdisciplinary approach to teaching entomology. *American Entomologist* 55:95–105.

Baldwin, D. H., J. A. Spromberg, T. K. Collier, and N. L. Scholz. 2009. A fish of many scales: extrapolating sublethal pesticide exposures to the productivity of wild salmon populations. *Ecological Applications* 19:2004–2115.

Baxter, C. V., K. D. Fausch, and W. C. Saunders. 2005. Tangled webs: reciprocal flows of invertebrate prey link streams and riparian zones. *Freshwater Biology* 50:201–220.

Bebi, P., D. Kilakowski, and T. T. Veblen. 2003. Interactions between fire and spruce beetles in a subalpine Rocky Mountain forest landscape. *Ecology* 84:362–371.

Belle-Isle, J., and D. Kneeshaw. 2007. A stand and landscape comparison of the effects of a spruce budworm (*Choristoneura fumiferana* (Clem.)) outbreak to the combined effects of harvesting and thinning on forest structure. *Forest Ecology and Management* 246:163–174.

Belovsky, G. E., and J. B. Slade. 2000. Insect herbivory accelerates nutrient cycling and increases plant production. *Proceedings of the National Academy of Sciences USA* 97:14412–14417.

Brown, M., T. A. Black, Z. Nesic, V. N. Foord, D. L. Spittlehouse, A. L. Fredeen, N. J. Grant, P. J. Burton, and J. A. Trofymow. 2010. Impact of mountain pine beetle on the net ecosystem production of lodgepole pine stands in British Columbia. *Agricultural and Forest Meteorology* 150:254–264.

Cerritos, R., and Z. Cano-Santana. 2008. Harvesting grasshoppers *Sphenarium purpurascens* in Mexico for human consumption: a comparison with insecticidal control for managing pest outbreaks. *Crop Protection* 27:473–480.

Chapman, S. K., S. C. Hart, N. S. Cobb, T. G. Whitham, and G. W. Koch. 2003. Insect herbivory increases litter quality and decomposition: an extension of the acceleration hypothesis. *Ecology* 84:2867–2876.

Christenson, L. M., G. M. Lovett, M. J. Mitchell, and P. M. Groffman. 2002. The fate of nitrogen in gypsy moth frass deposited to an oak forest floor. *Oecologia* 131:444–452.

Clark, D.B. and D.A. Clark. 1985. Seedling dynamics of a tropical tree: impacts of herbivory and meristem damage. *Ecology* 66:1884–1892.

Clark, K. L., N. Skowronski and J. Hom. 2010. Invasive insects impact forest carbon dynamics. *Global Change Biology* 16:88–101.

Classen, A. T., S. C. Hart, T. G. Whitham, N. S. Cobb, and G. W. Koch. 2005. Insect infestations linked to changes in microclimate: important climate change implications. *Soil Science Society of America Journal* 69:2049–2057.

Claudianos, C., H. Ranson, R. M. Johnson, S. Biswas, M. A. Schuler, M. R. Berenbaum, R. Feyereisen, and J. G. Oakeshott. 2006. A deficit of detoxification enzymes: pesticide sensitivity and environmental response in the honey bee. *Insect Molecular Biology* 15:615–636.

Clausen, L. W. 1954. *Insect Fact and Folklore*. MacMillan, New York.

Coleman, D. C., D. A. Crossley, Jr., and P. F. Hendrix. 2004. *Fundamentals of Soil Ecology*, 2nd ed. Elsevier, Amsterdam.

Collins, S. 1961. Benefits to understory from canopy defoliation by gypsy moth larvae. *Ecology* 42:836–838.

Costanza, R., R. d'Arge, R. de Groot, S. Farber, M. Grasso, B. Hannon, K. Limburg, S. Naeem, R. V. O'Neill, J. Paruela, R. G. Raskin, P. Sutton, and M. van den Belt. 1997. The value of the world's ecosystem services and natural capital. *Nature* 387:253–260.

Crossley, D. A. Jr., and H. F. Howden. 1961. Insect–vegetation relationships in an area contaminated by radioactive wastes. *Ecology* 42:302–317.

Dasgupta, P., S. Levin, and J. Lubchenco. 2000. Economic pathways to ecological sustainability. *Bioscience* 50:339–345.

Davidson, D.W. 1993. The effects of herbivory and granivory on terrestrial plant succession. *Oikos* 68:23–35.

Dolch, R., and T. Tscharntke. 2000. Defoliation of alders (*Alnus glutinosa*) affects herbivory by leaf beetles on undamaged neighbors. *Oecologia* 125:504–511.

Downing, A. L., and M. A. Leibold. 2002. Ecosystem consequences of species richness and composition in pond food webs. *Nature* 416:837–841.

Downing, K. B., and W. R. Williams. 1978. Douglas-fir tussock moth: did it affect private recreational businesses in northeastern Oregon? *J. Forestry* 76:29–30.

Dungan, R. J., M. H. Turnbull, and D. Kelly. 2007. The carbon costs for host trees of a phloem-feeding herbivore. *Journal of Ecology* 95:603–613.

Edwards, E. P. 1982. Hummingbirds feeding on an excretion produced by scale insects. *Condor* 84:122.

Eshleman, K. N., R. P. Morgan II, J. R. Webb, F. A. Deviney, and J. N. Galloway. 1998. Temporal patterns of nitrogen leakage from mid-Appalachian forested watersheds: role of insect defoliation. *Water Resources Research* 34:2005–2116.

Feeley, K. J., and J. W. Terborgh. 2005. The effects of herbivore density on soil nutrients and tree growth in tropical forest fragments. *Ecology* 86:116–124.

Foley, J. A., M. H. Costa, C. Delire, N. Ramankutty, and P. Snyder. 2003. Green surprise? How terrestrial ecosystems could affect earth's climate. *Frontiers in Ecology and the Environment* 1:38–44.

Fonte, S. J., and T. D. Schowalter. 2004. Decomposition of greenfall vs. senescent foliage in a tropical forest ecosystem in Puerto Rico. *Biotropica* 36:474–482.

Fonte, S. J., and T. D. Schowalter. 2005. The influence of a neotropical herbivore (*Lamponius portoricensis*) on nutrient cycling and soil processes. *Oecologia* 146: 423–431.

Forbush, E. H., and C. H. Fernald. 1896. *The Gypsy Moth*. Massachusetts Board of Agriculture, Boston.

Frost, C. J., and M. D. Hunter. 2004. Insect canopy herbivory and frass deposition affect soil nutrient dynamics and export in oak mesocosms. *Ecology* 85:3335–3347.

Frost, C. J., and M. D. Hunter. 2007. Recycling of nitrogen in herbivore feces: plant recovery, herbivore assimilation, soil retention, and leaching losses. *Oecologia* 151:42–53.

Frost, C. J., and M. D. Hunter. 2008. Insect herbivores and their frass affect *Quercus rubra* leaf quality and initial stages of subsequent decomposition. *Oikos* 117:13–22.

Furniss, R. L., and V. M. Carolin. 1977. *Western Forest Insects*. USDA Forest Service Misc. Publ. 1339. Government Printing Office, Washington, DC.

Gatto, P., A. Zocca, A. Battisti, M. J. Barrento, M. Branco, and M. R. Paiva. 2009. Economic assessment of managing processionary moth in pine forests: a case study in Portugal. *Journal of Environmental Management* 90:683–691.

Gehring, C. A., and T. G. Whitham. 1995. Duration of herbivore removal and environmental stress affect the ectomycorrhizae of pinyon pine. *Ecology* 76:2118–2123.

Gutschick, V. P. 1999. Biotic and abiotic consequences of differences in leaf structure. *New Phytologist* 143:3–18.

Gutschick, V. P., and F. W. Wiegel. 1988. Optimizing the canopy photosynthetic rate by patterns of investment in specific leaf mass. *American Naturalist* 132:67–86.

Harmon, M. E., J. F. Franklin, F. J. Swanson, P. Sollins, S. V. Gregory, J. D. Lattin, N. H. Anderson, S. P. Cline, N. G. Aumen, J. R. Sedell, G. W. Lienkaemper, K. Cromack Jr., and K. W. Cummins. 1986. Ecology of coarse woody debris in temperate ecosystems. *Advances in Ecological Research* 15:133–302.

Hättenschwiler, S., and P. Gasser. 2005. Soil animals alter plant litter diversity effects on decomposition. *Proceedings of the National Academy of Sciences USA* 102:1519–1524.

Helson, J. E., T. L. Capson, T. Johns, A. Aiello, and D. M. Windsor. 2009. Ecological and evolutionary bioprospecting: using aposematic insects as guides to rainforest plants active against disease. *Frontiers in Ecology and the Environment* 7:130–134.

Hollinger, D. Y. 1986. Herbivory and the cycling of nitrogen and phosphorus in isolated California oak trees. *Oecologia* 70:291–297.

Hunter, M. D., C. R. Linnen, and B. C. Reynolds. 2003. Effects of endemic densities of canopy herbivores on nutrient dynamics along a gradient in elevation in the southern Appalachians. *Pedobiologia* 47:231–244.

Jenkins, M. J., E. Herbertson, W. Page, and C. A. Jorgensen. 2008. Bark beetles, fuels, fires and implications for forest management in the Intermountain West. *Forest Ecology and Management* 254:16–34.

Johnson, D. M., A. M. Liebhold, P. C. Tobin, and O. N. Bjørnstad. 2006. Allee effects and pulsed invasion by the gypsy moth. *Nature* 444:361–363.

Juang, J-Y., G. G. Katul, A. Porporato, P. C. Stoy, M. S. Sequeira, M. Detto, H-S. Kim, and R. Oren. 2007. Eco-hydrological controls on summertime convective rainfall triggers. *Global Change Biology* 13:887–896.

Karban, R., and I.T. Baldwin. 1997. *Induced Responses to Herbivory*. University of Chicago Press, Chicago.

Kiffney, P. M., J. S. Richardson, and J. P. Bull. 2003. Responses of periphyton and insects to experimental manipulation of riparian buffer width along forest streams. *Journal of Applied Ecology* 40:1060–1076.

Klock, G. O., and B. E. Wickman. 1978. Ecosystem effects. In *The Douglas-fir Tussock Moth: A Synthesis* (M. H. Brookes, R. W. Stark and R. W. Campbell, Eds.), pp. 90–95. US Forest Service Tech. Bull. 1585. US Forest Service, Washington, DC.

Knapp, A. K., and T. R. Seastedt. 1986. Detritus accumulation limits productivity of tallgrass prairie. *Bioscience* 36:662–668.

Koenig, W. D., and A. M. Liebhold. 2005. Effects of periodical cicada emergences on abundances and synchrony of avian populations. *Ecology* 86:1873–1882.

Kolb, T. E., K. A. Dodds, and K. M. Clancy. 1999. Effect of western spruce budworm defoliation on the physiology and growth of potted Douglas-fir seedlings. *Forest Science* 45: 280–291.

Kremen, C., N. M. Williams, and R. W. Thorp. 2002. Crop pollination from native bees at risk from agricultural intensification. *Proceedings of the National Academy of Science USA* 99: 16812–16816.

Kudo, G., T. Y. Ida, and T. Tani. 2008. Linkages between phenology, pollination, photosynthesis, and reproduction in deciduous forest understory plants. *Ecology* 89:321–331.

Kurz, W. A., C.C. Dymond, G. Stinson, G. J. Rampley, E. T. Neilson, A. L. Carroll, T. Ebata, and L. Safranyik. 2008. Mountain pine beetle and forest carbon feedback to climate change. *Nature* 452:987–990.

Letourneau, D. K., and L. A. Dyer. 1998. Density patterns of *Piper* ant-plants and associated arthropods: top-predator trophic cascades in a terrestrial system? *Biotropica* 30:162–169.

Letourneau, D. K., J. A. Jedlicka, S. G. Bothwell, and C. R. Moreno. 2009. Effects of natural enemy biodiversity on the suppression of arthropod herbivores in terrestrial ecosystems. *Annual Review of Ecology, Evolution and Systematics* 40:573–592.

Leuschner, W. A. 1980. Impacts of the southern pine beetle. In *The Southern Pine Beetle* (R.C. Thatcher, J. L. Searcy, J. E. Coster, and G. D. Hertel, Eds.), pp. 137–151. US Forest Service Tech. Bull. 1631. US Forest Service, Washington, DC.

Louda, S. M., K. H. Keeler, and R. D. Holt. 1990. Herbivore influences on plant performance and competitive interactions. In *Perspectives on Plant Competition* (J. B. Grace and D. Tilman, Eds.), pp. 413–444. Academic Press, San Diego.

Lovett, G. M., and A. E. Ruesink. 1995. Carbon and nitrogen mineralization from decomposing gypsy moth frass. *Oecologia* 104:133–138.

Lovett, G., and P. Tobiessen. 1993. Carbon and nitrogen assimilation in red oaks (*Quercus rubra* L.) subject to defoliation and nitrogen stress. *Tree Physiology* 12:259–269.

Lovett, G. M., L. M. Christenson, P. M. Groffman, C. G. Jones, J. E. Hart, and M. J. Mitchell. 2002. Insect defoliation and nitrogen cycling in forests. *Bioscience* 52:335–341.

Marquis, R. J., and C. J. Whelan. 1994. Insectivorous birds increase growth of white oak through consumption of leaf-chewing insects. *Ecology* 75:2007–2014.

Marschner, H. 1995. *The Mineral Nutrition of Higher Plants*, 2nd ed. Academic Press, San Diego.

Mattson, W. J., and N. D. Addy. 1975. Phytophagous insects as regulators of forest primary production. *Science* 190:515–522.

Mattson, W. J., and R. A. Haack. 1987. The role of drought in outbreaks of plant-eating insects. *Bioscience* 37:110–118.

Mbata, K. J., E. N. Chidumayo, and C. M. Lwatula. 2002. Traditional regulation of edible caterpillar exploitation in the Kopa area of Mpika district in northern Zambia. *Journal of Insect Conservation* 6:115–130.

McCullough, D. G., R. A. Werner, and D. Neumann. 1998. Fire and insects in northern and boreal forest ecosystems of North America. *Annual Review of Entomology* 43:107–127.

McNaughton, S. J. 1979. Grazing as an optimization process: grass–ungulate relationships in the Serengeti. *American Naturalist* 113:691–703.

Meentemeyer, V. 1978. Macroclimate and lignin control of litter decomposition rates. *Ecology* 59: 465–472.

Meher-Homji, V. M. 1991. Probable impact of deforestation on hydrological processes. *Climate Change* 19:163–173.

Menninger, H. L., M. A. Palmer, L. S. Craig, and D. C. Richardson. 2008. Periodical cicada detritus impacts stream ecosystem metabolism. *Ecosystems* 11:1306–1317.

Michalson, E. L. 1975. Economic impact of mountain pine beetle on outdoor recreation. *Southern Journal of Agricultural Economics* 7(2):43–50.

Millenium Ecosystem Assessment. 2005. *Ecosystems and Human Well-Being: Biodiversity Synthesis*. World Resources Institute, Washington, DC.

Mooney, K. A. 2007. Tritrophic effects of birds and ants on a canopy food web, tree growth, and phytochemistry. *Ecology* 88:2005–2014.

Moore, J. R., and D. A. Maguire. 2005. Natural sway frequencies and damping ratios of trees: influence of crown structure. *Trees* 19:363–373.

North, M., J. Innes, and H. Zald. 2007. Comparison of thinning and prescribed fire restoration treatments to Sierran mixed-conifer historic conditions. *Canadian Journal of Forest Research* 37:331–342.

Pedigo, L. P., S. H. Hutchins, and L. G. Higley. 1986. Economic injury levels in theory and practice. *Annual Review of Entomology* 31:341–368.

Peltzer, D. A., R. B. Allen, G. M. Lovett, D. Whitehead, and D. A. Wardle. 2010. Effects of biological invasions on forest carbon sequestration. *Global Change Biology* 16:732–746.

Perlman, F., E. Press, J. A. Googins, A. Malley, and H. Poarea. 1976. Tussockosis: reactions to Douglas fir tussock moth. *Annals of Allergy* 36:302–307.

Pfadt, R. E., and D. M. Hardy. 1987. A historical look at rangeland grasshoppers and the value of grasshopper control programs. In *Integrated Pest Management on Rangeland* (J. L. Capinera, ed.), pp. 183–195. Westview Press, Boulder, CO.

Pray, C. L, W. H. Nowlin, and M. J. Vanni. 2009. Deposition and decomposition of periodical cicadas (Homoptera: Cicadidae: Magicicada) in woodland aquatic ecosystems. *Journal of the North American Benthological Society* 28:181–195.

Progar, R., T. D. Schowalter, J. J. Morrell, and C. M. Freitag. 2000. Respiration from coarse woody debris as affected by moisture and saprotroph functional diversity in western Oregon. *Oecologia* 124:426–431.

Ramos-Elorduy, J. 2009. Anthro-entomophagy: cultures, evolution and sustainability. *Entomological Research* 39:271–288.

Regal, R. J. 1982. Pollination by wind and animals: ecology of geographic patterns. *Annual Review of Ecology and Systematics* 13:497–424.

Reynolds, B. C., D. A. Crossley Jr., and M. D. Hunter. 2003. Response of soil invertebrates to forest canopy inputs along a productivity gradient. *Pedobiologia* 47:127–139.

Riley, C. V. 1878. *First Annual Report of the United States Entomological Commission for the Year 1877 Relating to the Rocky Mountain Locust and the Best Methods of Preventing Its Injuries and of Guarding Against Its Invasions, in Pursuance of an Appropriation Made by Congress for This Purpose.* US Department of Agriculture, Washington, DC.

Ritchie, M. E., D. Tilman, and J. M. H. Knops. 1998. Herbivore effects on plant and nitrogen dynamics in oak savanna. *Ecology* 79:165–177.

Rodgers, H.L., M.P. Brakke, and J.J. Ewel. 1995. Shoot damage effects on starch reserves of *Cedrela odorata*. *Biotropica* 27:71–77.

Romme, W. H., D. H. Knight, and J. B. Yavitt. 1986. Mountain pine beetle outbreaks in the Rocky Mountains: regulators of primary productivity? *American Naturalist* 127:484–494.

Schowalter, T. D. 2008. Insect herbivore responses to management practices in conifer forests in North America. *Journal of Sustainable Forestry* 26:204–222.

Schowalter, T. D. 2011. *Insect Ecology: An Ecosystem Approach*, 3rd ed. Elsevier/Academic, San Diego.

Schowalter, T. D., and D. A. Crossley, Jr. 1983. Forest canopy arthropods as sodium, potassium, magnesium and calcium pools in forests. *Forest Ecology and Management* 7:143–148.

Schowalter, T. D., and T. E. Sabin. 1991. Litter microarthropod responses to canopy herbivory, season and decomposition in litterbags in a regenerating conifer ecosystem in western Oregon. *Biology and Fertility of Soils* 11:93–96.

Schowalter, T. D., and P. Turchin. 1993. Southern pine beetle infestation development: interaction between pine and hardwood basal areas. *Forest Science* 39:201–210.

Schowalter, T. D., T. E. Sabin, S. G. Stafford, and J. M. Sexton. 1991. Phytophage effects on primary production, nutrient turnover, and litter decomposition of young Douglas-fir in western Oregon. *Forest Ecology and Management* 42:229–243.

Schowalter, T. D., D. C. Lightfoot, and W. G. Whitford. 1999. Diversity of arthropod responses to host-plant water stress in a desert ecosystem in southern New Mexico. *American Midland Naturalist* 142:281–290.

Seastedt, T. R., and D. A. Crossley Jr. 1981. Microarthropod response following cable logging and clearcutting in the southern Appalachians. *Ecology* 62:126–135.

Seastedt, T. R., D. A. Crossley Jr., and W. W. Hargrove. 1983. The effects of low-level consumption by canopy arthropods on the growth and nutrient dynamics of black locust and red maple trees in the southern Appalachians. *Ecology* 64:1040–1048.

Sheppard, S., and P. Picard. 2006. Visual-quality impact of forest pest activity at the landscape level: a synthesis of published knowledge and research needs. *Landscape and Urban Planning* 77: 321–342.

Shewchuk, B. A., and W. A. Kerr. 1993. Returns to grasshopper control on rangelands in southern Alberta. *Journal of Range Management* 46:458–462.

Singh, K. P., and R. S. Jayasomu. 2002. *Bombyx mori* – a review of its potential as a medicinal insect. *Pharmaceutical Biology* 40:28–32.

Smith, R. C. 1954. An analysis of 100 years of grasshopper populations in Kansas (1854 to 1954). *Transactions of the Kansas Academy of Science* 57:397–433.

Smith, R. H. 2007. *History of the Boll Weevil in Alabama*. Alabama Agricultural Experiment Station Bulletin 670. Alabama Agricultural Experiment Station, Auburn.

Smith, S., T. E. Reagan, J. L. Flynn, and G. H. Willis. 1983. Azinphosmethyl and fenvalerate runoff loss from a sugarcane–insect IPM system. *Journal of Environmental Quality* 12:534–537.

Stachurski, A., and J. R. Zimka. 1984. The budget of nitrogen dissolved in rainfall during its passing through the crown canopy in forest ecosystems. *Ekologia Polska* 32:191–218.

Stadler, B., and T. Müller. 1996. Aphid honeydew and its effect on the phyllosphere microflora of *Picea abies* (L.) Karst. *Oecologia* 108:771–776.

Stadler, B., B. Michalzik, and T. Müller. 1998. Linking aphid ecology with nutrient fluxes in a coniferous forest. *Ecology* 79:1514–1525.

Swank, W. T., J. B. Waide, D. A. Crossley Jr., and R. L. Todd. 1981. Insect defoliation enhances nitrate export from forest ecosystems. *Oecologia* 51:297–299.

Taylor, S. L., and D. A. MacLean. 2009. Legacy of insect defoliators: increased wind-related mortality two decades after a spruce budworm outbreak. *Forest Science* 55:256–267.

Thompson, D. C., and K. T. Gardner. 1996. Importance of grasshopper defoliation period on southwestern blue grama–dominated rangeland. *Journal of Range Management* 49:494–498.

Ticehurst, M., and S. Finley. 1988. An urban forest integrated pest management program for gypsy moth: an example. *Journal of Arboriculture* 14:172–175.

Torrell, L. A., J. H. Davis, E. W. Huddleston, and D. C. Thompson. 1989. Economic injury levels for interseasonal control of rangeland insects. *Journal of Economic Entomology* 82:1289–1294.

Treseder, K. K. 2008. Nitrogen additions and microbial biomass: a meta-analysis of ecosystem studies. *Ecology Letters* 11:1111–1120.

Trlica, M. J., and L. R. Rittenhouse. 1993. Grazing and plant performance. *Ecological Applications* 3:21–23.

Trumble, J. T., D. M. Kolodny-Hirsch, and I. P. Ting. 1993. Plant compensation for arthropod herbivory. *Annual Review of Entomology* 38:93–119.

Turlings, T. C. J., J. H. Loughrin, P. J. McCall, U. S. R. Röse, W. J. Lewis, and J. H. Tumlinson. 1995. How caterpillar-damaged plants protect themselves by attracting parasitic wasps. *Proceedings of the National Academy of Sciences USA* 92:4169–4174.

Van Bael, S. A., A. Aiello, A. Valderrama, E. Medianero, M. Samaniego, and S. J. Wright. 2004. General herbivore outbreak following an El Niño–related drought in a lowland Panamanian forest. *Journal of Tropical Ecology* 20:625–633.

White, W. B., and N. F. Schneeberger. 1981. Socioeconomic impacts. In *The Gypsy Moth: Research Toward Integrated Pest Management* (C.C. Doane and M.L. McManus, Eds.), pp. 681–694. US Forest Service Tech. Bull. 1584. US Forest Service, Washington, DC.

Whitford, W. G., V. Meentemeyer, T. R. Seastedt, K. Cromack Jr., D. A. Crossley Jr., P. Santos, R. L. Todd and J. B. Waide. 1981. Exceptions to the AET model: deserts and clear-cut forest. *Ecology* 62:275–277.

Wickman, B. E., 1980. Increased growth of white fir after a Douglas-fir tussock moth outbreak. *Journal of Forestry* 78:31–33.

Williamson, S. C., J. K. Detling, J. L. Dodd, and M. I. Dyer. 1989. Experimental evaluation of the grazing optimization hypothesis. *Journal of Range Management* 42:149–152.

Wood, T. E., D. Lawrence, D. A. Clark, and R. L. Chazdon. 2009. Rain forest nutrient cycling and productivity in response to large-scale litter manipulation. *Ecology* 90:109–121.

Wright, L. C., A. A. Berryman, and B. E. Wickman. 1986. Abundance of the fir engraver, *Scolytus ventralis*, and the Douglas-fir beetle, *Dendroctonus pseudotsugae*, following tree defoliation by the Douglas-fir tussock moth, *Orgyia pseudotsugata*. *Canadian Entomologist* 116:293–305.

Yen, A. L. 2009. Entomophagy and insect conservation: some thoughts for digestion. *Journal of Insect Conservation* 13:667–670.

Zenk, M. H., and M. Juenger. 2007. Evolution and current status of the phytochemistry of nitrogenous compounds. *Phytochemistry* 68:2757–2772.

Zimmerman, K. M., J. A. Lockwood, and A. V. Latchininsky. 2004. A spatial, Markovian model of rangeland grasshopper (Orthoptera: Acrididae) population dynamics: do long-term benefits justify suppression of infestations? *Environmental Entomology* 33:257–266.

Part IV

Genetics and Evolution

13

Evidence for Outbreaks from the Fossil Record of Insect Herbivory

Conrad C. Labandeira

Insect Outbreaks Revisited, First Edition. Edited by Pedro Barbosa, Deborah K. Letourneau and Anurag A. Agrawal.
© 2012 Blackwell Publishing Ltd. Published 2012 by Blackwell Publishing Ltd.

13.1   Introduction

The topic of insect outbreaks is virtually nonexistent in the paleoecological literature of plant–insect associations. Of the few examples explicitly studied to date, all are of Holocene age and occurred within the past 11 000 years, after the great megafaunal mass extinctions of the Americas and Eurasia. Evidence for these subfossil insect outbreaks largely is confined to North America and involves evidence for unusually high levels of plant consumption by insects in forest ecosystems. Some of the best examples consist of evidence for multiple rounds of elevated herbivory by two species of lepidopteran larvae in subfossil pollen floras affecting conifer and woody dicot hosts in Eastern Deciduous Forest or in related, ecologically nonanalog communities. These Holocene outbreak studies are extensions of modern plant–insect associational dynamics to the subfossil record. Such shallow-time studies involve prominent insect herbivores that periodically express outbreaks on modern tree hosts (Johnson and Lyons 1991, Metcalf and Metcalf 1993). By contrast, examination of older events provide better understanding of how insect outbreaks may have originated and occurred in floras and plant communities that lack ecologic parallels in the modern world. Very little is known about the earlier, pre-Holocene record of insect outbreaks extending to the rest of the Cenozoic, the Mesozoic, and even the mid- to late Paleozoic when insect herbivory initially was launched on land (Labandeira 2006a, 2006b). Pre-Holocene examples of insect outbreaks historically have not been a concern of paleobotanists, paleoentomologists, or, except lately, even investigators of fossil plant–insect associations. However, there are examples of unusually elevated levels of insect herbivory in the older fossil record that targeted particular plant hosts, often at the species level, but also consumption of groups of closely related species within encompassing host taxa. These examples of elevated insect host specificity on particular hosts have rarely been termed "insect outbreaks" in the paleobiological literature, although now would be the time to address the issue and ascertain the similarities that these intriguing examples have to modern insect outbreak phenomena.

   In this chapter, I will discuss initially Holocene events, referred to as "shallow-time" examples of insect outbreaks, followed by four more extensive examples of earlier, "deep-time" outbreaks. Then I will provide an operational definition of how insect outbreaks may be detected in the fossil record. I will end the chapter in discussing the paleobiological and macroevolutionary significance of outbreak phenomena for both plant hosts and their insect herbivores. It is the hope that this discussion will further the detection of insect outbreaks in the fossil record and determine if insect behavioral or physiological correlates of outbreak phenomena can be assessed directly in the fossil insect or plant–insect associational record.

13.2   A broad operational definition of insect outbreaks
in the fossil record

I shall not today attempt further to define the kinds of material I understand to be embraced … but I know it when I see it.

Supreme Court Justice Potter Stewart's
concurrence in the *Jacobellis v. Ohio*
obscenity case: 378 US 184 (1964)

One of the challenges in examining the variety and levels of insect herbivory in the fossil record is recognizing intuitively that there are instances of plants being targeted and consumed apparently by a single insect herbivore. However, often it is uncertain as to what metric to use for measuring the aftermath. This assuredness in intuitively recognizing outbreak levels of herbivory, but indecision in using the most appropriate measure for documentation, is reminiscent of Justice Stewart's celebrated statement in attempting to define obscenity in an issue facing the US Supreme Court in 1964. Similarly, presented below are several issues that indicate the presence of insect outbreaks in the fossil record, ranging from the late Paleozoic to the Holocene.

The first issue is the relevant unit of measurement. Holocene material typically is collected as peat cores to depths from 1 to 8 m that contain several types of microfossil information. Most important are pollen, charcoal, and fungal sporangia and spores that are used principally for plant community and habitat reconstruction. Seeds, leaf fragments, and wood fragments often provide ancillary data. For the Holocene insect outbreak record, accumulations of fecal pellets, related fungal material, insect head capsules, other disarticulated arthropod sclerites, and coprolites are used as proxy data to infer intervals of elevated herbivore intensity. These subfossil constituents are individually displayed as frequency abundance profiles, with time as the vertical axis, and data expressed as counts of the units of interest, such as pollen and spores itemized at the genus or form-genus level of resolution. For example, in each analyzed stratum or level, coprolites (fossilized fecal pellets) or head capsule abundance frequencies are used to infer insect herbivore intensity, often standardized to an equal volumetric measure to record the density of occurrence. (For additional details, see Chapter 7, this volume.)

When compared to Holocene deposits, the older, pre-Holocene fossil record employs different measures to indicate herbivore activity. This deep-time (pre-Holocene) approach instead focuses on items of consumption – typically foliage and seeds – whereas the shallow-time (Holocene) tack is to emphasize rather the bodily effects of the insects themselves, such as coprolites, head capsules, and other frass used to assess herbivory. Given this shift in focus, there are three commonly used metrics in the deep-time record that document the variety and intensity of insect herbivore feeding. The first metric is feeding diversity, or tallying the number of different, diagnosable damage types (DTs) that occur on a particular plant host (Labandeira *et al.* 2007). The second metric is the frequency of attack, or the proportion of leaves that are herbivorized, revealed by one or more DTs. The third metric is a measure the amount of herbivorized leaf surface area, typically as cm^2, as a proportion of total foliar surface area. These three metrics can be used to evaluate herbivory levels at the organ level for a plant species in a flora, such as external feeding on foliage (Prevec *et al.* 2009), or as comparisons of species both within (Wilf *et al.* 2006, Wappler *et al.* 2009) or among (Wilf *et al.* 2001, 2005; Labandeira *et al.* 2002b) floras, or spatiotemporally among species within plant lineages (Labandeira *et al.* 2007, Currano *et al.* 2010). Such approaches have been used in floras ranging in age from the Early Permian (Beck and Labandeira 1998, Labandeira and Allen 2007) to those at the Oligocene–Miocene boundary (Wappler 2010), and represent worldwide plant taxa from many localities and habitats.

Second, what is the herbivory level that must exceed a particular background level to warrant designation as an insect outbreak? The measures mentioned above represent several methods that could be used to indicate a significant increase of herbivory on a particular plant taxon relative to other taxa within a bulk flora, or to a bulk flora compared with other such bulk floras. In general, at least a twofold or threefold (preferably higher) increase in some measure of herbivory above the ambient level is desirable. The next section provides examples from the fossil record where significant increases in damage diversity, frequency of attack, or herbivorized surface area relative to total foliage area, appear to reach insect out-break levels. However, it is difficult to know if such outbreaks occur at very local spatiotemporal scales, are broadly regional phenomena, or represent early attempts at targeted and elevated herbivory. These herbivory levels hopefully establish an operational and more quantitative approach in determining fossil insect outbreaks, and result in an improvement to the "I know it when I see it" approach that was used in the rendering the Supreme Court's definition of obscenity.

Thirdly, how similar are data from the fossil record to modern methods of data collection? In other words, do separate norms need to be established to demon-strate premodern insect outbreaks? Interestingly, data from the deep-time and shallow-time parts of the fossil record frequently are qualitatively and quantita-tively identical to modes of data collection for the present-day record (e.g., Coley and Barone, 1996; Wilf *et al.* 2001; Currano *et al.* 2010). Measures such as herbivorized foliar surface area as a percentage of the total surface area available, or frequency of attack on leaves, either for target host plant species or for bulk floras, have very close parallels with similar studies on modern plant taxa. Less commonly and more difficult to quantify are three-dimensional measures, partic-ularly abundances of distinctive coprolite populations that originate from a single, above-ground insect producer, and accumulate on the surface of plant litter or peat to be incorporated volumetrically in a fossil deposit, such as Late Carboniferous coal ball deposits (Labandeira and Phillips 1996). Yet this category of data also is methodologically equivalent to studies in modern insect outbreaks that record the rate of fecal pellet deposition (Rinker and Lowman 2001), or other measures of frass accumulation on forest floors (Frost and Hunter 2008). With appropriate caveats regarding the differential preservational potential of various types of leaves, coprolite types, and other evidence for insect consumption of plants, the fossil record of insect herbivory should offer an opportunity toward measuring the extent and intensity of insect outbreaks at macroevolutionary time scales.

13.3   The curious case of the discovery, outbreaks, and extinction of the rocky mountain grasshopper

The Rocky Mountain grasshopper, *Melanoplus spretus* Walsh (Acrididae), was the most devastating agricultural pest in the western North America following settlement by Europeans until about 1900, when the last living specimen was found and the species became extinct. The history of this species occurred during the brief 80 years that represented its modern discovery, several major outbreaks, and sudden extinction of probably the most infamous insect of nineteenth-century North America (Table 13.1). The Rocky Mountain grasshopper has been

Table 13.1 Examples of Insect Outbreaks from the Fossil Record of Plant–Insect Associations.

			Holocene (11 000 BP to Present)				
Affected region	*Deposit, age, and figure herein*	*Plant community*	*Dominant plant taxa*	*Folivorized plant host*	*Outbreak insect folivore*	*Outbreak and ambient levels*	*References*
Western Great Plains from 100°W to Rocky Mtn. Crest; from southern Saskatchewan to northeastern-most Mexico	Glaciers, possibly alluvium; early 1800s to 1900 CE from discovery to species extinction (Figure 13.1)	Grassland and savanna; major river floodplains	Poaceae, Aster-aceae; riparian associated dicot trees and shrubs	Widely polyphagous; esp Poaceae, Brassicaceae, Apiaceae, Rosaceae, Salicaceae, Juglandaceae, Anacardiaeae, Asteraceae, etc.	Orthoptera: *Melanoplus spretus* Walsh; (Rocky Mountain Grasshopper; probably margin feeder, DTs 12,13,14,15, 26,143	Very high but quantitatively unrecorded; "they can sweep and clean a field" (Riley 1877, 85; inferred > 95% of all ambient foliage; also fruits and stored products	Riley 1877; Gurney 1953; Lockwood and Debrey 1990; Lockwood 2001; Lockwood et al. 1994;
Mid Holocene Hemlock Forest; Northern United States to Great Lakes Region, north to southern Quebec and Ontario	Unconsolidated peat; 5700–3200 yr BP for mid-Holocene hemlock decline	Northern Eastern Deciduous Forest	Betulaceae, Sapindaceae, Fagaceae, Pinaceae, Rosaceae, Salicaceae	*Tsuga canadensis* L. (hemlock)	Lepidoptera: Tortricidae: *Choristoneura fumiferana* (Clemens), spruce budworm; Geometridae: *Lambdina fiscal-lana* Gven., East. hemlock looper	Up to eightfold decrease in hemlock pollen; approx. sixfold increase in spruce budworm feces; about 10-fold increase in lepidopteran head capsules	Davis 1981; Bhiry and Filion 1996; Fuller 1998; Foster et al. 2006; Simard et al. 2006

(Continued)

Table 13.1 (Cont'd).

				Pre-Holocene (280 Ma to 64.0–64.7 Ma)			
Flora, locality, and repository[1]	Strata and age	Inferred plant community	Dominant plant taxa[2]	Folivorized plant host(s)[3]	Inferred outbreak insect-feeding type	Outbreak and ambient levels[4]	References
Taint (USNM49689); Redbed Sequence, NW Baylor Co., National Museum of Natural History	Waggoner Ranch Fm.; Early Permian, Artinskian Stage; c. 280 Ma (Plate 13.1)	Dry Riparian Woodland	Gigantopterids, peltasperms, cycadophyte, indeterminate broadleaf sp. conifers	Zeilleropteris sp., Cathyasiopteris sp. Gigantopteridium sp. (all Gigantopteridaceae)	Orthoptera; external foliage feeder (several margin, hole and surface feeding DTs)	Gigantopterid folivory exceeds the bulk floral value by 2×; gigantopterid DTs much more diverse	Beck and Labandeira 1998; Labandeira et al. 2007
Aasvoëlberg 311 SW Karoo Basin, Western Cape, So. Africa. Bernard Price Institute of Palaeontology, Johannesburg	Molteno Fm.; Late Triassic, early Carnian Stage; c. 228 Ma (Plate 13.2)	Heidiphyllum Thicket	Voltzialean conifer; hamshawvialean ginkgophyte, petriellalean ginkgophyte, equisetalean, corystosperm	Heidiphyllum elongatum (Morris) Retallack (Voltziaceae)	Coleoptera, possibly Buprestoidea; leaf miner (DT71)	Herbivory frequency on H. elongatum is 7.2% (10 358 total leaves versus 741 mined) compared to levels–<1% elsewhere in other Molteno floras	Anderson and Anderson 1989; Scott et al. 2004; Labandeira 2006b; pers. obs.; Labandeira et al. 2007

Site[1]	Age/Formation	Habitat	Plant taxa[2]	Most abundant plant	Insect herbivory[3]	Notes[4]	References
Somebodys Garden (DMNH2203); Williston Basin, Slope Co., North Dakota. Denver Museum of Nature and Science a	Hell Creek Fm.; Late Cretaceous, latest Maastrichtian Stage; 54.0 m below K–Pg boundary; circa 66.0 Ma (Plate 13.3)	Herbaceous Meadow	Urticaceae sp. 1, Cercidiphyllaceae, Rosidae, Urticaceae sp. 2 Cercidiphyllaceae, ?Rosidae, dicot indet.	Undescribed leaf morphotype HC81 (Urticaceae)	Coleoptera: Chrysomelidae, Curculionidae; hole feeder (DT57, DT50)	Comprises 50% of flora but 61.8% of folivory; elevation of about 1.62 × that of bulk floral value	Johnson 2002; Labandeira et al. 2002a, 2002b, 2007
Mexican Hat (USNM 42090, DMNH1251) Powder River Basin, Custer Co., MT. National Museum of Natural History	Fort Union Fm., Lebo Mbr.; early Paleogene, early Danian Stage, age 64.0–64.7 Ma (Plate 13.4)	Mesic Riparian Woodland	Platanaceae, Juglandaceae, dicot indet. 1, Lauraceae, ?Salicaceae, dicot indet 2, dicot indet. 3	Platanus raynoldsi Newberry (Platanaceae)	Diptera: Phytomyzites biliapchaensis Winkler, Labandeira and Wilf, leaf miner (DT104)	Attack frequency of mined interval (68.0%) is approx 2 × that of entire deposit for this host plant (37.9%), which already is high	Lang 1996; Wilf et al. 2006; Labandeira et al. 2007; Winkler et al. 2010

[1] USA Museum repository abbreviations: USNM, National Museum of Natural History, Washington, DC; and DMNH, Denver Museum of Nature and Science, Denver, CO.
[2] In rank order of the seven most abundant taxa in the flora.
[3] In order of folivorized abundance for multiple species in the same taxonomic clade.
[4] Various indices to measure folivory are used to assess outbreak versus ambient levels.

the subject of considerable popular interest (Lockwood 2004), including studies of the causes of migratory outbreak cycles and factors responsible for its extinction (Lockwood and Debrey 1990, Lockwood *et al*. 1994), paralleling interest in the passenger pigeon, *Ectopistes migratorius* L., which became extinct in the wild at about the same time. The earliest records of the Rocky Mountain grasshopper are recounted by Riley (1877), who mentions early "breakouts" of the grasshopper in the Great Plains during the 1830s–1860s, describing the outbreaks as a series of "locust plagues." The Rocky Mountain grasshopper consumed vast amounts of forage, typically grasses and associated forbs in Short Grass Prairie that constituted much of the habitat extending from southern Canada of Saskatchewan and Alberta, through the United States between the 100th meridian and the crest of the Rocky Mountains, and southward to extreme northeastern Mexico. This widely polyphagous insect additionally consumed riparian trees, fruits, stems, other insects, and exhibited cannibalistic tendencies (Riley 1877). Although consumption levels of the Rock Mountain grasshopper were not precisely quantified (Lockwood 2004), anecdotal historical references indicate that practically all photosynthetic tissue and even structural tissue such as bark and seed tests were consumed. This suggests foliar frequency of attack levels near 100% and herbivorized leaf-area removal rates approaching 100%. Such comprehensive herbivory would mimic vertebrate herbivore feeding, presumably as such near-complete levels of foliage removal would not provide items to enter the fossil record, except for the occasional leaf fragments with resistant veinal stringers typical of intense free feeding (Coulson and Witter 1984). In addition, historical records suggest that these regional feeding events were regulated by a periodical control mechanism, similar to that during the twentieth century occurring along the northern African to south Eurasian belt and extending eastward to the savanna and grasslands of India. Even if foliar items could enter the fossil record, detection of periodic outbreaks from decadal to century cycles of recurrence would be below stratigraphic resolution of the fossil record, except under such exceptional circumstances of uninterrupted accumulation as fine-grained strata in basins such as shallow lakes. In the rare case of preservation of an outbreak event attributable to a widely polyphagous insect, such as a migratory locust, one would expect the same recurring assemblage of externally feeding DTs across all plant taxa, representing a unique, highly stereotyped, but not specialized feeding pattern that would be conspicuous in the fossil record. Such a pattern would not be expected with typical monophagous insects whose feeding patterns would be reflected by single DT feeding marks on a single or group of closely related host plants.

13.4 Insect outbreaks in shallow time: The holocene spruce budworm and eastern hemlock looper

For some time, palynologists have examined Holocene (circa 11 000 yr BP to present) pollen records from subfossil peat and other organically rich sediment cores to document the periodic demise of hemlock, *Tsuga canadensis* L. A principal observation from North American, mid-Holocene core data has been multiple episodes of decline, beginning at 8200 yr BP, but especially during 5800–4800 yr BP in which there was dramatic range restriction of hemlock. The driver for these pulses of biogeographical contraction and expansion has been attributed to major climatic changes, particularly regional drought and fire, resulting from deglaciation (Lavoie 2001; Foster *et al.* 2006; Zhao *et al.* 2010), perhaps biotically stressing hemlock populations and causing susceptibility to fungal pathogen attack (Davis 1981; Haas and McAndrews 2000). Alternatively, some have suggested that repeated rounds of hemlock decline were caused by insect outbreaks involving the spruce budworm, *Choristoneura fumiferana* (Clemens) (Lepidoptera: Tortricidae) (Bhiry and Filion, 1996; Filion *et al.* 2006); the eastern hemlock looper, *Lambdina fiscellaria* (Guenée) (Lepidoptera: Geometridae) (Simard *et al.* 2002); and possibly the hemlock wooly adelgid, *Aldeges tsugae* Annand (Heard and Valente 2009). Both contending hypotheses – climatic change and insect outbreaks – postulate major regional events of degradation in hemlock decline, the apparent common denominator of which was introduction of a pathogen responsible for the severe reduction of the species (Allison *et al.* 1986). The possibility remains that both attributed causes are linked at some level, a phenomenon long appreciated by studies of North American wood-boring beetles and their conifer hosts (e.g., Bonello *et al.* 2003).

Several types of data from soil cores have been used to buttress hypotheses of climate change, insect outbreaks, or indirectly pathogens as explanations accounting for hemlock decline (Simard *et al.* 2002). These data include frequency–abundance profiles of distinctive coprolite types to distinguish background levels of herbivory from the peak levels during insect outbreaks. Informative contents from coprolites provide an opportunity to diagnose source plant species, a technique used by foresters to assess tree vigor in modern forests (Weiss 2005). Not only coprolite abundance levels of spruce budworm and hemlock looper are checked against the hemlock decline intervals in postglacial northern North America, but so also are the corresponding head capsule abundance profiles of these species (Simard *et al.* 2006). Coprolite, head capsule, and charcoal frequency profiles often are used separately as internal checks for validation of data, to indicate that drought and fire exhibit a strong correspondence to degradation of forests.

13.5　Insect outbreaks in deep time: focused folivory from four fossil floras

Four fossil deposits have been identified that contain particular plant taxa with exceptionally elevated levels of herbivory (Table 13.1). These case studies document collections of bulk floras from quarries measuring up to a few square meters in surface extent and traversing 0.4–2.0 m of stratal section. In these four deposits, all recognizable foliar material greater than a minimum established size of circa 1.0 cm^2 in surface area was collected, evaluated, and analyzed, regardless of preservational state, plant taxonomic status, or the presence or absence of insect damage. This approach provides an unbiased sample of all available plant fossils retrieved from the preserved, original plant community. In addition, this collection strategy includes evaluation of all multiple foliar or other plant organs occurring on each shale or siltstone slab, often of considerable surface extent, indicating an accurate capture of dispersed plant taxa at the time when they were originally incorporated in a stratum of sediment.

The four deposits collectively represent a variety sedimentologic facies within the fluvial depositional environment. These deposits overwhelmingly or exclusively contained plant material, as nonplant fossils such as insects were frequently absent, and vertebrates were more rarely preserved. The material is housed in museum repositories, and, with the exception of Aasvoëlberg 311, has been published using a variety of methods to indicate herbivory, especially the frequency and diversity analyses derived from the presence or absence of data of damage types (DTs) recorded from each fossil flora or plant host morphotype (Labandeira *et al.* 2007). In each of the four floras discussed in this section, seven of the most abundant plant taxa are listed, including the most highly herbivorized plant taxon. Interestingly, the most abundant plant taxon in each of the four floras, as measured by the frequency of specimens, is the presumed outbreak taxon. The proposed insect outbreak taxa in two of the floras are exophytically feeding mandibulate insects, and the other two floras are attacked by endophytically feeding leaf miners. Other examples of different DTs with conspicuously high levels of herbivory occur on other plant hosts from similar basinal deposits within the same regional megaflora. For example, some localities sample the Paleocene–Eocene Thermal Maximum, a brief 2×10^5 yr long interval when global temperatures were significantly elevated, consisting of an interval of time when North American temperate floras became subtropical, accompanied by considerably elevated herbivory levels (Currano *et al.* 2008, 2010). These patterns suggest that insect outbreaks may be a common phenomenon in the fossil history of insect herbivory.

13.5.1 Gigantopterids at Taint: Early Permian of the redbed sequence, northern Texas

The informal designation "Taint" was given to a site in Baylor County, in north-central Texas (USNM loc. 40689), United States, of Early Permian (Artinskian) strata corresponding to circa 280 Ma. This site was retrieved from the long-studied redbed sequence that historically has produced various synapsid reptiles, especially pelycosaurs. The Taint locality sampled 15 m of horizontal outcrop that consisted of five adjacent trenches from 0.4 to 1.0 m in depth and approximately 0.5 m in width for which all plant material was collected, identified to the genus level, analyzed, and published (Beck and Labandeira 1998). The results of this study indicated that 2.55% of the all foliar surface area in this deposit was removed by insect herbivory, a seemingly high mean value for a Paleozoic flora (Labandeira and Allen 2007). More importantly, there was considerable variation in the level of herbivory with some taxa lacking or having a trace amount of damage (*Asterophyllites* sphenopsids, *Brachyphyllum* conifers, and *Taeniopteris* cycadophytes), whereas most ferns and seed ferns (*Callipteris*, *Comia*, *Sphenophyllum*, and *Wattia*) displayed low to intermediate levels of herbivory. However, it was the gigantopterid seed ferns (*Gigantopteridium*, *Cathaysiopteris*, and *Zeilleropteris*) that received about twice the amount of damage compared to the mean for the bulk flora (Table 13.1, Figure 13.1), having a much greater

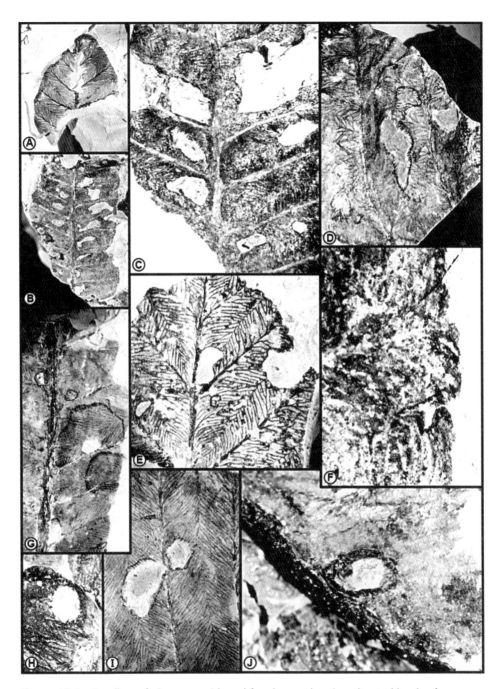

Figure 13.1 A gallery of gigantopterid seed fern leaves showing elevated levels of herbivory from the Early Permian (Artinskian Stage, about 280 Ma) redbed Taint locality (USNM loc. 40689), of the Waggoner Ranch Formation in north–central Texas, United States. The most prominent examples of insect feeding on the Gigantopteridaceae are medium to large sized, polylobate hole feeding in (A–C), (E), and (H–J). Less pronounced margin feeding occurs in (E) and (F). Significant herbivory also is present as surface abrasion, shown in (D) and (G). Based on these distinctive feeding patterns, the same or perhaps several orthopteroid insect taxa probably were responsible for this spectrum of plant damage. This material is housed in the Department of Paleontology, National Museum of Natural History, in Washington, DC. Further details are provided in Beck and Labandeira (1998).

variety of damage than any other plant taxa. Insect folivory of gigantopterids included overwhelmingly hole-feeding damage types (DTs 3, 4, 5, 77, and 113), as well as margin feeding (DT 12) and surface abrasion (DT 30 and 31) (Labandeira *et al.* 2007). This is different from the highly stereotyped damage in the more recent sites, of which the outbreak host plant is represented by a single DT. Taint damage consists of a distinctive spectrum of external foliage-feeding DTs (Figure 13.1), and is significant, particularly for a compression–impression flora occurring early in the launching of late Paleozoic, Euamerican herbivory on land. This elevated level of herbivory may not constitute an outbreak given current definitions, but it does represent the first documented instance in a fossil compression flora of the extensive targeting of a group of closely related plants by a recognizable syndrome of external foliage feeding.

13.5.2   *Heidiphyllum elongatum* at Aasvoëlberg 311: Late Triassic of the Karoo Basin, South Africa

In the stratigraphically lowest sequence of the Molteno Formation, in the Northern Cape Province of South Africa, seven Aasvoëlberg sites represent lake deposits that contain diverse assemblages of foliar material, seeds and other reproductively associated structures, stems, and roots (Anderson and Anderson 1989; Scott *et al.* 2004; Labandeira 2006b). These sites are of Upper Triassic age and are assignable to the lower Carnian Stage, corresponding to circa 228 Ma. One of these sites at Aasvoëlberg, Aas 311, consists of 11 677 specimens representing 15 taxa of plants, and is dominated by the voltzialean conifer leaf *Heidiphyllum elongatum* (Plate 13.1). This broad, parallel-veined leaf, similar to modern *Agathis* (Araucariaceae), consists of 88.7% of all specimens, and 91.9% of the material assigned to this whole-plant taxon when *Dordrechtites elongatus*, its dispersed seed, is added. The remaining seven most abundant and identifiable taxa, in decreasing rank order, are the ginkgophyte *Spenobaiera schenckii* (1.1%), three species of the ginkgophyte *Kannaskoppifolia* (0.5%) considered together, the sphenopsid *Equisetites kanensis* (0.5%), the corystosperm *Dicroidium crassinervis* (0.5%), the moss *Muscites guescelini* (0.3%), the liverwort *Marchantites* (0.3%), and the voltzialean conifer *Rissikia media* (<0.1%), all of which total to 95.1% of the flora. The remaining flora consists of very rare ferns, lycopods, unaffiliated seeds, and unidentifiable foliage. Notably, at Aas 311, 7.2% of the 10 357 *H. elongatum* specimens are leaf mined (DT71), indicating an exceptionally high level of herbivory, particularly as this plant is one of the earliest known leaf mined hosts in the fossil record (Table 13.1, Plate 13.1). Other Molteno sites with leaf-mined *H. elongatum*, other seed plants, ferns, and sphenopsids are at least an order of magnitude less abundant, suggesting by comparison that the leaf-miner at Aas 311 was an outbreak species.

13.5.3   HC81 at Somebody's Garden: Latest Cretaceous of the Williston Basin, North Dakota

Eighty localities from outcrops across the Williston Basin in southwestern North Dakota, United States, represent the best published example worldwide of plants and their insect associations above and below the Cretaceous–Paleogene (K–Pg) boundary (Lang 1996, Johnson 2002, Labandeira *et al*. 2002a, 2002b). The K–Pg boundary consists of the Hell Creek Formation below and Fort Union Formation above, and defines a major extinction event that affected major clades of terrestrial animals, including dinosaurs, as well as a considerable plethora of marine life, such as mosasaurs, rudistid clams, and ammonites. Also affected were terrestrial plants and their associations with herbivorous insects (Labandeira *et al*. 2002a, 2002b), although the effect may have been considerably dampened in Eurasia (Wappler *et al*. 2009). One locality, "Somebody's Garden" (DMNH loc. 2203), positioned 54.2 m below the boundary, consists of almost entirely herbaceous species, dominated by dicot angiosperms, and occupying an extensive sand bar or similar ephemeral habitat associated with a stream. From this locality was collected 412 plant specimens, of which almost half (207) are of the intensely attacked herbaceous dicot plant, morphotype HC81, an unnamed nettle (Urticaceae) (Table 13.1, Plate 13.2). The other half of the flora consists of 20 plant taxa, of which the seven most abundant and identifiable taxa are, in decreasing rank order, are HC229 (Cercidiphyllaceae indet.), HC224 (Rosidae indet.), HC228 (Urticaceae, sp. 2), HC212 (*Cercidiphyllum ellipticum*), HC90 (?Rosidae indet.), HC92 ("*Cinnamomum*" *lineafolia*, fam. indet.), and HC226 (Ranunculaceae indet.). The most important aspect of this flora, other than its overwhelming herbaceousness, is the distinctive herbivorized pattern on HC81, a palmately veined mesic leaf consisting of extensive hole feeding at primary vein angles (DT57) and associated circular to ovoidal feeding DTs 1, 2, and 3 elsewhere on the leaf blade (Labandeira *et al*. 2002a; Plate 13.2). Approximately half (51.2%) of the HC81 specimens exhibit one or more examples of this syndrome of plant damage, frequently with several instances of damage per leaf. When considered from a bulk floral perspective based on presence or absence data, 61.8% of all damage in the flora is on HC81, representing a 1.62 increase in damage level above the ambient level of attacked plant hosts other than HC81. Of the four candidate sites for insect outbreaks, though visually arresting, the weakest case perhaps lies with HC81 folivory at Somebody's Garden.

13.5.4 *Platanus raynoldsi* at Mexican Hat: Early Paleocene of the Powder River Basin, Montana

Mexican Hat (USNM loc. 42090, DMNH loc. 1251) is an early Paleocene site from the Lebo Member of the Fort Union Formation, dated between 64.0 and 64.7 Ma (Belt *et al*. 2004), in the Powder River Basin of eastern Montana, United States. This site historically has been known for the presence of pollen and small vertebrates occurring approximately within a million years after the K–Pg event, and provides a distinct contrast with the pre-event, more subtropical Cretaceous Somebody's Garden site mentioned in Section 13.5.3 (Wilf *et al*. 2006). Characterized by 15 dicot taxa totaling to 2218 leaves, Mexican Hat is dominated by the sycamore *Platanus raynoldsi* (Platanaceae), which consists of 52.9% of all leaves recorded from the deposit (Plate 13.3). The next seven most abundant leaves, in decreasing rank order, are *Juglandiphyllites glabra* (Juglandaceae), *Zizyphus flabella* (affiliation unknown), MH–Lauraceae sp. 2, "*Populus*" *nebrascensis* (?Salicaceae), *Dicotylophyllum anomalum* (affiliation unknown), "*Ficus*" *artocarpoides* (Lauraceae), and *Ternstroemites aureavallis*

(Theaceae). When *P. raynoldsi* is considered throughout the entire deposit, 37.9% of the leaves are attacked, exhibiting one or more occurrences of a broad spectrum of 28 DTs. This value is at the upper end of herbivory values for taxa in the flora, ranging from 18.8% for *J. glabra* to 43.6% for *Z. flabella* for all leaves with a rank of 7 or above (N≥84). However, when only the *P. raynoldsi* leaf-mined interval is analyzed, which is defined as the relatively thin stratum between the first and last appearances of leaf mine DT104 on *P. raynoldsi* (Winkler *et al.* 2010), the herbivory incidence rate almost doubles to 68.0% of all leaves. In this confined stratigraphic layer there are 32 leaf mines, and each *P. raynoldsi* leaf hosts a mean of 2.15 mines per leaf, some harboring up to 6 mines that obliterate much of the foliar mesophyll and other parenchymatic tissue (Table 13.1, Plate 13.3). While *P. raynoldsi* is clearly the dominant leaf and most heavily attacked at Mexican Hat, it is the stratally confined leaf-mined interval that exhibits qualities of an insect outbreak, particularly when damage frequencies and the extensive removal of inner leaf tissue are taken into account.

13.6 The macroevolutionary significance of insect outbreaks

There are important implications in any attempt to document insect outbreaks in deep time. A basic issue is whether insect outbreaks can be documented at all in the fossil record. Once an insect outbreak is known to occur, identification of culprits may be possible only if the evidence surrounding the damage is compelling. If the damage is distinctive and sufficiently abundant, and given an appropriate plant host or geochronological context, it may provide primary data to address major insect phylogenetic events. Most important, though, the fossil record of insect outbreaks can provide insight into the paleobiology of how insects have had a major effect on their plant resource environment and when they acquired the behaviors and physiologies that would allow outbreaks, at least as the term currently is understood.

13.6.1 How limited are fossil data on past insect outbreaks?

It is difficult to distinguish in the fossil record the difference between an insect outbreak of broad regional scope versus a more localized area containing a plant host or group of hosts that has been intensively herbivorized. This issue is related to the areal dimensions and nature of the examined, surviving outcrop. For example, fluvial sedimentary rocks representing certain restricted habitats, such as exposed sandy point bars or finer grained overbank deposits that form particular floodplain deposits, may be annually colonized by particular herbaceous or shrubby plants and subsequently attacked by host–specialist herbivores. But such ephemeral, restricted habitats do not necessarily represent regional insect outbreaks. Although fluvial deposits are abundant in the terrestrial sedimentological record and may provide equivocal evidence for fossil insect outbreaks, more regional outbreaks would more likely be preserved in extensive lacustrine deposits, particularly if exact stratigraphic correlations can be made across distant outcrops.

Another, related issue is the portion of the original insect outbreak region that is recoverable from the fossil record. Fossiliferous deposits in the terrestrial record are sporadic in time; but more important for insect outbreak phenomena, in space as well. Consequently, a deposit with plants providing evidence for an insect outbreak invariably preserves only a small area of what would have been an originally extensive region of a plant host (or hosts) with anomalously high levels of herbivory. This is attributable to two processes. Firstly, plant organs from regional plant host species or entire floras may have been deposited minimally in the first place, and thus preserved outcrops would provide a limited snapshot of what was once present. Most areas of the world, today as in the past, occur in landscapes of net erosion and are uncommonly deposited in basins of sediment accumulation. Secondly, many basinal deposits are uplifted, tectonically ravaged, eroded, or otherwise taphonomically altered such that they either preserve an even smaller fraction of the original outbreak record, or otherwise disappear entirely from the fossil record. Consequently, it is difficult to determine the extent of the original, two-dimensional regional surface that once contained an outbreak, even assuming that stratigraphic correlation among outcrops is not an issue.

13.6.2   Can culprits ever be assigned to fossil insect outbreaks?

For the deep-time part of the fossil record, generalized insect DTs, particularly those made by exophytic consumers such as external foliage feeders, are rarely assignable to specific insect lineages. By contrast, stereotyped DTs that are produced by endophytic, host-specialist consumers, such as leaf miners and gallers, frequently produce plant damage that allows taxonomic assignment to well-identified culprits. In addition, those DTs occurring in the more recent part of the fossil record, where a descendent association still may persist, typically are more amenable to reliable assignment than those from the more distant past. As a result, identification of insect culprits responsible for insect outbreaks in the more remote part of the fossil record is variable: rather poor for exophytic feeders and occasionally good for endophytic feeders. This is in distinct contrast with Holocene occurrences, where relevant data – fecal pellets, larval head capsules, and DT distinctiveness – frequently are present on the same hosts living in the same region, thus providing high certainty for an identification. In addition, the biology of many important Holocene associations in forests often is documented in the economic entomological literature, revealing much about their current biology.

13.6.3   What is the effect of phylogenetic constraints on herbivory patterns?

The recognition of specific features of insect herbivore behavior is detectible in fossil floral material, such as ovipositional probing by leaf miner flies (Winkler *et al.* 2010). One prominent telltale damage pattern involves stereotyped foliar galls on particular organs and tissues that are found on particular plant lineages, such as tenthredinid sawflies attacking the Salicaceae (willows and poplars) (Price 2003). The role of plant secondary chemistry in determining gall placement is crucial, in part because of considerable control that gall-forming insects have on the tissue localization of host plant secondary compounds such as phenolics (Nyman and Tiitto 2000), which frequently lead to galling intensities that reach outbreak status (Price 2003). Such patterns, including gall outbreaks and host switching, should be detectable in the better documented Paleogene and Neogene fossil record where well-preserved and identifiable galls and their host plants are often linked to extant lineages.

Similarly, more deep-time, coarser grained opportunities for recognition of the effect of phylogenetic events on herbivory patterns, including outbreaks, should be detectable in the fossil record. One example, broached earlier, would be the extensiveness of outbreak patterns in dominantly gymnosperm-dominated floras predating the Jurassic Parasitoid Revolution, and conversely in angiosperm-dominated floras subsequent to this event (Labandeira 2002). The Parasitoid Revolution, with major trophic implications for herbivores, was launched during the Early Jurassic with the appearance of aculeate and other parasitoid clades in the Hymenoptera and Diptera (Whitfield 1998, Eggleton and Belshaw 1992). Most major parasitoid lineages were established by the Early Cretaceous (Labandeira 2002). An important test for the influence of the Parasitoid Revolution on herbivory would be the contrast of herbivory intensity levels, plant concealment patterns, and other examples of herbivore defense (e.g., Krassilov *et al.* 2008), whether outbreak or otherwise, between ecologically analogous pre-Jurassic and post-Jurassic floras. Such an investigation could reveal whether earlier bottom→up regulation of herbivory driven by primary producers was dominant, versus later top→down regulation by parasitoids.

13.6.4 Do snapshots of insect outbreaks inform us about biotic communities from the deep past?

Insect herbivore specialization on particular plant hosts undoubtedly occurred early in terrestrial ecosystem evolution (Labandeira 2006a), recognizing that the trace-fossil record of insect-damaged plants typically occurs earlier in time than the corresponding body-fossil record of the insect culprits themselves (Labandeira 2007). Thus, fossil documentation of insect-caused plant damage is the only record of the expansion of herbivory across landscapes during the late Paleozoic, in lieu of insect body fossils. Particular Euramerican plant hosts were preferentially targeted, such as virtually all tissues of marattialean tree ferns in Late Pennsylvanian coal swamp forests (Labandeira 2001, Labandeira and Phillips, 1996), and subsequently leaves of gigantopterid seed ferns in Early Permian open woodland communities dominated by seed plants (Beck and Labandeira 1998; Glasspool *et al.* 2003). In Gondwana, certain species of glossopterid seed ferns were preferentially subjected to high levels of oviposition and margin feeding (Prevec *et al.* 2009). A crucial question is at what point in this strengthening of plant–insect associational nexus of relationships did outbreak behavior in insects originate? Gigantopterids may be a reasonable early candidate, as they are an early seed fern lineage that had angiosperm-like features (Hilton and Bateman 2006) and were extensively herbivorized. An answer to this question involves identification of those insect lineages responsible for the recorded high levels of damage. In the case of the earliest, late Paleozoic example, it appears that insects responsible for gigantopterid damage at Taint, in northern Texas, probably were external foliage-feeding orthopteroids. The Early Permian age of this deposit suggests the absence of other mandibulate herbivore lineages that would create the suite of DTs occurring on Taint gigantopterids, such as phytophagous adult beetles or herbivorous larvae from several early holometabolous clades. During this time, when particular plant resources soon became available as food and while most enemy-free space was bereft of top-down regulation by parasitoids (Labandeira 2002), insect outbreaks would be expected.

Perhaps a better example of an outbreak would be the endophytic leaf miner, probably a polyphagan beetle, at the Aas 311 site from the Late Triassic of South Africa. This elevated level of leaf mining represents even better defended enemy-free space than that of the Early Permian example, as it not only precedes the Jurassic Parasitoid Revolution (Labandeira 2002), but also acquired histological isolation of the larvae by adopting an endophytic feeding habit, probably as protection against active insect predators. As in the Late Permian case, there is high but not exact host plant specificity: the Molteno leaf miner was overwhelmingly on *H. elongatum*, but the same DT71 mine occurs on unrelated seed fern and ginkgophyte taxa. Analogously, the Taint external foliage feeder was consuming four confamilial genera within the Gigantopteridaceae.

During the latest Cretaceous, long after the Jurassic Parasitoid Revolution was established (Labandeira 2002), a different type of insect outbreak emerges at Somebody's Garden, in the latest Cretaceous from the Williston Basin of North Dakota. The intensely herbivorized nettle, HC81 (Urticaceae), was highly defended by dense arrays of resinous glands and integument-penetrating trichomes, indicating use of host secondary chemicals and physical defenses as major feeding deterrents to insect herbivores, which likely were chrysomeloid beetles (Labandeira 2002a). Once the chemical and physical defenses of the nettle host were breached, additional enemy-free space would have been purchased through consumed secondary chemicals occurring in herbivore tissues to ward off parasitoids and predators. Unlike the other deep-time examples, this episode of herbivory targets a herbaceous host within a habitat of few, if any, woody plants which grew in a disturbed, periodically reforming community, suggesting rapid colonization by potential plant hosts and their herbivore host specialists. Under these ephemeral conditions, limited insect outbreaks could be expected and would last as long as the habitat would experience the next flooding event.

After the K–Pg extinction event, the low-diversity Paleocene "dead zone" (Labandeira *et al*. 2002a), was dominated by mire communities, predominant throughout the Western Interior of North America (Johnson 2002). At Mexican Hat, from the early Paleocene in the Powder River Basin of Montana, the presence of a thin stratum of fossil leaves was overwhelmingly dominated by heavily mined *P. raynoldsi* sycamore leaves, strongly suggesting a heavily targeted host by an agromyzid leaf miner. This relationship was probably similar to modern agromyzid leaf miners of *Phytomyza* on the holly *Ilex* (Aquifoliaceae) (Scheffer and Wiegmann 2000), which occasionally achieve outbreak status (Eber *et al*. 2001). Like the analogous Molteno leaf miner, the Paleocene example of an endophytic habit on woody seed-plant foliage displays host monospecificity and co-evolved features such as leaf surface probes by an ovipositor for evidently assessing host-plant palatability. If this early Paleocene Agromyzidae–Platanaceae system is like other tight leaf miner associations in the fossil record (Lopez-Vaamonde *et al.* 2006), there is reason to infer a significantly earlier origin, presumably during the Late Cretaceous, that would be predicted by the sequential evolution hypothesis (Jermy 1976).

In addition to these examples from the compression–impression fossil record of foliage, a second source of fossil evidence for insect outbreaks is the record of amber. Evidence from Dominican amber, of Early Miocene age (Penney 2010), and considerably older New Jersey amber, of mid-Late Cretaceous age (Grimaldi and Nascimbene 2010), indicates that a relevant fraction of tree resin was produced as a response to beetle infestation, suggested by the presence of associated bark and ambrosia beetle as inclusions, and particularly by elevated δC^{13} values that provide evidence for physiological stress through the production of secondary metabolites (McKellar *et al.* 2011). This relationship has been demonstrated by modern mountain pine beetle, *Dendroctonus ponderosae*, a scolytine curculionid that infests considerable swaths of coniferous forest in western North America, responsible for elevated δC^{13} values (Kurz *et al.* 2009). While amber provides the best type of deep-time evidence for insect outbreaks on coniferous trees, it has serious limitations in addressing folivore-based outbreaks in the fossil record. The amber record does not preserve considerable coverage of foliage surface area, necessary for quantification of herbivory levels, and also amber with culprit wood borer inclusions extends only to the mid=Early Cretaceous, rendering as inaccessible the older two thirds of the deep-time record.

13.7   Summary and conclusions

There are seven major points that can be summarized from these studies. This summary not only provides ways of detecting insect outbreaks in the fossil record, but also indicates the important role of phylogenetic control of insect herbivore macroecology in site-specific increases in herbivory rate. It is anticipated that this contribution will goad paleoecologists to look for insect outbreak evidence in future studies of herbivory in the fossil record.

1. The definition of what constitutes an insect outbreak in the fossil record should remain flexible in terms of the metric used, but always should provide for a distinct, quantitative increase above the ambient herbivory level compared either to most other herbivorized host species within the same source flora, or alternatively to other hosts in spatiotemporally proximal floras.
2. It is unclear if an insect outbreak, such as a "locust plague," whereby virtually all photosynthetic (and other) tissue is rampantly consumed within one or a few annual seasons, would be preserved in the fossil record. If preserved, special observational and analytic skills would be required for its detection.
3. There are two principal modes of recognizing insect outbreaks from fossils. First is to examine the biological effects of the insect herbivore, such as the

accumulation of fecal pellets and head capsules in organically rich, three-dimensional deposits in the more recent part of the fossil record. The second is to assess herbivory levels on leaf surfaces from bedding planes of compression–impression deposits using existing methodology that has been applied to the more deep-time part of the record.

4. Exceptional increases in insect herbivory punctuate the fossil record and provide obvious examples of host specialist targeting of particular plant hosts. These exceptional events are spatiotemporally confined and suggest that host plant defenses are occasionally breached by exophytic and endophytic insect herbivores that can result in distinct outbreaks.

5. Because of the nature of the fossil record, the true extent of an insect outbreak may never be known because of imperfection in the preservation process and the vagaries of erosion that destroy much of what was deposited.

6. In general, the older the deposit, the more unlikely that the culprit of an insect outbreak is satisfactorily identifiable. This mostly is attributable to more accumulated extinctions, in the deeper time part of the fossil record, or to sufficiently modern plant–insect associational analogs for comparison.

7. The fossil history of insect outbreak phenomena can provide evolutionary biologists, particularly those who examine plant–insect associations, a better sense of the evolution of insect outbreaks. It is clear that insect outbreaks are a pattern of extensive herbivory that is not confined only to the present day or the Holocene, but extends into deep time, and probably to the late Paleozoic.

Acknowledgments

I am grateful to Pedro Barbosa for inviting me to contribute a deep-time perspective of insect outbreaks to this volume. I thank Anurag Agrawal and two anonymous reviewers for comments that greatly improved the manuscript. I thank Finnegan Marsh for rendering the figures. This is contribution 177 of the Evolution of Terrestrial Ecosystems Consortium at the National Museum of Natural History in Washington, DC.

References

Allison, T. D., R. E. Mueller, and M. B. Davis. 1986. Pollen in laminated sediments provides evidence for a mid-Holocene forest pathogen outbreak. *Ecology* 67:1101–1105.

Anderson, J. M., and H. M. Anderson. 1989. *Palaeoflora of Southern Africa: Molteno Formation (Triassic). Volume 2: Gymnosperms (excluding Dicroidium)*. Balkema, Rotterdam.

Beck, A. L., and C. C. Labandeira. 1998. Early Permian folivory on a gigantopterid-dominated riparian flora from north-central Texas. *Palaeogeography, Palaeoclimatology, Palaeoecology* 142:139–173.

Bhiry, N., and L. Filion. 1996. Mid-Holocene hemlock decline in Eastern North America linked with a phytophagous insect activity. *Quaternary Research* 45:312–320.

Bonello, P., A. J. Storer, W. R. McNee, T. R. Gordon, D. L. Wood, and W. Heller. 2003. Systemic effects of *Heterobasidion annosum* (Basidiomycotina) infection on the phenolic metabolism of ponderosa pine and the feeding behavior of *Ips paraconfusus* (Coleoptera: Scolytidae). *Journal of Chemical Ecology* 29:1167–1182.

Coley, P. D., and J. A. Barone. 1996. Herbivory and plant defenses in tropical forests. *Annual Review of Ecology and Systematics* 27:305–335.

Coulson, R. A., and J. A. Witter. 1984. *Forest Entomology: Ecology and Management*. Wiley Interscience, New York.

Currano, E. D., C. C. Labandeira, and P. Wilf. 2010. Fossil insect folivory tracks paleotemperature for six million years. *Ecological Monographs* 80:547–567.

Currano, E. D., P. Wilf, S. L. Wing, C. C. Labandeira, E. C. Lovelock, and D. L. Royer. 2008. Sharply increased insect herbivory during the Paleocene–Eocene Thermal Maximum. *Proceedings of the National Academy of Sciences USA* 105:1960–1964.

Davis, M. B. 1981. Outbreaks of forest pathogens in Quaternary history. Pages 216–227 *in* D. Bharadwaj, Vishnu-Mittre, and H. Maheshwari (editors), *Proceedings of the Fourth International Palynological Conference*, vol. 3. Birbal Sahni Institute of Palaeobotany, Lucknow, India.

Eber, S., H. P. Smith, R. K. Didham, and H. V. Cornell. 2001. Holly leaf-miners on two continents: what makes an outbreak species? *Ecological Entomology* 26:124–132.

Eggleton, P., and R. Belshaw. 1992. Insect parasitoids: an evolutionary overview. *Philosophical Transactions of the Royal Society of London B* 337:1–20.

Filion, L., S. Payette, R. E. Delwaide, and A. C. Lemieux. 2006. Insect-induced tree dieback and mortality gaps in high altitude balsam fir forests of northern New England and adjacent areas. *Écoscience* 13: 275–287.

Foster, D. R., W. W. Oswald, E. K. Faison, D. D. Doughty, and B. C. S. Hansen. 2006. A climatic driver for abrupt mid-Holocene vegetation dynamics and the hemlock decline in New England. *Ecology* 87:2959–2966.

Frost, C. J., and M. D. Hunter. 2008. Insect herbivores and their frass affect *Quercus rubra* leaf quality and initial stages of subsequent litter decomposition. *Oikos* 117:13–22.

Glasspool, I., J. Hilton, and M. Collinson. 2003. Foliar herbivory in Late Palaeozoic Cathaysian gigantopterids. *Review of Palaeobotany and Palynology* 127:125–132.

Grimaldi, D., and P. C. Nascimbene 2010. Raritan (New Jersey) amber. Pages 167–191 *in* D. Penney (editor), *Biodiversity of Fossils in Amber from the Major World Deposits*. Siri Scientific Press, Manchester, UK.

Haas, J. N., and H. McAndrews. 2000. The summer drought related hemlock (*Tsuga canadensis*) decline in eastern North America 5700 to 5100 years ago. Pages 81–88 *in* K. McManus (editor), *Proceedings: Symposium on Sustainable Management of Hemlock Ecosystems in Eastern North America*, Durham, NH, 1999. US Forest Service General Technical Report NE-267. US Forest Service, Washington, DC.

Heard, M. J., and M. J. Valente. 2009. Fossil pollen records forecast response of forests to hemlock wooly adelgid invasion. *Ecography* 32:881–887.

Hilton, J., and R. M. Bateman. 2006. Pteridosperms are the backbone of seed-plant phylogeny. *Journal of the Torrey Botanical Society* 133:119–168.

Jermy, T. 1976. Insect–host plant relationships – coevolution or sequential evolution? *Symposium Biologicae Hungaricae* 16:109–113.

Johnson, K. R. 2002. The megaflora of the Hell Creek and lower Fort Union formations in the western Dakotas: vegetational response to climate change, the Cretaceous–Tertiary boundary event, and rapid marine transgression. *Geological Society of America Special Paper* 361:329–391.

Johnson, W. T., and H. H. Lyons. 1991. *Insects That Feed on Trees and Shrubs*, 2nd ed. Cornell University Press, Ithaca, NY.

Krassilov, V., N. Silantieva, and Z. Lewy. 2008. Traumas on fossil leaves from the Cretaceous of Israel. Pages 7–187 *in* V. Krassilov and A. Rasnitsyn (editors), *Plant-Arthropod Interactions in the Early Angiosperm History: Evidence from the Cretaceous of Israel*. Pensoft Publishers, Sofia, and Brill Publishers, Leiden.

Kurz, W. A., C. C. Dymond, G. Stinson, G. J. Rampley, E. T. Nielson, A. L. Carroll, T. Ebata., and L. Safranyik. 2009. Mountain pine beetle and forest carbon feedback to climate change. *Nature* 452:987–990.

Labandeira, C. C. 2001. Rise and diversification of insects. Pages 82–88 *in* D. E. G. Briggs and P. W. Crowther (editors), *Palaeobiology II*. Blackwell Science, London.

Labandeira, C. C. 2002. Paleobiology of predators, parasitoids, and parasites: accommodation and death in the fossil record of terrestrial invertebrates. Pages 211–250 *in* M. Kowalewski and P. H. Kelley (editors), *The Fossil Record of Predation*. Paleontological Society Special Papers, no. 8. Paleontological Society, Boulder, CO.

Labandeira, C. C. 2006a. The four phases of plant–arthropod associations in deep time. *Geologica Acta* 4:409–438.

Labandeira, C. C. 2006b. Silurian to Triassic plant and hexapod clades and their associations: new data, a review, and interpretations. *Arthropod Systematics and Phylogeny* 64:53–94.

Labandeira, C. C. 2007. Assessing the fossil record of plant–insect associations: ichnodata versus body-fossil data. Pages 9–26 *in* R. Bromley, L. Buatois, M.G. Mángano, J.S. Genise, and R. Melchor (editors), *Ichnology at the Crossroads: A Multidimensional Approach to the Science of Organism–Substrate Interactions*. Society of Economic Paleontologists and Mineralogists Special Publication, vol. 88. Society of Economic Paleontologists and Mineralogists, Tulsa, OK.

Labandeira, C. C., and E. G. Allen. 2007. Minimal insect herbivory for the Lower Permian Coprolite Bone Bed site of north-central Texas, USA, and comparison to other Late Paleozoic floras. *Palaeogeography, Palaeoclimatology, Palaeoecology* 247:197–219.

Labandeira, C. C., K. R. Johnson, and P. Lang. 2002a. Preliminary assessment of insect herbivory across the Cretaceous–Tertiary boundary: major extinction and minimum rebound. Pages 297–327 *in* J. H. Hartman, K. R. Johnson and D. J. Nichols (editors), *The Hell Creek Formation of the Northern Great Plains*. Geological Society of America Special Paper, vol. 361. Geological Society of America, Boulder, CO.

Labandeira, C. C., K. R. Johnson, and P. Wilf. 2002b. Impact of the terminal Cretaceous event on plant–insect associations. *Proceedings of the National Academy of Sciences USA* 99:2061–2066.

Labandeira, C. C., and T. L. Phillips. 1996. A carboniferous insect gall: insight into early ecologic history of the Holometabola. *Proceedings of the National Academy of Sciences USA* 93:8470–8474.

Labandeira, C. C., P. Wilf, K. R. Johnson, and F. Marsh. 2007. *Guide to Insect (and Other) Damage Types on Compressed Plant Fossils*, v. 3.0 – Spring 2007. Smithsonian Institution, Washington, DC. http://paleobiology.si.edu/pdfs/insect/DamageGuide.pdf

Lang, P. J. 1996. Fossil evidence for patterns of leaf-feeding from the Late Cretaceous and Early Tertiary. PhD thesis, University of London.

Lavoie, C. 2001. Reconstructing the late-Holocene history of subalpine environment (Charlevoix, Québec) using fossil insects. *The Holocene* 11:89–99.

Lockwood, J. A. 2004. *Locust: The Devastating Rise and Mysterious Disappearance of the Insect That Shaped the American Frontier*. Basic Books, New York.

Lockwood, J. A., and L. D. Debrey. 1990. A solution for the sudden and unexplained extinction of the Rocky Mountain grasshopper (Orthoptera: Acrididae). *Environmental Entomology* 19:1194–1205.

Lockwood, J. A., L. D. Debrey, C. D. Thompson, C. M. Love, R. A. Nunamaker, S. R. Shaw, S. P. Schell, and C.R. Bamar. 1994. Preserved insect fauna of glaciers of Fremont County in Wyoming: insights into the ecology of the extinct Rocky Mountain Locust. *Environmental Entomology* 23:220–235.

Lopez-Vaamonde, C. N. Wikström, C. C. Labandeira, H. C. H. Godfray, S. J. Goodman, and J. M. Cook. 2006. Fossil-calibrated molecular phylogenies reveal that leaf-mining moths radiated millions of years after their host plants. *Journal of Evolutionary Biology* 19:1314–1326.

McKellar, R. C., A. P. Wolfe, K. Muehlenbachs, R. Tappert, M. S. Engel, T. Cheng, and G. A. Sánchez-Azofeifa. 2011. Insect outbreaks produce distinctive carbon isotope signatures in defensive resins and fossiliferous ambers. *Proceedings of the Royal Society B*. doi:10.1098/rspb.2011.0276

Metcalf, R. L., and R. A. Metcalf. 1993. *Destructive and Useful Insects: Their Habits and Control*. McGraw-Hill, New York.

Nyman, T., and R. J. Tiitto. 2000. Manipulation of the phenolic chemistry of willows by gall-inducing sawflies. *Proceedings of the National Academy of Sciences USA* 97:13184–13187.

Penney, D. 2010. Dominican amber. Pages 22–41 *in* D. Penney (editor), *Biodiversity of Fossils in Amber from the Major World Deposits*. Siri Scientific Press, Manchester, UK.

Prevec, R., C. C. Labandeira, J. Neveling, R. A. Gastaldo, C. V. Looy, and M. Bamford. 2009. A portrait of a Gondwanan ecosystem: a new late Permian locality from KwaZulu–Natal, South Africa. *Review of Palaeobotany and Palynology* 156:454–493.

Price, P. W. 2003. *Macroevolutionary Theory on Macroecological Patterns*. Cambridge University Press, Cambridge.

Riley, C. 1877. *The Locust Plague in the United States, Being More Particularly a Treatise on the Rocky Mountain Locust or So-Called Grasshopper, as It Occurs East of the Rocky Mountains, with Practical Recommendations for Its Destruction*. Rand, McNally and Co., Chicago.

Rinker, H. B., and M. D. Lowman. 2001. Literature review: canopy herbivory and soil ecology, the top-down impact of forest processes. *Selbyana* 22:225–231.

Scheffer, S. J., and B. M. Wiegmann. 2000. Molecular phylogenetics of the holly leafminers (Diptera: Agromyzidae: *Phytomyza*): species limits, speciation, and dietary specialization. *Molecular Phylogenetics and Evolution* 17:244–254.

Scott, A. C., J. M. Anderson, and H. M. Anderson. 2004. Evidence of plant–insect interactions in the Upper Triassic Molteno Formation of South Africa. *Journal of the Geological Society of London* 161:401–410.

Simard, I., H. Morin, and C. Lavoie. 2006. A millennial-scale reconstruction of spruce budworm abundance in Saguenay, Québec, Canada. *The Holocene* 16:31–37.

Simard, I., H. Morin, and B. Potelle. 2002. A new paleoecological approach to reconstructing long-term history of spruce budworm outbreaks. *Canadian Journal of Forestry Research* 32:428–438.

Wappler, T. 2010. Insect herbivory close to the Oligocene–Miocene transition—a quantitative analysis. *Palaeogeography, Palaeoclimatology, Palaeoecology* 292:540–550.

Wappler, T., E. D. Currano, P. Wilf, J. Rust, and C. C. Labandeira. 2009. No post-Cretaceous ecosystem depression in European forests? Rich insect-feeding damage on diverse middle Palaeocene plants, Menat, France. *Proceedings of the Royal Society B* 276:4271–4277.

Weiss, M. R. 2005. Defecation behavior and ecology of insects. *Annual Review of Entomology* 51:635–661.

Whitfield, J. B. 1998. Phylogeny and evolution of host–parasitoid interactions in Hymenoptera. *Annual Review of Entomology* 43:129–151.

Wilf, P., C. C. Labandeira, K. R. Johnson, P. D. Coley, and A. D. Cutter. 2001. Insect herbivory, plant defense, and early Cenozoic climate change. *Proceedings of the National Academy of Sciences USA* 98:6221–6226.

Wilf, P., C. C. Labandeira, K. R. Johnson, and N. R. Cúneo. 2005. Richness of plant–insect associations in Eocene Patagonia: a legacy for South American biodiversity. *Proceedings of the National Academy of Sciences USA* 102:8944–8948.

Wilf, P., C. C. Labandeira, K. R. Johnson, and B. Ellis. 2006. Decoupled plant and insect diversity after the end-Cretaceous extinction. *Science* 313:1112–1115.

Winkler, I. S., C. C. Labandeira, T. Wappler, and P. Wilf. 2010. Distinguishing Agromyzidae (Diptera) leaf mines in the fossil record: new taxa from the Paleogene of North America and Germany and their evolutionary implications. *Journal of Paleontology* 84:933–952.

Zhao, Y., Z-C. Yu, and C. Zhao. 2010. Hemlock (*Tsuga canadensis*) declines at 9800 and 5300 cal. yr. BP caused by Holocene climatic shifts in northeastern North America. *The Holocene* 20: doi:10.1177/0959683610365932

14

Implications of Host-Associated Differentiation in the Control of Pest Species

Raul F. Medina

14.1   Introduction

Genetic structure of insect populations can reflect their geographic distribution but may also reflect their long-term associations with particular plant or arthropod hosts. The latter is known as "host-associated differentiation" (HAD). Theoretically, HAD may explain the staggering species diversity of insects (Stireman *et al.* 2005, Forbes *et al.* 2009), and practically, HAD may have important implications for pest control in agroecosystems. In this chapter I will discuss the applied angle of HAD by providing some ideas on how this type of genetic structure may relate to insect outbreaks in agroecosystems and affect the practices used to control pest species. But before we delve into that discussion, let me provide a brief background on HAD, which has been found in herbivorous insects and in insect parasitoids associated with both agricultural and wild ecosystems.

Host-associated differentiation is the formation of genetically divergent lineages or subpopulations of parasites (i.e., organisms that obtain nourishment

Insect Outbreaks Revisited, First Edition. Edited by Pedro Barbosa, Deborah K. Letourneau and Anurag A. Agrawal.
© 2012 Blackwell Publishing Ltd. Published 2012 by Blackwell Publishing Ltd.

and shelter on or in another organism) when they occur in association with different hosts (Bush 1969, Abrahamson *et al.* 2003). HAD can be considered a special case of ecological speciation wherein hosts, whether they be plants or animals, function as differentially selective environments (Schluter 2001, Rundle and Nosil 2005, Matsubayashi *et al.* 2010). HAD can be detected (1) when parasites associated with different host species possess distinct genotypes despite co-occurring in the same geographic area; and (2) when, based on their similarity, parasite genotypes can be grouped by their host associations rather than by their geographic origin (Berlocher and Feder 2002, Stireman *et al.* 2006, Scheffer and Hawthorne 2007). Ultimately, the degree of HAD depends on the strength of ecological barriers to gene flow between parasitic individuals on different host species. In other words, HAD is the result of partial or total reproductive isolation among populations associated with different host species. If reproductive isolation among parasites associated with different host species is sustained over time, HAD can result in speciation even if the hosts involved occur sympatrically (Dres and Mallet 2002).

Ecological barriers may originate due to different selection pressures experienced by polyphagous insects feeding on different host plant species (Bernays 1991, Carroll and Boyd 1992, Cronin and Abrahamson 2001, Pappers *et al.* 2001, Nosil 2007). Strong differential selection on parasites feeding on different host species (and/or the presence of pre-adapted traits for the exploitation of novel hosts) may promote reproductive isolation between parasites associated with different hosts. Once differential selection generates ecologically and reproductively isolated populations, genetic differences among these populations can accumulate due to genetic drift. If selection is strong enough, these reproductively isolated populations can be maintained in sympatry due to allochrony (Pashley *et al.* 1992, Feder *et al.* 1993, Groman and Pellmyr 2000, Thomas *et al.* 2003), host fidelity (Feder *et al.* 1993, Craig *et al.* 2001), reinforcement (Nosil 2007, Joyce *et al.* 2010b), and/or selection against migrants to non-natal host species (Via *et al.* 2000, Matsubayashi *et al.* 2010, Dickey and Medina 2011a).

As Forbes *et al.* (2009) eloquently expressed it, "a major cause of biodiversity may be biodiversity itself." That is, specialization on different hosts by phytophagous insects may generate reproductive isolation, ultimately resulting in an increase in biodiversity (Feder *et al.* 1988, Abrahamson *et al.* 2001, Dres and Mallet 2002). Thus, different host species may represent distinct selective landscapes to which their parasites adapt. Two of the best-studied cases of HAD involve insects associated with species of goldenrods (Waring *et al.* 1990, Brown *et al.* 1996, Abrahamson and Weis 1997, Craig *et al.* 1997, Cronin and Abrahamson 2001, Nason *et al.* 2002, Stireman *et al.* 2005, Dorchin *et al.* 2009) or with apples and hawthorns (Bush 1969, Feder *et al.* 1988, Berlocher and Feder 2002, Linn *et al.* 2003, 2005). In both of these case studies, HAD occurs in multiple insect species associated with two different host plant species. In both study systems, HAD is not limited to herbivorous insect species; rather, HAD has cascaded up to the third trophic level where natural enemies have displayed a similar host-associated pattern. In addition to these systems, HAD has been found in several other insect species inhabiting wild and agricultural ecosystems (Pashley 1986, Goyer *et al.* 1995, Funk 1998, Via 1999, Martel *et al.* 2003, Thomas *et al.* 2003, Sword *et al.* 2005, Vialatte *et al.* 2005, Nosil 2007, Rosas-Garcia *et al.* 2010).

Recent empirical evidence indicates that HAD is more common than previously thought (Stireman *et al.* 2005), altering the concept of generalist herbivores. Several insects traditionally considered to be generalist species have been found to be composed of genetically distinct and reproductively isolated populations of specialist individuals (Fox and Morrow 1981, Rausher 1982, Futuyma and Peterson 1985, Bernays 1988, Mopper and Strauss 1998, Via *et al.* 2000, Nyman *et al.* 2002, Brunner *et al.* 2004) or even cryptic species (Hebert *et al.* 2004, Condon *et al.* 2008). The presence of HAD within an insect species is not uniformly distributed across its geographic distributions; some populations within the geographic distribution of an insect may experience HAD, while populations at different locations may not (Abrahamson *et al.* 2001). Thus, as has been proposed by Thompson (2005) regarding co-evolution, HAD could be more a population trait than a species trait, resulting in a geographic mosaic of HAD.

HAD can occur relatively quickly, with a species developing reproductive or ecological isolation and/or morphological divergence among populations specializing on different host plants in hundreds of years. The speed at which HAD can create genetically distinct host-associated population depends on several variables such as the amount of genetic variation present in host-associated populations, the feeding mode, the mode of reproduction, the strength of selection, and so on. Thus, the timing at which HAD can create genetically distinct populations on different host species will vary on different parasite species. The lower bound for the date at which the first apple maggot, *Rhagoletis pomonella*, switched from its native hawthorn species to introduced apples is 1620, when the first apples were introduced from Europe (Bush 1969, Bush and Smith 1998, Feder 1998, Feder *et al.* 2003). Similarly, the lower bound for the date at which the first European corn borer, *Ostrinia nubilalis*, started attacking corn in Europe is in the 1500s, and the lower bound for the date at which the first fall army worm, *Spodoptera frugiperda*, started attacking rice and sugar cane in the Americas is in the 1600s. Both of these Lepidoptera species are now composed of genetically distinct lineages when attacking some of their host plant species (Pashley 1986, Dres and Mallet 2002, Thomas *et al.* 2003). Although we don't know the precise length of time it took for these crop pests to develop HAD, we know that such a change can occur within decades. For example, the rhopalid soapberry bug, *Jadera haematoloma*, has evolved different beak lengths to exploit the seeds of introduced plant species in just 20 years (Carroll and Boyd 1992).

Agro-ecosystems may represent rich systems for assessing the commonality of HAD and the rapidity with which it can develop. Today most of the plant species grown for human consumption are non-natives in much of their current ranges (Sauer 1993, Vaughan and Geissler 2009). Many food crops in the Americas were imported by colonists between ~500 and 200 years ago (Sauer 1993). Thus, enough time may have elapsed for herbivorous insect pests to show HAD in most cultivated plants worldwide. Indeed, evidence from agricultural systems indicates that HAD in agroecosystems is not rare (Pashley 1986, Via 1999, Thomas *et al.* 2003, Vialatte *et al.* 2005, Peccoud *et al.* 2009). However, the question of how common is HAD in agroecosystems remains unanswered. To answer this question, more studies testing HAD in agroecosystems are needed.

Although speciation in insects is thought to occur over millennia, genetic differentiation of specific traits, due to strong selection, may appear in the course of few generations. For example, insecticide resistance in insect pest populations can be selected in just a couple of generations (Georghiou 1972). Similarly, strong selection by different host species may create genetic differentiation on specific parasite traits relatively fast (Carroll and Boyd 1992). However, for HAD to occur, the parasite traits selected by different hosts need to be associated with assortative mating. For example, the goldenrod gall fly *Eurosta solidaginis* shows strong HAD in two goldenrod species, *Solidago altissima* and *S. gigantea* (Abrahamson and Weis 1997), and prefers to mate with individuals originating from the same goldenrod species (Craig *et al*. 1993, 2001). Similarly, the apple maggot *R. pomonella* shows strong HAD in apples and hawthorns (Feder *et al*. 1988) and prefers to mate on its respective fruits (Prokopy *et al*. 1971, 1972). Alternatively, pre-adaptation scenarios could initiate the process of HAD when native insects switch to non-native plants. For example, insect herbivores may consist of several genetically distinct populations at certain locations. These genetically distinct populations may overlap in space but not in time (Feder *et al*. 1993) or may be ecologically isolated from each other (Abrahamson *et al*. 2001, Nosil and Crespi 2006a). Some of these genetically distinct populations could be pre-adapted to feed on non-native plants, allowing some native herbivorous insect to readily attack certain novel crops soon after their introduction. For example, the native north American Colorado potato beetle was first noticed in North American potatoes ~200 years after potatoes were introduced from South America (Hare 1990, Hajek 2004). More dramatically, in Brazil, the native Neotropical pentatomid, *Euchistus heros*, has become the most abundant pest of introduced soybean in just 30 years (Panizzi and Oliveira 1998). This suggests that pre-adapted individuals already present in native insect populations can readily colonize non-native plants (Kibota and Courtney 1991). If this pre-adapted individuals are able to keep their associations with their novel hosts due to any sort of selective advantage (e.g., better host nutritional quality, higher host abundance, and lower number of competitors or natural enemies on specific hosts), then HAD may occur if individuals associated with the novel host reproduce with each other rather than with individuals associated with the native hosts. Thus, generalist insect herbivores may often represent conglomerates of diversified genotypes, some of which have the potential to become host-associated populations if particular allelic combinations allow them to attack novel plant species and if different host species exert differential selection on traits promoting assortative mating.

Host-associated differentiation may generate pest outbreaks as a result of native herbivorous insects using non-native crops in which they encounter few or no competitors and/or natural enemies (Jaenike 1990, Feder 1995, Thomas *et al*. 2003). Thus, HAD may allow herbivorous insects feeding on non-native plants to be released of their natural enemies not at a geographic (Crawley 1997, Elton 2000) but at an ecological scale (i.e., native pest species will be "protected" based on their new association with a non-native host plant). The presence of enemy-free space may explain why several genetically distinct host-associated insect lineages remain feeding on novel but often nutritionally suboptimal plant species (Brown *et al*. 1995, 1996, Berdegue *et al*. 1996, Gratton and Welter 1999, Berlocher and Feder 2002, Heard *et al*. 2006, Dorchin *et al*. 2009).

14.2　Host-associated differentiation in herbivorous insect pests

In agroecosystems, field crops and surrounding wild vegetation represent two distinct selective regimes (Dres and Mallet 2002). For example, monocultures offer less structural, morphological, and chemical variation than wild vegetation. In addition, domesticated plants usually present reduced chemical and morphological defenses (Rosenthal and Dirzo 1997, Massei and Hartley 2000). Furthermore, differences in disturbance levels between crops and wild vegetation or variation in management practices (e.g., insecticide use and tillage levels) can further exacerbate the contrast between these selective environments. Thus, agroecosystems present several opportunities for divergent selection to initiate the process of host-associated differentiation as reflected in the examples of HAD in agroecosystems that we can find in the literature.

Most of the reported pests showing HAD consist of mites (Navajas *et al.* 2000, Magalhaes *et al.* 2007) or insects in the orders Hemiptera and Lepidoptera, and fewer in the orders Diptera and Thysanoptera (Pashley 1986, Feder *et al.* 1988, Martel *et al.* 2003, Brunner *et al.* 2004, Sharma *et al.* 2008, Rosas-Garcia *et al.* 2010). Among the Hemiptera, aphids seem to be prone to HAD (Vanlerberghe-Masutti and Chavigny 1998, Via 1999, Ruiz-Montoya *et al.* 2003, Vialatte *et al.* 2005, Carletto *et al.* 2009, Lombaert *et al.* 2009, Peccoud *et al.* 2009). There are reports of host-associated aphid genotypes in at least 18 aphid species (Dickey and Medina 2010), the majority of which are pest species. Most of the hemipteran and thysanopteran pests (e.g., *Thrips tabaci*) for which HAD is reported are parthenogenetic and exophytic, while most of the dipteran and lepidopteran pests reported to present HAD are endophytic and nonparthenogenetic (Pashley 1986, Feder *et al.* 1988, Via 1999, Thomas *et al.* 2003, Brunner *et al.* 2004, Vialatte *et al.* 2005, Peccoud *et al.* 2009).

There are several ways to be parthenogenetic (e.g., thelytokous-apomictic, thelytokous- automictic, and arrhenotokous-automictic). Although all forms of parthenogenesis reduce recombination rates compared to obligate sexual reproduction, not all do so equally. Recombination is highest in arrhenotokous insects, followed by automictic thelytokous insects (which perform meiosis) and then followed by apomictic thelytokous (which perform mitosis). In apomictic thelytokous insects, the level of recombination varies depending on the number of parthenogenetic generations they have prior to a sexually reproducing generation. For example, a cynipid wasp or a univoltine phylloxeran would have only two parthenogenetic generations prior to reproducing sexually, but yellow pecan aphids can have as many as 30 parthenogenetic generations prior to sexual reproduction. Lack of recombination may preserve key host-selected loci (e.g., loci for host preference, for synchronization of parasites' life cycles to their hosts' life cycles, and for parasites' physiological adaptations to their hosts) grouped together. Recombination may alter the right allele combinations needed to exploit specific hosts and to preserve HAD. Thus, the occurrence of HAD might increase along a continuum of decreasing recombination rate (Dickey 2010). Therefore, species with different kinds of parthenogenesis may differ in their likelihood to show HAD (Figure 14.1).

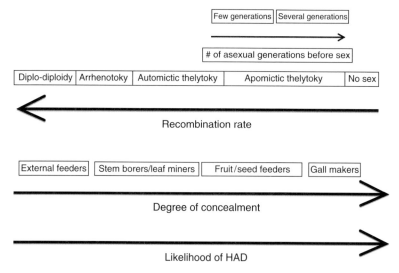

Figure 14.1　Hypothetically, the likelihood of HAD in insects increases with lower levels of recombination and with greater degrees of concealment. Thus, HAD is expected to be the most likely in asexual gall makers or in gall makers with several asexual generations before sexual reproduction (e.g., in some species of *Phylloxera*). HAD is expected to be least likely in diplo-diploid organisms feeding externally (e.g., exophytic Lepidoptera).

Similarly, the degree of association of parasites with their hosts may select for not only one but also a suite of traits needed to successfully exploit specific hosts. The more intimately associated parasites are to their hosts, the more traits are needed to be able to survive in association with their hosts. Thus, an "intimacy" continuum may correlate with the propensity for HAD. That is, within insects that share the same mode of reproduction (e.g., the same kind of parthenogenesis, or in sexually reproducing insects) we would expect HAD to be more common in gall makers, less common in fruit and seed feeders, even less common in stem borers and leaf miners, and least common in exophagous insects (Figure 14.1). The degree of "intimacy" between insect herbivores and their host plants depends on three components: variation in insect and plant physiology, insect and plant phenology, and the degree to which plants may act as insect refuges against their natural enemies. For endophytic insects, phenological and physiological factors may be critical drivers of HAD. Few herbivores are as intimately associated to their host plants as gall makers, which often satisfy all of the intimacy criteria listed above. Gall formation on different plant species may require adaptations such as the up-regulation of specific plant genes or the production of specific kinds of phytohormones. Gall makers can engineer their host plant tissues to serve as shelter from natural enemies (Weis and Abrahamson 1986). In addition, galls made by the same insect on different plant species or even on different plant parts may vary significantly in size, providing their inhabitants with differential protection against natural enemies (Brown *et al*. 1995, Stireman *et al*. 2008). Another intimate interaction between endophagous insects and their host plants can be found in insects that feed within the reproductive structures of their host plants, such as within fruits or seeds. Like gall makers, fruit and seed feeders are not only feeding but also living and in many instances mating within or on the surfaces of their host plants. In addition, endophagous herbivores associated with fruits and seeds must exhibit precise synchrony with the phenologies of their host plants, so that the appropriate food is available for each instar at the right time (Feder *et al*. 1993). A less intimate interaction between herbivorous insects and their host plants can be found in stem borers and leaf miners. These insects do not engage in dramatic genetic manipulations of their host plants, and synchronization of stem borers and leaf miners with their host plants can be less precise than that of fruit and seed feeders, because the plant organs used by leaf miners and stem borers are available for relatively long periods of time.

14.3 Host-associated differentiation in parasitoids

If the degree of intimacy between parasitic organisms and their hosts influences the likelihood of HAD, then we would expect to find several cases of HAD in parasitoid species due to the close association between parasitoids and their hosts (Tauber and Tauber 1989). Parasitoids may differentiate by the host insects they attack, by the host plants used by their insect hosts, or by both (Vaughn and Antolin 1998, Althoff and Thompson 2001, Kankare *et al*. 2005, Smith *et al*. 2006, Stireman *et al*. 2006, Forbes *et al*. 2009). Few studies have explored HAD in parasitoids, and even fewer have explored HAD in predators (Eubanks *et al*. 2003). Most of the studies of HAD in parasitoids have been done by laboratories working originally on HAD in their herbivorous hosts. For example, HAD has been shown to occur in parasitoids of herbivores associated with both of the most well-known HAD systems: goldenrods and the apple–hawthorn system. The parasitoid wasps, *Platygaster variabilis* and *Copidosoma gelechiae*, both show HAD when parasitizing their insect hosts (i.e., the gall midge *Rhopalomyia solidaginis* and the gelechiid moth *Gnorimoschema gallaesolidaginis*, respectively) on *Solidago altissima* and *Solidago gigantea* (Stireman *et al*. 2006, Kolaczan *et al*. 2009). Similarly, *Diachasma alloeum*, a braconid parasitoid of apple maggots, also shows HAD when attacking *R. pomonella* in apples and hawthorns and *R. mendax* in blueberries (Stelinski and Liburd 2005, Forbes *et al*. 2009). The occurrence of HAD in natural enemies of insect herbivores that show HAD has been termed "sequential radiation" (Abrahamson and Blair 2008). However, Cronin and Abrahamson (2001) have suggested that specialization and subsequent diversification by hymenopteran parasitoids may not be dependent on the specialization and subsequent diversification by their insect hosts. Traits of both plants and herbivore hosts influence parasitoid fitness in some cases (Price *et al*. 1980, Souissi and Le Rü 1998), and in others, plant traits alone may directly influence parasitoid fitness. For instance, differences in plant morphology and architecture as well as in plant chemistry, may explain the differential parasitism experienced by pest populations of the same pest species when on different host plants (Van Lenteren *et al*. 1995, Benrey *et al*. 1997, Barbosa and Benrey 1998, Fujiwara *et al*. 2000, Liu and Jiang 2003). Thus, parasitoids' adaptations to specific plant morphological, chemical, and/or architectural features may promote HAD in the third trophic level without the need of HAD in their herbivore hosts. Thus, sequential radiation (i.e., the presence of HAD in parasitoids as a consequence of HAD in their insect hosts) may be an artifact of the logistics involved in the exploration of HAD. However, there are no examples of HAD in parasitoids of hosts lacking HAD. The attempts made by our laboratory to search for such examples (in parasitoids of the green clover worm in alfalfa and soybean and in parasitoids of the processionary moth in different pine species) have both failed to detect HAD.

Parasitoid species are important natural enemies of insect pests. Thus, an understanding of the mechanisms promoting HAD in parasitoids and their hosts will improve our understanding and implementation of biological control (see the "HAD and Biocontrol" section). As will be discussed in Section 14.4, several pest control practices should consider the implications of HAD for their efficient application.

14.4   Impact of host-associated differentiation in agricultural practices

Host-associated populations of pests and natural enemies may differ in phenotypic traits relevant to pest control (Hufbauer and Roderick 2005). Several pest populations are phenotypically distinct when associated with different host plant species (Walsh 1864, Guttman *et al.* 1981, Gillham and Claridge 1994, Adams and Funk 1997, Pappers *et al.* 2001, Diegisser *et al.* 2004). *Helicoverpa zea*, for example, differs in larval developmental time, pupal weight, and survival when feeding on different host plant species (Martin *et al.* 1976, Gore *et al.* 2003). Similarly, the pea aphid, *Acyrthosiphon pisum*, has differential survival, fecundity, and resistance against parasitism when feeding on two different sympatric host plant species (Via 1991a, b, Henter and Via 1995). Apple maggots associated with apples and hawthorns differ significantly in adult body size, adult emergence dates, and morphological traits such as ovipositor length and number of postorbital bristles (Bush 1969). Fitness differences of pests associated with different host plant species (Harvey *et al.* 1995, Hunter 2003), as well as the differential "apparency" (detectability) of pests to parasitoids when on different host plants (Benrey *et al.* 1997, Barbosa and Caldas 2004), may make pests more or less susceptible to their natural enemies.

14.4.1   HAD and biocontrol

Host-associated populations of pests may differ in their vulnerability to natural enemies when on different host plant species (Mackauer 1976, Stansly *et al.* 1997, Barbosa and Benrey 1998, Nwanze *et al.* 1998, Fujiwara *et al.* 2000, Joyce and Bellows 2000, Le Corff *et al.* 2000, Barbosa *et al.* 2001, Billqvist and Ekbom 2001, Lill *et al.* 2002). Consequently, the fitness of their parasitoids may also differ when attacking the same pest species on different host plants (Verkerk *et al.* 1998). If sequential radiation is common (Stireman *et al.* 2006, Abrahamson and Blair 2008, Forbes *et al.* 2009), parasitoids associated with different plant species may be as compartmentalized as their herbivorous hosts. HAD in natural enemies may translate into parasitoids' failure to oviposit and kill pest species (Lozier *et al.* 2008, 2009, Medina and Barbosa 2008). Thus, failure to consider HAD within biocontrol programs may hamper the success of this practice.

In classical biocontrol, for example, it would be advisable to obtain natural enemy populations associated not only with the right pest species but also with the right host plant species. Ideally, parasitoids that attack the target pest on the same or at least on a closely related host plant species to the crop of interest should be preferred over parasitoids attacking the target pest on plants unrelated to the crop of interest. Omission of the host plant component in classical biocontrol programs may reduce the effectiveness of imported biocontrol agents to the point of rendering them useless.

The processes leading to HAD (e.g., strong divergent selection on characters involved in reproduction, and genetic drift on small ecologically and reproductively isolated populations) may also affect augmentation biological control. Since some insect adaptations may occur relatively rapidly, insects used in augmentative biological control could in theory, become adapted to the rearing facility conditions in which they are kept. Such lab-adapted lineages may not perform as efficient control agents in the field (Hopper *et al.* 1993). Several studies have reported deterioration of behavioral traits during laboratory rearing (Hopper *et al.* 1993). Rearing conditions could be selecting for traits involved in reproduction that although useful in the rearing environment may not be the best fit in the field. For example, different kinds of rearing materials could be selecting for differences in traits involved in reproduction. Different rearing materials have been shown to affect substrate-born vibrations used as mating signals by the braconid parasitoid *Cotesia flavipes* (Joyce *et al.* 2010a). Parasitoid colonies reared for several generation on these different materials could experience divergent selection on the traits involved in mating signal emission and/or reception, creating populations that, when released, will be reproductively isolated from field populations (Joyce *et al.* 2008, 2010a). The number of generations it may take for lab-associated lineages to evolve depends on the strength of the selective forces exerted by laboratory conditions, the genetic variability of the insect species involved, and perhaps the mode of reproduction (e.g., parthenogenesis) and feeding mode (e.g., endophagy), among other factors. Thus, it becomes crucial to know which species are prone to experience selection on reproductive traits in augmentative biological control programs so we can prevent laboratory or rearing facility adaptation. One way to prevent lab adaptation is by re-introducing field populations regularly into laboratory or rearing facility colonies to disrupt any kind of selection for rearing conditions. Species likely to show HAD may be more likely to be adapted to artificial rearing conditions. Thus, knowledge on which species are likely to present HAD will justify the implementation of practices aimed to prevent laboratory or rearing facility adaptation in such species.

HAD may also affect the success of conservation biological control practices. The use of wild vegetation as a source of natural enemies in conservation biological control assumes natural enemies move from wild vegetation to crops (Barbosa

1998, Landis *et al.* 2000). However, as data from parasitoids of *R. pomonella* and *E. solidaginis* have shown (Stireman *et al.* 2006, Forbes *et al.* 2009), natural enemies associated with different host plant species, although able to disperse among different host plant species, may be reproductively isolated. Thus, parasitoids associated with one host plant species may not oviposit on their insect hosts if they occur on alternative plant species or if they do, they could experience low fitness. If this is a common occurrence in parasitoids, the role of wild vegetation as a source of natural enemies may be unfounded. Some studies have found that parasitoids associated with different host plant species differ in their search behaviors. For example, parasitoids of *Greya* moths differ in oviposition behavior when ovipositing in different host plant species (Althoff and Thompson 2001). Differences in search and oviposition behaviors may reflect differential selection at specific genes involved in the expression of these traits or alternatively could be due to phenotypic plasticity (Daza-Bustamante *et al.* 2002). If genetic, strong differential selection on traits involved in mating and oviposition on different host plant species may contribute to HAD by promoting assortative mating of insects associated with different plant species.

More studies on the occurrence of HAD in parasitoids are needed. However, results from the few studies that have been completed suggest that not all parasitoid species seem to show HAD. Even when insect hosts show strong HAD, their parasitoids may be able to oviposit on them without showing any evidence of HAD themselves. For example, *Aphelinus perpallidus*, a parasitoid of the yellow pecan aphid, *Monelliopsis pecanis*, shows no HAD (Dickey and Medina 2011b), even though its insect host shows strong HAD (Dickey and Medina 2010). Similarly, no HAD has been detected in the parasitoid *Aphidius transcaspicus*, even though its aphid hosts show HAD (Lozier *et al.* 2009). So far, sequential radiation has been found in parasitoids of endophagous insects but not in parasitoids of exophagous herbivores (Dickey and Medina 2011b). Perhaps parasitoids of endophagous herbivores are more prone to evidence sequential radiation than parasitoids of exophagous herbivores, due to the extra layer of adaptations involved in the evolution of ovipositors suitable for different plant species.

Thus, knowing which parasitoid species in which kind of systems show HAD will refine our considerations in the design of successful biocontrol strategies. In order to be able to do so, we need to diversify our research on HAD by including data from a wide variety of parasitoids and hosts (e.g., idiobionts and koinobionts, parthenogenetic and non parthenogenetic parasitoids and/or hosts, and parasitoids of endophagous and exophagous hosts) and in as many kinds of agroecosystems as possible.

14.4.2   HAD and the movement of pests among host plant species

In agroecosystems, it is usually assumed that when pest species are observed on one host plant species (e.g., a crop or uncultivated host plant), and later observed on another species, the pest has moved among host plants. However, the observation of individuals from a presumed pest population, appearing on different host plant species at different times of the year, should not be interpreted as a demonstration of the generalist nature of a particular pest population. For example, although pea aphids, *Acyrthosiphon pisum*, associated with forage crops move to vegetable crops after harvest of forage crops, their fitness in vegetable crops is low (Losey and Eubanks 2000). Survival and fecundity of pea aphids from forage crops are significantly lower on vegetable crops, and mortality is significantly higher. For example, pea aphids associated with alfalfa fail to feed on soybean (Losey and Eubanks 2000). Thus, it seems that aphids immigrating from forage to vegetable crops do not establish permanent damaging populations. Similarly, the grain aphid, *Sitobion avenae*, is observed in uncultivated and cultivated cereals. However, individuals associated with cultivated and uncultivated cereals belong to distinct populations (Vialatte *et al.* 2005). Thus, uncultivated host plants in the margins adjacent to cereal plants seem to play little role as reservoirs for grain aphid populations colonizing cereal crops, and can be ignored when designing integrated pest management strategies against this pest (Vialatte *et al.* 2005). Such a dichotomy has also been found in the European corn borer, *O. nubilalis*. European corn borer populations from corn are genetically distinct from those on mugwort, *Artemisia vulgaris,* which was previously thought to be a refuge for this moth (Martel *et al.* 2003, Thomas *et al.* 2003). Similarly, the apple maggot, *R. pomonella*, is observed on apples early in the season and later on hawthorns. However, as we have discussed earlier, apple maggots on apples and hawthorns belong to two distinct and reproductively isolated populations (Feder and Filchak 1999). Thus, in agroecosystems, several polyphagous pests that are thought to move from one host plant species to another during the year in reality may represent a conglomerate of distinct populations of host plant specialists whose biology is finely tuned to the phenologies of their specific host plant species. In such cases, control or management of adjacent vegetation to crops may be unnecessary.

14.4.3   HAD and insecticide resistance

Some insect pests use the same enzymatic pathways for the detoxification of host plant allelochemicals and pesticides (Brattsten *et al.* 1984, Muehleisen *et al.* 1989, Li *et al.* 2000). Thus, individuals from the same pest species exposed to plant secondary metabolites from different plant species may differ in their degree of tolerance and/or resistance to insecticides (Hunter *et al.* 1994, Riley and Tan 2003, Behere *et al.* 2007, Liang *et al.* 2007). For instance, some of the monooxygenases or cytochrome pigment 450 enzymes (CYP450) and glutathione-S-transferases induced by plant allelochemicals are known to be highly effective at detoxifying synthetic insecticides (Yu 1982, 1983, Danielson *et al.* 1996). Plant defense compounds such as jasmonate and salicylate can increase expression levels of enzymes involved in detoxification in *H. zea* such as *CYP6B8*, *CYP6B28*, *CYP6B9*, and *CYP6B27* (Li *et al.* 2002). Some of these enzymes metabolize insecticides. For example, *CYP6B8* in *H. zea* metabolizes neurotoxic insecticides such as alpha-cypermethrin, aldrin, and diazinon (Li *et al.* 2004, Rupasinghe *et al.* 2007). Differences in insecticide tolerance among individuals of the same pest species associated with different host plant species have been observed in the laboratory and in the field. For example, pests consuming artificial diets containing different plant allelochemicals show differences in levels of insecticide tolerance (Castle *et al.* 2009). In the field, *Bemisia tabaci* in several locations (i.e., China, Florida, Texas, Arizona, and Georgia) is more tolerant to some pyrethroids when feeding on broccoli, cauliflower, and squash than when feeding on cantaloupe or cotton (Sivasupramaniam *et al.* 1997, Wolfenbarger *et al.* 1998, Riley and Tan 2003, Castle *et al.* 2009). Differences in insecticide resistance among different host-associated populations may require chemical control measures tailored to specific crops.

14.4.4   HAD and pest resistance to natural enemies

If pest species on different host plant species belong to different lineages, it is possible that they differ in resistance against their natural enemies (Kraaijeveld and Van Alphen 1994, Kraaijeveld and Godfray 1999, Hufbauer and Roderick 2005). Pests' differential resistance to natural enemies when associated with different host plant species may be explained by differences in apparency (Nosil and Crespi 2006b), pest immune defense against natural enemies (Hunter 2003), and the effect on parasitoids of different plant tissues fed on by their hosts (Hunter 2003). Shlichta and Barbosa (unpublished data), for example, have found that caterpillars associated with different tree species show differences in encapsulation ability (a phenomenon they called "host-associated encapsulation"). Differences such as this one may occur even in the absence of HAD (when host associated populations are selected for such traits at specific loci without the occurrence of reproductive isolation).

14.4.5   HAD and transgenic crops

An IPM practice that could be seriously affected by the presence of HAD involves the use of genetically modified crops. When genetically modified crops possessing insecticidal traits are used (such as *Bt* crops, for example), refuges are needed to dilute the appearance of resistance (Gould 1998). These refuges consist of plants without the genetic modification of interest. On plants without the genetic modification conferring insecticidal properties, susceptible pests can build up their populations. These susceptible individuals are expected to mate randomly with resistant individuals associated with the genetically modified crops, diluting, in this way, their resistance. If, however, HAD is present between resistant and nonresistant crop varieties, no reproduction will occur among individuals associated with them and resistance to the genetically modified plant will appear relatively quickly. In theory it is possible that individuals associated with genetically modified and unmodified varieties of the same crop plant species could be subjected to strong disruptive selection by the different selection scenarios offered by these two varieties.

　　If the pest in question is a generalist, noncrop plants can also be used as refuges. Such is the case of *Helicoverpa armigera* in Asia (Tabashnik and Carriere 2008). If HAD exists in such a system, then pest populations associated with a genetically modified crop and with noncrop refuges will not mate. If mating between resistant (i.e., individuals resistant to the genetically modified crop) and susceptible (i.e., individuals associated to noncrop refuges) individuals does not occur, resistance to the genetically modified crop will evolve. Thus, it is crucial to consider the possibility of HAD for species that use noncrop species as refuges.

14.5   Conclusions

HAD may influence pest management practices in ways that have heretofore been ignored or undervalued. The frequency of HAD varies across taxa. Some insect groups appear to be more likely to show HAD than others (e.g., parthenogenetic and endophagous insects). IPM programs that aim to be realistic and sustainable should consider this phenomenon more explicitly. We currently do not know how common HAD is in pests and natural enemies in agroecosystems. However, recent evidence indicates that HAD may be more common than previously thought. Thus, it is important to be able to predict which agroecosystems, which pests, and which natural enemies are prone to HAD when designing IPM programs. It is possible that differences in an insect's mode of reproduction may determine the likelihood of HAD. For example, cyclically parthenogenetic species could be more prone to HAD than strictly sexually reproducing species. Similarly, the degree of association of insect species with their resources (e.g., exophytic versus endophytic or endoparasitic versus ectoparasitic habit) may also influence the likelihood of HAD. The present chapter has discussed some of the factors that may promote HAD and has delineated the agricultural practices that may be influenced by this phenomenon. Future studies should explore these issues further. Knowing when HAD should be considered would improve the design and long-term sustainability of biological control.

Acknowledgments

Special thanks go to Apurba Barman, Aaron Dickey, Emilie Hartfield, Andrea Joyce, and Mauro Simonato for fruitful discussions on the role of HAD in the biology of herbivorous insects and their parasitoids. I also want to thank Patricia Pietrantonio for the exchange of ideas that produced the section on the influence of HAD on the response of insects to plant chemical defenses and insecticides, and to thank Andrew Forbes and three anonymous reviewers for their suggestions and encouragement.

References

Abrahamson, A. G., and C. P. Blair. 2008. Sequential radiation through host-race formation: herbivore diversity leads to diversity in natural enemies. Pages 188–202 *in* K. J. Tilmon (editor), *Specialization, Speciation and Radiation: The Evolutionary Biology of Herbivorous Insects*. University of California Press, Berkeley.

Abrahamson, W. G., C. P. Blair, M. D. Eubanks, and S. A. Morehead. 2003. Sequential radiation of unrelated organisms: the gall fly Eurosta solidaginis and the tumbling flower beetle *Mordellistena convicta*. *Journal of Evolutionary Biology* 16:781–789.

Abrahamson, W. G., M. D. Eubanks, C. P. Blair, and A. V. Whipple. 2001. Gall flies, inquilines, and goldenrods: a model for host-race formation and sympatric speciation. *American Zoologist* 41:928–938.

Abrahamson, W. G., and A. E. Weis. 1997. *Evolutionary Ecology across Three Trophic Levels: Goldenrods, Gallmakers, and Natural Enemies*. Princeton University Press, Princeton.

Adams, D. C., and D. J. Funk. 1997. Morphometric inferences on sibling species and sexual dimorphism in neochlamisus bebbianae leaf beetles: multivariate applications of the thin-plate spline. *Systematic Biology* 46:180–194.

Althoff, D. M., and J. N. Thompson. 2001. Geographic structure in the searching behaviour of a specialist parasitoid: combining molecular and behavioural approaches. *Journal of Evolutionary Biology* 14:406–417.

Barbosa, P. 1998. *Conservation Biological Control*. Academic Press, New York.

Barbosa, P. and B. Benrey. 1998. The influence of plants on insect parasitoids: implications for conservation biological control. Pages 55–82 *in* P. Barbosa (editor), *Conservation Biological Control*. Academic Press, San Diego.

Barbosa, P., and A. Caldas. 2004. Patterns of parasitoid–host associations in differentially parasitized macrolepidopteran assemblages on black willow *Salix nigra* (Marsh) and box elder *Acer negundo* L. *Basic and Applied Ecology* 5:75–85.

Barbosa, P., A. E. Segarra, P. Gross, A. Caldas, K. Ahlstrom, R. W. Carlson, D. C. Ferguson, E. E. Grissell, R. W. Hodges, P. M. Marsh, R. W. Poole, M. E. Schauff, S. R. Shaw, J. B. Whitfield, and N. E. Woodley. 2001. Differential parasitism of macrolepidopteran herbivores on two deciduous tree species. *Ecology* 82:698–704.

Behere, G. T., W. T. Tay, D. A. Russell, D. G. Heckel, B. R. Appleton, K. R. Kranthi, and P. Batterham. 2007. Mitochondrial DNA analysis of field populations of *Helicoverpa armigera* (Lepidoptera: Noctuidae) and of its relationship to *H. zea*. *Bmc Evolutionary Biology* 7:117–126.

Benrey, B., R. F. Denno, and L. Kaiser. 1997. The influence of plant species on attraction and host acceptance in *Cotesia glomerata* (Hymenoptera:Braconidae). *Journal of Insect Behavior* 10:619–630.

Berdegue, M., J. T. Trumble, J. D. Hare, and R. A. Redak. 1996. Is it enemy-free space? The evidence for terrestrial insects and freshwater arthropods. *Ecological Entomology* 21:203–217.

Berlocher, S. H., and J. L. Feder. 2002. Sympatric speciation in phytophagous insects: moving beyond controversy? *Annual Review of Entomology* 47:773–815.

Bernays, E. A. 1988. Host specificity in phytophagous insects: selection pressure from generalist predators. *Entomologia Experimentalis Et Applicata* 49:131–140.

Bernays, E. A. 1991. Evolution of insect morphology in relation to plants. *Philosophical Transactions of the Royal Society of London Series B-Biological Sciences* 333:257–264.

Billqvist, A., and B. Ekbom. 2001. Effects of host plant species on the interaction between the parasitic wasp *Diospilus capito* and pollen beetles (*Meligethes* spp.). *Agricultural and Forest Entomology* 3:147–152.

Brattsten, L. B., C. K. Evans, S. Bonetti, and L. H. Zalkow. 1984. Induction by carrot allelochemicals of insecticide-metabolizing enzymes in the southern armyworm (*Spodoptera eridania*). *Comparative Biochemistry and Physiology C-Pharmacology Toxicology and Endocrinology* 77:29–37.

Brown, J. M., W. G. Abrahamson, R. A. Packer, and P. A. Way. 1995. The role of natural enemy escape in a gallmaker host-plant shift. *Oecologia* 104:52–60.

Brown, J. M., and P. A. Way. 1996. Mitochondrial DNA phylogeography of host races of the goldenrod ball gallmaker, *Eurosta solidaginis* (Diptera: Tephritidae). *Evolution* 50:777–786.

Brunner, P. C., E. K. Chatzivassiliou, N. I. Katis, and J. E. Frey. 2004. Host-associated genetic differentiation in *Thrips tabaci* (Insecta; Thysanoptera), as determined from mtDNA sequence data. *Heredity* 93:364–370.

Bush, G. L. 1969. Sympatric host race formation and speciation in frugivorous flies of genus *Rhagoletis* (Diptera: Tephritidae). *Evolution* 23:237–251.

Bush, G. L., and J. J. Smith. 1998. The genetics and ecology of sympatric speciation: a case study. *Researches on Population Ecology* 40:175–187.

Carletto, J., E. Lombaert, P. Chavigny, T. Brevault, L. Lapchin, and F. Vanlerberghe-Masutti. 2009. Ecological specialization of the aphid *Aphis gossypii* Glover on cultivated host plants. *Molecular Ecology* 18:2198–2212.

Carroll, S. P., and C. Boyd. 1992. Host race radiation in the soapberry bug: natural-history with the history. *Evolution* 46:1052–1069.

Castle, S. J., N. Prabhaker, T. J. Henneberry, and N. C. Toscano. 2009. Host plant influence on susceptibility of *Bemisia tabaci* (Hemiptera: Aleyrodidae) to insecticides. *Bulletin of Entomological Research* 99:263–273.

Condon, M. A., S. J. Scheffer, M. L. Lewis, and S. M. Swensen. 2008. Hidden neotropical diversity: greater than the sum of its parts. *Science* 320:928–931.

Craig, T. P., J. D. Horner, and J. K. Itami. 1997. Hybridization studies on the host races of *Eurosta solidaginis*: implications for sympatric speciation. *Evolution* 51:1552–1560.

Craig, T. P., J. D. Horner, and J. K. Itami. 2001. Genetics, experience, and host-plant preference in *Eurosta solidaginis*: implications for host shifts and speciation. *Evolution* 55:773–782.

Craig, T. P., J. K. Itami, W. G. Abrahamson, and J. D. Horner. 1993. Behavioral evidence for host-race formation in *Eurosta solidaginis*. *Evolution* 47:1696–1710.

Crawley, M. J. 1997. *Plant Ecology*. Blackwell Science, Cambridge.

Cronin, J. T., and A. G. Abrahamson. 2001. Do parasitoids diversify in response to host-plant shifts by herbivorous insects? *Ecological Entomology* 26:347–355.

Danielson, P. B., S. L. Gloor, R. T. Roush, and J. C. Fogleman. 1996. Cytochrome P450-mediated resistance to isoquinoline alkaloids and susceptibility to synthetic insecticides in Drosophila. *Pesticide Biochemistry and Physiology* 55:172–179.

Daza-Bustamante, P., E. Fuentes-Contreras, L. C. Rodriguez, C. C. Figueroa, and H. M. Niemeyer. 2002. Behavioural differences between *Aphidius ervi* populations from two tritrophic systems are due to phenotypic plasticity. *Entomologia Experimentalis Et Applicata* 104:321–328.

Dickey, A. M. 2010. *Host-Associated Differentiation in an Insect Community*. Texas A&M University, College Station.

Dickey, A. M., and R. F. Medina. 2010. Testing host-associated differentiation in a quasi-endophage and a parthenogen on native trees. *Journal of Evolutionary Biology* 23:945–956.

Dickey, A. M., and R. F. Medina. 2011a. Immigrant inviability in yellow pecan aphid. *Ecological Entomology* 36:526–531.

Dickey, A. M., and R. F. Medina. 2011b. Lack of sequential radiation in a parasitoid of a host-associated aphid. *Entomologia Experimentalis Et Applicata* 139:154–160.

Diegisser, T., J. Johannesen, C. Lehr, and A. Seitz. 2004. Genetic and morphological differentiation in Tephritis bardanae (Diptera: Tephritidae): evidence for host-race formation. *Journal of Evolutionary Biology* 17:83–93.

Dorchin, N., E. R. Scott, C. E. Clarkin, M. P. Luongo, S. Jordan, and W. G. Abrahamson. 2009. Behavioural, ecological and genetic evidence confirm the occurrence of host-associated differentiation in goldenrod gall-midges. *Journal of Evolutionary Biology* 22:729–739.

Dres, M., and J. Mallet. 2002. Host races in plant-feeding insects and their importance in sympatric speciation. *Philosophical Transactions of the Royal Society of London Series B-Biological Sciences* 357:471–492.

Elton, C. S. 2000. *The Ecology of Invasions by Animal and Plants*. University of Chicago Press, Chicago.

Eubanks, M. D., C. P. Blair, and W. G. Abrahamson. 2003. One host shift leads to another? Evidence of host-race formation in a predaceous gall-boring beetle. *Evolution* 57:168–172.

Feder, J. L. 1995. The effects of parasitoids on sympatric host races of *Rhagoletis pomonella* (Diptera, Tephritidae). *Ecology* 76:801–813.

Feder, J. L. 1998. The apple maggot fly, *Rhagoletis pomonella*: flies in the face of conventional wisdom about speciation? Pages 130–144 *in* D. J. Howard and S. H. Berlocher (editors), *Endless Forms: Species and Speciation*. Oxford University Press, Oxford.

Feder, J. L., S. H. Berlocher, J. B. Roethele, H. Dambroski, J. J. Smith, W. L. Perry, V. Gavrilovic, K. E. Filchak, J. Rull, and M. Aluja. 2003. Allopatric genetic origins for sympatric host-plant shifts and race formation in *Rhagoletis*. *Proceedings of the National Academy of Sciences of the USA* 100:10314–10319.

Feder, J. L., C. A. Chilcote, and G. L. Bush. 1988. Genetic differentiation between sympatric host races of the apple maggot fly *Rhagoletis pomonella*. *Nature* 336:61–64.

Feder, J. L., and K. E. Filchak. 1999. It's about time: the evidence for host plant–mediated selection in the apple maggot fly, *Rhagoletis pomonella*, and its implications for fitness trade-offs in phytophagous insects. *Entomologia Experimentalis Et Applicata* 91:211–225.

Feder, J. L., T. A. Hunt, and L. Bush. 1993. The effects of climate, host-plant phenology and host fidelity on the genetics of apple and hawthorn infesting races of *Rhagoletis pomonella*. *Entomologia Experimentalis Et Applicata* 69:117–135.

Forbes, A. A., T. H. Q. Powell, L. L. Stelinski, J. J. Smith, and J. L. Feder. 2009. Sequential sympatric speciation across trophic levels. *Science* 323:776–779.

Fox, L. R., and P. A. Morrow. 1981. Specialization: species property or local phenomenon. *Science* 211:887–893.

Fujiwara, C., J. Takabayashi, and S. Yano. 2000. Effects of host-food plant species on parasitization rates of *Mythimna separata* (Lepidoptera : Noctuidae) by a parasitoid, *Cotesia kariyai* (Hymenoptera: Braconidae). *Applied Entomology and Zoology* 35:131–136.

Funk, D. J. 1998. Isolating a role for natural selection in speciation: host adaptation and sexual isolation in *Neochlamisus bebbianae* leaf beetles. *Evolution* 52:1744–1759.

Futuyma, D. J., and S. C. Peterson. 1985. Genetic variation in the use of resources by insects. *Annual Review of Entomology* 30:217–238.

Georghiou, G. P. 1972. The evolution of resistance to pesticides. *Annual Review of Ecology and Systematics* 3:133–168.

Gillham, M. C., and M. F. Claridge. 1994. A multivariate approach to host-plant associated morphological variation in the polyphagous leafhopper, Alnetoidia-Alneti (Dahlbom). *Biological Journal of the Linnean Society* 53:127–151.

Gore, J., B. R. Leonard, and R. H. Jones. 2003. Influence of agronomic hosts on the susceptibility of *Helicoverpa zea* (Boddie) (Lepidoptera: Noctuidae) to genetically engineered and non-engineered cottons. *Environmental Entomology* 32:103–110.

Gould, F. 1998. Sustainability of transgenic insecticidal cultivars: integrating pest genetics and ecology. *Annual Review of Entomology* 43:701–726.

Goyer, R. A., T. D. Paine, D. P. Pashley, G. J. Lenhard, J. R. Meeker, and C. C. Hanlon. 1995. Geographic and host-associated differentiation in the fruittree leafroller (Lepidoptera, Tortricidae). *Annals of the Entomological Society of America* 88:391–396.

Gratton, C., and S. C. Welter. 1999. Does "enemy-free space" exist? Experimental host shifts of an herbivorous fly. *Ecology* 80:773–785.

Groman, J. D., and O. Pellmyr. 2000. Rapid evolution and specialization following host colonization in a yucca moth. *Journal of Evolutionary Biology* 13:223–236.

Guttman, S. I., T. K. Wood, and A. A. Karlin. 1981. Genetic differentiation along host plant lines in the sympatric *Enchenopa binotata* Say complex (Homoptera: Membracidae). *Evolution* 35:205–217.

Hajek, A. E. 2004. *Natural Enemies: An Introduction to Biological Control*. Cambridge University Press, Cambridge.

Hare, J. D. 1990. Ecology and management of the Colorado potato beetle. *Annual Review of Entomology* 35:81–100.

Harvey, J. A., I. F. Harvey, and D. J. Thompson. 1995. The effect of host nutrition on growth and development of the parasitoid wasp *Venturia canescens*. *Entomologia Experimentalis Et Applicata* 75:213–220.

Heard, S. B., J. O. Stireman, J. D. Nason, G. H. Cox, C. R. Kolacz, and J. M. Brown. 2006. On the elusiveness of enemy-free space: spatial, temporal, and host-plant-related variation in parasitoid attack rates on three gallmakers of goldenrods. *Oecologia* 150:421–434.

Hebert, P. D. N., E. H. Penton, J. M. Burns, D. H. Janzen, and W. Hallwachs. 2004. Ten species in one: DNA barcoding reveals cryptic species in the neotropical skipper butterfly *Astraptes fulgerator*. *Proceedings of the National Academy of Sciences of the USA* 101:14812–14817.

Henter, H. J., and S. Via. 1995. The potential for coevolution in a host–parasitoid system. 1. Genetic variation within an aphid population in susceptibility to a parasitic wasp. *Evolution* 49:427–438.

Hopper, K. R., R. T. Roush, and W. Powell. 1993. Management of genetics of biological-control introductions. *Annual Review of Entomology* 38:27–51.

Hufbauer, R. A., and G. K. Roderick. 2005. Microevolution in biological control: mechanisms, patterns, and processes. *Biological Control* 35:227–239.

Hunter, M. D. 2003. Effects of plant quality on the population ecology of parasitoids. *Agricultural and Forest Entomology* 5:1–8.

Hunter, M. D., D. J. Biddinger, E. J. Carlini, B. A. Mcpheron, and L. A. Hull. 1994. Effects of apple leaf allelochemistry on tufted apple bud moth (Lepidoptera, Tortricidae) resistance to azinphosmethyl. *Journal of Economic Entomology* 87:1423–1429.

Jaenike, J. 1990. Host specialization in phytophagous insects. *Annual Review of Ecology and Systematics* 21:243–273.

Joyce, A. L., M. Aluja, J. Sivinski, S. B. Vinson, R. Ramirez-Romero, J. S. Bernal, and L. Guillen. 2010a. Effect of continuous rearing on courtship acoustics of five braconid parasitoids, candidates for augmentative biological control of *Anastrepha* species. *Biocontrol* 55:573–582.

Joyce, A. L., and T. S. Bellows. 2000. Field evaluation of *Amitus bennetti* (Hymenoptera: Platygasteridae), a parasitoid of *Bemisia argentifolii* (Hemiptera: Aleyrodidae), in cotton and bean. *Biological Control* 17:258–266.

Joyce, A. L., J. S. Bernal, S. B. Vinson, R. E. Hunt, F. Schulthess, and R. F. Medina. 2010b. Geographic variation in male courtship acoustics and genetic divergence of populations of the *Cotesia flavipes* species complex. *Entomologia Experimentalis Et Applicata* 137:153–164.

Joyce, A. L., R. E. Hunt, J. S. Bernal, and S. B. Vinson. 2008. Substrate influences mating success and transmission of courtship vibrations for the parasitoid *Cotesia marginiventris*. *Entomologia Experimentalis Et Applicata* 127:39–47.

Kankare, M., S. Van Nouhuys, and I. Hanski. 2005. Genetic divergence among host-specific cryptic species in *Cotesia melitaearum* aggregate (Hymenoptera : Braconidae), parasitoids of checkerspot butterflies. *Annals of the Entomological Society of America* 98:382–394.

Kibota, T. T., and S. P. Courtney. 1991. Jack of one trade, master of none: host choice by *Drosophila magnaquinaria*. *Oecologia* 86:251–260.

Kolaczan, C. R., S. B. Heard, K. A. Segraves, D. M. Althoff, and J. D. Nason. 2009. Spatial and genetic structure of host-associated differentiation in the parasitoid *Copidosoma gelechiae*. *Journal of Evolutionary Biology* 22:1275–1283.

Kraaijeveld, A. R., and H. C. J. Godfray. 1999. Geographic patterns in the evolution of resistance and virulence in *Drosophila* and its parasitoids. *American Naturalist* 153:S61–S74.

Kraaijeveld, A. R., and J. J. M. Van Alphen. 1994. Geographical variation in resistance of the parasitoid *Asobara tabida* against encapsulation by *Drosophila melanogaster* larvae: the mechanism explored. *Physiological Entomology* 19:9–14.

Landis, D. A., S. D. Wratten, and G. M. Gurr. 2000. Habitat management to conserve natural enemies of arthropod pests in agriculture. *Annual Review of Entomology* 45:175–201.

Le Corff, J., R. J. Marquis, and J. B. Whitfield. 2000. Temporal and spatial variation in a parasitoid community associated with the herbivores that feed on Missouri Quercus. *Environmental Entomology* 29:181–194.

Li, X. C., J. Baudry, M. R. Berenbaum, and M. A. Schuler. 2004. Structural and functional divergence of insect CYP6B proteins: from specialist to generalist cytochrome P450. *Proceedings of the National Academy of Sciences of the USA* 101:2939–2944.

Li, X. C., M. A. Schuler, and M. R. Berenbaum. 2002. Jasmonate and salicylate induce expression of herbivore cytochrome P450 genes. *Nature* 419:712–715.

Li, X. C., A. R. Zangerl, M. A. Schuler, and M. R. Berenbaum. 2000. Cross-resistance to alpha-cypermethrin after xanthotoxin ingestion in *Helicoverpa zea* (Lepidoptera: Noctuidae). *Journal of Economic Entomology* 93:18–25.

Liang, P., J. Z. Cui, X. Q. Yang, and X. W. Gao. 2007. Effects of host plants on insecticide susceptibility and carboxylesterase activity in *Bemisia tabaci* biotype B and greenhouse whitefly, *Trialeurodes vaporariorum*. *Pest Management Science* 63:365–371.

Lill, J. T., R. J. Marquis, and R. E. Ricklefs. 2002. Host plants influence parasitism of forest caterpillars. *Nature* 417:170–173.

Linn, C., J. L. Feder, S. Nojima, H. R. Dambroski, S. H. Berlocher, and W. Roelofs. 2003. Fruit odor discrimination and sympatric host race formation in Rhagoletis. *Proceedings of the National Academy of Sciences of the USA* 100:11490–11493.

Linn, C. E., H. Dambroski, S. Nojima, J. L. Feder, S. H. Berlocher, and W. L. Roelofs. 2005. Variability in response specificity of apple, hawthorn, and flowering dogwood-infesting *Rhagoletis* flies to host fruit volatile blends: implications for sympatric host shifts. *Entomologia Experimentalis Et Applicata* 116:55–64.

Liu, S. S., and L. H. Jiang. 2003. Differential parasitism of *Plutella xylostella* (Lepidoptera: Plutellidae) larvae by the parasitoid *Cotesia plutellae* (Hymenoptera: Braconidae) on two host plant species. *Bulletin of Entomological Research* 93:65–72.

Lombaert, E., J. Carletto, C. Piotte, X. Fauvergue, H. Lecoq, F. Vanlerberghe-Masutti, and L. Lapchin. 2009. Response of the melon aphid, *Aphis gossypii*, to host-plant resistance: evidence for high adaptive potential despite low genetic variability. *Entomologia Experimentalis Et Applicata* 133:46–56.

Losey, J. E., and M. D. Eubanks. 2000. Implications of pea aphid host-plant specialization for the potential colonization of vegetables following post-harvest emigration from forage crops. *Environmental Entomology* 29:1283–1288.

Lozier, J. D., G. K. Roderick, and N. J. Mills. 2008. Evolutionarily significant units in natural enemies: identifying regional populations of *Aphidius transcaspicus* (Hymenoptera: Braconidae) for use in biological control of mealy plum aphid. *Biological Control* 46:532–541.

Lozier, J. D., G. K. Roderick, and N. J. Mills. 2009. Molecular markers reveal strong geographic, but not host associated, genetic differentiation in *Aphidius transcaspicus*, a parasitoid of the aphid genus Hyalopterus. *Bulletin of Entomological Research* 99:83–96.

Mackauer, M. 1976. Genetic problems in production of biological-control agents. *Annual Review of Entomology* 21:369–385.

Magalhaes, S., M. R. Forbes, A. Skoracka, M. Osakabe, C. Chevillon, and K. D. Mccoy. 2007. Host race formation in the Acari. *Experimental and Applied Acarology* 42:225–238.

Martel, C., A. Rejasse, F. Rousset, M. T. Bethenod, and D. Bourguet. 2003. Host-plant-associated genetic differentiation in northern French populations of the European corn borer. *Heredity* 90:141–149.

Martin, P. B., P. D. Lingren, and G. L. Greene. 1976. Relative abundance and host preferences of cabbage-looper Lepidoptera-Noctuidae, soybean looper Lepidoptera-Noctuidae, tobacco budworm Lepidoptera-Noctuidae, and corn-earworm Lepidoptera-Noctuidae, on crops grown in northern Florida. *Environmental Entomology* 5:878–882.

Massei, G., and S. E. Hartley. 2000. Disarmed by domestication? Induced responses to browsing in wild and cultivated olive. *Oecologia* 122:225–231.

Matsubayashi, K. W., I. Ohshima, and P. Nosil. 2010. Ecological speciation in phytophagous insects. *Entomologia Experimentalis Et Applicata* 134:1–27.

Medina, R. F., and P. Barbosa. 2008. The role of host plant species in the phenotypic differentiation of sympatric populations of *Aleiodes nolophanae* and *Cotesia marginiventris*. *Entomologia Experimentalis Et Applicata* 128:14–26.

Mopper, S., and S. Y. Strauss. 1998. *Genetic Structure and Local Adaptation in Natural Insect Populations*. Chapman and Hall, London.

Muehleisen, D. P., J. H. Benedict, F. W. Plapp, and F. A. Carino. 1989. Effects of cotton allelochemicals on toxicity of insecticides and induction of detoxifying enzymes in bollworm (Lepidoptera, Noctuidae). *Journal of Economic Entomology* 82:1554–1558.

Nason, J. D., S. B. Heard, and F. R. Williams. 2002. Host-associated genetic differentiation in the golden-rod elliptical-gall moth, *Gnorimoschema gallaesolidaginis* (Lepidoptera: Gelechiidae). *Evolution* 56:1475–1488.

Navajas, M., A. Tsagkarakov, J. Lagnel, and M. J. Perrot-Minnot. 2000. Genetic differentiation in *Tetranychus urticae* (Acari: Tetranychidae): polymorphism, host races or sibling species? *Experimental and Applied Acarology* 24:365–376.

Nosil, P. 2007. Divergent host plant adaptation and reproductive isolation between ecotypes of *Timema cristinae* walking sticks. *American Naturalist* 169:151–162.

Nosil, P., and B. J. Crespi. 2006a. Ecological divergence promotes the evolution of cryptic reproductive isolation. *Proceedings of the Royal Society B-Biological Sciences* 273:991–997.

Nosil, P., and B. J. Crespi. 2006b. Experimental evidence that predation promotes divergence in adaptive radiation. *Proceedings of the National Academy of Sciences of the USA* 103:9090–9095.

Nwanze, K. F., Y. V. R. Reddy, F. E. Nwilene, K. G. Kausalya, and D. D. R. Reddy. 1998. Tritrophic interactions in sorghum, midge (*Stenodiplosis sorghicola*) and its parasitoid (*Aprostocetus* spp.). *Crop Protection* 17:165–169.

Nyman, T., T. Ylioja, and H. Roininen. 2002. Host-associated allozyme variation in tree cambium miners, *Phytobia* spp. (Diptera: Agromyzidae). *Heredity* 89:394–400.

Panizzi, A. R., and E. D. M. Oliveira. 1998. Performance and seasonal abundance of the neotropical brown stink bug, *Euschistus heros* nymphs and adults on a novel food plant (pigeonpea) and soybean. *Entomologia Experimentalis Et Applicata* 88:169–175.

Pappers, S. M., H. van Dommelen, G. van der Velde, and N. J. Ouborg. 2001. Differences in morphology and reproductive traits of *Galerucella nymphaeae* from four host plant species. *Entomologia Experimentalis Et Applicata* 99:183–191.

Pashley, D. P. 1986. Host-associated genetic differentiation in fall armyworm (Lepidoptera: Noctuidae): a sibling species complex. *Annals of the Entomological Society of America* 79:898–904.

Pashley, D. P., A. M. Hammond, and T. N. Hardy. 1992. Reproductive isolating mechanisms in fall army-worm host strains (Lepidoptera, Noctuidae). *Annals of the Entomological Society of America* 85:400–405.

Peccoud, J., A. Ollivier, M. Plantegenest, and J. C. Simon. 2009. A continuum of genetic divergence from sympatric host races to species in the pea aphid complex. *Proceedings of the National Academy of Sciences of the USA* 106:7495–7500.

Price, P. W., C. E. Bouton, P. Gross, B. A. McPheron, J. N. Thompson, and A. E. Weis. 1980. Interactions among 3 trophic levels: influence of plants on interactions between insect herbivores and natural enemies. *Annual Review of Ecology and Systematics* 11:41–65.

Prokopy, R. J., E. W. Bennett, and G. L. Bush. 1971. Mating behavior in *Rhagoletis pomonella* (Diptera:Tephritidae). 1. Site of assembly. *Canadian Entomologist* 103:1405–1409.

Prokopy, R. J., G. L. Bush, and E. W. Bennett. 1972. Mating hehavior in *Rhagoletis pomonella* (Diptera: Tephritidae). 2. Temporal organization. *Canadian Entomologist* 104:97–104.

Rausher, M. D. 1982. Population differentiation in Euphydryas-Editha butterflies – larval adaptation to different hosts. *Evolution* 36:581–590.

Riley, D. G., and W. J. Tan. 2003. Host plant effects on resistance to bifenthrin in silverleaf whitefly (Homoptera: Aleyrodidae). *Journal of Economic Entomology* 96:1315–1321.

Rosas-Garcia, N. M., S. L. Sarmiento-Benavides, J. M. Villegas-Mendoza, S. Hernandez-Delgado, and N. Mayek-Perez. 2010. Genetic differentiation among *Maconellicoccus hirsutus* (Hemiptera: Pseudococcidae) populations living on different host plants. *Environmental Entomology* 39:1043–1050.

Rosenthal, J. P., and R. Dirzo. 1997. Effects of life history, domestication and agronomic selection on plant defence against insects: evidence from maizes and wild relatives. *Evolutionary Ecology* 11:337–355.

Ruiz-Montoya, L., J. Nunez-Farfan, and J. Vargas. 2003. Host-associated genetic structure of Mexican populations of the cabbage aphid *Brevicoryne brassicae* L. (Homoptera: Aphididae). *Heredity* 91:415–421.

Rundle, H. D., and P. Nosil. 2005. Ecological speciation. *Ecology Letters* 8:336–352.

Rupasinghe, S. G., Z. Wen, T. L. Chiu, and M. A. Schuler. 2007. *Helicoverpa zea* CYP6B8 and CYP321A1: different molecular solutions to the problem of metabolizing plant toxins and insecticides. *Protein Engineering Design & Selection* 20:615–624.

Sauer, J. D. 1993. *Historical Geography of Crop Plants*. CRC Press, Boca Raton, FL.

Scheffer, S. J., and D. J. Hawthorne. 2007. Molecular evidence of host-associated genetic divergence in the holly leafminer *Phytomyza glabricola* (Diptera: Agromyzidae): apparent discordance among marker systems. *Molecular Ecology* 16:2627–2637.

Schluter, D. 2001. Ecology and the origin of species. *Trends in Ecology & Evolution* 16:372–380.

Sharma, R. K., V. K. Gupta, J. Jindal, and V. K. Dilawari. 2008. Host associated genetic variations in whitefly, *Bemisia tabaci* (Genn.). *Indian Journal of Biotechnology* 7:366–370.

Sivasupramaniam, S., S. Johnson, T. F. Watson, A. A. Osman, and R. Jassim. 1997. A glass-vial technique for monitoring tolerance of *Bemisia argentifolii* (Homoptera: Aleyrodidae) to selected insecticides in Arizona. *Journal of Economic Entomology* 90:66–74.

Smith, M. A., N. E. Woodley, D. H. Janzen, W. Hallwachs, and P. D. N. Hebert. 2006. DNA barcodes reveal cryptic host-specificity within the presumed polyphagous members of a genus of parasitoid flies (Diptera: Tachinidae). *Proceedings of the National Academy of Sciences of the USA* 103:3657–3662.

Souissi, R., and B. Le Rü. 1998. Influence of the host plant of the cassava mealybug *Phenacoccus manihoti* (Hemiptera: Pseudococcidae) on biological characteristics of its parasitoid *Apoanagyrus lopezi* (Hymenoptera: Encyrtidae). *Bulletin of Entomological Research* 88:75–82.

Stansly, P. A., D. J. Schuster, and T. X. Liu. 1997. Apparent parasitism of *Bemisia argentifolii* (Homoptera: Aleyrodidae) by aphelinidae (Hymenoptera) on vegetable crops and associated weeds in south Florida. *Biological Control* 9:49–57.

Stelinski, L. L., and O. E. Liburd. 2005. Behavioral evidence for host fidelity among populations of the parasitic wasp, *Diachasma alloeum* (Muesebeck). *Naturwissenschaften* 92:65–68.

Stireman, J. O., E. M. Janson, T. G. Carr, H. Devlin, and P. Abbot. 2008. Evolutionary radiation of *Asteromyia carbonifera* (Diptera: Cecidomyiidae) gall morphotypes on the goldenrod *Solidago altissima* (Asteraceae). *Biological Journal of the Linnean Society* 95:840–858.

Stireman, J. O., J. D. Nason, and S. B. Heard. 2005. Host-associated genetic differentiation in phytophagous insects: general phenomenon or isolated exceptions? Evidence from a goldenrod insect-community. *Evolution* 59:2573–2587.

Stireman, J. O., J. D. Nason, S. B. Heard, and J. M. Seehawer. 2006. Cascading host-associated genetic differentiation in parasitoids of phytophagous insects. *Proceedings of the Royal Society B-Biological Sciences* 273:523–530.

Sword, G. A., A. Joern, and L. B. Senior. 2005. Host plant-associated genetic differentiation in the snakeweed grasshopper, *Hesperotettix viridis* (Orthoptera: Acrididae). *Molecular Ecology* 14:2197–2205.

Tabashnik, B. E., and Y. Carriere. 2008. Evolution of insect resistance to transgenic plants. In K. J. Tilmon (editor), *Specialization, Speciation and Radiation*. University of California Press, Berkeley.

Tauber, C. A., and M. J. Tauber. 1989. Sympatric speciation in insects: perception and perspective. Pages 307–345 *in* D. Otte and J. A. Endler (editors), *Speciation and Its Consequences*. Sinauer Associates, Sunderland, MA.

Thomas, Y., M. T. Bethenod, L. Pelozuelo, B. Frerot, and D. Bourguet. 2003. Genetic isolation between two sympatric host-plant races of the European corn borer, *Ostrinia nubilalis* Hubner. I. Sex pheromone, moth emergence timing, and parasitism. *Evolution* 57:261–273.

Van Lenteren, J. C., Z. H. Li, J. W. Kamerman, and R. M. Xu. 1995. The parasite–host relationships between *Encarsia formosa* (Hym., Aphelinidae) and *Trialeurodes vaporariorum* (Hom., Aleyrodidae). XXVI. Leaf hairs reduce the capacity of Encarsia to control greenhouse whitefly on cucumber. *Journal of Applied Entomology* 119:553–559.

Vanlerberghe-Masutti, F., and P. Chavigny. 1998. Host-based genetic differentiation in the aphid *Aphis gossypii* Glover, evidenced from RAPD fingerprints. *Molecular Ecology* 7:905–914.

Vaughan, J., and C. Geissler. 2009. *The New Oxford Book of Food Plants*. Oxford University Press, New York.

Vaughn, T. T., and M. F. Antolin. 1998. Population genetics of an opportunistic parasitoid in an agricultural landscape. *Heredity* 80:152–162.

Verkerk, R. H. J., S. R. Leather, and D. J. Wright. 1998. The potential for manipulating crop-pest–natural enemy interactions for improved insect pest management. *Bulletin of Entomological Research* 88:493–501.

Via, S. 1991a. The genetic-structure of host plant adaptation in a spatial patchwork – demographic variability among reciprocally transplanted pea aphid clones. *Evolution* 45:827–852.

Via, S. 1991b. Specialized host plant performance of pea aphid clones is not altered by experience. *Ecology* 72:1420–1427.

Via, S. 1999. Reproductive isolation between sympatric races of pea aphids. I. Gene flow restriction and habitat choice. *Evolution* 53:1446–1457.

Via, S., A. C. Bouck, and S. Skillman. 2000. Reproductive isolation between divergent races of pea aphids on two hosts. II. Selection against migrants and hybrids in the parental environments. *Evolution* 54:1626–1637.

Vialatte, A., C. A. Dedryver, J. C. Simon, M. Galman, and M. Plantegenest. 2005. Limited genetic exchanges between populations of an insect pest living on uncultivated and related cultivated host plants. *Proceedings of the Royal Society B-Biological Sciences* 272:1075–1082.

Walsh, B. D. 1864. On phytophagic varieties and phytophagic species. *Proceedings of the Entomological Society of Philadelphia* 3:403–430.

Waring, G. L., W. G. Abrahamson, and D. J. Howard. 1990. Genetic differentiation among host-associated populations of the gallmaker *Eurosta solidaginis* (Diptera: Tephritidae). *Evolution* 44:1648–1655.

Weis, A. E., and W. G. Abrahamson. 1986. Evolution of host-plant manipulation by gall makers: ecological and genetic factors in the *Solidago-Eurosta* system. *American Naturalist* 127:681–695.

Wolfenbarger, D. A., D. G. Riley, C. A. Staetz, G. L. Leibee, G. A. Herzog, and E. V. Gage. 1998. Response of silverleaf whitefly (Homoptera: Aleyrodidae) to bifenthrin and endosulfan by vial bioassay in Florida, Georgia and Texas. *Journal of Entomological Science* 33:412–420.

Yu, S. J. 1982. Host plant induction of glutathione S-transferase in the fall armyworm. *Pesticide Biochemistry and Physiology* 18:101–106.

Yu, S. J. 1983. Induction of detoxifying enzymes by allelochemicals and host plants in the fall armyworm. *Pesticide Biochemistry and Physiology* 19:330–336.

Part V

Applied Perspectives

15

Disasters by Design: Outbreaks along Urban Gradients

Michael J. Raupp, Paula M. Shrewsbury, and Dan A. Herms

15.1   Introduction

Urbanization has dramatically changed the character of land as forests, deserts, prairies, and croplands transform into cities and suburbs and their attendant buildings, roads, airports, waterways, gardens, and parks (McDonnell and Pickett 1990, Faeth *et al.* 2005, Grimm *et al.* 2008). The trend shows no sign of slowing. By 2100, urban areas in the United States are predicted to increase over current levels by 74–164%, and suburban areas by 59–154% (Bierwagen *et al.* 2010). The juxtaposition of developed areas to natural ones creates a unique ecological gradient along which patterns of richness, abundance, and ecosystem processes can be studied in a context of human culture, economics, and politics (Frankie and Ehler 1978, Dreistadt *et al.* 1990, McDonnell and Pickett 1990, McIntyre 2000, Rickman and Conner 2003, Shochat *et al.* 2006, McKinney 2008). Biota shifts from communities of plants and animals at various stages of succession to communities associated with collections of ornamental plants, remnants of natural habitats, and species capable of thriving in man-altered systems (McDonnell and Pickett 1990, McIntyre 2000, Faeth *et al.* 2005, Shochat *et al.* 2006).

Insect Outbreaks Revisited, First Edition. Edited by Pedro Barbosa, Deborah K. Letourneau and Anurag A. Agrawal.
© 2012 Blackwell Publishing Ltd. Published 2012 by Blackwell Publishing Ltd.

Many herbivorous arthropods found on woody plants in urban environments attain dramatically greater abundance that rarely, if ever, occurs in natural areas (Frankie and Ehler 1978, Dreistadt *et al.* 1990, Watson *et al.* 1994). Key features of urban ecosystems include contrived plant communities resulting from widespread use of exotic plants, and often dramatic reductions in the density, diversity, and complexity of vegetation. Natural processes are disrupted by impervious surfaces and anthropogenic maintenance practices such as pulsed inputs of fertilizers, water, and pesticides. These factors create opportunities for populations of insects and mites to increase due to alterations in host quality and accessibility (bottom-up factors); natural enemy abundance and activity (top-down factors); available microhabitats, including heat islands; and matrixes that disrupt movement and colonization of herbivores and natural enemies (Herms *et al.* 1984, Thomas 1989, Shrewsbury and Raupp 2006, Shochat *et al.* 2006, Raupp *et al.* 2010).

This chapter focuses on outbreaks of arthropod pests along one or more features of urban gradients. "Outbreaks" are dramatic increases in the abundance of arthropods occurring in relatively short periods (Berryman 1987) that are often associated with feeding injury to plants exceeding economic or aesthetic thresholds. Clearly, patterns of species richness and abundance track key ecological features such as productivity, patch size, and habitat quality (Shochat *et al.* 2006, McKinney 2008), and some taxa reach greatest richness and abundance at intermediate or low levels of urbanization (Rickmann and Conner 2003, Shochat *et al.* 2006, McKinney 2008, Raupp *et al.* 2010). However, this chapter examines the subset of arthropods that reach their greatest densities with increasing urbanization.

The objective of this review is to identify key features and processes underlying elevated populations. We draw on studies of phytophagous mites and insects due to the number and rich diversity of accounts documenting changes in abundance of these types of organisms associated with features of urban landscapes. A wide array of literature dealing with population dynamics of herbivorous arthropods in forest ecosystems provides relevant mechanistic explanations for outbreaks in cities. We present examples of empirical studies on the diversity and abundance of herbivores and their natural enemies along urban gradients, and explore how features and properties of urbanization affect arthropod abundance and ecosystem processes in general. We hope to identify gaps in our knowledge of herbivores in urban landscapes and identify hypotheses worthy of further investigation.

15.2 Case studies of arthropod outbreaks along urban gradients

Table 15.1 provides a summary of more than 20 studies of arthropod abundance in urban landscapes. It includes several feeding guilds such as suckers, chewers, gall makers, leafminers, and borers, and species with different modes of dispersal including scale insects and mites that are wind dispersed and others such as wood-boring and folivorous Lepidoptera that are highly mobile. It summarizes cases that demonstrate elevated densities of pests in the urban settings that exceed densities in natural or less urbanized habitats. In many case studies, definitions

Table 15.1 Case Studies of Arthropods with Greater Abundance in Urban Settings.

Herbivore	Host	Contrast	Pest abundance[a]	Reference
Tetranychidae				
Oligonychus sp.	*Quercus* sp.	Shopping plaza	174/site	Ehler and Frankie (1979)
		Natural stand	17/site	
Oligonychus subnudus	*Pinus radiata*	Roadside	~100 – 1400/ sample	Koehler and Frankie (1968)
		Watershed	~20 – 90/ sample	
Tetranychus sp.	*Tilia cordata*	Street	13/cm^2	Balder *et al.* (1999)
		Park	2/cm^2	
Eotetranychus tiliarum	*Tilia* × *europaea*	Sunny side street	17.6/cm^2	Schneider *et al.* (2000)
		Shady side street	4.5/cm^2	
Eotetranychus tiliarum	*Tilia* × *europaea*	Asphalt present	8.5/cm^2	Schneider *et al.* (2000)
		Gardens present	0/cm^2	
Eotetranychus tiliarum	*Tilia* sp.	Street	Outbreak	Kropczynska *et al.* (1986)
		Park	<0.1 leaf	
Eotetranychus tiliarum	*Tilia* spp.	Street	Severe attack (43%)	Fostad and Pederson (1997)
		Park	Detectable attack (1%)	
Platytetranychus multidigituli	*Gleditsia triacanthos*	Street	More[b]	Cranshaw and Hart (1988)
		Landscapes	Less	
Platytetranychus multidigituli	*Gleditsia triacanthos*	Impervious surface	More[c]	Sperry *et al.* (2001)
		Pervious surface	Less	
Platytetranychus multidigituli	*Gleditsia triacanthos*	Low-density host	More[c]	Sperry *et al.* (2001)
		High density host	Less	
Miridae				
Diaphnocoris chlorionis	*Gleditsia triacanthos*	Low-diversity landscape	More[c]	Sperry *et al.* (2001)
		High-diversity landscape	Less	
Tingidae				
Stephanitis pyrioides	*Rhododendron* sp.	Simple	600/m^2	Shrewsbury and Raupp (2000)
		Complex	1/m^2	

(Continued)

Table 15.1   (*Cont'd*).

Herbivore	Host	Contrast	Pest abundancea	Reference
Aphididae				
Aphis nerii	*Nerium oleander*	Urban	61/terminal	Hall and Ehler (1980)
		Rural	0.1/terminal	
Aphis pomi	*Cretaeagus* sp.	Motorway	234–489/ shoot	Braun and Flückiger (1984a, b)
		Natural	5–46/shoot	
Coccidae				
Pulvinaria regalis	*Aesculus hippocastanum*	Impervious surface	More[c]	Speight *et al.* (1998)
	Tilia cordata *Acer pseudoplatanus*	Pervious surface	Less	
Ceroplastes rubens	*Schefflera actinophylla*	Roadsides	1.15/leaf	Loch and Zaluchi (1996)
		Gardens	0.26/leaf	
Diaspididae				
Pseudaulacaspis pentagona	*Morus alba*	Landscape	3/cm^2	Hanks and Denno (1993)
		Forest	0.001/ cm^2	
Chionaspis pinifoliae	*Pinus* sp.	Impoverished	60/search	Tooker and Hanks (2000)
		Wooded	2/search	
Fiorinia externa	*Tsuga diversifolia*	Cultivated	37/1000 needles	McClure (1986)
		Natural	0/1000 needles	
Fiorinia externa	*Tsuga sieboldii*	Cultivated	22/1000 needles	McClure (1986)
		Natural	2/1000 needles	
Melanaspis obscura	*Quercus* sp.	Homes and parks	More[b]	Stoetzal and Davidson (1971)
		Forests	Less	
Nuculaspis tsugae	*Tsuga diversifolia*	Cultivated	22/1000 needles	McClure (1986)
		Natural	6/1000 needles	
Nuculaspis tsugae	*Tsuga sieboldii*	Cultivated	14/1000 needles	McClure (1986)
		Natural	6/1000 needles	
Nuculaspis californica	*Pinus* sp. (large trees)	Disturbed (dust)	~1.4/cm	Edmunds (1973)
		Undisturbed	>0.025/cm	

Table 15.1 (*Cont'd*).

Herbivore	Host	Contrast	Pest abundance[a]	Reference
Sesiidae				
Podosesia syringae	*Fraxinus pennsylvanicus*	City	7.3/tree	Cregg and Dix (2001)
		Park-like	0/tree	
Plutellidae				
Homadaula anisocentra	*Gleditsia triacanthos*	Urban forest	More[b]	Hart *et al.* (1986)
		Natural forest	Less	
Homaduala anisocentra	*Gleditsia triacanthos*	Impervious surface	More[d]	Sperry *et al.* (2001)
		Pervious surface	Less	
Lymantriidae				
Euproctis similis	*Crateagus monogyna*	Central reserve	~8/bush	Port and Thompson (1980)
		Verge	~3/ bush	
Brassolidae				
Brassolis sophorae	*Arecaceae*	Hardscape area	More[c]	Ruszszyk (1996)
		Vegetated area	Less	
Agromyzidae				
Phytomyza ilicicola	*Ilex opaca*	Cultivated	More[b]	Potter (1985)
		Forest	Less	
Phytomyza ilicicola	*Ilex opaca*	Urban	12/100 leaves	Kahn (1988), Kahn and Cornell (1989)
		Forest	1/100 leaves	
Pinyonia edulicola	*Pinus edulis*	Urban	162–15 179/4 min	Frankie *et al.* (1987)
		Natural	72–329 /4 min	
Cynipidae				
Disholcaspis cinerosa	*Quercus* spp.	Shopping mall	~10–100/ 0.5 m	Frankie and Morgan (1992)
		Natural	~1/0.5 m	

[a] Where multiple measures of abundance were reported, concordant spatial or temporal maxima were included in the table.
[b] Qualitative observations.
[c] Results of regression analyses preclude reporting direct measures of abundance. More indicates a positive relationship between increasing urbanization and pest abundance and less indicates the opposite.
[d] Effect not seen at some spatial scales examined.

of contrasting habitats or features along the urban gradient were clear. Plants were sampled in natural areas, parks, residential areas, or cities. Other studies measured patterns of herbivore abundance along specific features of the urban gradient related to plant density and diversity, vegetational complexity, patterns of land use, or amounts of impervious surface. In some cases, but not all, elevated densities of herbivores were referred to as "outbreaks." Many studies discussed causes of variation in abundance, but most did not examine features of urbanization independently. Habitat variables were often confounded. In very few studies were mechanisms underlying patterns convincingly documented. Generally, it was difficult or impossible to completely separate processes associated with urbanization, such as deterioration in habitat quality, from those associated with habitat loss, fragmentation, productivity, and isolation (Nuckols and Conner 1995, Rickman and Connor 2003, McKinney 2008).

Mites, especially tetranychid mites, dominated the taxa of arthropods that outbreak in urban habitats. In this regard, mites were followed closely by scale insects, particularly diaspidid scales. Sap-sucking Hemiptera more commonly reached eruptive populations levels in urban settings than mandibulate herbivores such as Lepidoptera, Coleoptera, or Hymenoptera. Several species, such as *Eotetranychus tiliarum*; *Pulvinaria regalis*; red wax scale, *Ceroplastes rubens*; elongate hemlock scale, *Fiorinia externa*; *and* white peach scale, *Pseudaulacaspis pentagona* are polyphagous. Others such as honeylocust spider mite, *Platytetranychus multidigituli*; azalea lace bug, *Stephanitis pyrioides*; apple aphid, *Aphis* pomi; ash borer, *Podosesia syringae*; *Brassolis sophorae*; native holly leafminer, *Phytomyza ilicicola*; pinyon spindlegall midge, *Pinyonia edulicola*; and *Disholcaspis cinerosa* specialize at the level of plant family, genus, or species.

15.3 Features and mechanisms contributing to outbreaks

Several features of urbanized landscapes alter the biological attributes of plants, herbivores, and natural enemies, and their interactions, and ultimately affect the abundance of herbivorous insects and mites. We recognize that phenomena such as drought, global change, and epizootics can affect population dynamics of herbivorous arthropods simultaneously over broad geographic areas in natural areas such as forests and human-altered habitats. The following discussions focus on elements unique to urban landscapes that link pest outbreaks with underlying mechanisms.

15.3.1 Low biodiversity and catastrophic loss in cities

The use of woody plants in human-altered ecosystems dates back thousands of years to Egyptian, Greek, and Chinese civilizations (Koch 2000). Urban forests developed in three common ways in North America (McBride and Jacobs 1976, Dreistadt *et al.* 1990). Firstly, the clearing of forested land for agriculture and settlement allowed some native trees and their vegetative progeny to remain. Forest remnants and second growth forests are common in urban areas and suburbs in the Northeast, Midwest, and Pacific Northwest regions of the United States. Secondly, contrived urban forests arose in areas lacking large stands of native trees such as in the Great Plains and arid regions in the Southwestern

United States. Arthropods colonize these habitats by moving with their hosts, being transported inadvertently or deliberately by humans, or by crossing inhospitable matrixes that lacked suitable hosts, refuges, or other vital resources. Finally, in both historical contexts, woody plants establish and reproduce independently of human intervention.

Areas once forested and later cleared for agriculture with subsequent urban development suffer from habitat fragmentation and degradation (McDonnell and Pickett 1990, McIntyre 2000, Rickman and Connor 2003, Shochat *et al.* 2006, McKinney 2008). At some intermediate levels of urbanization along the gradient where primary productivity may reach a maximum due to the addition of plants and anthropogenic inputs, richness of some taxa may also be highest (Shochat *et al.* 2006, McKinney 2008, Raupp *et al.* 2010). However, there is general agreement that at the highest levels of urbanization, the richness of many taxa of plants and animals declines dramatically (McKinney 2002, 2008, Shochat *et al.* 2006, Raupp *et al.* 2010).

Erosion of floristic diversity can have serious consequences with respect to the sustainability of the urban forest when evolutionarily naïve plants confront exotic pests for which they lack resistance (Herms 2002a, Gandhi and Herms 2010). Dutch elm disease (DED) is the quintessential example of how a lack of floristic diversity conspired with an exotic pathogen, *Ophiostoma ulmi*, to create catastrophic loss in an urban setting. American elm was one of the most commonly used street trees in the United States before the introduction of DED from Europe in the 1930s in logs to be used for veneer. Since its introduction, DED has killed more than 40 million elm trees, imposed enormous economic loss on cities faced with removal and replacement of trees, and left many city streets barren in the United States (Campana and Stipes 1981, Dreistadt *et al.* 1990). In the last two decades, two important and devastating insect pests, the Asian longhorned beetle, *Anoplophora glabripennis* (ALB), and the emerald ash borer, *Agrilus planipennis* (EAB), appear poised to supplant DED as the most significant killers of trees in urban landscapes in the United States (Nowak *et al.* 2001, Poland and McCullough 2006, Raupp *et al.* 2006, Kovacs *et al.* 2010). A recent survey of street trees in 12 major cities in eastern North American found generic diversity of street trees to be surprisingly low. Maple and ash, highly suitable hosts for ALB and EAB, respectively dominated urban forest canopies. This analysis revealed that 29–70% of the street trees found in these cities were at risk to these two pests (Raupp *et al.* 2006). Urban landscapes continue to be dominated by few species or genera of woody plants that may lack genetic variation in resistance due to clonal propagation, predisposing cities to catastrophic loss due to pests. This fact signals an ongoing need for diversification.

15.3.2 Vegetational texture and top-down regulation

Vegetational texture varies along urban gradients. Vegetational texture has been defined in terms of plant density, patch size, and vegetational diversity (Kareiva 1983, Denno and Roderick 1991, Shrewsbury and Raupp 2000). Plant density is the distance between individual plants. Patch size is the geographical extent of the host plant stand. A host plant patch can consist of all plants of the same species, or plants of different species, but all are hosts of a common herbivore species. Vegetational diversity is the frequency and species composition of nonhost plants in association with the host plants of an herbivore (Kareiva 1983, Denno and Roderick 1991). In addition to these, several other components reflect the vegetational texture of a habitat. These include the structural complexity of the habitat; plant species diversity, evenness, and richness; plant growth form; color contrasts; and volatile plant compounds (Letourneau 1990).

Reductions in vegetational diversity and complexity may contribute to important reductions in the abundance and richness of natural enemies and biological control services they provide. Linden trees lining broad city streets in Berlin often housed 10-fold increases in the numbers of *Eotetranychus tiliarium* and other spider mites compared to lindens residing on narrower side streets amid gardens or in parks or natural areas where plants in several taxa were arrayed in multiple strata, thereby elevating vegetational complexity. Elevated abundance of spider mites was attributed to the relative rarity of predators in trees along hot, wide boulevards compared to side streets where vegetation surrounded trees (Balder *et al.* 1999, Schneider *et al.* 2000). Abundance of the native holly leaf miner was 10 times greater in urban landscapes than in forests (Kahn 1988), possibly because of lower rates of parasitism and predation in urban sites (Kahn and Cornell 1989). Populations of apple aphid were significantly greater on crabapple trees along motorways in Basel, Switzerland, compared to trees at a distance from motorways (Braun and Flückiger 1984). Cages were used to exclude predators on branches of trees along motorways and trees ~200 m away from the motorway. After three weeks, on trees near motorways, the ratio of aphids exposed to predators compared to unexposed was about 1:3. On trees that were a distance from motorways, the ratio was about 1:54, suggesting much lower levels of predation near motorways. In a study of the population dynamics of the white peach scale on mulberry trees, scales were three orders of magnitude more abundant on trees in mesic open landscapes than on trees in nearby forest remnants (Hanks and Denno 1993). Abundant natural enemies in forested sites reduced survival of immatures and enabled white peach scale to persist at high densities only in mesic sites with few natural enemies (Hanks and Denno 1993). Elevated density of pine needle scale, *Chionaspis pinifoliae*, on exotic pines was similarly attributed to low diversity and abundance of natural enemies in urban landscapes, nurseries, and Christmas tree plantations (Eliason and McCullough 1997, Tooker and Hanks 2000).

Several studies have found that urban habitats with greater vegetational diversity or complexity supported greater abundance (Hanks and Denno 1999, Frank and Shrewsbury 2004a, Shrewsbury *et al.* 2004, Shrewsbury and Raupp 2006) or richness (Tooker and Hanks 2000) of natural enemies, especially generalist predators. In a test of mechanisms underlying these patterns, vegetationally complex residential landscapes had greater abundance and retention of generalist predators, notably of a hunting spider, *Anyphaena celer*, than did landscapes with reduced complexity. Greater abundance of predators was attributed to more abundant alternative prey. Enhanced top-down pressure reduced the abundance of a key pest, the azalea lace bug. Elevated populations of lace bugs were observed only in landscapes with low levels of complexity despite the fact that host quality was superior in complex landscapes (Shrewsbury and Raupp 2006).

Habitats with increased vegetational diversity, structural complexity, and productivity provide more favorable microhabitats and refuges for predators, alternative food resources such as pollen and nectar, and greater diversity and abundance of alternative prey for generalist predators (Hanks and Denno 1993, Landis *et al.* 2000, Langellotto and Denno 2004, Shochat *et al.* 2004, Shrewsbury *et al.* 2004, Shrewsbury and Raupp 2006, Tooker and Hanks 2005) and refuge for parasitoids (Frankie *et al.* 1992). A review by Langellotto and Denno (2004) found that increasing habitat structure resulted in significant increases in natural enemy abundance in seven of nine guilds examined. Greater abundance of resident predators provides a mechanism for dampening herbivore outbreaks in diverse and complex habitats (Langellotto and Denno 2004, Shrewsbury and Raupp 2006).

Other factors relating to vegetational texture such as isolation and fragmentation of patches may also influence herbivore – natural enemy dynamics. In other human-influenced habitats such as agricultural systems, isolation of habitat fragments and lack of connectivity resulted in reduced colonization by parasitoids and a concomitant release from top-down suppression by natural enemies (Kruess and Tscharntke 1994). Along urban gradients, isolated trees may provide herbivores with temporary refugia from their natural enemies, thereby uncoupling predator–prey and host–parasitoid interactions. For example, galls produced by the pinyon spindle gall midge were rare in natural settings, but extremely abundant in some urban sites. Gall midges were thought to colonize urban sites in advance of their natural enemies, which likely created temporal windows of relaxed top-down regulation enabling gall midge populations to increase temporarily in urban settings (Frankie *et al.* 1987). A cynipid gall wasp, *Disholcaspis cinerosa*, attained much higher densities in urban sites than in natural areas. Newly transplanted trees were the most heavily attacked, with increased host quality and temporal escape from parasitoids implicated as contributing factors (Frankie and Morgan 1984, Frankie *et al.* 1992). Other studies suggest land use patterns or productivity may be important along urban gradients. For example, the trophic dynamics of epigeal arthropods varied greatly across four human-altered sites that varied in habitat structure and land use, including residential landscapes, farms, industrial sites, and desert remnants (McIntyre *et al.* 2001). Predators, herbivores, and detritivores were most abundant in agricultural sites, while omnivores were equally abundant across all land use types. Clearly, variation in habitat complexity along urban gradients may differentially affect trophic dynamics and ecosystem function enabling insects and mites to escape natural enemies in time and space, thereby increasing the likelihood of outbreaks.

15.3.3   Alien plants and herbivores

The proportion of alien plants in landscapes often increases with increasing levels of urbanization, thus creating a complex matrix that includes exotic arthropods encountering endemic hosts and endemic arthropods encountering alien hosts (Owen 1983, McDonnell and Pickett 1990, McIntyre 2000). Furthermore, during the last two centuries, the number of exotic arthropods that have established on woody plants in North America has increased dramatically (Liebhold *et al*. 1995, Gandhi and Herms 2010). The impact of these events on the diversity and abundance of herbivores in urban settings appears to be mixed, with evidence indicating negative, positive, and no effects (McIntyre 2000, Tallamy 2004, McKinney 2008, Burghardt *et al*. 2009).

The enemy release hypothesis proposes an explanation for the successful invasion of natural systems by alien plants, predicting that alien plants should support fewer herbivores than natives (Keane and Crawley 2002, Tallamy 2004) because many herbivores are restricted to coevolved hosts (Janieke 1990). In support of this hypothesis, Burghardt *et al*. (2009) found the species richness and abundance of Lepidoptera in suburban landscapes dominated by native plants to be three and four times greater, respectively, than the richness and abundance of Lepidoptera in landscapes dominated by alien plants. Compounding the historical constraint is the fact that many alien plants used in urban landscapes have been selected specifically for their resistance to key pests (Raupp *et al*. 1992, Herms 2002a, Tallamy 2004). Therefore, use of alien plants may lead to reduced richness and abundance of herbivores in urban landscapes. If this is indeed the case, then herbivore-poor alien plant communities would also support fewer natural enemies (Tallamy 2004). Consequently, adapted herbivores arriving in alien-dominated landscapes may experience eruptive outbreaks more frequently or intensely due to relaxed top-down regulation.

Although many herbivores are narrowly restricted to hosts with which they share a long historical association (Janieke 1990), others are not. This may mitigate to some degree the loss of ecosystem services associated with alien plants in urban landscapes. If alien host plants are not well defended from endemic herbivores owing to lack of a evolutionary history, then release from bottom-up regulation may be just as important as relaxation of top-down supression in the population dynamics of invasive pests as they proliferate in "defense free space" (Gandhi and Herms 2010). For example, fecundity and survival of the pine needle scale, which is endemic to North America, was much lower on its native hosts than on Eurasian *Pinus mugo* and *P. sylvestris* (Nielsen and Johnson 1973, Glynn and Herms 2004). Bronze birch borer, a native of North America, is consistently lethal to Eurasian birches, but North American birches are much more tolerant (Miller *et al*. 1991, Herms 2002a).

In an evolutionary *quid quo pro*, some native plants are far more susceptible to alien pests than their exotic congeners. Prominent examples include hemlock woolly adelgid, *Adelges tsugae*, and eastern North American hemlocks (Havill 2006); balsam wooly adelgid, *Adelges piceae*, and North American firs (Witter and Ragenovich 1986); beech bark scale, *Cryptococcus fagisuga*, and North American beeches (Houston 1994); and emerald ash borer and North American ashes (Rebek *et al*. 2008). The pattern, however, is not universal, as the fecundity of two species of Asian diaspidid scales was higher on hemlock hosts from their native range than on naïve North American hemlocks (McClure 1983). Futhermore, many native generalist herbivores readily incorporate introduced plants into their diet (Owen 1983, Keane and Crawley 2002, Agrawal and Kotanen 2003, Tallamy 2004, Gaston *et al*. 2007). Tent caterpillars, *Malacosoma* spp.; fall webworm, *Hyphantria cunea*; and bagworm, *Thyridopteryx ephemeraeformis* are classic examples of North American folivores that welcome aliens from several plant families into their diets. For example, eastern tent caterpillar, *Malacosoma americanum* regularly feeds on *Prunus* spp. from Europe and Asia; fall webworm, *Hyphantria cunea* regularly feeds on European and Asian hosts in North America and in its invaded range in Europe and Asia; and bagworm consumes many species of deciduous and evergreen hosts from Europe and Asia (Johnson and Lyon 1991).

Further exacerbating the problems caused by alien plants in urban landscapes is the fact that specialist herbivores from the endemic range of alien plants often are cointroduced with their host into new regions and subsquently become key pests. The exotic azalea lace bug and euonymus scale, *Unaspis euonymi*, are prominent examples. Invasion of North American cites by the alien tree, *Ailanthus altissima*, has allowed its specialist cynthia moth, *Samia cynthia*, to persist in urban environments (Pyle 1975). When their natural enemies are lacking, exotic herbivores accompanying alien plants may be more prone to outbreaks, due to relaxation of supression from predators and parasitoids (Liebhold *et al.* 1995, Tallamy 2004).

The addition of alien plants can also lead to an increase in plant diversity along some portions of urban gradients and may affect invertebrate richness or abundance (McKinney 2008, Goddard *et al.* 2009). Species richness of phytophagous and predatory mites on trees lining avenues and populating parks was greater than richness on trees in wooded suburbs of Como, Italy (Rigamonti and Lozzia 1999). On avenues and in parks, the diverse combination of resident native trees and exotic trees elevated arthropod diversity at two trophic levels. In examining the contribution of urban gardens in the United Kingdom, Gaston *et al.* (2007) conclude that vegetational structure in three-dimensional space and diversity of plant species are more important for maintaining invertebrate species richness than plant origin.

15.3.4 Impervious surfaces contribute to outbreaks

One of the hallmarks of urbanization is the domination of land by human-made structures. Impervious surfaces, also known as "hardscape," create an inhospitable matrix that includes buildings and infrastructure associated with transit, parking areas, and sidewalks. The proportion of impervious surfaces, which vary dramatically across urban gradients, fragment habitats, affect plant density, alter thermal regimes and plant–water relations, and disrupt interactions between herbivores and their natural enemies (Oke 1989, McDonnell and Pickett 1990, Speight *et al.* 1998, McIntyre 2000, Sperry *et al.* 2001, McKinney 2002, Arnfield 2003, Rickman and Conner 2003).

Water Infiltration and Drought Stress
Impervious surfaces dramatically reduce water infiltration and disrupt hydrological cycles in urban environments (Bierwagon *et al.* 2010). In cities dominated by hardscape, water limitations may exacerbate drought stress, thereby imposing strong bottom-up effects on arthropods. For instance, a large body of mostly anecdotal evidence has led to the proposition that drought stress

can trigger insect outbreaks by enhancing host quality (Mattson and Haack 1987). The stress hypothesis proposes that water deficits increase the nutritional quality and/or weaken the natural defenses of plants, thereby increasing herbivore fecundity and survival (Rhoades 1983, White 1984) and the likelihood of outbreaks. Indeed, stress has been proposed as a key cause of insect outbreaks in urban forests (Dreistadt *et al*. 1990, McIntyre 2000). However, reviews of empirical studies challenge this generality, concluding that effects of stress on host quality are highly variable (Larsson 1989, Koricheva *et al*. 1998, Herms 2002a, Huberty and Denno 2004).

Only a few studies have actually examined the relationship between urban stress and insect infestation, and results vary. Oaks growing in a downtown site experienced more severe stress, and associated lace bugs and aphids were generally more numerous than on trees growing on a nearby campus (Cregg and Dix 2001). In another study, density of horse chestnut scale, *Pulvinaria regalis*, was highest where water and nutrient infiltration were inhibited by impermeable surfaces, and the authors concluded that stress enhanced host quality (Speight *et al*. 1998). A similar study of a red wax scale in Australia revealed elevated densities on trees along roads compared to those in gardens, and proposed superior host suitability or elevated colonization as potential underlying mechanisms (Loch and Zalucki 1996).

Other studies of sucking insects support the alternative notion that drought stress decreases host quality of woody plants for sucking insects (Koricheva *et al*. 1998, Huberty and Denno 2004). For example, survival of the armored scale *P. pentagona* on white mulberry, *Morus alba*, was lower on water-stressed trees in urban environments than on trees in forested and urban sites that experienced less stress (Hanks and Denno 1993).

Effects of drought stress on folivores have been more variable. Water stress decreased folivore performance in some cases, but other studies reached the opposite conclusion or found effects to be minimal (Herms 2002a). Studies with bark beetles have also reached conflicting conclusions (Herms 2002a). It has been proposed that effects of drought stress on secondary metabolism and host plant resistance may be nonlinear, but are instead quadratic, with moderately drought stressed trees more resistant to herbivores than either severely stressed or rapidly growing trees (Mattson and Haack 1987, Larsson 1989, Herms 1999) which provides a potential explanation for otherwise contradictory results. Although evidence for the quadratic effects of drought on host quality is limited, there is some (Herms 1999, 2002a).

The evidence is stronger that drought-stressed trees are more susceptible to wood borers (Larsson 1989, Herms 2002a, Huberty and Denno 2004). For example, trees of green ash, *Fraxinus pennsylvanica*, planted in a downtown urban environment experienced more severe drought stress and suffered higher levels of borer damage than did trees planted on a park-like campus (Cregg and Dix 2001). Similarly, flowering dogwood, *Cornus florida*, an understory species that does not tolerate midday water stress, was much more susceptible to colonization by the dogwood borer, *Synanthedon scitula*, when planted in full sun relative to trees planted in at least partial shade (Potter and Timmons 1981).

Water stress also increased the susceptibility of eucalypts to the eucalyptus longhorned borer, *Phoracantha semipunctata* (Hanks *et al*. 1999).

Huberty and Denno (2004) suggested that intermittent, as opposed to chronic, drought stress may have underappreciated effects on host quality for herbivores, especially phloem sap suckers. Their pulsed stress hypothesis proposes that bouts of drought, followed by recovery of turgor, increase nitrogen content of phloem sap, which favors sucking insects. This hypothesis may have particular relevance to the urban forest because trees and shrubs are often planted in medians, planters, and other confined spaces where they experience more intense water deficits than trees growing in extensive soil profiles (Krizek and Dubik 1987, Lindsey and Bassuck 1991). The frequency of water deficit may also increase because, for example, restricted soil volumes dry faster during dry periods, and saturate more rapidly when it rains. Attempts to alleviate drought stress by periodic irrigation may further exacerbate this problem and enhance susceptibility of trees in urban sites. If so, then the pulsed stress hypothesis predicts that the effects of host quality on population dynamics of sap-sucking arthropods will differ markedly across the urban-rural gradient. Evidence supporting the pulsed stress hypothesis in urbanized areas can be found in studies by Hall and Ehler (1980) and Hanks and Denno (1993). Oleander in rural sites had more of the senescent terminals and less succulent growth that are favored by aphids compared to urban sites where intermittent irrigation and pruning produced abundant succulent growth (Hall and Ehler 1980). In non-irrigated urban sites, moisture deficit suppressed outbreaks of white peach scale, but in intermittently mesic urban landscapes, scales achieved dramatically elevated densities (Hanks and Denno 1993).

Heat islands

Impervious surfaces displace vegetation, including trees, shrubs, and ground cover (Taha 1997, Nowak and Dwyer 2000, Grimm *et al*. 2008, Bierwagen *et al*. 2010). Vegetation greatly reduces the amount of solar radiation reaching the ground and buildings where it is absorbed. This in turn reduces the amount of energy reradiated as heat (Oke 1989). Trees can reduce incoming solar radiation by 90%, and vegetation further cools urban areas through evapotranspiration (Oke 1989, Taha 1997, Nowak and Dwyer 2000). Furthermore, rapid runoff from impervious surfaces reduces water available for transpiration (Taha 1997). Collectively, these factors result in "heat islands." Cities can be as much as 10°C warmer than surrounding suburban and natural areas (Oke 1989, Kim 1992, Baker *et al*. 2003, Arnfield 2003, Stone 2007, Grimm *et al*. 2008). With respect to climate change, Bale *et al*. (2002) suggest that nondiapausing, multivoltine herbivores such as aphids are especially likely to benefit from increased temperatures. For example, a 2.8°C increase in average temperature increased the reproductive output and survival of the aphid *Acyrthosiphon svalbardicum*, resulting in an 11-fold increase in the number of overwintering eggs (Strathdee *et al*. 1993).

Referencing data along an urban gradient near Phoenix, Arizona, Baker *et al*. (2003) suggest that warming near cities creates a "thermal window" supporting greater abundance of arthropods. Given a strongly theoretical base, it is not surprising that outbreaks of several species of phytophagous herbivores in urban settings have been linked to elevated temperatures. The most common taxon of arthropods achieving outbreaks in urban areas appears to be spider mites (Table 15.1). Elevated temperatures have dramatic effects on the survival, development, and fecundity of these small multivoltine herbivores. Doubling the temperature from 15°C to 30°C reduced the developmental period from egg to adult of polyphagous twospotted spider mite, *Tetranychus urticae*, by about 80%, from 36 to 7 days (Sabilis 1981). Schneider *et al.* (2000) reported lifetime fecundity of *Eotetranychus tiliarium* to be 13 eggs per female at 12°C, but at 20°C fecundity increased to 47 eggs per female. Kropczynska *et al.* (1988) also documented increased lifetime fecundity (89 vs. 77 eggs per female), enhanced survival (92% vs. 61%), and faster development (14 vs. 19 days egg to adult) for *E. tiliarium* raised at higher versus lower temperatures, respectively.

Lepidoptera may also benefit from heat island effects. Reynolds *et al.* (2007) found caterpillar abundance in natural forest stands to be positively correlated with the thermal sum of daily summer temperatures. Because cities may be warmer than surrounding natural areas, some species may be able to expand their range along elevational and latitudinal gradients. Defoliation by mimosa webworm, *H. anisocentra*, was lower following cold winters when overwinter pupae experienced supercooling minima (i.e., the temperature below which webworms fail to survive) (Hart *et al.* 1986). Hart *et al.* (1986) suggested that cities provide thermal refuges for mimosa webworm in the northern part of its range where it would be unable to survive in natural forest stands. Elevated temperatures may also mitigate direct mortality associated with the inability of small insect larvae to recover from unusually cold temperatures. Fordyce and Shapiro (2003) found significantly greater mortality of small, slowly developing swallowtail larvae compared to faster growing larvae during chilly weather. Slower growing larvae succumbed to cold temperatures in a state of torpor called a "chill coma," whereas larger, faster growing larvae were less likely to die. Warmer temperatures in urban areas may enable insects to pass through thermally vulnerable stages more rapidly, thereby elevating survival.

Phenological synchrony of host and herbivore

The phenological window hypothesis (Feeny 1976) proposes that herbivores are best adapted to a narrow window of time when host traits are most suitable for their growth and development, and that herbivore performance will decline when phenological synchrony between host and herbivore is disrupted. For spring-feeding herbivores, tight synchrony with the appearance of nutritious young foliage directly enhances survival, development, and fecundity (Feeny 1970, Raupp and Denno 1983, Mattson and Scriber 1987, Hunter and Elkinton 2000, Dukes *et al.* 2009). Conversely, phenological asynchrony has been shown to decrease performance when herbivores were confronted with foliage of lower nutritional value and/or higher concentrations of secondary metabolites (Meyer and Montgomery 1987, Hunter and Elkinton 2000).

One outcome of global climate change, namely, warming in the Northern Hemisphere, has resulted in seasonal advancement of several plant phenological events, including early production of leaves in spring (Schwartz *et al.* 2006,

Cleland *et al.* 2007, Schwartz and Hanes 2009). If the developmental response of herbivores to variation in temperature differs from that of their host plant, then elevated temperatures characteristic of urban centers may alter the phenological synchrony between herbivores and their hosts (Ayres 1993, Watt and McFarlane 2002). Such effects can be positive as well as negative. For example, warmer temperatures in cities may further advantage herbivores if developmental rates of insects and mites are more responsive to elevated temperatures than maturation rates of host leaves, thereby expanding the phenological window (e.g., Ayers 1993, Dukes *et al.* 2009). A broadened phenological window of host suitability may also reduce the risk of starvation of early season folivores due to untimely emergence. However, effects of close synchrony with plant phenology on population dynamics may be mitigated by natural enemies in some early season leaf-feeding Lepidoptera such as gypsy moth, *Lymantria dispar*, and autumnal moth, *Epirrita autumnata* (Hunter and Elkinton 2000, Klemola *et al.* 2003). For example, delayed emergence of gypsy moth larvae resulted in higher levels of dispersal thereby reducing local densities and mortality inflicted by density-dependent natural enemies (Hunter and Elkinton 2000). The effects of phenological asynchrony among trophic levels arising from heat island effects may have important effects on population dynamics of herbivores in urban environments. However, these impacts have yet to be explored.

Effects on predator–prey interactions

Higher temperatures in urbanized environments may indirectly affect herbivore mortality by natural enemies. The slow growth–high mortality hypothesis proposes that host quality traits influencing the development rate of insects in turn influence the duration herbivores are vulnerable or exposed to natural enemies (Benrey and Denno 1997). Similarly, factors other than host quality such as elevated temperatures common in urban centers may alter developmental rates of arthropods. Many insects have specific windows of vulnerability that increase susceptibility to attack by natural enemies. For example, five instars of the black cutworm, *Agrotis ipsilon*, differed in their susceptibility to attack by 12 species of predators (carabids, staphylinids, spiders) of varying body size (Frank and Shrewsbury 2004b). In general, smaller predators were unable to eat larger prey, but larger predators consumed larvae of most stages, demonstrating that once prey obtained a certain size they were no longer accessible to a subset of predators (Frank and Shrewsbury 2004b). Early instars of azalea lace bug were more susceptible to predation by larvae of the lacewing *Chrysoperla carnea* than were older, more active nymphs and adults, which have behavioral defenses (Lepping 2003). Moreover, lace bugs attained their less vulnerable adult stage three days faster in sunny habitats than in shady ones (Lepping 2003).

Theoretical predications and empirical evidence regarding indirect effects of urban warming on predator–prey interactions are mixed. Berggren *et al.* (2009) suggest that elevated temperatures may enhance top-down regulation of herbivores when poikilothermic predators are more sensitive to warming than their cold-blooded prey. Studies of host–parasite interactions by Campbell *et al.* (1974) and Virtanen and Neuvonen (1999) support this notion. However, empirical evidence links elevated temperatures to outbreaks of spider mites in cities

(Table 15.1). Kropczynska *et al.* (1988) demonstrated a curious asymmetry between spider mites and their phytoseiid predators with respect to the effects of temperature on survival, development, and fecundity. When raised at elevated temperatures, the spider mite *E. tiliarium* developed 23% faster, experienced 80% less mortality, and laid 13% more eggs. Over the same range of temperatures the predatory phytoseiids, *Amblyseius findlandicus, A. aberrans, Paraseius soleiger, Phytoseius micropilis,* and *Typhlodromus pyri* demonstrated no changes in survival or fecundity. Two of these predators, *A. aberrans* and *Ph. micropolis,* developed 3% and 28% more rapidly, respectively, at elevated temperatures. A similar study by Sabelis (1981) found no such asymmetry in developmental rate between *T. urticae* and its predator *Ph. persimilis.* Perhaps flighted natural enemies that depend on rapid thermal warming to seek prey such as the parasitoids studied by Campbell *et al.* (1974) and Virtanen and Neuvonan (1999) are differentially favored by elevated temperatures in cities, while leaf-bound predators such as predatory mites are outpaced by their herbivorous prey.

A major determinant of patterns in herbivore abundance is the relative rates at which herbivores and natural enemies colonize plants (Thomas 1989). Herbivores and natural enemies have distinct modes of dispersal, levels of mobility, and searching behaviors for hosts or prey. Movement of herbivores and natural enemies can be differentially affected by matrix composition which includes areas separating habitat patches like impervious surfaces and other factors including patch isolation, habitat quality, plant density, and richness (Thomas 1989, Cronin 2003, Lee and Heimpel 2005, Diekotter *et al.* 2007). Impervious surfaces fragment urban habitats and may affect movement of herbivorous arthropods and natural enemies. Although numerous studies have examined aspects of matrix composition on predator–prey interactions, to our knowledge no studies have directly tested the effects of impervious surfaces on herbivore–natural enemy movement, colonization dynamics, and their impact on pest outbreaks in urbanized environments.

At present, only a few studies offer tantalizing evidence that impervious surfaces uncouple predator–prey interactions in cities. For example, the lymantriid moth *Euproctis similis* reached outbreak densities along the central reserve (median strip of a highway) but not on the verge (edge) of motorways (Port and Thompson 1980). Predatory wasps killed a greater proportion of larvae along the verge than along the central reserve. Although the presence of impervious surface may have played a role, Port and Thompson (1980) attributed the major difference in abundance of *E. similis* to inputs of oxides of nitrogen from exhaust that elevated nitrogen contents of plants, thereby increasing nutrition for herbivores. Locations with more hardscape may affect herbivore–natural enemy dynamics by providing refuge for herbivores from their natural enemies. In studying mortality of caterpillars along urban gradients in São Paulo, Brazil, Ruszczyk (1996) found a positive relationship between the land area covered by buildings and successful eclosion of an urban palm caterpillar, *Brassolis sophorae.* Ruszczyk (1996) reasoned that human-made features such as mailboxes, garages, closets, and skirting boards (wooden boards that cover the lowest part of a wall) provided partial refuges for pupating larvae to avoid parasitoids resulting in higher caterpillar densities in cities.

15.3.5   Anthropogenic inputs

Fertilization and Pest Outbreaks

Ornamental landscapes are commonly fertilized (Braman *et al.* 1998) based in part on the rationale that it enhances host plant resistance (Caldwell and Funk 1999). However, numerous studies have shown that fertilization generally decreases the resistance of woody plants to sucking arthropods including aphids, adelgids, scales, psyllids, plant bugs, and spider mites; folivores including caterpillars, sawflies, and leaf beetles; as well as subcortical-feeding shoot and stem borers (Kytö *et al.* 1996, Herms 2002b). Elevated levels of fertilization have been implicated in outbreaks of spider mites in urban sites such as shopping malls (Ehler and Frankie 1979). Thus, fertilization has the potential to increase herbivore populations, especially species such as mites, scales, and adelgids that complete multiple generations on the same plant (Herms 2002b).

Positive effects of fertilization on host quality have been attributed to the effects of nitrogen on enhanced nutritional quality and decreased concentrations of carbon-based secondary metabolites (Kytö *et al.* 1996, Herms 2002b). Nitrogen is the nutrient most often limiting to trees in urban soils (Harris 1992), and phosphorus fertilization has little effect on secondary metabolism of woody plants (Koricheva *et al.* 1998b). The growth and reproduction of phytophagous insects generally increase with the nitrogen concentration of their host (Mattson 1980), which is a nearly universal response of woody plants to nitrogen fertilization (Koricheva *et al.* 1998b). The growth/differentiation balance hypothesis (GDBH) attributes fertilizer effects on chemical defense to a physiological trade-off that results in decreased secondary metabolite concentrations as plant growth rate increases (Herms and Mattson 1992, Herms 2002b). Recent tests of GDBH have reinforced the generalization that increased nutrient availability decreases secondary metabolism and insect resistance of woody plants (Glynn *et al.* 2003, 2007, Hale *et al.* 2005, Lloyd *et al.* 2006).

High concentrations of secondary metabolites are maintained in nutrient-deficient plants because photosynthesis per unit leaf area is less sensitive to moderate nutrient limitation than is growth (Luxmoore 1991), which can increase the availability of carbon to support secondary metabolism when growth is limited (Glynn *et al.* 2007). In contrast, when nutrient limitation is severe enough to decrease photosynthesis, GDBH predicts that secondary metabolism will also be decreased as all plant functions are carbon limited, thus resulting in a nonlinear, parabolic pattern of secondary metabolism across a nutrient gradient (Herms and Mattson 1992).

Urban soils are often characterized by extreme nutrient limitation (Scharenbroch and Lloyd 2004). Hence, it could be hypothesized that fertilization increases pest resistance in such sites. However, because of physiological buffering mechanisms that include increased root–shoot ratios and nutrient use efficiency (Aerts and Chapin 2000), even severe nitrogen deficiency rarely induces chlorosis in established trees (Harris 1992). Rather, chlorosis is induced and photosynthesis decreased in response to nitrogen deficiency only when a plant cannot adjust to sudden changes in its internal nutrient balance. This can occur, for example, when trees are transplanted from high- to low-nutrient sites

to which they will eventually acclimate (Harris 1992). For example, in two species of willow, secondary metabolism was decreased only temporarily in response
to extreme nutrient limitation and increased over time as plants acclimated to
nutrient stress (Glynn *et al.* 2007). In conclusion, theory and empirical studies
suggest few situations in which fertilization increases tree resistance to insects.

Pollution and host quality

Detrimental effects of man-made pollutants on natural and agricultural
ecosystems are well known and have been the focus of entire treatises (Heliövaara
and Väisänen 1993). Pollutants also affect plant and insect interactions in urban
ecosystems. Ozone is an important stressor in urban forests (Bobbink 1998,
Taylor *et al.* 1994). It has been proposed that ozone stress generally decreases
tree resistance to insects (Hain 1997). Accordingly, a recent meta-analysis found
that ozone generally increased host quality for insect herbivores (Taylor *et al.*
1994). However, the pattern was variable. For example, experimental ozone
fumigation increased the quality of *Populus tremuloides* for four species of
outbreak Lepidoptera, including gypsy moth (Herms *et al.* 1996), but had no
effect on the quality of *Acer saccharum* or hybrid poplar as hosts for gypsy moth
(Lindroth 1993), or on the host quality of *Betula papyrifera* for *Orgyia
leucostigma* (Kopper *et al.* 2001). Similarly, exposure of *Populus deltoides* to
ozone had no effect on the aphid *Chaitophorus populicola* (Coleman and Jones
1988a), but decreased fecundity of the imported willow leaf beetle, *Plagiodera
versicolora* (Coleman and Jones 1988b).

Deposition of atmospheric nitrogen originating from fossil fuel combustion
provides a potential mechanism by which nitrogen enrichment may influence
the population dynamics of phytophagous insects in urban environments on a
wide scale. The magnitude of nitrogen deposition can be substantial, exceeding
$30 kg ha^{-1}$ in many urban and rural regions of Europe and North America
(Taylor *et al.* 1994, Bobbink 1998). Chronic nitrogen loading alters patterns of
nutrient cycling and increases nitrogen uptake in forest ecosystems, with the
potential to increase herbivore populations and alter herbivore community
structure through effects on host quality (Throop and Lerdae 2004). In southern California, nitrogen deposition and ozone exposure interacted to increase
pine susceptibility to bark beetle infestation during drought (Jones *et al.* 2004).
Nitrogen deposition associated with vehicle exhaust was also implicated in elevating the quality of crabapple trees for the lymantriid *Euproctis similis* that
resulted in elevated populations of the moth on trees near roadways (Port and
Thompson 1980). Rose bushes exposed to ambient air in central Munich were
superior hosts, and enhanced growth rates of rose aphids, *Microsiphon rosae*, by
20% compared to roses grown in chambers supplied with filtered air to remove
pollutants (Dohmen 1985). By planting beans and viburnums along motorways,
exposing them to filtered and unfiltered air, and measuring population growth
of black bean aphid, *Aphis fabae*, Bolsinger and Flückinger (1987) demonstrated
elevated levels of foliar nitrogen in both plants and a twofold increase in density
of aphid populations on plants exposed to atmospheric pollutants.

De-icing salts are commonly applied in cities where snow accumulates on
streets. The use of de-icing salts in European cities has been implicated in

improving the quality of hosts for herbivores and contributing to dramatically elevated populations of spider mites and aphids along roadways compared to more natural areas such as parks and natural areas at a distance from motorways (Kropczynska *et al.* 1986, Braun and Flückiger 1984, Fostad and Pederson 1997).

Pollutants affect the quality of host plants for natural enemies as well as herbivores. Airborne particulates such as dust from roadways can disrupt activities of natural enemies on foliage and may result in elevated populations of herbivores near thoroughfares. Greater abundance of an armored scale, *Nuculaspis californica*, has been associated with increased levels of dust along roadways compared to natural areas (Edmunds 1973).

Pesticides

Pesticides disrupt ecosystem processes at several spatial scales. Government agencies may inadvertently create outbreaks of insects and mites in urban settings while conducting area-wide attempts to eradicate invasive exotic pests. An eradication program for a localized infestation of Japanese beetle, *Popillia japonica*, in California resulted in dramatic increases of citrus red mite, *Panonychus citri*; woolly whitefly, *Aleurothrixus floccosus*; and purple scale, *Lepidosaphes beckii*, and associated mortality of citrus trees occurred in residential landscapes (DeBach and Rose 1977). Disruption of natural enemy communities and a relaxation of top-down forces contributed to outbreaks of these pests. Eradication programs for Mediterranean fruit fly, *Ceratitis capitata*, resulted in outbreaks of several species of sucking insects and mites in urban areas (Dreistadt and Dahlsten 1986). An endemic gall midge that normally occurs at very low levels was 90 times more abundant in areas treated insecticides (Ehler *et al.* 1984). Szczepaniec *et al.* (2011) documented dramatic increases of in populations of the Schoenei spider mite, *Tetranychus schoenei*, on elm trees treated with insecticides as part of an eradication program for Asian longhorned beetle in Central Park, New York.

Where municipalities attempt to manage biting and filth flies, secondary pest outbreaks may occur. In a resort community in California, prolonged applications of malathion to control adult mosquitoes resulted in elevated and damaging levels of pine needle scale on lodgepole and Jeffery pines (Dahlsten *et al.* 1969). A similar case occurred on Mackinaw Island, Michigan, where European fruit lecanium, *Parthenolecanium corni*, attained high densities on trees lining streets following weekly applications of dimethoate to control filth-flies (Merritt *et al.* 1983). A program to control flies in breeding sites and the cessation of sprays were followed by rapid declines in scale populations. Mortality associated with parasitoids that colonized infested trees from untreated areas nearby was the mechanism believed to underlie the decline (Merritt *et al.* 1983).

At the spatial scale of individual residences, Raupp *et al.* (2001) found that broad-scale applications of insecticides increased the diversity and abundance of scale insects in residential landscapes. Seventy-five percent of chronically treated residential landscapes experienced elevated populations of scales, but only 40% of sites sprayed at intervals of fewer than four years had serious problems with scale insects (Raupp *et al.* 2001). Recurrently treated sites had been colonized by nine species of Diaspididae and three species of Coccidae, whereas only four species of Diaspididae and no Coccidae were found in landscapes that were rarely sprayed (Raupp *et al.* 2001). Residual organophosphate insecticides reduced the abundance and activity of several taxa of natural enemies. Dimethoate applied to entire or partial canopies of eastern hemlock, *Tsuga canadensis*, infested with elongate hemlock scale resulted in a dramatic resurgence in scale insect populations (McClure 1977). Scale insects on treated trees developed faster and were more fecund due to an improvement in host quality following an initial reduction in scale density. Moreover, the insecticide virtually elimi-nated an important parasitoid and three species of common predators from the lower crowns of partially treated trees (McClure (1977).

Some systemic insecticides such as imidacloprid have been associated with elevated populations of honeylocust spider mite on honeylocust (Sclar *et al.* 1998); spruce spider mite, *Oligonychus ununguis*, and hemlock rust mite, *Nalepella tsugifolia*, on Canadian hemlocks (Raupp *et al.* 2004); twospotted spider mite on *Rosa* and euonymus plants, *Euonymus alatus* (Gupta and Krischik 2007, Chiriboga 2009); and Schoenei spider mites on American elm (Raupp *et al.* 2008). Loss of top-down regulation by natural enemies, improved host quality, and enhanced mite fecundity have been implicated in these outbreaks (Gupta and Krischik 2007, Creary 2008, Chiriboga 2009, Raupp *et al.* 2010, Szczepaniec *et al.* 2011). It is noteworthy that in several pesticide-related out-breaks, cessation of pesticide applications was followed by dramatic declines in herbivore populations as mortality caused by parasitoids and predators increased (Luck and Dahlsten 1975, Merritt *et al.* 1983, Ehler and Kinsey 1991).

15.4 Conclusions

Case studies indicate that many herbivorous taxa in several feeding guilds, with diverse life histories, modes of dispersal, and host specificities, outbreak along urban gradients. Small sucking arthropods like scale insects and mites dominate eruptive taxa. These multivoltine pests with short generation times may be advantaged by elevated temperatures in cities. Strong theoretical evidence and several empirical studies indicate that many taxa benefit from enhanced resource quality which is linked to periods of intermittent drought and pulsed inputs of water, nutrients, and pollutants. Relaxation of top-down suppression associated with reduced richness or abundance of natural enemies are implicated in outbreaks of herbivores in cities, which may lack critical resources such as abundant alternative prey for generalist natural enemies. Physical structures in cities provide spatial refuge for pests from natural enemies, and elevated temperatures may provide temporal escape from predators, parasitoids, and unfavorable thermal regimes. Pesticides often eliminate natural enemies, impair their activities, alter plant quality, or stimulate reproduction in pests, thereby contributing to outbreaks in urban areas.

Further research is necessary before a robust predictive framework emerges regarding the relative importance of bottom-up and top-down effects on population dynamics of phytophagous arthropods in urban environments. Clearly, the effects of abiotic stress on host quality are variable and complex and depend on the type, timing, and intensity of stress; the physiological response of the plant; and the behavior, physiology, and life history of the herbivore (Jones and Coleman 1991). Few studies have adequately quantified these variables. Much remains to be learned about the relative contributions of natural enemies and the effects of urban matrixes on herbivore–natural enemy population dynamics and movement as they relate to pest outbreaks. Many important questions remain unresolved regarding the impact of alien plants on community structure, ecosystem processes, and pest outbreaks in urban environments (Goddard *et al.* 2009). The gradient from natural areas to cities will continue to provide unique opportunities to study features of the natural world, such as patterns of community structure and function, in a context of human culture, economics, and politics in an increasingly urbanized world.

Acknowledgments

We are grateful to P. Barbosa, D. Letourneau, A. Agrawal, and two reviewers whose helpful comments improved the manuscript. The National Research Initiative Competitive Grants Program of USDA 2005-00915 and International Society of Arboriculture Tree Fund grants to M.J.R. and the National Research Initiative Competitive Grants Program of USDA 2002-0002577 to P.M.S. helped support this research.

References

Aerts, R., and F. S. Chapin III. 2000. The mineral nutrition of wild plants revisited: a re-evaluation of processes and patterns. *Advances in Ecological Research* 30:1–67.

Agrawal, A. A., and P. M. Kotanen. 2003. Herbivores and the success of exotic plants: a phylogenetically controlled experiment. *Ecology Letters* 6:712–715.

Arnfield, A. J. 2003. Two decades of urban climate research: a review of turbulence, exchanges of energy and water, and the urban heat island. *International Journal of Climatology* 23:1–26.

Ayres, M. 1993. Plant defense herbivory and climate change. Pages 75–94 *in* P. M. Kareiva, J. G., Kingsolver, and R. B. Huey (editors), *Biotic Interactions and Global Change*. Sinauer Associates, Sunderland, MA.

Baker, L., A. Brazel, N. Selover, C. Martin, N. McIntyre, F. Steiner, A. Nelson, and L. Musacchio. 2003. Urbanization and warming of Phoenix (Arizona, USA): impacts, feedbacks and mitigation. *Urban Ecosystems* 6:183–203.

Balder, H., B. Jäckel, and B. Pradel. 1999. Investigations on the existence of beneficial organisms on urban trees in Berlin. Pages 189–195 *in* M. Lemattre, P. Lamettre, and F. Lamettre (editors), *Proceedings of the International Symposium on Urban Tree Health*. International Society for Horticultural Science, Brugge, Belgium.

Bale, J. S., G. J. Masters, I. D. Hodkinson, C. Awmack, T. M. Bezemer, V. Brown, J. Butterfield, A. Buse, J. C. Coulson, J. Farrar, J. G. Good, R. Harrington, S. Hartley, T. H. Jones, R. Lindroth, M. Press, I. Symrnioudis, A. D. Watt, and J. B. Whittaker. 2002. Herbivory in global climate change research: direct effects of rising temperature on insect herbivores. *Global Change Biology* 8:1–16.

Benrey, B., and R. F. Denno. 1997. The slow-growth–high-mortality hypothesis: a test using the cabbage butterfly. *Ecology* 78:987–999.

Berggren, Å., C. Björkman, H. Bylund, and M. P. Ayres. 2009. The distribution and abundance of animal populations in a climate of uncertainty. *Oikos* 118:1121–1126.

Berryman, A. A. 1987. The theory and classification of outbreaks. Pages 3–30 *in* P. Barbosa, and J. C. Shultz (editors), *Insect Outbreaks*. Academic Press, San Diego.

Bierwagen, B. G., D. M. Theobald, C. R. Pyke, A. Choate, P. Groth, J. V. Thomas, and P. Morefield. 2010. National housing and impervious surface scenarios for integrated climate impact assessments. *Proceedings of the National Academy of the Sciences of the USA* 107:20887–20892.

Bobbink, R. 1998. Impacts of tropospheric ozone and airborne nitrogenous pollutants on natural and seminatural ecosystems: a commentary. *New Phytologist* 139:161–168.

Bolsinger, M., and W. Flückiger. 1987. Enhanced aphid infestation at motorways: the role of ambient air pollution. *Entomologia Experimentalis et Applicata* 45:237–243.

Braman, S. K., J. G. Latimer, and C. D. Robacker. 1998. Factors influencing pesticide use and integrated pest management implementation in urban landscapes: a case study in Atlanta. *HortTechnology* 8:145–157.

Braun, S., and W. Flückiger. 1984. Increased population of aphid *Aphis pomi* at a motorway. Part 2 – the effect of drought and deicing salt. *Environmental Pollution Series A, Ecological and Biological* 36:261–270.

Burghardt, K., D. W. Tallamy, and G. Shriver. 2009. Impact of native plants on bird and butterfly biodiversity in suburban landscapes. *Conservation Biology* 23:219–224.

Caldwell, D., and R. Funk. 1999. Tree fertilization. *Arborist News* 8:24–26.

Campana, R. J., and R. J. Stipes. 1981. Dutch elm disease in North America with particular reference to Canada: Success or failure of conventional control methods. *Canadian Journal of Plant Pathology* 3:252–259.

Campbell, A., B. D. Frazer, N. Gilbert, A. P. Gutierrez, and M. Mackauer. 1974. Temperature requirements of some aphids and their parasites. *Journal of Applied Ecology* 11:431–438.

Chiriboga, A. 2009. Physiological responses of woody plants to imidacloprid formulations. Master's thesis, The Ohio State University, Columbus.

Cleland, E. E., I. Chuine, A. Menzel, H. A. Mooney, and M. D. Schwartz. 2007. Shifting plant phenology in response to global change. *Trends in Ecology and Evolution* 22:357–365.

Coleman, J. S., and C. G. Jones. 1988a. Acute ozone stress on eastern cottonwood (*Populus deltoides* Bartr.) and the pest potential of the aphid, *Chaitophorus populicola* Thomas (Homoptera: Aphididae). *Environmental Entomology* 17:207–212.

Coleman, J. S., and C. G. Jones. 1988b. Plant stress and insect performance: cottonwood, ozone and a leaf beetle. *Oecologia* 76:57–61.

Cranshaw, W. and E. Hart. 1988. Rogue's gallery. *American Nurseryman* 69–75.

Creary, S. 2009. Indirect effects of imidacloprid on two predators of spider mite on elms and boxwoods. Master's thesis, University of Maryland, College Park.

Cregg, B. M., and M. E. Dix. 2001. Tree moisture stress and insect damage in urban areas in relation to heat island effects. *Journal of Arboriculture* 27:8–17.

Cronin, J. T. 2003. Matrix heterogeneity and host–parasitoid interactions in space. *Ecology* 84:1506–1516.

Dahlsten, D. L., R. Garcia, J. E. Prine, and R. Hunt. 1969. Insect problems in forest recreation areas, pine needle scale and mosquitoes. *California Agriculture* 23:4–6.

DeBach, P., and M. Rose. 1977. Environmental upsets caused by chemical eradication. *California Agriculture* 23:4–6.

Denno, R.F. and G.K. Roderick 1991. Influence of patch size, vegetation texture, and host plant architecture on the diversity, abundance, and life history styles of sap feeding herbivores. Pages 169–196 *in* S.S. Bell, E.D. McCoy, and H.R. Mushinsky (editors), *Habitat Structure: The Physical Arrangement of Objects in Space*. Chapman and Hall, New York.

Dohmen, G. P. 1985. Secondary effects of air pollution: enhanced aphid growth. *Environmental Pollution Series A, Ecological and Biological* 39:227–234.

Dreistadt, S. H., and D. L. Dahlsten. 1986. Medfly eradication in California: lessons from the field. *Environment* 28:18–20, 40–44.

Dreistadt, S. H., D. L. Dahlsten, and G. W. Frankie. 1990. Urban forests and insect ecology. *BioScience* 40:192–198.

Dukes, J. S., J. Pontius, D. Orwig, J. R. Garnas, V. L. Rodgers, N. Brazee, B. Cooke, K. A. Theoharides, E. E. Stange, R. Harrington, J. Ehrenfeld, J. Gurevitch, M. Lerdau, K. Stinson, R. Wick, and M. Ayres. 2009. Responses of insect pests, pathogens, and invasive plant species to climate change in the forests of northeastern North America: what can we predict? *Canadian Journal of Forest Research* 39:231–248.

Edmunds, G. F. Jr. 1973. Ecology of black pineleaf scale (Homptera: Diaspididae). *Environmental Entomology* 2:765–777.

Ehler, L. E., P. C. Endicott, M. B. Hertlein, and B. Alvarado-Rodriquez. 1984. Medfly eradication in California. *Entomologia Experimentalis et Applicata* 36:201–208.

Ehler, L. E., and G. W. Frankie. 1979. Arthropod fauna of live oak in urban and natural stands in Texas. II. Characteristics of the mite fauna (Acari). *Journal of the Kansas Entomological Society* 52:86–92.

Ehler, L. E., and M. G. Kinsey. 1991. Ecological recovery of a gall midge and its parasite guild following disturbance. *Environmental Entomology* 20:1295–1300.

Eliason, E. A., and D. G. McCullough. 1997. Survival and fecundity of three insects reared on four varieties of Scotch pine Christmas trees. *Journal of Economic Entomology* 90:1598–1608.

Faeth, S. H., P. S. Warren, E. Schochat, and W. A. Marussich. 2005. Trophic dynamics in urban communities. *BioScience* 55:399–407.

Feeny, P. 1970. Seasonal changes in oak leaf tannins and nutrients as a cause of spring feeding by winter moth caterpillars. *Ecology* 51:565–581.

Feeny, P. 1976. Plant apparency and chemical defense. *Recent Advances in Phytochemistry* 10:1–40.

Fordyce, J. A., and A. M. Shapiro. 2003. Another perspective on the slow growth / high mortality hypothesis: chilling effects on swallowtail larvae. *Ecology* 84:263–268.

Fostad, O., and P. A. Pederson. 1997. Vitality, variation, and causes of decline in trees in Oslo center (Norway). *Journal of Arboriculture* 23:155–165.

Frank, S. D., and P. M. Shrewsbury. 2004a. Effect of conservation strips on the abundance and distribution of natural enemies and predation of *Agrotis ipsilon* (Lepidoptera: Noctuidae) on golf course fairways. *Environmental Entomology* 33:1662–1672.

Frank, S. D., and P. M. Shrewsbury. 2004b. Consumption of black cutworms, *Agrotis ipsilon* (Lepidoptera: Noctuidae), and alternative prey by common golf course predators. *Environmental Entomology* 33:1681–1688.

Frankie, G. W., W. Brewer, W. Cranshaw, and J. F. Barthell. 1987. Abundance and natural enemies of the spindle gall midge, *Pinyonia edulicola* Gagné, in natural and urban stands of pinyon pine in Colorado (Diptera: Cecidomyiidae). *Journal of the Kansas Entomological Society* 60:133–144.

Frankie, G. W., and L. E. Ehler. 1978. Ecology of insects in urban environments. *Annual Review of Entomology* 23:367–387.

Frankie, G. W. and D. L. Morgan. 1984. Role of the host plant and parasites in regulating insect herbivore abundance, with an emphasis on gall inducing insects. Pages 110–140 *in* P. W. Price, C. N. Slobodchikoff, and W. Gaud (editors), *A New Ecology: Novel Approaches to Interactive Systems*. John Wiley & Sons, Inc., New York.

Frankie, G. W., D. L. Morgan, and E. E. Grissell. 1992. Effects of urbanization on the distribution and abundance of the cynipid gall wasp, *Disholcaspis cinerosa*, on ornamental live oak in Texas, USA. Pages 258–279 *in* J. D. Shorthouse, and O. Rohfritsch (editors), *Biology of Insect-Induced Galls*. Oxford University Press, New York.

Gandhi, K. J. K., and D. A. Herms. 2010. Direct and indirect effects of invasive insect herbivores on ecological processes and interactions in forests of eastern North America. *Biological Invasions* 12:389–405.

Glynn, C., and D. A. Herms. 2004. Local adaptation in pine needle scale (*Chionaspis pinifoliae*): natal and novel host quality as tests for specialization within and among red and Scots pine. *Environmental Entomology* 33:748–755.

Glynn, C., D. A. Herms, M. Egawa, R. Hansen, and W. J. Mattson. 2003. Effects of nutrient availability on dry matter allocation, and constitutive and induced insect resistance of poplar. *Oikos* 101:385–397.

Glynn, C., D. A. Herms, C. M. Orians, R. C. Hansen, and S. Larsson. 2007. Testing the growth-differentiation balance hypothesis: dynamic responses of willows to nutrient availability. *New Phytologist* 176:623–634.

Griffin, J. M., G. M. Lovett, M. A. Arthur, and K. C. Weathers. 2003. The distribution and severity of beech bark disease in the Catskill Mountains, N.Y. *Canadian Journal of Forest Research* 33:1754–1760.

Grimm, N. B., S. H. Faeth, N. E. Golubiewski, C. L. Redman, J. Wu, X. Bai, and J. M. Briggs. 2008. Global change and the ecology of cities. *Science* 319:756–760.

Gupta, G., and V. V. Krischik. 2007. Professional and consumer insecticides for the management of adult Japanese beetle on hybrid tea rose. *Journal of Economic Entomology* 100:830–837.

Hain, F. P. 1987. Interactions of insects, trees, and air pollutants. *Tree Physiology* 3:93–102.

Hale, B. K., D. A. Herms, R. C. Hansen, T. P. Clausen, and D. A. Arnold. 2005. Effects of drought stress and nutrient availability on dry matter allocation, phenolic glycosides and rapid induced resistance of poplar to two lymantriid defoliators. *Journal of Chemical Ecology* 31:2601–2620.

Hall, R. W., and L. E. Ehler. 1980. Population ecology of *Aphis nerii* on oleander. *Environmental Entomology* 9:338–344.

Hanks, L. M., and R. F. Denno. 1993. Natural enemies and plant water relations influence the distribution of an armored scale insect. *Ecology* 74:1081–1091.

Hanks, L. M., T. D. Paine, J. G. Millar, C. D. Campbell, and U. K. Schuch. 1999. Water relations of host trees and resistance to the phloem-boring beetle *Phoracantha semipunctata* F. (Coleoptera: Cerambycidae). *Oecologia* 119:400–407.

Harris, R. W. 1992. Root–shoot ratios. *Journal of Arboriculture* 18:39–42.

Hart, J. H., F. D. Miller Jr., and R. A. Bastian. 1986. Tree location and winter temperature influence on mimosa webworm populations in a northern urban environment. *Journal of Arboriculture* 12:237–240.

Havill, N. P., M. E. Montgomery, G. Yu, S. Shiyake, and A. Caccone. 2006. Mitochondrial DNA from hemlock woolly adelgid (Hemiptera: Adelgidae) suggests cryptic speciation and pinpoints the source of the introduction to eastern North America. *Annals of the Entomological Society of America* 99:195–203.

Heliövaara, K., and R. Väisänen. 1993. *Insects and Pollution*. CRC Press, Boca Raton, FL.

Herms, D. A. 1999. Physiological and abiotic determinants of competitive ability and herbivore resistance. *Phyton* 39:53–64.

Herms, D. A. 2002a. Strategies for deployment of insect resistant ornamental plants. Pages 217–237 in M. R. Wagner, K. M. Clancy, F. Lieutier, and T. D. Paine (editors), *Mechanisms and Deployment of Resistance in Trees to Insects*. Kluwer Academic, Dordrecht, the Netherlands.

Herms, D. A. 2002b. Effects of fertilization on insect resistance of woody ornamental plants: reassessing an entrenched paradigm. *Environmental Entomology* 31:923–933.

Herms, D. A., and W. J. Mattson. 1992. The dilemma of plants: to grow or defend. *The Quarterly Review of Biology* 67:283–335.

Herms, D. A., R. C. Akers, and D. G. Nielsen. 1984. The ornamental landscape as an ecosystem: implications for pest management. *Journal of Arboriculture* 10:303–307.

Herms, D. A., W. J. Mattson, D. N. Karowe, M. D. Coleman, T. M. Trier, B. A. Birr, and J. G. Isebrands. 1996. Variable performance of outbreak defoliators on aspen clones exposed to elevated CO_2 and O_3. Pages 43–55 in J. Hom, R. Birdsey, and K. O'Brian (editors), *Proceedings of the 1995 North Global Change Program, Pittsburg, Pennsylvania*. US Forest Service General Technical Report NE-214. US Forest Service, Washington, DC.

Houston, D. R. 1994. Major new tree disease epidemics: beech bark disease. *Annual Review of Phytopathology* 32:75–87.

Huberty, A. F., and R. F. Denno. 2004. Plant water stress and its consequences for herbivorous insects: a new synthesis. *Ecology* 85:1383–1398.

Hunter, A. F., and J. S. Elkinton. 2000. Effects of synchrony with host plant on populations of a spring-feeding Lepidopteran. *Ecology* 81:1248–1261.

Jaenike, J. 1990. Host specialization in phytophagous insects. *Annual Review of Entomology* 21:243–273.

Johnson, W. T., and H. H. Lyon. 1994. *Insects That Feed on Trees and Shrubs*. Cornell University Press, Ithaca, NY.

Jones, C. G., and J. S. Coleman. 1991. Plant stress and insect herbivory: toward an integrated perspective. Pages 249–280 in H. A. Mooney, W. E. Winner, and E. J. Pell (editors), *Response of Plants to Multiple Stresses*. Academic Press, San Diego.

Jones, M. E., T. D. Paine, M. E. Fenn, and M. A. Poth. 2004. Influence of ozone and nitrogen deposition on bark beetle activity under drought conditions. *Forest Ecology and Management* 200:67–76.

Kahn, D. M. 1988. Population ecology of an insect herbivore: native holly leafminer, *Phytomyza ilicicola*. PhD dissertation, University of Delaware, Newark.

Kahn, D. M., and H. V. Cornell. 1989. Leafminers, early leaf abscission, and parasitoids: a tritrophic interaction. *Ecology* 70:1219–1226.

Kareiva, P. 1983. Influence of vegetation texture on herbivore populations: resource concentration and herbivore movement. Pages 259–289 in Denno, R.F. and M.S. McClure (editors), *Variable Plants and Herbivores in Natural and Managed Systems*. Academic Press, New York.

Keane, R. M., and M. J. Crawley. 2002. Exotic plant invasions and the enemy release hypothesis. *Trends in Ecology and Evolution* 17:164–170.

Kim, H. H. 1992. Urban heat-island. *International Journal of Remote Sensing* 13:2319–2336.

Klemola, T., K. Ruohomäki, M. Tanhuanpää, and P. Kaitaniemi. 2003. Performance of a spring-feeding moth in relation to time of oviposition and bud-burst phenology of different host species. *Ecological Entomology* 28:319–327.

Kruess, A., and T. Tscharntke. 1994 Habitat fragmentation, species loss, and biological control. *Science* 264:1581–1584.

Koch, J. 2000. The origins of urban forestry. Pages 1–10 *in* J. E. Kuser (editor), *Handbook of Urban and Community Forestry in the Northeast*. Academic, New York.

Koehler, C. S., and G. S. Frankie. 1968. Distribution and seasonal abundance of *Oligonychus subnudus* on Monterey pine. *Annals of the Entomological Society of America* 61:1500–1506.

Kopper, B. J., R. L. Lindroth, and E. V. Nordheim. 2001. CO_2 and O_3 effects on paper birch (Betulaceae: *Betula papyrifera*) phytochemistry and whitemarked tussock moth (Lymantriidae: *Orgyia leucostigma*) performance. *Environmental Entomology* 30:1119–1126.

Koricheva J., S. Larsson, and E. Haukioja. 1998a. Insect performance on experimentally stressed woody plants: a meta-analysis. *Annual Review of Entomology* 43:195–216.

Koricheva, J., S. Larsson, E. Haukioja, and M.Keinänen. 1998b. Regulation of woody plant secondary metabolism by resource availability: hypothesis testing by means of meta-analysis. *Oikos* 83:212–226.

Kovacs, K. F., R. G. Haight, D. G. McCullough, R. J. Mercader, N. W. Siegert, and A. M. Liebhold. 2010. Cost of potential emerald ash borer damage in U.S. communities, 2009–2019. *Ecological Economics* 69:569–578.

Krizek, D. T., and S. P. Dubik. 1987. Influence of water stress and restricted root volume on growth and development of urban trees. *Journal of Arboriculture* 13:47–55.

Kropczynska, D., M.van de Vrie, and A.Tomczyk. 1986. Woody ornamentals. Pages 385–393 *in* W. Helle, and M.W. Sabelis (editors), *Spider Mites: Their Biology, Natural Enemies and Control*, vol. 1B. Elsevier, Amsterdam.

Kropczynska, D., M. van de Vrie, and A. Tomczyk. 1988. Bionomics of *Eotetranychus tiliarium* and its phytoseiid predators. *Experimental and Applied Acarology* 5:65–81.

Kytö, M., P. Niemela, and S. Larsson. 1996. Insects on trees: population and individual response to fertilization. *Oikos* 75:148–159.

Landis, D. A., S. D. Wratten, and G. M. Gurr. 2000. Habitat management to conserve natural enemies of arthropod pests in agriculture. *Annual Review of Entomology* 45:175–201.

Langellotto, G. A., and R. F. Denno. 2004. Responses of invertebrate natural enemies to complex-structured habitats: a meta-analytical synthesis. *Oecologia* 139:1–10.

Larsson, S. 1989. Stressful times for the plant stress–insect performance hypothesis. *Oikos* 56:277–283.

Letourneau, D.K. 1990. Mechanisms of predator accumulation in a mixed crop system. *Ecological Entomology.* 15: 63–69.

Lee, J. C., and G. E. Heimpel. 2005. Impact of flowering buckwheat on lepidopteran cabbage pests and their parasitoids at two spatial scales. *Biological Control* 34:290–301.

Lepping, M. D. 2003. The influence of thermal environment on development and vulnerability to predation of the azalea lace bug, *Stephanitis pyrioides* (Heteroptera: Tingidae). Master's thesis, University of Maryland, College Park.

Liebhold, A. M., W. L. Macdonald, D. Bergdahl, and V. C. Mastro. 1995. Invasion by exotic forest pests: a threat to forest ecosystems. *Forest Science Monograph* 30:1–49.

Lindroth, R. L., P. B.Reich, M. G. Tjoelker, J. C. Volin, and J. Oleksyn. 1993. Light environment alters response to ozone stress in seedlings of *Acer saccharum* Marsh. and hybrid *Populus* L. *New Phytologist* 124:647–651.

Lindsey, P., and N. Bassuck. 1991. Specifying soil volumes to meet the water needs of mature urban street trees and trees in containers. *Journal of Arboriculture* 17:141–149.

Lloyd, J. E., D. A. Herms, J. V. Wagoner, and M. A. Rose. 2006. Fertilization rate and irrigation scheduling in the nursery influence growth, insect performance, and stress tolerance of 'Sutyzam' crabapple in the landscape. *HortScience* 41:442–445.

Loch, A. D., and M. P. Zalucki. 1996. Spatial distribution of pink wax scale, *Ceorplastes rubens* Maskell (Hemiptera: Coccidea), on umbrella trees in South-eastern Queensland: the pattern of outbreak. *Australian Journal of Zoology* 44:599–609.

Luck, R. F., and D. L. Dahlsten. 1975. Natural decline of a pine needle scale (*Chionaspis pinifoliae* (Fitch)) outbreak at South Lake Tahoe, California, following cessation of adult mosquito control with malathion. *Ecology* 56:893–904.

Luxmoore, R. J. 1991. A source-sink framework for coupling water, carbon, and nutrient dynamics of vegetation. *Tree Physiology* 9:267–280.

Mattson, W. J. 1980. Herbivory in relation to plant nitrogen content. *Annual Review of Ecology and Systematics* 11:119–161.

Mattson, W. J., and R. A. Haack. 1987. The role of plant water deficits in provoking outbreaks of phytophagous insects. Pages 365–407 *in* P. Barbosa and J.C. Shultz (editors), *Insect Outbreaks*. Academic, San Diego.

Mattson, W. J., and J. M. Scriber. 1987. Nutritional ecology of insect folivores of woody plants: nitrogen, water fiber and mineral considerations. Pages 105–146 *in* F. Slansky Jr. and J. G. Rodriguez (editors), *Nutritional Ecology of Insects, Mites, Spiders and Related Invertebrates*. John Wiley & Sons, Inc., New York.

McBride, J. and D. Jacobs. 1976. Urban forest development: a case study, Menlo Park, California. *Journal of Urban Ecology* 2:1–14.

McClure, M. S. 1977. Resurgence of the scale, *Fiorinia externa* (Homoptera: Diaspididae) on hemlock following insecticide application. *Environmental Entomology* 6:480–484.

McClure, M. S. 1983. Reproduction and adaptation of exotic hemlock scales (Homoptera: Diaspididae) on their new and native hosts. *Environmental Entomology* 12:1811–1815.

McClure, M. S. 1986. Population dynamics of Japanese hemlock scales: a comparison of endemic and exotic communities. *Ecology* 67:1411–1421.

McDonnell, M. J., and S.T.A. Pickett. 1990. Ecosystem structure and function along urban-rural gradients: an unexploited opportunity for ecology. *Ecology* 71:1232–1237.

McIntyre, N. E. 2000. Ecology of urban arthropods: a review and a call to action. *Annals of the Entomological Society of America* 93:825–835.

McIntyre, N. E., J. Rango, W. F. Fagan, and S. H. Faeth. 2001. Ground arthropod community structure in a heterogeneous urban environment. *Landscape and Urban Planning* 52:257–274.

McKinney, M. L. 2002. Urbanization, biodiversity, and conservation. *BioScience* 52:883–890.

McKinney, M. L. 2008. Effects of urbanization on species richness: a review of plants and animals. *Urban Ecosystems* 11:161–176.

Merritt, R.W., M. K. Kennedy, and E. F. Gersabeck. 1983. Integrated pest management of nuisance and biting flies in a Michigan resort: dealing with secondary pest outbreaks. Pages 277–299 *in* G.W. Frankie and C.S. Koehler (editors), *Urban Entomology: Interdisciplinary Perspectives*. Praeger, New York.

Meyer, G. A., and M. E. Montgomery. 1987. Relationships between leaf age and the food quality of cottonwood foliage for the gypsy moth, *Lymantria dispar. Oecologia* 72:527–532.

Miller, R. O., P. D. Bloese, J. W. Hanover, and R. A. Haack. 1991. Paper birch and European white birch vary in growth and resistance to bronze birch borer. *Journal of the American Society for Horticultural Science* 116:580–584.

Nielsen, D. G., and N. E. Johnson. 1973. Contributions to the life history and dynamics of the pine needle scale, *Phenacaspis pinifoliae*, in central New York. *Annals of the Entomological Society of America* 66:34–43.

Nowak, D. J., and J. F. Dwyer. 2000. Understanding the costs and benefits of urban forest ecosystems. Pages 11–25 *in* J.E. Kuser (editor), *Handbook of Community and Urban Forestry in the Northeast*. Kluwer, New York.

Nowak, D. J., J. E. Pasek, R. A. Sequeira, D. E. Crane, and V. C. Mastro. 2001. Potential effect of *Anoplophora glabripennis* (Coleoptera: Cerambycidae) on urban trees in the United States. *Journal of Economic Entomology* 94:16–22.

Nuckols, M. S., and E. F. Conner. 1995. Do trees in ornamental plantings receive more damage by insects than trees in natural forests? *Ecological Entomology* 20:253–260.

Oke, T. R. 1989. The micrometeorology of the urban forest. *Philosophical Transactions of the Royal Society, London, England. Serial B* 324:335–349.

Owen, J. 1983. Effects of contrived plant diversity and permanent succession on insects in English suburban gardens. Pages 395–422 *in* G.W. Frankie and C.S. Koehler (editors), *Urban Entomology: Interdisciplinary Perspectives*. Praeger, New York.

Poland, T. M., and D. G. McCullough. 2006. Emerald ash borer: invasion of the urban forest and the threat to North America's ash resource. *Journal of Forestry* 104:118–124.

Port, G. R., and J. R. Thompson. 1980. Outbreaks of insect herbivores on plants along motorways in the United Kingdom. *Journal of Applied Ecology* 17:649–656.

Potter, D. 1985. Population regulation of the native holly leaf miner, *Phytomyza ilicicola* Loew (Diptera: Agromyzidae), on American holly. *Oecologia* 66:499–505.

Potter, D. A., and G. M. Timmons. 1981. Factors affecting predisposition of flowering dogwood trees to attack by the dogwood borer. *HortScience* 16:677–679.

Pyle, R. M. 1975. Silk moth of the railroad yards. *Natural History* 84:43–51.

Raupp, M. J., and R. F. Denno. 1983. Leaf age as a predictor of herbivore distribution abundance. Pages 91–121 *in* R. E. Denno and M. S. McClure (editors), *Variable Plants and Herbivores in Natural and Managed Systems*. Academic Press, New York.

Raupp, M. J., J. J. Holmes, C. S. Sadof, P. M. Shrewsbury, and J. A. Davidson. 2001. Effects of cover spray and residual pesticides on scale insects and natural enemies in urban forests. *Journal of Arboriculture* 27:203–213.

Raupp, M. J., C. S. Koehler, and J. A. Davidson. 1992. Advances in implementing integrated pest management for woody landscape plants. *Annual Review of Entomology* 37:561–585.

Raupp, M. J., A. Szczepaniec, and A. Cumming Buckelew. 2008. Prophylactic pesticide applications and low species diversity: do they create pest outbreaks in the urban forest? Pages 59–61 *in Proceedings of the 18th USDA Interagency Research Forum on Gypsy Moth and Other Invasive Species 2007*. US Forest Service, Newton Square, PA.

Raupp, M. J., R. Webb, A. Szczepaneic, D. Booth, and R. Ahern. 2004. Incidence, abundance, and severity of mites on hemlocks following applications of imidacloprid. *Journal of Arboriculture* 30:108–113.

Raupp, M. J., A. Buckelew Cumming, and E. C. Raupp 2006. Street tree diversity in Eastern North America and its potential for tree loss to exotic pests. *Journal of Arboriculture* 32:297–304.

Raupp, M. J., P. M. Shrewsbury, and D. A. Herms. 2010. Ecology of herbivorous arthropods in urban landscapes. *Annual Review of Entomology* 55:19–38.

Rebek, E. J., D. A. Herms, and D. R. Smitley. 2008. Interspecific variation in resistance to emerald ash borer (Coleoptera: Buprestidae) among North American and Asian ash (*Fraxinus* spp.). *Environmental Entomology* 37:242–246.

Reynolds, L.V., M. P. Ayres, T. G. Siccama, and R.T. Holmes. 2007. Climatic effects on caterpillar fluctuations in northern hardwood forests. *Canadian Journal of Forest Research* 73:481–489.

Rhoades, D. F. 1983. Herbivore population dynamics and plant chemistry. Pages 155–220 *in* R.F. Denno and M.S. McClure (editors), *Variable Plants and Herbivores in Natural and Managed Systems*. Academic, New York.

Rickman. N. K., and E. F. Connor. 2003. The effect of urbanization on the quality of remnant habitats for leaf-mining Lepidoptera on *Quercus agrifolia*. *Ecography* 26:777–787.

Rigamonti, I. E., and G. C. Lozzia. 1999. Injurious and beneficial mites on urban trees in Northern Italy. Pages 177–182 *in* M. Lemattre, P. Lamettre, and F. Lamettre (editors), *Proceedings of the International Symposium on Urban Tree Health, Brugge, Belgium*.

Ruszczyk, A. 1996. Spatial patterns of pupal mortality in urban palm caterpillars. *Oecologia* 107:356–363.

Sabelis, M. W. 1981. *Biological Control of Twospotted Spider Mites Using Phytoseiid Predators*. Agriculture Research Reports 910. Pudoc, Wageningen, the Netherlands.

Scharenbroch, B. C., and J. E. Lloyd. 2004. A literature review of nitrogen availability indices for use in urban landscapes. *Journal of Arboriculture* 30:214–230.

Schneider, K., H. Balder, B. Jackel, and B.Pradel. 2000. Bionomics of *Eotatranychus tiliarum* as influenced by key factors. Pages 102–108 *in* G.F. Backhaus, H. Balder, and E. Idczak (editors), *International Symposium on Plant Health in Urban Horticulture*. Parey Buchverlag, Berlin, Germany.

Schwartz, M. D., R. Ahas, and A. Aasa. 2006. Onset of spring starting earlier across the Northern Hemisphere. *Global Change Biology* 12:343–351.

Schwartz, M. D., and J. M. Hanes. 2009. Intercomparing multiple measures of the onset of spring in eastern North America. *International Journal of Climatology* 30:1614–1626.

Sclar, D. C., D. Gerace, and W. S. Cranshaw. 1998. Observations of population increase and injury by spider mites (Acari: Tetranychidae) on ornamental plants treated with imidacloprid. *Journal of Economic Entomology* 91:250–255.

Shochat, E., W. L. Stefanov, M. E. A. Whitehouse, and S. J. Faeth. 2004. Urbanization and spider diversity: influences of human modification of habitat structure and productivity. *Ecological Applications* 14:268–280.

Shochat, E., P.S. Warren, S.H. Faeth, N.E. McIntyre, and D. Hope. 2006. From patterns to emerging processes in mechanistic urban ecology. *Trends in Ecology and Evolution* 4:186–191.

Shrewsbury, P. M., J. H. Lashomb, G. C. Hamilton, J. Zhang, J. M. Patt, and R. A. Casagrande. 2004. The influence of flowering plants on herbivore and natural enemy abundance in ornamental landscapes. *International Journal of Ecology and Environmental Sciences* 30:23–33.

Shrewsbury, P. M., and M. J. Raupp. 2000. Evaluation of components of vegetational texture for predicting azalea lace bug, *Stephanitis pyrioides* (Heteroptera: Tingidae), abundance in managed landscapes. *Environmental Entomology* 29:919–1026.

Shrewsbury, P. M., and M. J. Raupp. 2006. Do top-down or bottom-up forces determine *Stephanitis pyrioides* abundance in urban landscapes? *Ecological Applications* 16:262–272.

Speight, M. R., R. S. Hails, M. Gilbert, and A. Foggo. 1998. Horse chestnut scale (*Pulvinaria regalis*) (Homoptera: Coccidae) and urban host tree environment. *Ecology* 79:1503–1513.

Sperry, C. E., W. R. Chaney, G. Shao, and C. S. Sadof. 2001. Effects of tree density, tree species diversity, and percentage of hardscape on three insect pests of honeylocust. *Journal of Arboriculture* 27:263–271.

Stoetzal, M. B., and J. A. Davidson. 1971. Biology of the obscure scale, *Melanaspis obscura* (Homoptera: Diaspididae) on pin oak in Maryland. *Annals of the Entomological Society of America* 64:45–50.

Strathdee, A. T., J. S. Bale, W. C. Block, S. J. Coulson, I. D. Hodkinson, and N. R. Webb. 1993. Effects of temperature elevation on a field population of *Acyrthosiphon svalbardicum* (Hemiptera: Aphididae) on Spitsbergen. *Oecologia* 96:457–465.

Szczepaniec, A., S.F. Creary, K.L. Laskowski, J. P. Nyrop, and M. J. Raupp. 2011. Neonicotinoid insecticide imidacloprid causes outbreaks of spider mites on elm trees in urban landscapes. *PLoS One* 6:e20018. doi:10.1371/journal.pone.0020018

Taha, H. 1997. Urban climates and heat islands: albedo, evapotranspiration, and anthropogenic heat. *Energy and Buildings* 25:99–103.

Tallamy, D. W. 2004. Do alien plants reduce insect biomass? *Conservation Biology* 18:1689–1692.

Taylor, G. E. Jr., D. W. Johnson, and C. P. Andersen. 1994. Air pollution and forest ecosystems: a regional to global perspective. *Ecological Applications* 4:662–689.

Thomas, C. D. 1989. Predator–herbivore interaction and the escape of isolated plants from phytophagous insects. *Oikos* 55:291–298.

Throop H. L., and M. T. Lerdau. 2004. Effects of nitrogen deposition on insect herbivory: implications for community and ecosystem processes. *Ecosystems* 7:109–133.

Tooker, J. F., and L. M. Hanks. 2000. Influence of plant community structure on natural enemies of pine needle scale (Homoptera: Diaspididae) in urban landscapes. *Environmental Entomology* 29:1305–1311.

Virtanen, T., and S. Neuvonen. 1999. Performance of moth larvae on birch in relations to altitude, climate, host quality and parasitoids. *Oecologia* 120:92–101.

Watson, J. K., P. L. Lambdin, and K. Langdon. 1994. Diversity of scale insects (Homoptera: Coccoidea) in the Great Smoky Mountains National Park. *Annals of the Entomological Society of America* 87:225–230.

Watt, A. D., and A. M. McFarlane. 2002. Will climate change have a different impact on different trophic levels? Phenological development of winter moth *Operophtera brumata* and its host plants. *Ecological Entomology* 27:254–256.

White, T. C. R. 1984. The abundance of invertebrate herbivores in relation to the availability of nitrogen in stressed food plants. *Oecologia* 63:90–105.

Witter, J. A., and I. R. Ragenovich. 1986. Regeneration of Fraser fir at Mt. Mitchell, North Carolina, after depredations by the balsam woolly adelgid. *Forest Science* 32:585–594.

16

Resistance to Transgenic Crops and Pest Outbreaks

Bruce E. Tabashnik and Yves Carrière

16.1 Introduction

Transgenic crops represent one of the most controversial and rapidly adopted technologies in the history of agriculture (Tabashnik 2010). First grown commercially in 1996, transgenic crops covered 134 million hectares (ha) in 25 countries during 2009, with about half of the world's total grown in the United States (James 2009). To reduce reliance on insecticide sprays, scientists genetically engineered corn and cotton plants to make insecticidal proteins encoded by genes from the common bacterium *Bacillus thuringiensis* (Bt). These Bt proteins kill some devastating insect pests, but cause little or no harm to most other organisms including people (Mendelsohn *et al*. 2003, National Research Council 2010). Evolution of resistance to insecticides in more than 400 species of insects and the season-long presence of Bt toxins in transgenic crops raised concerns that rapid adaptation by pests could also reduce the efficacy of Bt crops and lead to pest outbreaks (Tabashnik 1994, Gould 1998, Onstad 2008). Although Bt crops remain effective against most pest populations (Wu *et al*. 2008, Hutchison *et al*. 2010, Tabashnik *et al*. 2010), strong evidence of field-evolved resistance to one or more Bt toxins in transgenic crops has been reported in six cases involving five species of pests (Ali *et al*. 2006, Ali and Luttrell 2007,

Insect Outbreaks Revisited, First Edition. Edited by Pedro Barbosa, Deborah K. Letourneau and Anurag A. Agrawal.
© 2012 Blackwell Publishing Ltd. Published 2012 by Blackwell Publishing Ltd.

Van Rensburg 2007, Tabashnik *et al.* 2008a, Bagla 2010, Downes *et al.* 2010a, Storer *et al.* 2010). Previous reviews on resistance to Bt crops have critically analyzed the evidence of field-evolved resistance, outlined application of evolutionary and ecological theory to understand and manage resistance, and tested the correspondence between the field evidence and theory (Tabashnik *et al.* 2008a, 2009a, Gassmann *et al.* 2009, Carrière *et al.* 2010, Tabashnik and Carrière 2010). Here we focus on the cases where strong evidence of field-evolved resistance has been reported and ask: has field-evolved pest resistance to Bt crops led to pest outbreaks? We begin with definitions of field-evolved resistance and outbreaks, then consider the evidence about resistance and outbreaks, and conclude by summarizing what we know and looking to the future.

16.2   Definitions: field-evolved resistance and outbreaks

Field-evolved (or field-selected) resistance is defined as a genetically based decrease in susceptibility of a population to a toxin caused by exposure of the population to the toxin in the field (National Research Council 1986, Tabashnik 1994, Tabashnik *et al.* 2009a). Insect populations often have natural genetic variation affecting response to a toxin, with some alleles conferring susceptibility and others conferring resistance. Alleles conferring resistance are typically rare in insect populations before the populations are exposed to a Bt toxin, with empirical estimates often close to one in a thousand (Tabashnik 1994, Gould *et al.* 1997, Burd *et al.* 2003, Tabashnik *et al.* 2008a, Downes *et al.* 2009, Huang *et al.* 2009, Gassmann *et al.* 2009, Carrière *et al.* 2010). Field-evolved resistance occurs when exposure of a field population to a toxin increases the frequency of alleles conferring resistance in subsequent generations. Hence, inherently low susceptibility of a species to a toxin does not signify field-evolved resistance. Likewise, merely detecting resistance-conferring alleles, without demonstrating that their frequency has increased, does not constitute evidence of field-evolved resistance. Methods for monitoring resistance are detailed in Tabashnik *et al.* (2009a).

We define "outbreak" as a marked increase in the population density of an insect pest (Capinera 2004). As with field-evolved resistance (Tabashnik *et al.* 2009a), both the magnitude and geographic area of outbreaks can vary widely. Pest outbreaks usually entail increases in pest population density that could cause economic losses and may trigger intervention with control tactics such as insecticide sprays to reduce damage. Evolution of resistance to Bt crops does not necessarily cause pest outbreaks. The relationship between resistance and outbreaks depends on many factors, including the frequency of resistance alleles and the extent to which pest population density is controlled by factors other than the Bt crop (Tabashnik *et al.* 2009a). Field-evolved resistance to a Bt crop is most likely to cause outbreaks of a pest when the Bt crop was the primary method of control for that pest before resistance occurred. Multitactic approaches used in integrated pest management can not only delay pest resistance (Tabashnik *et al.* 2010) but also reduce the likelihood and severity of outbreaks when resistance does occur.

16.3 Evidence: has resistance to Bt crops caused pest outbreaks?

Table 16.1 summarizes key information about the first six well-documented cases of field-evolved resistance to Bt toxins in transgenic crops, which involve five species of lepidopteran pests of corn and cotton on four continents. In five of these six cases, pest outbreaks are associated with pest resistance to Bt crops. As detailed below, quantitative data demonstrating field-evolved resistance are available in all six cases, but the evidence for outbreaks is largely anecdotal.

16.3.1 Pest resistance to Bt corn and outbreaks

Strong, undisputed evidence of field-evolved resistance to the Bt toxins in transgenic corn has been reported for some populations of two targeted noctuid moths: *Busseola fusca* and *Spodoptera frugiperda* (Table 16.1). Resistance of these two pests to Bt corn is linked with pest outbreaks, as described in this chapter.

Resistance of Spodoptera frugipdera *to Bt corn in Puerto Rico*

Field-evolved resistance of *S. frugiperda* (fall armyworm) to Bt corn producing Cry1F occurred in four years in the US territory of Puerto Rico (Storer *et al.* 2010), making this one of the fastest cases of field-evolved resistance to a Bt crop. This is also the first case of resistance leading to withdrawal of a Bt crop from the marketplace (Tabashnik *et al.* 2009a). In this case, anecdotal evidence of the pest outbreak preceded the documentation of resistance. The US Environmental Protection Agency (USEPA) required proactive resistance monitoring for three pests targeted by Cry1F corn, but not for *S. frugiperda*, which it considers a secondary pest of corn (USEPA 2005, 2010a). The EPA has stated that "resistance to Bt corn is not likely" for *S. frugiperda* in the continental United States, because it overwinters only in the South (south Texas, south Florida, and the Caribbean) and migrates to the North in the winter, and because "corn is not necessarily a primary host" for this polyphagous insect (USEPA 2010a). The EPA notes, however, that "a resistance monitoring plan could be warranted" if >400 ha of Bt corn (particularly Cry1F corn targeting *S. frugiperda*) were planted in the areas where this pest overwinters (USEPA 2010a).

In Puerto Rico, where *S. frugiperda* is the most important corn pest, Cry1F corn was first commercially available in 2003 (Storer *et al.* 2010). In 2006, populations of *S. frugiperda* were "unusually large" and damage to Cry1F corn was unusually high on some commercial farms and seed company field stations (Storer *et al.* 2010). To determine if resistance to Cry1F was contributing to these unusual events, insects were sampled during 2007 and 2008 from four farms in Puerto Rico and used as founders of four lab-reared strains, each derived from one farm. In lab bioassays, larvae from these four strains from Puerto Rico and larvae from four strains from the continental United States were exposed to diet treated with Cry1F. For the Puerto Rico strains, no substantial mortality or growth inhibition occurred. In contrast, the four strains from the continental United States were much more susceptible to Cry1F (Storer *et al.* 2010). The

Table 16.1 Field-evolved resistance to Bt crops and outbreaks by insect pests.

Species	Country[a]	Crop	Bt toxin	Year Comm.[b]	Year Detected[c]	Outbreak	Reference(s)
Busseola fusca	South Africa	corn	Cry1Ab	1998	2006	Yes	Van Rensburg 2007, Kruger et al. 2009
Spodoptera frugiperda	USA	corn	Cry1F	2003	2007[d]	Yes	Storer et al. 2010
Helicoverpa zea	USA	cotton	Cry1Ac	1996	2002	Yes	Ali et al. 2006, Tabashnik et al. 2008a
Helicoverpa zea	USA	cotton	Cry2Ab	2003	2005	Yes	Ali and Luttrell 2007, Tabashnik et al. 2009, Tabashnik and Carrière 2010
Pectinophora gossypiella	India	cotton	Cry1Ac	2002[e]	2008	Yes	Dhurua and Gujar 2011, Monsanto 2010
Helicoverpa punctigera	Australia	cotton	Cry2Ab	2004–05	2008–09	No	Downes et al. 2010a

[a] Only some populations were resistant in the countries listed.
[b] First year Bt crop was grown commercially in the location where resistance was detected.
[c] First year resistant insects were sampled from the field.
[d] Unusually large populations were seen in 2006; resistance was detected in insects collected in 2007.
[e] First year of legal planting; illegal planting was detected in 2001 and probably also occurred sooner.

virtually complete lack of susceptibility to Cry1F of the four resistant strains from Puerto Rico makes it difficult to quantify the magnitude of resistance. However, the highest concentration of Cry1F tested (10 000 ng of Cry1F per cm^2 of diet) – which had no effect on resistant strains – was more than 700 times greater than the concentration that caused 50% growth inhibition in a strain derived from Georgia in 2007 (Storer *et al.* 2010).

Although the data summarized above are consistent with field-evolved resistance, an alternative hypothesis is that Puerto Rico populations of *S. frugiperda* were not susceptible to Cry1F before Bt corn was introduced there. Storer *et al.* (2010) present field efficacy data refuting this hypothesis. A field trial in 1999, before Bt corn was commercialized, showed that 99% of Cry1F corn plants had little or no damage from *S. frugiperda*, while all non-Bt corn plants suffered extensive damage. In 2009, a similar field trial showed that the *S. frugiperda* damage to both Bt and non-Bt corn was extensive, with both types of corn receiving identical mean damage ratings (7.0 on a scale where 0 is no damage and 9 is total destruction of corn whorls). The 2009 data indicate that the Cry1F toxin in Bt corn provided no protection against *S. frugiperda*. These results imply that even two years after commercial sales of Cry1F corn had stopped, close to 100% of the pest population was still resistant to Cry1F.

The data from lab bioassays and field trials provide powerful evidence of field-evolved resistance of *S. frugiperda* to Cry1F corn in Puerto Rico. Although this resistance certainly contributed to pest outbreaks on Bt corn, the contributions of other factors cannot be ruled out. In particular, Storer *et al.* (2010) assert that a severe drought in Puerto Rico from October 2006 to April 2007 reduced the quality of alternative host plants, concentrating *S. frugiperda* populations on irrigated corn plants. Because the evolution of resistance coincided with the drought, the relative importance of these two factors in causing the pest outbreak cannot be teased apart readily. However, if a similar drought had occurred while the Bt corn remained effective and no outbreak occurred, this would demonstrate that drought alone did not cause the outbreak.

Resistance of Busseola fusca *to Bt corn in South Africa*

Experiments initiated in 2006 provided the first strong evidence of field-evolved resistance of *B. fusca* (maize stem borer) in South Africa to Bt corn producing Cry1Ab (Van Rensburg 2007, Tabashnik *et al.* 2009a). Resistance of *B. fusca* to Bt corn producing Cry1Ab has some striking parallels with *S. frugiperda* resistance to Bt corn producing Cry1F. With *B. fusca*, as with *S. frugiperda*, anecdotal evidence of the pest outbreak preceded documentation of resistance. Monitoring data show that field-evolved resistance to Bt corn producing Cry1Ab occurred in *B. fusca* in eight years or less (Van Rensburg 2007). The area of Cry1Ab corn planted in South Africa increased from <3% of corn (50 000 ha) in 1998, the first year of commercialization, to 35% (943 000 ha) in 2006, with higher percentages in some areas (James 2007). During the 2004–2005 season, severe damage to Bt corn by *B. fusca* occurred at some locations (Van Rensburg 2007). All of the damaged fields were irrigated and had a history of continuous Bt corn production. The damage was seen on various Bt corn hybrids from different seed companies and ELISA tests confirmed the presence of Bt toxin in the plants (Van Rensburg 2007).

In 2006, Van Rensburg (2007) collected diapausing *B. fusca* larvae from stubble in a damaged Cry1Ab corn field (R strain) near Christiana, in the North West province of South Africa, and in a non-Bt corn field (S strain) near Ventersdorp, 250 km to the northeast in the same province, where Bt corn had not been adopted. The larvae from both fields were reared to adults in the laboratory and their neonate progeny were used to infest corn plants in the field. Survival after 18 days was significantly higher for the R strain than the S strain on Bt corn, but did not differ between strains on non-Bt corn (Tabashnik *et al.* 2009a). Survival on Bt corn relative to non-Bt corn was 43–64% for the R strain versus 0% for the S strain (Tabashnik *et al.* 2009a). These data provide strong evidence that field-evolved resistance increased larval survival on Cry1Ab corn in the field.

Sixty km southwest from the site of the first resistant population, in the Vaalharts area of the Northern Cape province of South Africa, researchers found a second resistant population of *B. fusca* where farmers had reported increased damage to Bt corn (Kruger *et al.* 2009). The percentage of Vaalharts farmers reporting medium or severe damage to Bt corn from stem borers rose from 2.5% in the 2005–2006 growing season to 59% in the 2007–2008 growing season. In contrast to previous years when insecticides were not used for stem borer control on Bt corn, 55% of farmers applied insecticides to both Bt corn and non-Bt corn for stem borer control during the 2007–2008 growing season. Furthermore, Monsanto, a biotechnology company that developed some types of Bt corn grown in South Africa, covered spaying costs when stem borer density exceeded the economic threshold on early-planted Bt corn (Kruger *et al.* 2009).

As with the case in Puerto Rico, resistance to Bt corn in South Africa contributed to outbreaks of *B. fusca*, but the role of other factors cannot be excluded. In South Africa, agronomic practices including irrigation, staggered planting of corn, and cultivation methods that allow overwintering survival of larvae may have also contributed to outbreaks of *B. fusca* (Kruger *et al.* 2009).

16.3.2 Pest resistance to Bt cotton and outbreaks

In contrast to the undisputed evidence of resistance to Bt corn and associated pest outbreaks described above, insect resistance to Bt cotton has generated more controversy (Tabashnik *et al.* 2008b, Tabashnik and Carrière 2010). Other reviews have summarized cases where the evidence of resistance to Bt cotton is ambiguous, particularly *Helicoverpa armigera* resistance to Cry1Ac produced by Bt cotton in China and India (Tabashnik *et al.* 2008a, 2009a). We focus here on the cases where strong evidence of an increased frequency of resistance to Bt toxins in transgenic cotton has been reported: *Helicoverpa zea* in the United States, *Pectinophora gossypiella* in India, and *Helicoverpa punctigera* in Australia.

Resistance and cross-resistance of Helicoverpa zea *to Bt cotton in the United States*

Researchers saw the initial evidence of *Helicoverpa zea* (bollworm) resistance to Bt cotton producing Cry1Ac (Bollgard) in the southeastern United States in 2002, six years after this type of Bt cotton was commercialized (Luttrell *et al.* 2004, Ali *et al.* 2006, Tabashnik *et al.* 2008a, Tabashnik and Carrière 2010). This was the first documented case of field-evolved resistance to a Bt crop, and it generated considerable controversy (Tabashnik *et al.* 2008b, Tabashnik and Carrière 2010). Monsanto's US registration for Bollgard cotton expired in September 2009, removing this product from the market in the United States (USEPA 2010b). In the United States, Bollgard cotton was replaced primarily by Bt cotton that produces two Bt toxins, Cry1Ac and Cry2Ab (Bollgard II).

Before Bt cotton was commercialized, Luttrell *et al.* (1999) conducted field sampling and bioassays to establish the baseline susceptibility of *H. zea* field populations to Cry1Ac. After Bt cotton was commercialized, they con- ducted yearly resistance monitoring to determine if susceptibility to Cry1Ac had decreased and to check if such resistance was responsible for "occasional field reports of damaging bollworm populations on Bollgard cotton" (Luttrell and Ali 2007). The extensive monitoring data of Luttrell and his colleagues pro- vide strong evidence of field-evolved resistance of *H. zea* to Bt cotton producing Cry1Ac (Luttrell *et al.* 2004, Ali *et al.* 2006, Allen and Luttrell 2008, Tabashnik *et al.* 2008a, b, 2009a, Tabashnik and Carrière 2010). From 2002 to 2006, they found 14 field populations of *H. zea* in the southeastern United States with more than 100-fold increases in the concentration of Cry1Ac needed to kill 50% of larvae (LC_{50}) in lab bioassays (Ali *et al.* 2006, Luttrell *et al.* 2007).

Although widespread outbreaks were not reported, sporadic outbreaks of *H. zea* associated with its resistance to Cry1Ac are well documented (Tabashnik and Carrière 2010 and references therein). For example in 2002, Luttrell *et al.* (2004) sampled *H. zea* larvae surviving on Cry1Ac cotton plants from a "prob- lem field with unacceptable levels of boll damage" in Arkansas and from a second "problem field" in Mississippi. Lab bioassays with Cry1Ac incorporated in diet demonstrated that the strains derived from these two fields with unusually high boll damage were the most resistant strains found in 2002 (Ali *et al.* 2006). Relative to the LC_{50} value of larvae from a susceptible strain, the LC_{50} values of the strains from the problem fields in Arkansas and Mississippi were 40 and 22 times higher, respectively (Ali *et al.* 2006). Furthermore, larval survival on Bt cotton leaves relative to non-Bt cotton leaves was greater for both of these strains than for a susceptible strain (Luttrell *et al.* 2004). This pattern was confirmed in 2006 using similar experiments with two additional field-derived resistant strains (Luttrell and Ali 2007, Tabashnik *et al.* 2008b).

Tabashnik *et al.* (2008a) suggested that despite documented resistance to Cry1Ac of *H. zea*, widespread outbreaks of this pest did not occur for several reasons. First, most populations tested were not resistant. Second, many growers sprayed insecticides on Bollgard cotton fields from the outset because the con- centration of Cry1Ac produced by this transgenic cotton was not sufficient to control high-density populations of *H. zea* (Jackson *et al.* 2004a). Third, in greenhouse experiments, Bollgard caused 48–60% mortality of larvae from strains of *H. zea* with moderate levels of resistance (Jackson *et al.* 2004b). These results suggest that even for field populations of *H. zea* with moderate levels of resistance, Bollgard may have contributed to control. Finally, Bt cotton produc- ing Cry1Ac and Cry2Ab was registered in the United States in December 2002 (USEPA 2010b) and quickly began replacing Bt cotton that produced only Cry1Ac

(Tabashnik *et al.* 2008a). As noted above, the US registration for Bt cotton producing only Cry1Ac expired in 2009 (USEPA 2010b).

As with Cry1Ac, lab bioassays with Cry2Ab in diet documented field-evolved resistance to this Bt toxin in *H. zea* field populations from the southeastern United States (Ali and Luttrell 2007, Tabashnik *et al.* 2009a, Tabashnik and Carrière 2010). Annual resistance monitoring surveys show that the percentage of *H. zea* populations resistant to Cry2Ab rose from 0% in 2002 to 50% in 2005 (Ali and Luttrell 2007, Tabashnik *et al.* 2009a). In this analysis, a population was deemed resistant if its LC_{50} exceeded the diagnostic concentration of toxin (150 µg Cry2Ab per ml diet) and was at least 10 times higher than the LC_{50} for a standard susceptible strain tested in the same year (Tabashnik *et al.* 2009a). Although the combination of Cry1Ac and Cry2Ab was chosen for Bt cotton plants because cross-resistance between these two toxins was considered unlikely (Hernández-Rodriguez *et al.* 2008), Ali and Luttrell (2007) found that the LC_{50} values for these two toxins were positively correlated across 61 strains of *H. zea* derived from the field during 2002 to 2006 ($r = 0.32$, df $= 59$, $P = 0.01$) (Ali and Luttrell 2007). Moreover, Jackson *et al.* (2006) reported a positive genetic correlation between larval growth on diet treated with Cry1Ac and Cry2Ab in field populations of *H. zea* sampled during 2001 and 2002, before cotton with both toxins was registered in December 2002. These results suggest that selection with Cry1Ac probably caused some cross-resistance to Cry2Ab (Tabashnik *et al.* 2009a, b).

Similar to data cited above revealing that bollworm resistance to Cry1Ac increased survival on leaves of Bt cotton producing Cry1Ac, survival on Bt cotton leaves producing Cry1Ac and Cry2Ab was positively associated with the LC_{50} for Cry1Ac (Luttrell and Ali 2007, Tabashnik *et al.* 2008b). Furthermore, for strains started from larvae surviving on plants in the field during 2002 to 2005, the mean LC_{50} was more than double for strains from Bollgard II cotton relative to strains from non-Bt crops ($t = 2.65$, df $= 32$, $P = 0.01$) (Ali and Luttrell 2007).

In addition, field data from seven sites in North Carolina during 2000–2002 indicate that the mean density of bollworm adults was 1697 ± 682 per hectare on Bollgard II cotton, which was 3.6% relative to non-Bt cotton (Jackson *et al.* 2004b). These results likely reflect survival of susceptible larvae on the two-toxin cotton, because no genes conferring substantial resistance to Cry1Ac or Cry2Ab were detected in screening of 1252 families of bollworm from North Carolina in 2001 and 2002 (Jackson *et al.* 2006). Given that some susceptible individuals completed development in the field on cotton with both toxins, and that resistance to Cry1Ac in diet tests was associated with increased survival on cotton leaves containing Cry1Ac and Cry2Ab (Luttrell and Ali 2007), we expect that the observed increased survival on diet treated with diagnostic concentrations of Cry1Ac and Cry2Ab is linked with increased survival in the field on cotton plants producing these toxins. We do not know of published data that directly test this hypothesis.

Based on the number of male moths caught in pheromone traps, Adamczyk and Hubbard (2006) reported declines in *H. zea* populations in Mississippi from 2001 to 2005. They proposed that several factors other than Bt cotton caused this trend. However, during 2010 in Mississippi, dramatic increases in the abundance of *H. zea* surviving on Bollgard II cotton plants led to additional insecticide sprays by growers (Jeff Gore, personal communication). Bioassays will be needed to determine if these outbreaks in Mississippi during 2010 are associated with resistance. Other potential causes include changes in cropping patterns and weather.

Resistance of Pectinophora gossypiella *to Bt cotton in India*

Resistance of *Pectinophora gossypiella* (pink bollworm) to Bt cotton producing Cry1Ac was first detected in the offspring of samples collected from the field in 2008 in the state of Gujarat in western India (Dhurua and Gujar 2011). Bt cotton hybrids produced by crossing a Bt cotton cultivar with local non-Bt cotton cultivars were commercialized in India in 2002, but illegal planting of Bt cotton hybrids had started at least as early as 2001 in Gujarat (Lalitha *et al.* 2009, Showalter *et al.* 2009). Dhurua and Gujar (2011) tested the progeny of field-collected insects using lab bioassays with Cry1Ac incorporated in artificial diet. Their results show that a diagnostic concentration of Cry1Ac (1 μg toxin per ml diet) killed 96–100% of larvae from four sites in India, but only 24–31% of the larvae from the resistant population from the Amreli district of Gujarat. The LC_{50} of Cry1Ac was more than 40 times higher for the resistant Amreli population than for the most susceptible field population.

In March 2010, Monsanto (2010) reported that lab testing of samples collected during 2009 confirmed pink bollworm resistance to Cry1Ac in four districts of Gujarat: Amreli and the adjacent districts of Bhavnagar, Junagarh, and Rajkot. Monsanto (2010) also detected unusually high abundance of pink bollworm larvae on Cry1Ac cotton plants in the field during 2009 in Gujarat. According to the May 2010 report from India's Genetic Engineering Approval Committee (2010), Dr. K. R. Kranthi, director of India's Central Institute for Cotton Research (CICR), indicated that the number of pink bollworm moths captured in CICR pheromone traps was about 10 times higher in the region where resistance was detected compared with the 12 other areas monitored in India.

Kranthi reportedly asserted, however, that weather favoring pest survival in Gujarat, and not resistance, caused the unusually high survival of pink bollworm in the field (GEAC 2010). Although weather could have contributed to the outbreak in the field, weather can be excluded as a factor affecting the lab bioassay data demonstrating resistance to Cry1Ac (Dhurua and Gujar 2011, Monsanto, 2010). Kranthi has also challenged the bioassay data on the grounds that resistance monitoring should be based only on insects collected from non-Bt cotton, whereas Monsanto collected larvae from Bt cotton plants (GEAC 2010). Monsanto has appropriately countered this criticism by stating that their resistance monitoring based on insects collected from Bt cotton in India is "standard practice" (Bagla 2010). Indeed, sampling and testing survivors from Bt plants are essential components of resistance monitoring (Tabashnik *et al.* 2008a, b, 2009a). Ironically, some of the specious arguments disputing Monsanto's report of pink bollworm resistance to Cry1Ac-producing Bt cotton in India mirror those offered by Monsanto and others attempting to challenge documentation of *H. zea* resistance to Cry1Ac-producing Bt cotton in the United States (Tabashnik and Carrière 2010).

In sum, the laboratory data of Dhurua and Gujar (2011) and Monsanto (2010) provide strong evidence of pink bollworm resistance to Cry1Ac in Gujarat, India. The reports of unusually high abundance of pink bollworm in Gujarat in 2009, based on both larval and adult abundance (Monsanto 2010, GEAC 2010), imply that an outbreak was associated with resistance to Cry1Ac cotton in Gujarat.

Resistance of Helicoverpa punctigera *to Bt cotton in Australia*

As in the United States, Bt cotton producing only Cry1Ac was first grown commercially in Australia in 1996, followed by Bollgard II cotton that produces Cry1Ac and Cry2Ab (Downes *et al*. 2010b). Since the 2004–2005 growing season, Bollgard II cotton has constituted at least 80% of Australia's cotton (Downes *et al*. 2010a). In contrast to the situation in the United States, where overlap between Bollgard II and Bt cotton producing only Cry1Ac occurred for seven years, Bollgard II cotton completely replaced Cry1Ac cotton in the 2004–2005 growing season in Australia (Downes *et al*. 2010a, b). Also unlike the United States, where *H. zea* evolved resistance to Cry1Ac (as discussed in this chapter), the frequency of alleles conferring resistance to Cry1Ac has not increased in the two main target pests of Bt cotton in Australia, *Helicoverpa armigera* and *H. punctigera* (Mahon *et al*. 2007, Downes *et al*. 2009). However, based on F2 screen data for *H. punctigera* in Australia, the frequency of alleles conferring resistance to Cry2Ab increased sevenfold from 0.0018 (2002–2003 to 2006–2007 pooled) to 0.012 in 2008–2009 (Downes *et al*. 2010a). Moreover, based on F1 screen data for *H. punctigera*, the Cry2Ab resistance allele frequency tripled from 2007–2008 to 2008–2009 (Downes *et al*. 2010b). Also, during 2008–2009, the Cry2Ab resistance allele frequency was eight times higher in areas where Bt cotton was grown than in non-agricultural areas (Downes *et al*. 2010b). Although the Cry2Ab resistance alleles detected in *H. punctigera* do not confer resistance to the Cry1Ac toxin in Bollgard II, the concentration of Cry1Ac in Bollgard II typically declines at the end of the growing season, which may increase survival on Bollgard II of larvae that are resistant only to Cry2Ab (Mahon and Olsen 2009, Downes *et al*. 2010a).

Downes *et al*. (2010a) refer to their results as evidence of "incipient resistance" of *H. punctigera* to Cry2Ab and conclude that the increased frequency of resistance alleles has not caused pest outbreaks. They point out that the highest estimated Cry2Ab resistance allele frequency is 0.048, based on F1 screen data from 2008–2009. Noting that the Cry2Ab resistance alleles they found are recessive, and assuming Hardy–Weinberg equilibrium, they estimate the frequency of individuals resistant to Cry2Ab was only 0.002 (0.048^2) in 2008–2009. This frequency of resistant individuals would have little effect on population dynamics.

16.4　Conclusion

In five of the six cases reviewed here, pest outbreaks were associated with strong evidence of field-evolved resistance to Bt toxins produced by transgenic crops (Table 16.1). In three of these five cases, anecdotal evidence of increases in pest population density preceded the evidence of resistance. With both cases of resistance to Bt corn, *S. frugiperda* in Puerto Rico and *B. fusca* in South Africa, anecdotal observations of pest outbreaks spurred the bioassay experiments that documented resistance. Although researchers conducted routine monitoring for resistance of *H. zea* to Bt toxin Cry1Ac, problems controlling some populations of this pest in fields of Bt cotton producing Cry1Ac were observed before researchers documented resistance of these populations to Cry1Ac with bioassays. In the remaining two cases where outbreaks were linked with strong evidence of resistance to Bt crops, *H. zea* resistance to Cry2Ab in the United States and *P. gossypiella* resistance to Cry1Ac in India, bioassays detected resistance before increases in pest population density were reported.

The exceptional case, where evidence of resistance was found without any evidence of increased pest population density, is the "incipient" resistance of *H. punctigera* to Cry2Ab in Australia. Here the data show an increase in the recessive resistance allele frequency to at most 0.048, which implies a frequency of resistant individuals of about 0.002. At this frequency of resistant individuals, no detectable effect on pest population dynamics is expected. Options for responding to the early warning of increased Cry2Ab resistance allele frequency for *H. punctigera* include no action, limiting the area planted to Bollgard II, requiring more area planted to refuges of non-Bt host plants, increasing insecticide sprays, and deploying cotton with a novel toxin that kills larvae resistant to Cry2Ab (Downes *et al*. 2010a). It will be intriguing to see how the Australian cotton industry and the insects respond.

We think that transgenic crops producing insecticidal toxins from Bt and other sources will be increasingly important for the future of agriculture. For example, while this review focuses on lepidopteran pests, Bt corn effective against coleopteran pests such as *Diabrotica* spp. (corn rootworms) was first registered in the United States in 2003 (USEPA 2010b) and is now planted widely. Rapid evolution of resistance by *Diabrotica vergifera vergifera* to transgenic corn producing the coleopteran-active Bt toxin Cry3Bb would not be surprising, based on the relevant theory and empirical data (Meihls *et al*. 2008, Tabashnik 2008). More widespread use of Bt crops, a wider array of toxins, a broader range of targeted pests, and more years of pest exposure to Bt crops will provide a larger data set for testing ideas about the effects of Bt crops on pest population dynamics.

In the five cases reviewed here where pest outbreaks were associated with resistance to Bt crops, resistance probably contributed to the outbreaks. Nonetheless, we cannot exclude the effects of other factors, particularly changes in weather and cropping patterns. In contrast to these cases of resistance to Bt crops, we note that in several situations, sustained efficacy of Bt crops against some pests is associated with long-term, large-scale declines in the pests' population density (Carrière *et al*. 2003, 2010, Wu *et al*. 2008, Hutchison *et al*. 2010, Tabashnik *et al*. 2010). Whereas extensive data document these pest population declines, evidence of pest outbreaks associated with resistance to Bt crops is largely anecdotal. Although it will remain difficult to prove a causal connection between pest resistance to Bt crops and pest outbreaks, we encourage researchers to carefully track pest population density, spatial and temporal patterns of Bt crop deployment, and pest susceptibility to Bt toxins. By analyzing such data along with information on weather and cropping patterns, scientists can more rigorously test hypotheses about the effects of Bt crops on pest population dynamics.

Acknowledgments

We thank Mark Sisterson for comments. Portions of this chapter are reprinted from Tabashnik *et al.* (2009a) with permission from the Entomological Society of America. This work was supported by USDA-Agriculture and Food Research Initiative Grant 2008-35302-0390.

References

Adamczyk, J. J. Jr., and D. Hubbard. 2006. Changes in populations of *Heliothis virescens* (F.) (Lepidoptera: Noctuidae) and *Helicoverpa zea* (Boddie) (Lepidoptera: Noctuidae) in the Mississippi Delta from 1986 to 2005 as indicated by adult male pheromone traps. *Journal of Cotton Science* 10:155–160.

Allen, K. C., and R. G. Luttrell. 2008. Influence of surrounding landscape on response of *Helicoverpa zea* (Boddie) to Cry1Ac in diet-incorporated assays. *Southwestern Entomology* 33:265–279.

Ali, M. I., and R. G. Luttrell. 2007. Susceptibility of bollworm and tobacco budworm (Noctuidae) to Cry2Ab2 insecticidal protein. *Journal of Economic Entomology* 100:921–931.

Ali, M. I., R. G. Luttrell, and S. Y. Young. 2006. Susceptibilities of *Helicoverpa zea* and *Heliothis virescens* (Lepidoptera: Noctuidae) populations to Cry1Ac insecticidal protein. *Journal of Economic Entomology* 99:164–175.

Bagla, P. 2010. Hardy cotton-munching pests are latest blow to GM crops. *Science* 327:1429.

Burd, A. D., F. Gould, J. R. Bradley, J. W. Van Duyn, and W. J. Moar. 2003. Estimated frequency of non-recessive Bt resistance genes in bollworm, *Helicoverpa zea* (Boddie) (Lepidoptera: Noctuidae) in eastern North Carolina. *Journal of Economic Entomology* 96:137–142.

Capinera, J. L. 2004. Outbreak. In J. L. Capinera (ed.), *Encyclopedia of Entomology*, vol. 2, 1611. Kluwer Academic Publishers, Dordrecht, Netherlands.

Carrière, Y., C. Ellers-Kirk, M. Sisterson, L. Antilla, M. Whitlow, T. J. Dennehy, and B. E. Tabashnik. 2003. Long-term regional suppression of pink bollworm by *Bacillus thuringiensis* cotton. *Proceedings of the National Academy Sciences USA* 100:1519–1523.

Carrière, Y., D. W. Crowder, and B. E. Tabashnik. 2010. Evolutionary ecology of insect adaptation to Bt crops. *Evolutionary Applications* 3:561–573.

Downes, S., T. L. Parker, and R. J. Mahon. 2009. Frequency of alleles conferring resistance to the *Bacillus thuringiensis* toxins Cry1Ac and Cry2Ab in Australian populations of *Helicoverpa punctigera* (Lepidoptera: Noctuidae) from 2002 to 2006. *Journal of Economic Entomology* 102:733–742.

Downes, S., T. Parker, and R. Mahon. 2010a. Incipient resistance of *Helicoverpa punctigera* to the Cry2Ab Bt toxin in Bollgard II. *PLoS One* 5:e12567. doi:10.1371/journal.pone.0012567

Downes, S., R. J. Mahon, L. Rossiter, T. Leven, G. Fitt, and G. Baker. 2010b. Adaptive management of pest resistance by *Helicoverpa* species (Noctuidae) in Australia to the Cry2Ab Bt toxin in Bollgard II cotton. *Evolutionary Applications* 3:574–584.

Dhurua, S., and G. T. Gujar. 2011. Field-evolved resistance to Bt toxin Cry1Ac in pink bollworm, *Pectinophora gossypiella* (Saunders) (Lepidoptera: Gelechiidae) from India. *Pest Management Science* 67:898–903.

Gassmann, A. J., Y. Carrière, and B. E. Tabashnik. 2009. Fitness costs of insect resistance to *Bacillus thuringiensis*. *Annual Review of Entomology* 54:147–163.

Genetic Engineering Approval Committee. 2010. Decisions taken in the 100th meeting of the Genetic Engineering Approval Committee (GEAC), December 5. http://www.envfor.nic.in/divisions/csurv/geac/decision-may-100.pdf

Gould, F. 1998. Sustainability of transgenic insecticidal cultivars: integrating pest genetics and ecology. *Annual Review of Entomology* 43:701–726.

Gould, F., A. Anderson, A. Jones, D. Sumerford, D. G. Heckel, J. Lopez, S. Micinski, R. Leonard, and M. Laster. 1997. Initial frequency of alleles for resistance to *Bacillus thuringiensis* toxins in field populations of *Heliothis virescens*. *Proceedings of the National Academy of Sciences USA* 94:3519–3523.

Hernández-Rodríguez, C. S., A. Van Vliet, N. Bautsoens, J. Van Rie, and J. Ferré. 2008. Specific binding of *Bacillus thuringiensis* Cry2 insecticidal proteins to a common site in the midgut of *Helicoverpa* species. *Applied and Environmental Microbiology* 74:7654–7659.

Huang, F., R. Parker, R. Leonard, Y. Yong, and J. Liu. 2009. Frequency of resistance alleles to *Bacillus thuringiensis*–corn in Texas populations of sugarcane borer, *Diatraea saccharalis* (F.) (Lepidoptera: Crambidae). *Crop Protection* 28:174–180.

Hutchison, W. D., E. C. Burkness, P. D. Mitchell, R. D. Moon, T. W. Leslie, J. Fleischer, M. Abrahamson, K. L. Hamilton, K. L. Steffey, M. E. Gray, R. L. Hellmich, L. V. Kaster, T. E. Hunt, R. J. Wright, K. Pecinovsky, T. L. Rabaey, B. R. Flood, and E. S. Raun. 2010. Areawide suppression of European corn borer with *Bt* maize reaps savings to non-*Bt* maize growers. *Science* 330:222–225.

Jackson, R. E., J. R. Bradley Jr., J. W. Van Duyn, and F. Gould. 2004a. Comparative production of *Helicoverpa zea* (Lepidoptera: Noctuidae) from transgenic cotton expressing either one or two *Bacillus thuringiensis* proteins with and without insecticide oversprays. *Journal of Economic Entomology* 97:1719–1725.

Jackson, R. E., J. R. Bradley Jr., and J. W. Van Duyn. 2004b. Performance of feral and Cry1Ac-selected *Helicoverpa zea* (Lepidoptera: Noctuidae) strains on transgenic cottons expressing one or two *Bacillus thuringiensis* spp. *kurstaki* proteins under greenhouse conditions. *Journal of Entomological Science* 39:46–55.

Jackson, R. E., F. Gould, L. Bradley, and J. Van Duyn. 2006. Genetic variation for resistance to *Bacillus thuringiensis* toxins in *Helicoverpa zea* (Lepidoptera: Noctuidae) in eastern North Carolina. *Journal of Economic Entomology* 99:1790–1797.

James, C. 2007. Global Status of Commercialized Biotech/GM Crops: 2006. ISAAA Brief No. 35. International Service for the Acquisition of Ag-biotech Applications, Ithaca, NY.

James, C. 2009. Global Status of Commercialized Biotech/GM Crops: 2008. ISAAA Brief No. 39. International Service for the Acquisition of Ag-biotech Applications, Ithaca, NY.

Kruger, M., J. B. J. Van Rensburg, and J. Van den Berg. 2009. Perspective on the development of stem borer resistance to Bt maize and refuge compliance at the Vaalharts irrigation scheme in South Africa. *Crop Protection* 28:684–689.

Lalitha, N., B. Ramaswami, and P. K. Viswanathan. 2009. India's experience with Bt cotton: case studies from Gujarat and Maharashtra. *In* R. Tripp (ed.), *Biotechnology and Agricultural Development: Transgenic Cotton, Rural Institutions and Resource-Poor Farmers*, pp. 135–167. Routledge, New York.

Luttrell, R. G., and M. I. Ali. 2007. Exploring selection for Bt resistance in heliothines: results of laboratory and field studies. *In* D. A. Richter (ed.), *Proceedings, 2007 Beltwide Cotton Conferences, 9–12 January 2007, New Orleans, LA*, pp. 1073–1086. National Cotton Council of America, Memphis, TN.

Luttrell, R. G., L. Wan, and K. Knighten. 1999. Variation in susceptibility of noctuid (Lepidoptera) larvae attacking cotton and soybean to purified endotoxin proteins and commercial formulations of *Bacillus thuringiensis*. *Journal of Economic Entomology* 92:21–32.

Luttrell R. G., I. Ali, K. C. Allen, S. Y. Young III, A. Szalanski, K. Williams, G. Lorenz, C. D. Parker Jr., and C. Blanco. 2004. Resistance to Bt in Arkansas populations of cotton bollworm. *In* D. A. Richter (ed.), *Proceedings, 2004 Beltwide Cotton Conferences, 5–9 January 2004, San Antonio, TX*, pp. 1373–1383. National Cotton Council of America, Memphis, TN.

Mahon, R. J., and K. M. Olsen. 2009. Limited survival of a Cry2Ab-resistant strain of *Helicoverpa armigera* (Lepidoptera: Noctuidae) on Bollgard II. *Journal of Economic Entomology* 102:708–716.

Mahon, R. J., K. Olsen, S. J. Downes, and S. Addison. 2007. Frequency of alleles conferring resistance to the Bt toxins Cry1Ac and Cry2Ab in Australian populations of *Helicoverpa armigera* (Hübner) (Lepidoptera: Noctuidae). *Journal of Economic Entomology* 100:1844–1853.

Meihls, L. N., M. L. Higdon, B. D. Siegfried, N. J. Millerd, T. W. Sappington, M. R. Ellersieke, T. A. Spencer, and B. E. Hibbard. 2008. Increased survival of western corn rootworm on transgenic corn within three generations of on-plant greenhouse selection. *Proceedings of the National Academy of Sciences USA* 105:19177–19182.

Mendelsohn, M., J. Kough, Z. Vaituzis, and K. Matthews. 2003. Are Bt crops safe? *Nature Biotechnology* 21:1003–1009.

Monsanto. 2010. Cotton in India. http://www.monsanto.com/monsanto_today/for_the_ record/india_pink_ bollworm.asp

National Research Council. 1986. *Pesticide Resistance: Strategies and Tactics for Management*. National Academy Press, Washington, DC.

National Research Council. 2010. *The Impact of Genetically Engineered Crops on Farm Sustainability in the United States*. National Academies Press, Washington, DC.

Onstad, D. W. 2008. *Insect Resistance Management: Biology, Economics, and Prediction*. Academic Press, London.

Showalter, A. M., S. Heuberger, B. E. Tabashnik, and Y. Carrière. 2009. A primer for the use of insecticidal transgenic cotton in developing countries. *Journal of Insect Science* 9:22. http://www.insectscience.org/9.22

Storer, N. P., J. M. Babcock, M. Schlenz, T. Meade, and G. D. Thompson. 2010. Discovery and characterization of field resistance to Bt maize: *Spodoptera frugiperda* in Puerto Rico. *Journal of Economic Entomology* 103:1031–1038.

Tabashnik, B. E. 1994. Evolution of resistance to *Bacillus thuringiensis*. *Annual Review of Entomology* 39:47–79.

Tabashnik, B. E. 2008. Delaying insect resistance to transgenic crops. *Proceedings of the National Academy of Sciences USA* 105:19029–19030.

Tabashnik, B. E. 2010. Communal benefits of transgenic corn. *Science* 330:189–190.

Tabashnik, B. E., and Y. Carrière. 2010. Field-evolved resistance to Bt cotton: *Helicoverpa zea* in the U.S. and pink bollworm in India. *Southwestern Entomologist* 35:417–424.

Tabashnik, B. E., A. J. Gassmann, D. W. Crowder, and Y. Carrière. 2008a. Insect resistance to Bt crops: evidence versus theory. *Nature Biotechnology* 26:199–202.

Tabashnik, B. E., A. J. Gassmann, D. W. Crowder, and Y. Carrière. 2008b. Field-evolved resistance to Bt toxins. *Nature Biotechnology* 26:1074–1076.

Tabashnik, B. E., J. B. J. Van Rensburg, and Y. Carrière. 2009a. Field-evolved insect resistance to Bt crops: definition, theory, and data. *Journal of Economic Entomology* 102:2011–2025.

Tabashnik, B. E., G. C. Unnithan, L. Masson, D. W. Crowder, X. Li, and Y. Carrière. 2009b. Asymmetrical cross-resistance between *Bacillus thuringiensis* toxins Cry1Ac and Cry2Ab in pink bollworm. *Proceedings of the National Academy of Sciences USA* 16:11 889–11 984.

Tabashnik, B. E., M. S. Sisterson, P. C. Ellsworth, T. J. Dennehy, L. Antilla, L. Liesner, M. Whitlow, R. T. Staten, J. A. Fabrick, G. C. Unnithan, A. J. Yelich, C. Ellers-Kirk, V. S. Harpold, X. Li, and Y. Carrière. 2010. Suppressing resistance to Bt cotton with sterile insect releases. *Nature Biotechnology* 28:1304–1307.

US Environmental Protection Agency (USEPA). 2005. Biopesticide Registration Action Document *Bacillus thuringiensis* Cry1F corn, August 2004 (updated 2005). http://www.epa.gov/pesticides/biopesticides/ingredients/tech_docs/brad_006481.pdf

US Environmental Protection Agency (USEPA). 2010a. Biopesticide Registration Action Document Cry1Ab and Cry1F *Bacillus thuringiensis* (Bt) Corn Plant-Incorporated Protectants. http://www.epa.gov/opp00001/biopesticides/pips/cry1f-cry1ab-brad.pdf

US Environmental Protection Agency (USEPA). 2010b. Current & Previously Registered Section 3 PIP Registrations. http://www.epa.gov/pesticides/biopesticides/pips/ pip_list.htm

Van Rensburg, J. B. J. 2007. First report of field resistance by stem borer, *Busseola fusca* (Fuller) to Bt-transgenic maize. *South African Journal of Plant and Soil* 24:147–151.

Wu, K-M., Y-H. Lu, H-Q. Feng, Y-Y. Jiang, and J-Z. Zhou. 2008. Suppression of cotton bollworm in multiple crops in China in areas with Bt toxin-containing cotton. *Science* 321:1676–1678.

17

Natural Enemies and Insect Outbreaks in Agriculture: A Landscape Perspective

J. Megan Woltz, Benjamin P. Werling, and Douglas A. Landis

17.1 Introduction

Planting over a vast extent of country certain plants to the exclusion of others, offers to the insects which live at the expense of these plants conditions eminently favorable to their excessive multiplication.

Paul Marchal (1908)

In an early treatise on the use of insects for biological control in agriculture, the French scientist Paul Marchal noted that in modifying agricultural landscapes, humans frequently favor insect outbreaks (Marchal 1908). Moreover, he observed that in order to use natural enemies to prevent or suppress insect outbreaks, we must not destroy them through inappropriate cultural practices. While significant progress in understanding these phenomena has been made in the intervening

Insect Outbreaks Revisited, First Edition. Edited by Pedro Barbosa, Deborah K. Letourneau and Anurag A. Agrawal.
© 2012 Blackwell Publishing Ltd. Published 2012 by Blackwell Publishing Ltd.

100 years, insect ecologists and applied entomologists have yet to fully realize the potential of landscape management to suppress insect outbreaks in agriculture. In this chapter, we summarize a modern understanding of the ways that natural enemies interact with landscape structure to reduce herbivore outbreaks, and outline approaches to managing landscapes to improve the reliability and effectiveness of this pest suppression service.

Natural enemies provide a critical ecosystem service (Zhang *et al.* 2007, Power 2010) by preventing many if not most potential insect outbreaks (Swinton *et al.* 2006). This service is in turn directly influenced by landscape composition and configuration (Thies and Tscharntke 1999, Thies *et al.* 2003, Tscharntke *et al.* 2005) and can be disrupted by the simplification and intensification of agricultural landscapes (Gardiner *et al.* 2008). Because changes in landscape structure may be gradual, losses in pest suppression may be imperceptible over short time frames. For example, as humans initially begin to appropriate natural habitats for crop and livestock production, agricultural landscapes begin to differ from the natural landscapes from which they develop (Curtis 1956, Landis 1994). At first, agricultural crops will represent a small proportion of the overall landscape with the native habitat remaining as the landscape matrix. In such situations, pest suppression may remain high. However, on productive soils, agricultural uses frequently come to dominate, replacing natural and seminatural habitats (Medley *et al.* 1995). Such landscape changes can reduce the abundance and diversity of natural enemies and, ultimately, pest suppression. The scale at which landscape change begins to influence pest suppression will depend on the biology of functionally important pests and natural enemies. By more fully understanding the relationship between landscape structure and natural enemy biology, we may be able to better manage landscapes to conserve natural enemies and prevent or reduce the severity of pest outbreaks.

17.2 Landscape influences on natural enemies and herbivore suppression

Landscapes are described in different ways and at different scales depending on the particular processes being studied. In general, Turner *et al.* (2001) define a landscape as an "area that is spatially heterogeneous in at least one factor of interest." As we will discuss in this chapter, the "factor of interest" is often land cover or habitat type, and studies of landscape effects on herbivore outbreaks frequently focus on how heterogeneity in land cover – generally over scales of several square kilometers – influences natural enemy communities. *Landscape composition* describes the identity and amount of different land covers that are present in a landscape and is typically measured using the absolute area of each cover type (e.g., hectares of forest or wetland) or the proportion of the landscape they occupy (e.g., percent cover of forest or wetland). In contrast, *landscape configuration* describes the spatial arrangement of habitats on the landscape, and is a function of the size, shape, aggregation, and dispersion of patches of different land cover types (Cushman *et al.* 2008). *Landscape diversity* is a function of composition, increasing with the variety and evenness of cover types present in an area; it is often quantified by applying Simpson's or Shannon's diversity index at the landscape scale (e.g., Thies and Tscharntke 1999, Steffan-Dewenter *et al.* 2002, Roschewitz *et al.* 2005). The concept of *scale*, which is characterized by extent and grain, is central to understanding the measurement and implications of a given landscape change (Turner *et al.* 2001). *Extent* describes the total area under study, while *grain* describes the resolution, or smallest identifiable unit, of a landscape data set. In practice, many insect ecologists use the term scale to refer to the extent of the landscape that influences an organism, measuring composition in successively larger, often circular regions (i.e., extents) centered on the sampled area (e.g., Steffan-Dewenter *et al.* 2002). This approach to testing hypotheses about landscape structure has been used in a variety of studies (Bianchi *et al.* 2006), including many of those discussed in this chapter.

17.2.1 Landscape composition: importance of diverse noncrop habitat for natural enemies

Landscape composition can affect natural enemies at the individual, population, and community levels, influencing their ability to suppress outbreaks of herbivores (Figure 17.1). Because different habitats may provide different levels of requisite resources for natural enemies, landscape composition can ultimately influence the suppression of herbivore outbreaks. In agricultural landscapes, perennial noncrop habitats may play a particularly important role by providing natural enemies with pollen, nectar, and alternative prey that are often scarce in crops (Thies *et al.* 2003). Additionally, noncrop habitats can provide natural enemies with overwintering sites and refuge from disturbances including tillage and harvest (Landis *et al.* 2000) or insecticide use (Lee *et al.* 2001). The relative abundance of noncrop habitats can also affect the fitness of individual natural enemies. For example, predatory ground beetles, *Pterostichus cupreus*, were larger (i.e., had consumed more, used less energy in foraging, or both) and females had higher fecundity in landscapes with low proportions of annual crop cover (Bommarco 1998). Furthermore, landscape composition can have population- and community-level effects on natural enemies. Specifically, the proportion of noncrop area in the landscape has been positively related to predator abundance (Gardiner *et al.* 2009), species richness (Schmidt *et al.* 2005, Schmidt *et al.* 2008), and subsequently suppression of pests like soybean aphid, *Aphis glycines* (Gardiner *et al.* 2009), and rape pollen beetle, *Meligethes aeneus* (Thies and Tscharntke 1999, Thies *et al.* 2003).

In addition to the proportion of noncrop habitat within the landscape, the overall diversity of habitat types influences natural enemy communities. Because different habitat types can support different predatory and parasitic taxa, landscape diversity can directly influence natural enemy diversity (Bianchi *et al.* 2006), which can lead to higher rates of pest suppression (Kruess 2003). Diverse landscapes can also promote the recovery of herbivore suppression after disturbances (Tscharntke *et al.* 2005) and provide "spatial insurance" (Loreau *et al.* 2003) by supporting different natural enemies with complementary dispersal abilities. In contrast, simplified landscapes often

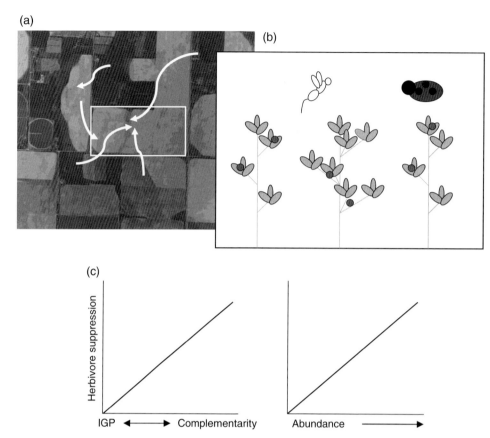

Figure 17.1 Traits of natural enemies and crop habitats interact at multiple levels to influence biocontrol outcomes in agricultural fields. (a) At the landscape level, both the dispersal ability of natural enemies and the abundance and configuration of resource habitats determine which natural enemies arrive in crop fields. (b) Within crop fields, plant architecture, prey distribution, and field management interact with natural enemy search efficiency and consumption capacity to determine the efficacy of individual natural enemies. (c) Within the natural enemy community, the overall abundance of natural enemies as well as the degree of intraguild predation or complementarity between natural enemy species determine whether the presence of diverse natural enemies interferes with or augments herbivore suppression.

support lower natural enemy diversity (Bianchi *et al.* 2006) and reduced suppression of pest herbivores (Andow 1983, Kareiva 1987, Thies and Tscharntke 1999, Symondson *et al.* 2002, Tscharntke *et al.* 2005). For example, Gardiner *et al.* (2009) showed that suppression of soybean aphid outbreaks was greater in landscapes with a high diversity of habitat types compared to agriculturally dominated landscapes in the north-central United States (Figure 17.2). In many cases, the effects of noncrop abundance and habitat diversity can be difficult to separate, as these two variables are often highly correlated in agricultural landscapes (e.g., Thies and Tscharntke 1999, Steffan-Dewenter *et al.* 2002, Roschewitz *et al.* 2005).

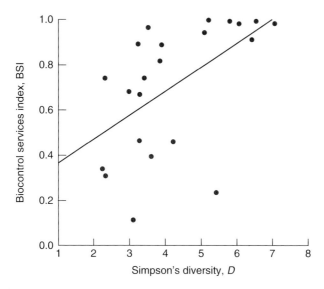

Figure 17.2   Studies in the Midwestern United States revealed a positive relationship between soybean aphid suppression (Biocontrol Services Index, or BSI) and landscape diversity (Simpson's diversity, D) at the 1.5 km scale surrounding focal soybean fields. Reprinted with permission from Gardiner *et al.* (2009).

Much of the research on agricultural landscape composition and pest suppression has been performed in the north temperate zone (e.g., Thies *et al.* 2005, Gardiner *et al.* 2009), where forests and grasslands comprise the majority of noncrop habitats. However, similar influences of landscape diversity on predator diversity have been found for desert habitats in arid agricultural regions of the Middle East (Pluess *et al.* 2010) and tropical rice-producing regions in Southeast Asia (Wilby *et al.* 2006). Additionally, managed perennial crops may provide some of the same benefits as noncrop habitats. For example, perennial pasture may contribute to the diversity of aphid predators in wheat fields (Elliott *et al.* 1999), and perennial crops have been related to lower rates of bird cherry-oat aphid, *Rhopalosiphum padi*, establishment (Östman *et al.* 2001) and higher activity density of linyphiid spiders (Oberg *et al.* 2007).

17.2.2   Landscape configuration: connectivity and patch size

The spatial arrangement, or configuration, of habitats on the landscape can also affect pest suppression. Increasing the connectivity of perennial noncrop habitats can increase natural enemy population growth (Benjamin *et al.* 2008) and dispersal rates (Corbett and Rosenheim 1996). However, the ways in which natural enemies perceive connectivity depend upon their biology. Generalist species may use multiple habitats and thus be less responsive to the connectivity of any single habitat type than specialists (Ryall and Fahrig 2006). A given landscape may also appear more connected to highly mobile species versus poor dispersers (deVries *et al.* 1996, Tscharntke *et al.* 2002). For example, in a study of coccinellid ability to track aphid populations in fragmented landscapes, the distribution of the pink lady beetle, *Coleomegilla maculata,* was more affected by habitat fragmentation than that of a better disperser, the multicolored Asian lady beetle, *Harmonia axyridis* (With *et al.* 2002). Furthermore, the scale at which connectivity influences natural enemies may depend on their dispersal abilities. For example, Perovíc *et al.* (2010) found that densities of smaller parasitoids were predicted by connectivity of noncrop habitats within 750 meters of a sample site, while larger predators like the melyrid beetle, *Dicranolaius bellulus,* responded significantly to the connectivity of noncrop habitat over 1.5–3.0 kilometers.

Average patch size can also influence herbivore suppression by natural enemies. For example, small fields associated with relatively large noncrop field borders influence parasitism of true armyworm, *Pseudaletia unipuncta,* in corn fields (Marino and Landis 1996) and predation of Colorado potato beetle, *Leptinotarsa decemlineata,* eggs (Werling and Gratton 2010). Similarly, a modeling exercise predicted higher lady beetle abundance and more effective aphid suppression in landscapes in which multiple, small patches of noncrop habitats were dispersed across the landscape instead of aggregated into fewer, larger patches, because this arrangement reduced the distance between field margins and crop fields and increased lady beetle colonization of crop fields (Bianchi and van der Werf 2003).

In practice, it is difficult to separate the effects of landscape composition and configuration on pest suppression, as landscapes are simultaneously simplified spatially and compositionally through agricultural intensification. This suggests that changes in both could combine to influence natural enemies as agricultural landscapes become simplified. For example, both the abundance of noncrop and perennial habitats and average field size have been related to predatory ground beetle, *Pterostichus cupreus,* fecundity (Bommarco 1998) and the species richness and diversity of aphid predators (Elliott *et al.* 1999).

17.3 Scales at which landscapes influence natural enemies and outbreaks

It is clear that diverse noncrop habitats provide critical resources to support natural enemy populations. These resources may be provided locally (i.e., within or adjacent to crop fields) or within the broader landscape. Noncrop vegetation planted adjacent to and within crop fields can support some parasitoid and predator populations (Rieux *et al.* 1999), contributing to the suppression of herbivore outbreaks (Hickman and Wratten 1996, Landis *et al.* 2000, Bianchi *et al.* 2006). Alternatively, insects may use field edges and adjacent habitats as refuges during times of crop disturbances (Duelli *et al.* 1999, Meek *et al.* 2002, Bianchi *et al.* 2006), and ground-dwelling predators are likely to disperse from field edges and hedgerows both after overwintering and during the summer growing season (Benjamin *et al.* 2008). However, the supply of natural enemies to these near- and within- crop habitats is still dependent upon

Table 17.1 Studies examining the relative effects of local and landscape traits on natural enemies and biocontrol services.

Types of local effects	Example studies
Field management (organic and conventional) Spider and carabid richness and activity density, as well as aphid parasitism rates, have been shown to be higher in landscapes with greater abundances of noncrop habitat and habitat diversity. In some (3/5) cases, density and richness also varied between organic and conventional fields, while in others (2/5) they did not.	Weibull *et al.* 2003; Schmidt *et al.* 2005; Purtauf *et al.* 2005; Roschewitz *et al.* 2005; Clough *et al.* 2005
Adjacent habitats Predator abundance and diversity and predation rates are higher in landscapes with more noncrop habitat. In some (2/3) cases, predators are also affected by the presence of beneficial resource habitats immediately adjacent to crop fields, while in others (1/3) they are not.	Prasifka *et al.* 2004; Werling and Gratton 2008, 2010
Within-field characteristics Both landscape and within-field factors like plant diversity and insecticide use explain variation in predator communities, but overall predator abundance was best explained by landscape composition and configuration variables.	Elliott *et al.* 1999; Drapela *et al.* 2008
Habitat type Arthropod abundances and community composition are simultaneously affected by the habitats in which they are sampled and characteristics of the surrounding landscape (2/5). In some cases (2/5) landscape effects are stronger; in others (1/5) the habitat has a greater influence.	Dauber *et al.* 2005; Aviron *et al.* 2005; Schweiger *et al.* 2005; Wilby *et al.* 2006; Oberg *et al.* 2007

the overall abundance of natural enemies in the broader landscape (Tscharntke *et al.* 2005, Clough *et al.* 2007). Thus, landscape traits can have an equal or greater influence on natural enemy communities than local factors (Table 17.1). For example, the diversity of carabid beetles in potato fields significantly increased with the amount of noncrop habitat in the landscape at 1.5 km surrounding potato fields, while the area of field margin adjacent to the potato fields had no effect on the composition or diversity of carabids in potato fields, despite the fact that the field margins supported diverse carabid communities (Werling and Gratton 2008).

The particular scale at which natural enemies respond to landscape characteristics will depend on taxon-specific traits. Specifically, the spatial scale at which a taxon responds to landscape composition and configuration can vary with its size and mobility (Roland and Taylor 1997) as well as its trophic level and degree of resource specialization (Tscharntke and Brandl 2004). For example, parasitoids with low dispersal capabilities are expected to respond at smaller scales, while more mobile generalist predators are affected at larger scales (Tscharntke *et al.* 2005). Gardiner *et al.* (2010) found that two groups of ground-dwelling predators, spiders and carabid beetles, responded to landscape structure at different scales, and concluded that such information is necessary to understanding taxon-specific responses to landscape structure. Furthermore, a single natural enemy taxon can respond to landscape traits at multiple scales. Spider species richness in agricultural crops is related to the occurrence of noncrop habitats at scales ranging from 95 m to 3 km (Schmidt *et al.* 2008) (Plate 17.1). In practice, natural enemy abundance and efficacy have been related to landscape characteristics at scales ranging from a few hundred meters up to 10 km (Table 17.2).

17.4 Interaction of predator biology and landscape traits

The dispersal abilities of particular natural enemies and the presence and configuration of source habitats for these species will determine which natural enemies coexist in agricultural fields (Bianchi *et al.* 2009). However, the ultimate impact of these communities on pest suppression depends on the traits of functionally important species and their interactions with landscape structure (Plate 17.1). Certain types of natural enemies may be more likely to respond to a particular aspect of landscape structure. For example, natural enemy responses to landscape diversity are primarily driven by organisms with high dispersal capabilities (Tscharntke *et al.* 2005). Furthermore, diverse landscapes may increase the effects of generalist predators in crop fields by subsidizing them with resources from other habitat types (Polis *et al.* 1997).

Landscape composition and configuration shape natural enemy communities within crop fields. However, their ultimate impact on pests will depend on subsequent interactions within these natural enemy communities. For example, changes in landscape structure can affect the species richness of predator communities, which may have neutral, positive, or negative effects on pest suppression depending on predator biology (Straub *et al.* 2008). Increased predator diversity could reduce biocontrol when predators interact via intraguild predation (Polis *et al.* 1989), as when predation on green lacewing, *Chrysoperla carnea*, larvae released cotton aphids, *Aphis gossypii*, from control (Rosenheim *et al.* 1993). In contrast, increased predator diversity could increase biocontrol when species use the prey resource base in a complementary fashion (Casula *et al.* 2006) by attacking different herbivores (Cardinale *et al.* 2003) or different life stages of the same herbivore (Wilby *et al.* 2005). Within crop fields, characteristics such as plant architecture and diversity can further influence the intensity of interactions among natural enemies or between natural enemies and prey, as complex plant architecture can provide a spatial refuge from intraguild predation (Finke and Denno 2002). Finally, the mechanisms through which natural enemy biodiversity influences pest suppression are likely to be different at different scales (Finke and Snyder 2010).

Table 17.2 Studies examining the scales of natural enemy response to landscape variables.

Natural enemies and landscape variables	Scales	Example studies
Parasitism		
Parasitism rates of pest aphids, lepidopterans, and beetles have been positively related to amount of noncrop habitat, perennial and forage crops, roadside vegetation, and habitat diversity in the landscape, and negatively related to the amount of agricultural land in the landscape.	0.3–10 km radius	Thies and Tscharntke 1999; Thies *et al*. 2003, 2005; Roschewitz *et al*. 2005; Bianchi *et al*. 2005, 2008
Predation		
Predation rates of pest aphids and lepidopterans have been positively related to habitat diversity and amount of wooded habitat in the landscape.	0.3–10 km radius	Bianchi *et al*. 2005; Gardiner *et al*. 2009
Spiders		
Spider richness, abundance, and activity density have been positively related to the amount of noncrop habitat, perennial crops, amount of roadside vegetation, and habitat diversity of the landscape.	88 m to 3.5 km radius	Schmidt and Tscharntke 2005; Clough *et al*. 2005; Schmidt *et al*. 2005, 2008; Isaia *et al*. 2006; Oberg *et al*. 2007; Drapela *et al*. 2008; Pluess *et al*. 2010; Gardiner *et al*. 2010
Carabid Beetles		
Carabid richness, abundance, activity density, and body size have been positively related to amount of noncrop habitat, grasslands, perennial crops, and habitat diversity of the landscape, as well as perimeter-to-area ratio of crop fields.	0.4–2.5 km radius; 1600 m² block	Bommarco 1998; Weibull *et al*. 2003; Purtauf *et al*. 2005; Batáry *et al*. 2007; Werling and Gratton 2008; Gardiner *et al*. 2010
Other Generalist Predators		
Abundance of generalist predators like coccinellid beetles, pirate bugs, and opiliones in crop fields have been positively related to amount of noncrop habitat in the landscape and landscape patchiness.	0.5–3.5 km radius; 2.56 km² and 23.04 km² blocks	Elliott *et al*. 1999; Prasifka *et al*. 2004; Gardiner *et al*. 2009, 2010

At larger scales, having a diversity of "insurance species" of natural enemies in the landscape that each respond differently to environmental conditions could assure stable pest suppression as conditions fluctuate over time and space (Tscharntke *et al.* 2005).

The timing of natural enemy dispersal can also modulate the effect of landscape composition and configuration on pest control (van der Werf 1995). It is not sufficient for a landscape to provide natural enemies to crop fields; rather, for effective biocontrol, natural enemies must be present in crop fields before pests reach high densities and achieve exponential population growth. For example, briefly delaying predation of aphids transmitting beet yellows virus (BYV) with predator exclusion devices resulted in season-long increases in BYV infection rates in sugar beets (Landis and van der Werf 1997). Similarly, reducing early-season populations of natural enemies in rice with insecticide resulted in much higher populations of pest herbivores in treated than untreated fields late in the season (Settle *et al.* 1996). The match between natural enemy dispersal ability and the distances between source and recipient habitats (i.e., crop fields) will determine the timing of natural enemy arrival in fields (Bianchi *et al.* 2009). In general, distances between source habitats and crop fields will be smaller when mean patch sizes within the landscape are smaller, allowing natural enemies to colonize fields earlier (Bianchi *et al.* 2009) which could explain higher biocontrol rates in landscapes consisting of small patches (Bianchi *et al.* 2006).

17.5 Managing agricultural landscapes to prevent insect outbreaks

Ultimately, a major goal of research regarding landscape influences on insect outbreaks is to develop better land management strategies for preventing or reducing the severity of such outbreaks. Any strategies will have to take into consideration how local management efforts are influenced by broader landscape patterns. Furthermore, due to the scales at which landscapes influence insect outbreaks, management strategies will often have to balance the needs and desires of a diverse group of stakeholders.

17.5.1 Unifying research from local and landscape scales

Since the 1960s, research on the role of noncrop habitats in pest suppression has focused on local "habitat management" (i.e., manipulating habitats in and around crop fields to enhance provision of food, shelter, and other resources to natural enemies) (Pickett and Bugg 1998, Landis *et al.* 2000). While many such studies have revealed increased abundance of natural enemies and increased rates of predation and parasitism in the vicinity of habitat management areas, these effects tended not to penetrate very far into associated crop fields, and resulted in effective pest control in relatively few cases (Gurr *et al.* 2000, 2004). In contrast, many studies since the mid-1990s have examined broader landscape effects on pest suppression, and it is well documented that natural pest control is enhanced in complex landscapes where it is positively associated with the occurrence of perennial herbaceous habitats (80% of cases), wooded habitats (71%), and landscape patchiness (70%) (Bianchi *et al*. 2006). A series of review and synthesis efforts have attempted to relate insights gleaned from these two approaches, focusing on the interaction of local- and landscape-scale approaches (Gardiner *et al*. 2008, Hunter 2008, Schellhorn *et al*. 2008). Tscharntke and others (2005) proposed that habitat management was unlikely to be effective in highly simplified, "cleared" landscapes due to a lack of natural enemies to respond to such habitats. Similarly, they hypothesized that complex landscapes would benefit little from habitat management. Rather, manipulating habitats near crop fields may be most effective where moderate landscape complexity assures a supply of natural enemies that can benefit from habitat management (Tscharntke *et al*. 2005), or when local habitats are connected to larger source pools of natural enemies in the landscape (Tscharntke 2007). This suggests that the effects of local management will be highly dependent on landscape structure. In any case, because the culmination of management actions on individual farms within the landscape ultimately results in the landscape characteristics that influence the population dynamics of pests and natural enemies, it will likely be advantageous to implement pest management and natural enemy conservation strategies at large spatial scales (Schellhorn *et al*. 2008).

17.5.2 Socio-economic considerations in managing landscapes for ecosystem services

Because ecosystem services, including pest suppression services, are often generated at scales larger than single-landowner holdings, management must take place at the landscape scale (Goldman *et al*. 2007). However, there are many challenges inherent in managing landscapes, including quantifying and valuing ecosystem services, analyzing cost and benefit trade-offs of alternative land uses, and creating and administering payments for ecosystem services (de Groot *et al*. 2010). While incentive programs can help increase land owner support for new land management policies (Goldman *et al*. 2007), promoting multifunctionality of landscapes (i.e., managing landscapes for the simultaneous production of ecosystem services and traditional agricultural products) could allow landowners to capitalize on increasing societal demand for sustainable agriculture without completely altering traditional land uses (Jordan and Warner 2010).

Steingröver *et al*. (2010) developed a framework for implementing landscape-scale management of ecosystem services based on bringing together a broad stakeholder base (including farmers and other land owners, conservationists, resource managers, legislators, and scientists), allowing them to take advantage of both expert and local knowledge to make a plan that was ecologically sound, area specific, and socially palatable. Working in a specific Dutch landscape, they were able to successfully create a landscape management plan that simultaneously preserved water purification and pest suppression services and had the support of multiple stakeholders (Steingröver *et al*. 2010). In contrast, developing landscape management plans may be more difficult in regions with no existing structural framework and/or with a long-established sociopolitical system resistant to changes in landscape use (Jordan and Warner 2010). In such cases, it is important to develop new partnerships that increase public understanding and support of ecosystem service production (Jordan and Warner 2010). Researchers must collaborate with other stakeholders to develop alternative scenarios for landscape management that incorporate landowner knowledge and are sensitive to stakeholder values (Cowling *et al.* 2008, de Groot *et al.* 2010, Steingröver *et al.* 2010). In all cases, managing landscapes for ecosystem services will involve feedbacks between the biophysical systems that generate these services and the socio-economic systems that value them. In other words, it is not sufficient to simply place a dollar value on ecosystem services like pest suppression; land owners must be provided with information about how to generate such services with their land use decisions, markets must exist in which land owners can profit from service generation, and public demand must support service markets (Jordan and Warner 2010).

17.6 Conclusions

Landscape structure profoundly influences the abundance and composition of natural enemy communities, and ultimately suppression of herbivores in agroecosystems. Managing landscapes to prevent herbivore outbreaks must involve careful consideration of the interactions between natural enemy biology, natural enemy community dynamics, and habitat characteristics. Furthermore, because of the large scales at which landscapes influence natural enemies, the needs and desires of a diverse group of landowners must be balanced in order to develop coordinated management plans. Recent efforts to integrate such diverse interests have demonstrated that stakeholder design of multifunctional landscapes is feasible, and these processes may be the model for future decision making. Additional studies will be required to determine if such landscapes can reliably reduce pest outbreaks while delivering other ecosystem services desired by society and if the benefits outweigh the real and perceived costs.

Acknowledgments

Support for this work was provided by the US Department of Agriculture (USDA) AFRI Program, the National Science Foundation (NSF) Long-Term Ecological Research Program, the US Department of Energy (DOE) Great Lakes Bioenergy Research Center (DOE BER Office of Science DE-FC02-07ER64494 and DOE OBP Office of Energy Efficiency and Renewable Energy DE-AC05-76RL01830), and Michigan State University (MSU) AgBioResearch.

References

Andow, D. 1983. The extent of monoculture and its effects on insect pest populations with particular reference to wheat and cotton. *Agriculture Ecosystems & Environment* 9:25–35.

Aviron, S., F. Burel, J. Baudry, and N. Schermann. 2005. Carabid assemblages in agricultural landscapes: impacts of habitat features, landscape context at different spatial scales and farming intensity. *Agriculture Ecosystems & Environment* 108:205–217.

Benjamin, R., G. Cedric, and I. Pablo. 2008. Modeling spatially explicit population dynamics of *Pterostichus melanarius* I11. (Coleoptera: Carabidae) in response to changes in the composition and configuration of agricultural landscapes. *Landscape and Urban Planning* 84:191–199.

Bianchi, F. J. J. A., N. A. Schellhorn, and W. van der Werf. 2009. Predicting the time to colonization of the parasitoid *Diadegma semiclausum*: the importance of the shape of spatial dispersal kernels for biological control. *Biological Control* 50:267–274.

Bianchi, F. J. J. A., C. J. H. Booij, and T. Tscharntke. 2006. Sustainable pest regulation in agricultural landscapes: a review on landscape composition, biodiversity and natural pest control. *Proceedings of the Royal Society B-Biological Sciences* 273:1715–1727.

Bianchi, F. J. J. A., P. W. Goedhart, and J. M. Baveco. 2008. Enhanced pest control in cabbage crops near forest in The Netherlands. *Landscape Ecology* 23:595–602.

Bianchi, F. J. J. A., and W. van der Werf. 2003. The effect of the area and configuration of hibernation sites on the control of aphids by *Coccinella septempunctata* (Coleoptera: Coccinellidae) in agricultural landscapes: a simulation study. *Environmental Entomology* 32:1290–1304.

Bianchi, F. J. J. A., W. K. R. E. van Wingerden, A. J. Griffioen, M. van der Veen, M. J. J. van der Straten, R. M. A. Wegman, and H. A. M. Meeuwsen. 2005. Landscape factors affecting the control of *Mamestra brassicae* by natural enemies in Brussels sprout. *Agriculture Ecosystems & Environment* 107:145–150.

Bommarco, R. 1998. Reproduction and energy reserves of a predatory carabid beetle relative to agroecosystem complexity. *Ecological Applications* 8:846–853.

Cardinale, B. J., C. T. Harvey, K. Gross, and A. R. Ives. 2003. Biodiversity and biocontrol: emergent impacts of a multi-enemy assemblage on pest suppression and crop yield in an agroecosystem. *Ecology Letters* 6:857–865.

Casula, P., A. Wilby, and M. B. Thomas. 2006. Understanding biodiversity effects on prey in multi-enemy systems. *Ecology Letters* 9:995–1004.

Clough, Y., A. Kruess, D. Kleijn, and T. Tscharntke. 2005. Spider diversity in cereal fields: comparing factors at local, landscape and regional scales. *Journal of Biogeography* 32:2007–2014.

Clough, Y., A. Kruess, and T. Tscharntke. 2007. Local and landscape factors in differently managed arable fields affect the insect herbivore community of a noncrop plant species. *Journal of Applied Ecology* 44:22–28.

Corbett, A., and J. A. Rosenheim. 1996. Impact of a natural enemy overwintering refuge and its interaction with the surrounding landscape. *Ecological Entomology* 21:155–164.

Cowling, R. M., B. Egoh, A. T. Knight, P. J. O'Farrell, B. Reyers, M. Rouget, D. J. Roux, A. Welz, and A. Wilhelm-Rechman. 2008. An operational model for mainstreaming ecosystem services for implementation. *Proceedings of the National Academy of Sciences USA* 105:9483–9488.

Curtis, J. T. 1956. The modification of mid-latitude grasslands and forests by man. *In* W. L. Thomas (editor), *Man's Role in Changing the Face of the Earth*, pp. 721–736. University of Chicago Press, Chicago.

Cushman, S. A., K. McGariyal, and M. C. Neel. 2008. Parsimony in landscape metrics: strength, universality, and consistency. *Ecological Indicators* 8:691–703.

Dauber, J., T. Purtauf, A. Allspach, J. Frisch, K. Voigtlander, and V. Wolters. 2005. Local vs. landscape controls on diversity: a test using surface-dwelling soil macroinvertebrates of differing mobility. *Global Ecology and Biogeography* 14:213–221.

de Groot, R. S., R. Alkemade, L. Braat, L. Hein, and L. Willemen. 2010. Challenges in integrating the concept of ecosystem services and values in landscape planning, management and decision making. *Ecological Complexity* 7:260–272.

deVries, H. H., P. J. denBoer, and T. S. vanDijk. 1996. Ground beetle species in heathland fragments in relation to survival, dispersal, and habitat preference. *Oecologia* 107:332–342.

Drapela, T., D. Moser, J. G. Zaller, and T. Frank. 2008. Spider assemblages in winter oilseed rape affected by landscape and site factors. *Ecography* 31:254–262.

Duelli, P., M. K. Obrist, and D. R. Schmatz. 1999. Biodiversity evaluation in agricultural landscapes: above-ground insects. *Agriculture Ecosystems & Environment* 74:33–64.

Elliott, N. C., R. W. Kieckhefer, J. H. Lee, and B. W. French. 1999. Influence of within-field and landscape factors on aphid predator populations in wheat. *Landscape Ecology* 14:239–252.

Finke, D. L., and R. F. Denno. 2002. Intraguild predation diminished in complex-structured vegetation: implications for prey suppression. *Ecology* 83:643–652.

Finke, D. L., and W. E. Snyder. 2010. Conserving the benefits of predator biodiversity. *Biological Conservation* 143:2260–2269.

Gardiner, M. M., A. K. Fiedler, A. C. Costamagna, and D. A. Landis. 2008. Integrating conservation biological control into IPM systems. *In* E. B. Radcliffe, W. D. Hutchison, and R. E. Cancelado (editors), *Integrated Pest Management: Concepts, Tactics, Strategies and Case Studies*, pp. 151–162. Cambridge University Press, Cambridge.

Gardiner, M. M., D. A. Landis, C. Gratton, C. D. DiFonzo, M. O'Neal, J. M. Chacon, M. T. Wayo, N. P. Schmidt, E. E. Mueller, and G. E. Heimpel. 2009. Landscape diversity enhances biological control of an introduced crop pest in the north-central USA. *Ecological Applications* 19:143–154.

Gardiner, M. M., D. A. Landis, C. Gratton, N. Schmidt, M. O'Neal, E. Mueller, J. Chacon, and G. E. Heimpel. 2010. Landscape composition influences the activity density of Carabidae and Arachnida in soybean fields. *Biological Control* 55:11–19.

Goldman, R. L., B. H. Thompson, and G. C. Daily. 2007. Institutional incentives for managing the landscape: inducing cooperation for the production of ecosystem services. *Ecological Economics* 64:333–343.

Gurr, G., S. Wratten, and M. A. Altieri. 2004. *Ecological Engineering for Pest Management: Advances in Habitat Manipulation for Arthropods*. Cornell University Press, Ithaca, NY.

Gurr, G. M., S. D. Wratten, and P. Barbosa. 2000. Success in conservation biological control of arthropods. *In* G. M. Gurr and S. D. Wratten (editors), *Biological Control: Measures of Success*, pp. 105–132. Kluwer Academic Publishers, Dordrecht, the Netherlands.

Hickman, J. M., and S. D. Wratten. 1996. Use of *Phacelia tanacetifolia* strips to enhance biological control of aphids by hoverfly larvae in cereal fields. *Journal of Economic Entomology* 89:832–840.

Hunter, M. D. 2008. *The Role of Landscape in Insect Ecology and Its Implications for Agriculture*. CAB Reviews: Perspectives in Agriculture, Veterinary Science, Nutrition, and Natural Resources 3. CABI, London.

Jordan, N., and K. D. Warner. 2010. Enhancing the multifunctionality of US agriculture. *Bioscience* 60:60–66.

Kareiva, P. 1987. Habitat fragmentation and the stability of predator prey interactions. *Nature* 326:388–390.

Kruess, A. 2003. Effects of landscape structure and habitat type on a plant–herbivore–parasitoid community. *Ecography* 26:283–290.

Landis, D. A. 1994. Arthropod sampling in agricultural landscapes: ecological considerations. *In* L. P. Pedigo and G. D. Buntin (editors), *Handbook of Sampling Methods for Arthropods in Agriculture*, pp. 15–31. CRC Press, Boca Raton, FL.

Landis, D. A., D. M. Carmona, and A. Pérez-Valdéz. 2000. Habitat management to enhance biological control in IPM. *In* G. G. Kennedy and T. B. Sutton (editors), *Enhancing Technologies for Integrated Pest Management*, pp. 226–239. APS Press, St. Paul.

Landis, D. A., and W. van der Werf. 1997. Early-season predation impacts the establishment of aphids and spread of beet yellows virus in sugar beet. *Entomophaga* 42:499–516.

Lee, J. C., F. B. Menalled, and D. A. Landis. 2001. Refuge habitats modify impact of insecticide disturbance on carabid beetle communities. *Journal of Applied Ecology* 38:472–483.

Loreau, M., N. Mouquet, and A. Gonzalez. 2003. Biodiversity as spatial insurance in heterogeneous landscapes. *Proceedings of the National Academy of Sciences USA* 100:12765–12770.

Marchal, P. 1908. The utilization of auxiliary entomophagous insects in the struggle against insects injurious to agriculture. *Popular Science Monthly* 72:352–370.

Marino, P. C., and D. A. Landis. 1996. Effect of landscape structure on parasitoid diversity and parasitism in agroecosystems. *Ecological Applications* 6:276–284.

Medley, K. E., B. W. Okey, G. W. Barrett, M. F. Lucas, and W. H. Renwick. 1995. Landscape change with agricultural intensification in a rural watershed, Southwestern Ohio, USA. *Landscape Ecology* 10:161–176.

Meek, B., D. Loxton, T. Sparks, R. Pywell, H. Pickett, and M. Nowakowski. 2002. The effect of arable field margin composition on invertebrate biodiversity. *Biological Conservation* 106: 259–271.

Oberg, S., B. Ekbom, and R. Bommarco. 2007. Influence of habitat type and surrounding landscape on spider diversity in Swedish agroecosystems. *Agriculture Ecosystems & Environment* 122:211–219.

Östman, O., B. Ekbom, and J. Bengtsson. 2001. Landscape heterogeneity and farming practice influence biological control. *Basic and Applied Ecology* 2:365–371.

Perovíc, D. J., G. M. Gurr, A. Raman, and H. I. Nicol. 2010. Effect of landscape composition and arrangement on biological control agents in a simplified agricultural system: a cost–distance approach. *Biological Control* 52:263–270.

Pickett, C. H., and R. L. Bugg. 1998. *Enhancing Biological Control: Habitat Management to Promote Natural Enemies of Agricultural Pests.* University of California Press, Berkeley.

Pluess, T., I. Opatovsky, E. Gavish-Regev, Y. Lubin, and M. H. Schmidt-Entling. 2010. Non-crop habitats in the landscape enhance spider diversity in wheat fields of a desert agroecosystem. *Agriculture Ecosystems & Environment* 137:68–74.

Polis, G. A., W. B. Anderson, and H. R. D. 1997. Toward and integration of landscape and food web ecology: the dynamics of spatially subsidized food webs. *Annual Review of Ecology and Systematics* 28:289–316.

Polis, G. A., C. A. Myers, and R. D. Holt. 1989. The ecology and evolution of intraguild predation: potential competitors that eat each other. *Annual Review of Ecology and Systematics* 20:297–330.

Power, A. G. 2010. Ecosystem services and agriculture: tradeoffs and synergies. *Philosophical Transactions of the Royal Society B: Biological Sciences* 365:2959–2971.

Prasifka, J. R., K. M. Heinz, and R. R. Minzenmayer. 2004. Relationships of landscape, prey and agronomic variables to the abundance of generalist predators in cotton (*Gossypium hirsutum*) fields. *Landscape Ecology* 19:709–717.

Rieux, R., S. Simon, and H. Defrance. 1999. Role of hedgerows and ground cover management on arthropod populations in pear orchards. *Agriculture Ecosystems & Environment* 73:119–127.

Roland, J., and P. D. Taylor. 1997. Insect parasitoid species respond to forest structure at different spatial scales. *Nature* 386:710–713.

Roschewitz, I., M. Hucker, T. Tscharntke, and C. Thies. 2005. The influence of landscape context and farming practices on parasitism of cereal aphids. *Agriculture Ecosystems & Environment* 108:218–227.

Rosenheim, J. A., L. R. Wihoit, and C. A. Armer. 1993. Influence of intraguild predation among generalist insect predators on the suppression of an herbivore population. *Oecologia* 96:439–449

Ryall, K. L., and L. Fahrig. 2006. Response of predators to loss and fragmentation of prey habitat: a review of theory. *Ecology* 87:1086–1093.

Schellhorn, N. A., S. Macfadyen, F. J. J. A. Bianchi, D. G. Williams, and M. P. Zalucki. 2008. Managing ecosystem services in broadacre landscapes: what are the appropriate spatial scales? *Australian Journal of Experimental Agriculture* 48:1549–1559.

Schmidt, M. H., I. Roschewitz, C. Thies, and T. Tscharntke. 2005. Differential effects of landscape and management on diversity and density of ground-dwelling farmland spiders. *Journal of Applied Ecology* 42:281–287.

Schmidt, M. H., C. Thies, W. Nentwig, and T. Tscharntke. 2008. Contrasting responses of arable spiders to the landscape matrix at different spatial scales. *Journal of Biogeography* 35:157–166.

Schmidt, M. H., and T. Tscharntke. 2005. Landscape context of sheetweb spider (Araneae: Linyphiidae) abundance in cereal fields. *Journal of Biogeography* 32:467–473.

Schweiger, O., J. P. Maelfait, W. Van Wingerden, F. Hendrickx, R. Billeter, M. Speelmans, I. Augenstein, B. Aukema, S. Aviron, D. Bailey, R. Bukacek, F. Burel, T. Diekotter, J. Dirksen, M. Frenzel, F. Herzog, J. Liira, M. Roubalova, and R. Bugter. 2005. Quantifying the impact of environmental factors on arthropod communities in agricultural landscapes across organizational levels and spatial scales. *Journal of Applied Ecology* 42:1129–1139.

Settle, W. H., H. Ariawan, E. T. Astuti, W. Cahyana, A. L. Hakim, D. Hindayana, A. S. Lestari, Pajarningsih, and Sartanto. 1996. Managing tropical rice pests through conservation of generalist natural enemies and alternative prey. *Ecology* 77:1975–1988.

Steffan-Dewenter, I., U. Munzenberg, C. Burger, C. Thies, and T. Tscharntke. 2002. Scale-dependent effects of landscape context on three pollinator guilds. *Ecology* 83:1421–1432.

Steingröver, E. G., W. Geertsema, and W. van Wingerden. 2010. Designing agricultural landscapes for natural pest control: a transdisciplinary approach in the Hoeksche Waard (The Netherlands). *Landscape Ecology* 25:825–838.

Straub, C. S., D. L. Finke, and W. E. Snyder. 2008. Are the conservation of natural enemy biodiversity and biological control compatible goals? *Biological Control* 45:225–237.

Swinton, S. M., F. Lupi, G. P. Robertson, and D. A. Landis. 2006. Ecosystem services from agriculture: looking beyond the usual suspects. *American Journal of Agricultural Economics* 88:1160–1166.

Symondson, W. O. C., K. D. Sunderland, and M. H. Greenstone. 2002. Can generalist predators be effective biocontrol agents? *Annual Review of Entomology* 47:561–594.

Thies, C., I. Roschewitz, and T. Tscharntke. 2005. The landscape context of cereal aphid–parasitoid interactions. *Proceedings of the Royal Society B-Biological Sciences* 272:203–210.

Thies, C., I. Steffan-Dewenter, and T. Tscharntke. 2003. Effects of landscape context on herbivory and parasitism at different spatial scales. *Oikos* 101:18–25.

Thies, C., and T. Tscharntke. 1999. Landscape structure and biological control in agroecosystems. *Science* 285:893–895.

Tscharntke, T., and R. Brandl. 2004. Plant–insect interactions in fragmented landscapes. *Annual Review of Entomology* 49:405–430.

Tscharntke, T., A. M. Klein, A. Kruess, I. Steffan-Dewenter, and C. Thies. 2005. Landscape perspectives on agricultural intensification and biodiversity – ecosystem service management. *Ecology Letters* 8:857–874.

Tscharntke, T., I. Steffan-Dewenter, A. Kruess, and C. Thies. 2002. Characteristics of insect populations on habitat fragments: a mini review. *Ecological Research* 17:229–239.

Tscharntke, T., Bommarco, R., Clough, Y., Crist, T.O., Kleihn, D., Rand, T.A., Tylianakis, J.M., van Nouhuys, S., and Vidal, S. 2007. Conservation biological control and enemy diversity on a landscape scale. *Biological Control* 43:16.

Turner, M. G., R. H. Gardner, and R. V. O'Neill. 2001. *Landscape Ecology in Theory and Practice: Pattern and Process*. Springer-Verlag, New York.

van der Werf, W. 1995. How do immigration rates affect predator/prey interactions in field crops? Predictions from simple models and an example involving the spread of aphid-borne viruses in sugar beet. *In* S. Toft, and W. Riedel (editors), *Arthropod Natural Enmies in Arable Land I: Density, Spatial Heterogeneity and Dispersal*, pp. 295–312. Aarhus University Press, Aarhus.

Weibull, A.C., O. Ostman, and A. Granqvist. 2003. Species richness in agroecosystems: the effect of landscape, habitat and farm management. *Biodiversity and Conservation* 12:1335–1355.

Werling, B. P., and C. Gratton. 2008. Influence of field margins and landscape context on ground beetle diversity in Wisconsin (USA) potato fields. *Agriculture Ecosystems & Environment* 128:104–108.

Werling, B. P., and C. Gratton. 2010. Local and broadscale landscape structure differentially impacts predation of two potato pests. *Ecological Applications* 20:1114–1125.

Wilby, A., L. P. Lan, K. L. Heong, N. P. D. Huyen, N. H. Quang, N. V. Minh, and M. B. Thomas. 2006. Arthropod diversity and community structure in relation to land use in the Mekong Delta, Vietnam. *Ecosystems* 9:538–549.

Wilby, A., S. C. Villareal, L. P. Lan, K. L. Heong, and M. B. Thomas. 2005. Functional benefits of predator species diversity depend on prey identity. *Ecological Entomology* 30:497–501.

With, K. A., D. M. Pavuk, J. L. Worchuck, R. K. Oates, and J. L. Fisher. 2002. Threshold effects of landscape structure on biological control in agroecosystems. *Ecological Applications* 12:52–65.

Zhang, W., T. H. Ricketts, C. Kremen, K. Carney, and S. M. Swinton. 2007. Ecosystem services and dis-services to agriculture. *Ecological Economics* 64:253–260.

18

Integrated Pest Management – Outbreaks Prevented, Delayed, or Facilitated?

Deborah K. Letourneau

18.1 Introduction

Integrated pest management (IPM) in the United States was initiated by a publication on "Integrated Control" by the late University of California (UC) entomologists Vern Stern, Ray Smith, Robert van den Bosh, and Ken Hagen (1959). At its inception, IPM was the theory-based antidote to the total reliance on synthetic insecticides in US agriculture that characterized conventional agriculture from 1945 to 1960. An integrated approach to pest management was prompted by an increasing acceptance that frequent insecticide applications increased the probability of selection for resistant pests (a limitation recognized at the onset of widespread use of DDT; see Ripper 1944). By the 1950s, there was clearly an overreliance on insecticides selected for resistant pest populations that were no longer controlled by the compounds that were previously effective (Kogan 1998).

Thus, there was a transition in the pre-IPM years, from occasional insect outbreaks to more severe outbreaks, with greater densities that persisted longer, sometimes

Insect Outbreaks Revisited, First Edition. Edited by Pedro Barbosa, Deborah K. Letourneau and Anurag A. Agrawal.
© 2012 Blackwell Publishing Ltd. Published 2012 by Blackwell Publishing Ltd.

devastating crops. In addition, outbreaks of secondary insect and mite pests that were previously not considered outbreak species began to occur. These secondary pest outbreaks were most often attributed to the suppression of arthropod natural enemies by the insecticides used to target the main pest outbreak (but see Hardin *et al*. 1995). In effect, pest control practices prevalent in the middle of the twentieth century were initiating new outbreaks and, in some cases, creating new insect pests.

An integrated control approach was proposed to re-establish the biological pest control that had been disrupted by frequent sprays of broad-spectrum pesticides meant to eliminate insect pests (Stern *et al*. 1959). Such a reliance on predators and parasitoids to prevent the majority of insect outbreaks would ensure that chemical control measures that disrupt beneficial organisms were reserved for when and where an outbreak was imminent. Insect outbreaks could be prevented or delayed through the mortality imposed by their natural enemies. Populations whose densities were monitored over time allowed for outbreaks to be anticipated, and truncated if necessary, through judicious use of selective insecticides that had a low disruptive capacity for natural enemies. Smith's (1962) concept enlisted the soil, the plant, its culture, weeds, arthropod pests and their competitors, and all of the conditioning factors within a cooperative IPM strategy unified toward profitable crop production without disrupting other parts of the ecosystem. Conditioning factors modify organisms and their activities. For example, the frequency of rainfall in one year may determine the depth of larvae feeding on root tissue in the soil, which makes them more or less accessible for oviposition by parasitoids, thus influencing their parasitism rates, which, in turn, modulate the size of their population in the next year. Thus, IPM was developed in those early years as a thoughtful strategy focused on preventing insect outbreaks, but encompassing all of the biotic and abiotic elements of the ecosystem.

In this chapter, I present an overview of IPM in the United States and an assessment of its progress as a rational and operational approach for managing insect outbreaks in agroecosystems. First, I summarize the history of IPM in the United States and the motivations for developing an IPM concept. My analysis appreciates IPM as a progressive strategy for reducing insect outbreaks and pesticide use in US agriculture, but recognizes the IPM concept as malleable and its practice as fraught with contingencies. Second, I consider the nature of insect outbreaks in agroecosystems. Are agricultural insect pest species more prone to outbreaks than other insects? To what degree are their eruptions artifacts of their agricultural setting? The observations discussed in this section help to identify measures needed to return agricultural insect outbreaks to occasional, short-term occurrences rather than facilitate or prolong them. Third, I consider the extent to which the IPM philosophy of its founders is practiced in twenty-first-century agriculture, and how new tools and strategies for regulating insect populations in agroecosystems fit into a sustainable agriculture framework for the future.

18.2   Historical development of IPM in the United States

During the first half of the twentieth century, economic entomologists relied primarily on their knowledge of pest biology to develop management schemes to avoid insect outbreaks in annual and perennial crop systems (e.g., Miles 1921), along with the application of insecticides such as sulfur, nicotine, or lead arsenate when an outbreak ensued despite preventative measures. After the discovery, during World WarII, that DDT was a highly effective insecticide on a wide variety of insects, organochlorines began to be applied prophylactically at scheduled intervals as insurance against insect outbreaks, and were widely adopted in agriculture. Inexpensive, persistent, synthetic insecticides eliminated insect outbreaks on a wide range of crops and increased yields and profits dramatically in the 1940s and 1950s. At the same time, insect outbreaks in agriculture were sporadically returning, and proving more difficult to suppress, owing primarily to adaptations that evolved in insect pest populations (selection for resistance traits) and ecological disruption of natural controls of insect population densities (resulting in pest resurgence and secondary pests).

In response to mounting failures of the new insecticides in eliminating insect outbreaks, Stern *et al.* (1959) presented a radical departure from the then prevalent notion of pest eradication. The ideal goal of "integrated control" was to reduce pest numbers below the economic injury level (EIL), *not* to eliminate them completely. In agricultural systems, insect densities at or above the EIL can serve as the definition of an agricultural insect outbreak. The EIL is the estimated density of pests at which the cost of intervention would be incrementally lower than the expected profit from suppressing insects to a level determined by the effectiveness of the intervention. That is, yield loss at the EIL makes it economically rational to invest in the cost of an insect pest control tactic (usually insecticide) that will drastically reduce their numbers. Thus, the concept of IPM articulated, for the first time, the strategy of preventing insect outbreaks by maintaining low levels of pest insects along with nonpest alternative hosts to support an effective community of natural enemies. Cost-effective monitoring schemes designed for target insects increased the feasibility and economy of field scouting, including using methods like sequential sampling, which adjusted the number of samples to the uncertainty of a population reaching the EIL. Insecticides would be applied only if pest levels reached the economic or action threshold, despite ecologically based preventative measures. These ecologically based measures included pest disruptive soil tillage practices, pest-resistant cultivars, temporal or spatial asynchrony between crop and insect pest, strategic weed removal, crop rotation, strip cropping, as well as augmentative, classical, and conservation biological control. Van den Bosch and Stern (1962) expressed the IPM philosophy in their statement: "if we are to expand our emphasis on the ecological approach to pest control, then no factor which impinges upon the agricultural ecosystem can be overlooked in the integrated control program." The IPM concept, then, was a dynamic, knowledge-intensive, strategic approach to managing insect outbreaks in agroecosystems, making use of a wide array of tactics, chosen first for their compatibility and environmental safety, while maintaining yields and profit.

By the 1970s and 1980s, IPM was becoming widely accepted in the academy and its extension services, and cost savings through lower expenditures for insecticides were demonstrated in experimental settings. Meanwhile, growers who responded to decreased effectiveness of their insecticide treatments by increasing spray frequency or using more concentrated formulations, thus increasing selection pressures on insect pest populations, were experiencing catastrophic yield losses

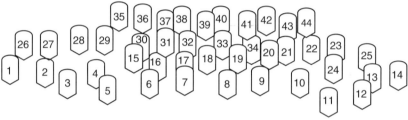

Figure 18.1 Division of Biological Control, University of California, Berkeley, February 2, 1979, with IPM Program Director Carl B. Huffaker (#7 on key), technical staff, visiting professors, postdoctoral researchers, and graduate students. Numbers on the key correspond to James Carey (1), Lowell Etzel (2), Nettie Mackay (3), Charles Kennett (4), Grace Leach (5), Denis Meals (6), Carl Huffaker (7), Barbara DeRochers (8), Richard Garcia (9), Jane Clarkin (10), Ying Wang (11), Gary Smith (12), John Andrews (13), Andrew Gutierrez (14), Cliff Kitayama (15), Kathleen Furneaux (16), Deborah Letourneau (17), Robert Pemberton (18), Ann Hajek (19), Sigurd Szerlip (20), David Rowney (21), Barbara Gemmill (22), Richard Hom (23), Paul Daley (24), Natalie Vandenberg (25), Leo Caltagirone (26), Johnnie Eaton (27), Peter Duelli (28), James Johnson (29), Pritam Singh (30), Ken Wahl (31), Robin Akers (32), Junji Hamai (33), William Voigt (34), Kenneth Hagen (35), John Perkins (36), David Williams (37), Hans Herren (38), Robert Nowierski (39), Richard Nathanson (40), Joanne Fox (41), Sean Swezey (42), Cliff Gold (43), and Frank Skinner (44). Professors Robert van den Bosch and Donald Dahlsten are not pictured.

due to sustained insect outbreaks coupled with higher costs of prevention. A major IPM research program headed by Carl B. Huffaker at the University of California, Berkeley (Figure 18.1) was funded by the National Science Foundation and the Environmental Protection Agency from 1972 to 1978, followed by the Consortium for Integrated Pest Management (1979–1985) and the US Department of Agriculture's Regional IPM Program in 1985. During the Clinton administration, the National IPM Initiative called for IPM on 75% of US croplands by the year 2000 (Dhaliwal *et al.* 2004). Food and Agriculture IPM Programs began to flourish during this period, and to develop into national policies promoting IPM (Matteson *et al.* 1994), leading to a strong push for IPM worldwide at the UN Conference on Environment and Development (UNCED) in Rio in 1992 (Dhaliwal *et al.* 2004, UNCED 1992). Agenda 21, Chapter 14 proclaimed that "Integrated pest management, which combines biological control, host plant resistance and appropriate farming practices and minimizes the use of pesticides, is the best option for the future, as it guarantees yields, reduces costs, is environmentally friendly and contributes to the sustainability of agriculture."

However, this hopeful trajectory leveled off in recent decades. Adoption of IPM, by almost any definition, even if defined as "using the minimum amount of pesticide to achieve an acceptable profit," was slower than expected (Trumble 1998). And even if the goal of reducing the number of pesticide sprays had been met in a vast number of cases because of effective scouting and better recognition of economic thresholds for potential pests, insufficient effort had been made in cropping systems to enhance, deploy, or integrate all of the various alternatives to pesticides – the actual foundations of an IPM approach. The prevention of insect outbreaks with the use of prophylactic management tactics involves creative ecological approaches, beginning with cultural control methods (Bajwa and Kogan 2004). Van den Bosch and Stern (1962) cautioned long ago against overlooking the fact that "it is the entire ecosystem and its components that are of primary concern and not a particular pest." Yet IPM on the ground was stagnating with its narrowed focus on curative controls of specific target insects with (albeit fewer) insecticide treatments (Ehler 2006). The rise in estimated crop losses from insect pests in the United States from 7% in 1945 to 13% four decades later (Pimentel *et al.* 1991) and the increase in pesticide use and crop losses from animal pests in North America since the 1960s (Oerke 2006) imply that IPM, as practiced, continued to facilitate insect outbreaks.

By the 40th anniversary of IPM in the late 1990s, there was widespread agreement that "there was very little 'I' in IPM" (Ehler and Bottrell 2000). Benbrook *et al.* (1996) estimated that only about 6% of US acreage qualified for the more expansive, multitactic definitions of IPM. In addition, USDA reported pesticide usage was somewhat higher in 2000 than in 1994, when the national IPM initiative was launched (Coble and Ortman 2009), indicating that as IPM was adopted, insect outbreaks were at best delayed, and possibly facilitated by those practices. Many recognized that "we can no longer afford to work within a vacuum of simple population effects and ignore multispecies interactions critical to the functioning of real agroecosystems and their surrounding habitats" (Letourneau and Andow 1999). A major report released during that period (NRC 1996) criticized IPM practices for managing individual components or organisms. It called for "a paradigm shift in pest-management theory … that examines processes, flows, and relationships among organisms." What had been lost, then, or at least rarely applied, were the gains made in understanding what causes pest outbreaks in agricultural systems in the first place, and the commitment to treat those systemic causes rather than the symptoms of insect outbreaks (Zorner 2000).

18.3 The nature of the beast

What are we up against in terms of eliminating or at least minimizing insect pest outbreaks in agriculture through IPM? Are they more frequent or more severe because of their agricultural setting and imposed management practices? Do natural enemies suppress insect outbreaks similarly in natural and agricultural systems, defining "natural systems," here, as ecosystems or habitat fragments that are not being managed? Many of the pests in agriculture are not known to reach outbreak proportions in natural systems, suggesting that at least some of the conditions in agricultural systems such as crop fields, plantations, or greenhouses promote insect outbreaks. Not only may an agricultural system promote plant exploitation by insect herbivores by the abundance of high-quality food compared to conditions in natural systems, but also the amount of direct damage, such as fruit infestation or defoliation, causes obvious dollar-value losses in a crop rather than a more indirect and even vague measure of harm, based on the integrity of a forest ecosystem, its aesthetic value, climate modulation, and so on.

18.3.1 Outbreaks in unmanaged systems versus agroecosystems

Are agricultural pest species atypical among insects? In non-agricultural outbreaks, there seems to be two main types of outbreak categories. First, there are very evident outbreaks among native insect herbivores (see Chapter 6, this volume) which typically live on perennial plants, are sometimes regionally synchronous, and exhibit population eruptions that persist for years, even in the face of insecticidal treatment, or outbreak sporadically in time, such as every 5–8 years (e.g., Jepsen *et al.* 2009). In many cases, the causes and underlying mechanisms of an insect outbreak in natural systems have been elusive (Dwyer *et al.* 2004). For example, the strong association between forest fragmentation and the duration of tent caterpillar outbreaks in boreal forest, suggested to increase moth oviposition or reduce tachinid parasitism (Roland 1993), was not upheld in other forests. The opposite factor, forest cover, was positively associated with tent caterpillar outbreak duration in a follow-up study (Wood *et al.* 2010). A combination of factors including the assortment of host plant quality, precipitation patterns, drought, warm temperatures, the presence of pathogens, the absence of natural enemies, synchrony, phenology, host plant concentration or density, and the amount of disturbance or spatial heterogeneity has been implicated as the causal force for insect outbreaks in unmanaged forests. Defoliation and tree death are indicators of insect outbreaks in forests, but the consequences of outbreaks on tree longevity, growth rate, or reproductive output are often not assessed. Outbreaks on unmanaged forbs and grasses may have a similar range of complex causes as those in forests. Measurable consequences have included reduced primary productivity (Coupe and Cahill 2003) and reproductive output (Root 1996), and increased plant mortality. Stands of thousands of lupine bushes may die off in periodic ghost moth outbreaks (Strong *et al.* 1995). Increases in herbivore density of several orders of magnitude, however, are not unusual in unmanaged systems, and may have no visible long-term effects on the plant. For example, tussock moth outbreaks on the same bush lupine cause repeated defoliation, sometimes for several years, but are highly localized, with plant recovery and foliage regrowth occurring after one growing season (Harrison and Maron 1995), and thus may be overlooked.

The second type of outbreak involves invasive species. Invasive species can reach very high densities in a very short time, and either lead to the virtual elimination of the host plant or more commonly reach some type of dynamic equilibrium over an extended period of time as susceptible plant individuals die out and/or as host plants in areas that are unsuitable for the pest, escape elimination (Kenis *et al.* 2009). A potential third type of outbreak in non-agricultural communities may be on the rise. This type would be indicative of directional system breakdowns or state changes, in which indirect stressors from human activities, perhaps associated with disruption of the ozone layer, global warming, pollutants, or fragmentation of habitats, fundamentally alter plants or whole ecosystems beyond the range of known natural variation, thus increasing their susceptibility to insect outbreaks or vulnerability to invasive insects. For example, bark beetle outbreaks may extend much farther north or be more frequent than expected based on historical records, if winter temperatures increase, trees in boreal forests are stressed from acid rain, and warmer summer temperatures cause drought stress.

Certainly agricultural insect pest outbreaks differ in many ways from natural systems outbreaks because of human intervention. Firstly, the criteria for defining an outbreak in agricultural production systems are likely to lead to more conservative assessments than in natural systems, because outbreak levels are tied to the loss of marketable crop yield rather than massive defoliation events or regional bouts of plant mortality (Figure 18.2a and 18.2b). If EILs indicate outbreak densities in an agroecosystem, then outbreaks are as much defined by the value of a bushel of apple or pound of grain and the cost of petroleum or pay rate for labor as they are by the amount of plant damage incurred. Despite the extrabiological criteria, the tolerable damage levels to trees, seeds, or fruits in natural forest are likely to be greater than, say, the damage levels that are tolerated in a commercial orchard without some kind of intervention, thus making insect outbreaks in agricultural systems more frequent. The number of insects suggested to signal an impending outbreak (economic threshold), and thus requiring rapid intervention, vary, of course, among species of insect and plant. In the Midwestern states, for instance, it might be 250 soybean aphids per soybean plant, one western bean cutworm egg mass per 20 field corn plants, or one adult potato leafhopper per sweep in a potato field. Economic thresholds are often lowered when insects transmit or facilitate infection by plant pathogens, and often involve conservative guesses, but careful field study can raise an economic threshold, even for vectors of pathogens (e.g., Brust and Foster 1999). Pest tolerance levels can be modified further at various stages along the market-based commodity chain from producer to consumer such that the EIL is lowest at the end of the growing season. Zero tolerance of internal insect pests or cosmetically flawless produce has been required for certain commodities, such as tart cherries, leafy greens, or sweet corn (Curtis 1998, Castle *et al.* 2009, Palumbo and Castle 2009).

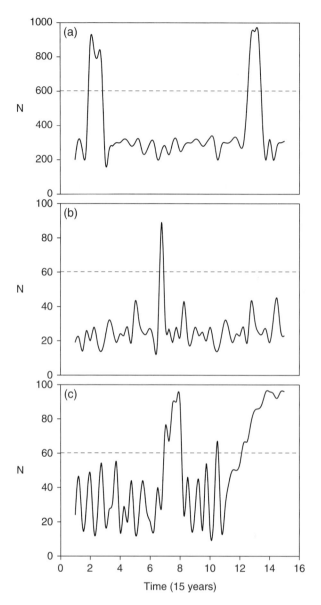

Figure 18.2 (a) Hypothetical 15-year period showing cyclic insect outbreaks lasting several years in an unmanaged system compared to (b) sporadic insect outbreaks with lower peak densities (y-axis) as defined by the economic injury level (EIL, dashed line) in high-value crops with integrated management for pests including tactics such as crop and resistant variety rotation, crop timing shifts, habitat preservation to enhance conservation biological control, monitoring, and target selective augmentation of nematode enemies compared to (c) sustained insect outbreaks occurring after a target insect population on a high-value crop develops resistance from frequent treatments with one and then another class of broad-spectrum insecticides that were previously effective in keeping its monitored densities below the EIL, and occasional pest resurgence episodes when, after treatment with insecticide, the target insect pest population rapidly increases to a level that exceeds that of an untreated population.

Under these circumstances, an insect outbreak is virtually defined as the presence of the insect in the field, thus minimizing the definition of an insect outbreak to its extreme in an agricultural setting. On the other hand, insecticides are also applied to insects that are not pests, but appear to be causing crop damage. For example, despite indications from physiological plant resource allocation models that a relatively high density of a crop-feeding insect is tolerable, since most are on crop organs destined to be aborted, conservative growers may choose to treat them. Incidentally, both outbreak determination and responses to it are made on a field-by-field basis in agroecosystems, possibly creating a more patchy effect to outbreak dynamics than is likely in an unmanaged system.

Secondly, agricultural insect outbreaks can be facilitated and maintained by the nature of the agroecosystem and how we manage it (Risch 1987). Crop fields or plantations typically have more homogeneous plant resources than do natural systems, because of their low plant diversity, including species richness, evenness, and genetic variability. Abundant, homogeneous plant resources may result in low diversity and high densities of herbivorous insects (e.g., Dalin *et al.* 2009). Furthermore, plant response mechanisms are limited in agroecosystems compared to unmanaged systems where individuals, neighbors, and offspring of target plants are subject to localized, predictable insect pest pressure. Trees, shrubs, grasses, and forbs in unmanaged systems exhibit physiological or morphological responses to insect exploitation that can reduce plant vulnerability to subsequent generations of insect herbivores through plastic responses or natural selection. In contrast, trees in orchards or plantations, cereals, and vegetable crops are reset at shorter intervals with "naïve" plants bred elsewhere. Furthermore, plants in managed systems are subject to artificial selection to maximize agronomic or marketability factors such as sweeter mesocarp tissue for peaches, size and color of the strawberry receptacle, endosperm weight for cereals, or closely packed aborted flower meristems for cauliflower. The overall nutritional quality or suitability of a cultivated plant can stimulate rapid population growth of insects when, compared to wild plants in their unmanaged habitats, it has a higher concentration of nitrogen in its tissues from fertilizer (Scriber 1984, Letourneau 1997), more succulence from irrigation or reduced plant–plant competition, or increased palatability and reduced chemical or physical antiherbivore defenses. Consequently, the amplitude of the outbreak in a cropping system may also be much smaller than in unmanaged systems if typical background levels of herbivores are relatively greater, making injurious population levels easier to reach in agroecosystems when conditions are favorable (Figure 18.2b), and outbreaks more frequent.

Third, the eruption of exotic insects occurs in both unmanaged and agricultural systems, causing direct damage to plants and indirect effects as serious as species extirpation in forests to crop quarantines and export embargoes in agriculture. The causal mechanism of the outbreak may be similar in unmanaged and agroecosystems if the insect herbivore is introduced into a new environment without the necessary complement of natural enemies that regulates its numbers in its center of origin (e.g., Haye *et al.* 2010) or if it is already explosive in its native habitat (e.g., Hu *et al.* 2009). A relative lack of host plant resistance in the new habitat can also cause the rapid spread of invasive insects that are comparatively contained in their native region. For example, the emerald ash borer (EAB), an exotic pest from Asia, is currently harming ashes in the United States, but the Manchurian ash from its native region is resistant to the borer. A host plant release mechanism is more likely in unmanaged habitats than agroecosystems if tight co-evolution is involved. Invasive insects in unmanaged systems are often one of the several exotic species in the ecosystem, whereas cropping systems can be dominated by non-native plants and herbivores.

The likelihood that exotic agricultural pests have been exposed to pesticides is higher than it is for an herbivore that feeds on noncrop vegetation, and thus they may exhibit pesticide resistance traits. For exotic insects that outbreak due to the absence of key natural enemies, classical biological control efforts can theoretically offer similar benefits (and costs, regarding nontarget hosts) in both types of systems.

Finally, crop protection efforts are designed to cut off the normal progression of an outbreak, through interventions such as the application of pesticides, chemical deterrents, attractive plants, inundative or inoculative releases of biological control agents, or even crop harvest. Population rebounds that persist over time are expected if these interventions facilitate the development of resistance to strong selection pressures (Figure 18.2c), insecticide-induced hormesis (Guedes *et al.* 2010), a combination of direct kills, sublethal effects, and starvation of natural enemies (Avila and Rodriguez-del-Bosque 2005), or through dispersal to more suitable habitat patches in an otherwise homogeneous crop landscape (Margosian *et al.* 2009).

18.3.2 Herbivore suppression by natural enemies in unmanaged systems versus agroecosystems

Insect outbreaks, whether in unmanaged or agricultural habitats, indicate that the available assemblage of predators, parasitoids, parasites, diseases, and other antagonists was not strong enough as a collection of top-down forces to regulate the herbivore; alternatively, the insect has escaped an otherwise suppressive natural enemy complex in time or space physically, physiologically, or behaviorally (e.g., Denno *et al.* 2002, Schott *et al.* 2010, Noma *et al.* 2010, Maron *et al.* 2001). For a number of reasons, it is more difficult in some cases to attribute the regulation of insect populations below outbreak densities solely to the action of natural enemies. Large-scale field studies are often based on correlative evidence rather than actual mortality measures; predator–prey models based on laboratory experiments lack the complexity of tritrophic interactions, apparent competition, and metapopulation dynamics. Further, experimental field studies are likely to be conducted at scales that can bias the results (Furlong and Zalucki 2010, Schott *et al.* 2010), under artificial conditions, or in simplified communities (Letourneau *et al.* 2009). Nevertheless, strong evidence of top-down control is provided by long-term correlative studies over large areas with measurements of multiple variables or in studies that corroborate outcomes with different approaches having different strengths and weaknesses. For example, population regulation by a combination of natural enemies explained long-term cyclic patterns in some forest insect outbreaks very well (e.g., Dwyer *et al.* 2004). Predator exclusion experiments verified the role of natural enemies in others (e.g., Turchin *et al.* 1999). In other cases, despite high parasitism rates at the local scale, parasitoids were not correlated with insect outbreak cycles on the larger spatial and temporal scale (Schott *et al.* 2010). Collections of "on-farm" data can be valuable measures of the extent to which natural enemies dampen insect outbreaks. For example, Walker *et al.* (2010) used the percentage of parasitized *Helicoverpa armigera* (Hübner) armyworms to adjust the economic threshold from one larva per plant to more than eight larvae per plant in tomato.

Mechanistic studies isolating the role of natural enemies in preventing insect outbreaks are less common than comparative mortality experiments at normal densities in part because a comparative non-outbreak control is often lacking (e.g., Hardin *et al.* 1995). Alternative approaches include comparing the enemy assemblage and attack rates of related species that do and do not erupt (Nothnagle and Schultz 1987, Eber *et al.* 2001) or simply inferring from measures of top-down forces on herbivore levels whether or not they are at high densities. None of these methods imply a systematically stronger role of enemies in unmanaged than agricultural systems or vice versa. Halaj and Wise (2001) found that predator abundance strongly affected herbivore densities in both agricultural and unmanaged systems. Using a meta-analysis on 66 studies in agriculture and 24 in unmanaged systems, Stireman *et al.* (2005) also found that the mean effect sizes (overall effect of natural enemies in reducing herbivore abundance) were similarly large in managed versus unmanaged ecosystems, despite very high variability in agricultural studies.

Because insect pest outbreaks in agriculture cause costly yield loss, it is relatively more critical in agroecosystems than in unmanaged systems for enemies to prevent outbreaks by maintaining low or moderate insect populations rather than to respond to them after they occur, increase the rate of population decline and reduce the duration of an outbreak. Prevention requires either density dependent regulation by key specialist enemies (parasitoids) efficient enough to keep peak pest densities below the EIL or a high enough predation pressure by a wider assemblage of enemies, including generalists, to maintain low enough densities. An assemblage of natural enemies reliably truncated outbreaks of *Aphis gossypii* on cotton before the production of harvestable yield, thus converting this aphid from a potential pest to a nonpest insect (Rosenheim *et al.* 1997). In this case, even though above-ground biomass was reduced by 45% on aphid-infested plants in the early season, cotton yield was not affected, because compensation took place before bolls were mature. Analyzing life tables of herbivores in unmanaged and managed systems, Hawkins *et al.* (1999) found that natural enemies are similarly important as key factors in both systems, but generalist predators are more common in unmanaged systems and specialist parasitoids are more likely to be key factors in managed systems.

With respect to enemy diversity, agriculture tends to reduce biodiversity on all trophic levels, and at increasingly larger scales as agricultural operations become more dominant in the landscape (Butler *et al.* 2007). Attwood *et al.* (2008) showed that the species richness of carnivorous arthropods was significantly greater in native vegetation than in agriculture, and in reduced-input cropping systems compared to conventional agriculture. Relatively simplified food webs in high input, monoculture agricultural systems may make them more responsive than unmanaged systems to small changes in enemy richness (see Stireman *et al.* 2005). The meta-analysis by Letourneau *et al.* (2009) of studies measuring the effect of enemy diversity on arthropod prey suppression is consistent with a higher sensitivity to enemy species addition or loss in agroecosystems. We found that increased enemy richness in unmanaged systems had inconsistent effects on herbivore suppression compared to a large and significant decrease in herbivore abundance with an increase in enemy richness in agriculture (Letourneau *et al.* 2009). Thus, mortality to natural enemies and community disruption caused by insecticides may put simplified agroecosystems in disproportionate jeopardy, causing an increase in the frequency of insect outbreaks beyond what is ever experienced in more complex unmanaged systems.

18.4 Integrating insect suppression tactics through IPM

The simplest vegetation management schemes such as improving natural enemy resources in field border vegetation or intercropping with plants that repel injurious insects are a rare phenomenon in mainstream US agriculture. Instead, insect suppression tactics that are adopted widely tend to be those that require the fewest changes in conventional practices and fit well with the associated infrastructure, such as monocultures of transgenic, insecticidal crops or insect pathogens delivered with standard spray equipment. The burgeoning field of insect resistance management (IRM) provides practical guidelines for varying the strength and type of selective pressures imposed on insect populations by pesticides. Weiss *et al.* (2009) seem to suggest that the sustainability of IPM programs can be assessed based on insecticide resistance rates alone. Whereas, IRM is designed to delay insect outbreaks due to pest resistance in IPM programs, management strategies that do not lead to insect outbreaks on the long term require more effective integration and implementation of a wider variety of methods and tools. The incentives for more comprehensive sustainability parameters for IPM practices are increasing as (1) pesticide cancellations through the Food Quality Protection Act accrue, (2) agroecology research and practices enter the policy sphere (e.g., De Schutter 2010), and (3) public support and demand increase for sustainable agriculture, slow food (an advocacy movement against fast food), and healthy diets (Merrigan 2000). In addition, the pesticide treadmill alternative, in which ineffective insecticides are replaced with ones that cause mortality in a different way, is becoming less and less feasible, with more than 500 species of insects and mites that are resistant to at least one class of insecticides (compared to the 70 noted by Stern *et al.* (1959)).

18.4.1 From facilitation to prevention of outbreaks with IPM

Castle *et al.* (2009) called the original, broader vision of integrated control (Stern *et al.* 1959) the most rational and knowledgeable approach to managing insect outbreaks. Many current descriptions, including the University of California definition of IPM, endorsed by the US Department of Agriculture, reflect this broad agenda:

> Integrated pest management (IPM) is an ecosystem-based strategy that focuses on long-term prevention of pests or their damage through a combination of techniques such as biological control, habitat manipulation, modification of cultural practices,

and use of resistant varieties. Pesticides are used only after monitoring indicates they are needed according to established guidelines, and treatments are made with the goal of removing only the target organism. Pest control materials are selected and applied in a manner that minimizes risks to human health, beneficial and nontarget organisms, and the environment. (UC-IPM 2008)

The feasibility of this truly integrative pest management may ultimately depend upon the structural redesign of agroecosystems to increase biodiversity on multiple scales, from genetic diversity among cultivars in a crop field to habitat diversity in the landscape (Birch *et al.* 2011). For the same 50 years since the original formulation of the integrated control concept, insect ecologists have been quantifying the suppressive effects of plant diversification schemes on insect pest populations that tend to reach outbreak proportions in monocultures (Pimentel 1961, Pimentel and Goodman 1978, Altieri and Letourneau 1984, Andow 1991, Nicholls and Altieri 2007). Monoculture seems to be a "lock-in" condition of conventional agriculture (*sensu* Vanloqueren and Baret 2009), such that the practice is entrenched and inflexible even when superior alternative approaches are available. If monoculture tends to make an agroecosystem fundamentally more favorable for insect pests than for their natural enemies, then biological control, the cornerstone of IPM, is not functioning as such. The ecological strategies called for in IPM are only possible in a functioning ecosystem, which is not likely to resemble conventional, industrial farming (including genotypic homogeneity, overfertilization, and breeding programs that select primarily for high yield under these conditions). Without integration into a multiple-resistance-factor, ecosystem-based IPM plan that builds durability, newer IPM tools such as transgenic Bt crops and selective insecticides, which reduce the disruption of natural enemies, are still destined for the same path of overexposure and pest resistance selection: the pesticide treadmill that IPM was meant to avoid.

Nicholls and Altieri (2007) have framed agroecosystem-based IPM as a set of three progressive steps: increased efficiency of pesticide inputs, input substitution with more benign materials or tactics, and then system redesign (Hill and MacRae 1995). System redesign focuses on restoring ecosystem structure and function, and emphasizes planned increases in biodiversity. It requires active integration and reliance on the foundational prevention of insect outbreaks using compatible tactics based on ecological knowledge. Goodell (2009) revisited one of the early IPM practices of strip-cropping alfalfa within cotton fields, showing how this trap cropping tactic and attention to alternate host source crops in the landscape prevented most *Lygus* bug outbreaks. Deguine *et al.* (2008) examine the history of insect pest management in cotton and paradigm shifts from traditional management to chemical pest control to IPM to area-wide eradication–suppression strategies coupled with more biologically based IPM programs. Whereas the IPM phase reduced the frequency of insecticide treatments and secondary pests and insect resistance management delayed pest resurgence and reduced the creation of secondary pests, the latter paradigm utilizes options such as pheromone lure-and-kill traps, mating disruption, cotton with stacked antifeedant and Bt genes with refugia to delay pest resistance, sterile male and beneficial insect releases, conservation biological control through intercropping and food sprays, locally adapted varieties, cover crops, living mulches, altered cropping geometries and sequences, and intercropping with safflower, tomato, or sorghum and selective biopesticides on intercalated trap crops.

Alternative options are required for many crops now that IPM as pesticide reduction and insect resistance management have failed to prevent insect outbreaks. For example, growing resistance and cross-resistance in thrips on onion, through its history of insecticide-based management, have led to sustained insect outbreaks documented by Diaz-Montano *et al.* (2011). They recognize an urgent need for an integrated array of options, among them the development of thrips-resistant cultivars and intercropping schemes to reduce the favorability of the onion crop for outbreaking insects. These historical examples illustrate that IPM as implemented in US agriculture has reduced the selection pressure that caused insect outbreaks compared to pre-IPM practices when up to 20 or more applications of insecticide were applied in a growing season (Smith and Reynolds 1972). IPM reduced the frequency of sprays, thus delaying the onset of insect outbreaks by reducing and varying the type of insecticide-induced mortality on the population. However, the same processes of resistance and cross-resistance development were still occurring, with IPM facilitating the eventual increase in population eruption of target insects and the emergence of secondary pests as in the previous era, albeit more slowly. For example, Frantz and Mellinger (2009) reported on the pest status of western flower thrips *Frankliniella occidentalis* (Pergande) in Florida, after it became established there in 1982. In the mid-1990s, these thrips occurred at very low numbers on tomatoes, peppers, and other vegetable crops. In the next decade, the western flower thrips persisted at outbreak densities two orders of magnitude above the economic threshold throughout most of the season in vegetables, causing devastating losses in marketable yield. Resistance to Spinosad (Spintor, Dow AgroScience) and disruption of natural enemies through frequent pyrethroid sprays are the likely cause in the development of these enduring outbreaks of thrips over a period of 11 years. A similar process may be developing with a reliance on insecticides coupled with Bt crops on cotton, with resistance development beginning and an increase in outbreaks of sucking insects. These case histories illustrate how a wide array of tactics embedded in redesigned cropping systems is the rational way to avoid a trajectory toward persistent insect outbreaks (Figure 18.2b) and provide enduring pest management with the longest possible intervals between insect outbreaks (Figure 18.2c), as sought in sustainable agroecosystems.

18.4.2 Letting nature take care of itself is not enough

Although preconceived culture-based assumptions about nature and its role in managing ecosystems are present to some degree in most researchers and practitioners (Hull *et al.* 2002, Perkins 1982), we are still struggling scientifically to understand how ecosystems function, both theoretically and practically. The notion that increasing biodiversity in agroecosystems alone will restore a critical balance and prevent insects from reaching outbreak proportions may be compelling, but it is less than convincing for a number of reasons. Firstly, insect outbreaks occur in unmanaged systems with their full complement of biodiversity. Reasons for outbreaks include interactions between plants and insects under certain abiotic conditions, disruption of biotic controls through asynchrony, indirect stressors from distant sources of pollution, and invasive organisms, all of which can also occur in agroecosystems. Secondly, if agronomically selected crops trade productivity and palatability for antiherbivore defense, then crops tend, theoretically, to be on the more susceptible end of the spectrum of their ancestors' natural variation (e.g., Rosenthal and Dirzo 1997). Thirdly, in their agricultural settings, crops are often more apparent (*sensu* Feeny 1976) than are their wild relatives in unmanaged habitats, thus individual plants may be colonized readily by herbivores, but perhaps the season is not long enough or vulnerable periods occur too early to develop a resident natural enemy community to curb populations before economic damage results. Also, natural enemy source pools may be limited in the landscape (Landis *et al.* 2000; Chapter 17, this volume). These factors may promote insect outbreaks through the disruption of insect–plant interactions despite a particular grower's efforts to diversify her or his cropping system (Macfadyen and Bohan 2010). Fourthly, "complexity begets stability" and "balance of nature" arguments, though commonly used metaphors or conceptual models, have not been upheld as general principles, either theoretically or empirically (Chapman 1939, May 1986, Walter 2008). Ecosystems are characterized more by dynamism than stasis and by chaos as much as by stability. In fact, allowing an agroecosystem to attain a static equilibrium, if this is possible in any system, may be counterproductive.

Consequently, thoughtful management is needed to achieve the "right kind of biodiversity" to support the goals of production agriculture (Landis *et al.* 2000, Letourneau and Bothwell 2008, Straub *et al.* 2008, Isaacs *et al.* 2009, Macfadyen *et al.* 2009, Letourneau *et al.* 2011). Critical investigations are underway in active research projects worldwide, including efforts to (1) distinguish between local host plants that promote pest or disease-carrying insects and those that primarily increase resources for natural enemies of insect pests (e.g., Barberi *et al.* 2010, Broatch *et al.* 2010, Schellhorn *et al.* 2010); (2) assess the importance of natural enemy species richness *per se* versus relative abundance, species identity, or species interaction networks for effective biological control services (Cardinale *et al.* 2006, Anjum-Zubair *et al.* 2010, Crowder *et al.* 2010, Moreno *et al.* 2010, Tylianakis *et al.* 2010); (3) understand the effects of landscape heterogeneity at different scales (Tscharntke *et al.* 2005, Werling and Gratton 2010, O'Rourke *et al.* 2011); (4) evaluate a wide range of natural enemies for complementary versus antagonistic effects on pest suppression (Johnson *et al.* 2010); (5) newly assess and quantitatively synthesize our current knowledge base (Poveda *et al.* 2008, Letourneau *et al.* 2009, Van Driesche *et al.* 2010); and, most comprehensively, (6) determine how best to apply this knowledge to redesign agroecosystems to be more resistant to insect outbreaks (Steingrover *et al.* 2010, Wood 2010) and potentially obtain many other benefits, even carbon credits for some of these practices (Gurr and Kvedaras 2010).

Other tactics that may operate synergistically with biological control to provide durable environmental resistance to insect pests of crops include mixed cultivars with an array of physical and chemical defenses, transgenic plants with multiple resistance mechanisms for each key pest expressed at critical times in targeted plant tissues, induced resistance mechanisms, and sterile insect techniques (Gurr and Kvedaras 2010), many of which are emerging agroscience topics (Lichtfouse *et al.* 2010). Integrating biological control into economic thresholds is essential because estimates of mortality due to predators, parasitoids, or pathogens more accurately determine if potential pest insects are under sufficient control, or if supplements are needed, such as an attractant like food sprays for enemies of honeydew-producing insects or a direct inoculative or inundative release of commercially available natural enemies. New-generation EILs include probabilistic EILs (PEIL) that allow growers to select among risk levels and environmental EILs (EEIL) that include environmental cost estimates into the managed cost parameter, though these have not replaced traditional EILs in the field (Higley and Peterson 2009). Given the cost, seasonality of workloads, and levels of facilities and expertise needed, Gurr and Kvedaras (2010) suggest that state departments of agriculture could become involved in providing services for informed pest management decisions. In addition, they call for developing molecular techniques, such as enzyme-linked immunosorbent assays (ELISA) or advanced DNA barcoding, so that field test kits available to growers, field scouts, or pest control advisors can provide an estimate of current levels of parasitoids or pathogens actively reducing pest reproductive capacities. Shared information and regional cooperation can cut costs for testing and alert growers to impending insect outbreaks.

Finally, in the case of insect outbreaks occurring within a well-designed agroecosystem, selective insecticides applied strategically (e.g., Spinosad with low toxicity to many natural enemies, granulosis viruses for specific hosts, or Bt var. tenebrionis for beetle outbreaks) will have a less pronounced effect on natural enemies and other beneficial insects than do the broad-spectrum materials that still dominate pest suppression tactics today. Naranjo and Ellsworth (2009) successfully combined treatment of *Bemesia tabaci* whiteflies with insect growth regulators and conservation biological control. Higher costs, more applications, the need for pest demographic data, and multiple individual pest decisions are some of the reasons why selective insecticides have not been adopted more widely (Castle and Naranjo 2009), as well as nonlethal effects on beneficials, but higher costs also reduce the likelihood of overapplication, and sophisticated knowledge is a commitment that must be made for the development of a new IPM era.

Instead of adopting the best fitting innovations into the existing industrial farming model, which promotes insect outbreaks, structural change is needed. This is not to say that insect outbreaks can be prevented completely in any ecosystem, especially agroecosystems, which tend to have more vulnerabilities and very low thresholds for insect outbreaks. However, the frequency, severity, and duration of insect outbreaks can be minimized using knowledge of the biology and ecology of the plants, herbivores, and natural enemies within the field, field margins, and larger landscape to manage population growth rates. The initiation of an outbreak can be delayed or prevented by decreasing reproductive success and crop colonization using knowledge of insect movement rates and colonization sources at the landscape level, mating disruption, timing of crop availability, repellents or attractants, and crop diversification schemes. Mortality rates are elevated through conserving and enhancing resident biological control agents and, if necessary, retarding herbivore growth and development, and augmenting pathogens and parasitoids. Whereas outbreaks of insects that are resistant to single tactics can persist through long time periods, outbreaks of shorter duration are likely with durable, multifactor strategies, requiring agroecosystem designs that enhance plant productivity by optimizing essential ecosystem processes and services.

Are insect outbreaks much more common in organic than in IPM or conventional agriculture? There are only a handful of cases for which insect outbreaks, relative pest densities, feeding damage and/or insect-vectored disease levels in organic crops have been compared to conventional or IPM crops in replicated field studies; some show no differences (Clark *et al.* 1998, Letourneau and Goldstein 2001, Lotter 2003). In those few cases, the prohibition of synthetic insecticides in organic farming did not result in devastating insect pest problems, despite the fact that the vast majority of public and private research and development in the United States has been focused on conventional agriculture. For example, cultivars bred for optimum performance under conventional agriculture conditions and inputs may not exhibit the same tolerance or resistance to insect outbreaks under organic management (Letourneau and van Bruggen 2006). Many organic farmers use the IPM concept (Zehnder *et al.* 2007) and contract pest control advisors (PCAs) to monitor insects and recommend management options. When damaging numbers of insect pests in organic crop fields do lower yields, the price premium for organic produce and sometimes lower input costs can help to preserve a reasonable profit margin (Tamis and van den Brink 1999, Lotter 2003, Swezey *et al.* 2007). Other production challenges, such as input costs for nutrients and weed management, are often reported as more severe than insect pest problems for organic growers.

18.5   Conclusions

Does IPM prevent insect outbreaks? Given that insect outbreaks occur at some point in almost every ecosystem, prevention is probably not possible. However, the potential for IPM to reduce the frequency or severity of insect outbreaks in managed systems is impossible to know, simply because IPM, as practiced, rarely represents an integrated, multitactic, expansive strategy with insecticides reserved as a last resort. For many reasons, including short-term economic risks, biological uncertainty, and power relations, hundreds of insects have developed resistance to one or more pesticides since IPM was proposed. Sustained target and secondary insect outbreaks contribute to the increased proportion of crop yield lost to insects during the IPM years. Will IPM practices improve? Not if, as Vanloqueren and Baret (2009) observed, scientists tend to envision the most *probable* rather than the most *desirable* future scenarios in agriculture, taking as a given the economic, social and political status quo. Prominent monoculture systems, global economic pressures, and pest management models in the United States provide many short-term arguments for plunging ahead with the "integrated *pesticide* management" type of IPM – a path that can truncate insect outbreaks for a decade or two, with a sequence of insecticides that work until they don't. Meanwhile, pesticide producers reap certain profits and growers, farmworkers, and consumers assume the risk when this path leads to the creation of resistant pests, secondary pest outbreaks, widespread crop losses, and health hazards, all of which are externalities not included in the cost of their production (Power 2010).

However, spatial econometrics of high-value ecosystem services associated with agricultural and unmanaged systems (e.g., Losey and Vaughan 2006, Power 2010) and high production estimates of alternative agricultural systems (Badgley *et al.* 2006) shed a different light on the relative costs and benefits of the status quo, and challenge formerly accepted assumptions with new data. Calls for fundamental changes arising from critiques of failed pesticide reduction policies (Zalom 1993, Geiger *et al.* 2010) seem to be fueling renewed commitments for innovations in agroecosystem engineering for integrative pest management (e.g., Birch *et al.* 2011). For example, the European Commission–funded ENDURE Network (ENDURE 2009) promotes the redesign of cropping systems to make them less vulnerable to pests. Part of this process includes legislation adopted in 2009 with increased restrictions on pesticides and the mandatory adoption of IPM principles before 2014. Lamine and Ricci's (2010) summary statements on the network's findings point out that (1) optimizing today's farming practices or substituting today's technologies and inputs for new ones will not sufficiently improve farming systems made vulnerable by years of reliance on pesticides; (2) some reconception and redesign of farming systems and innovations that go beyond the framework of reference currently adopted for the improvement of crop protection methods are needed; and (3) transitions toward sustainable crop protection strategies cannot be driven by farmers alone, but require a coordinated involvement of the entire agrifood system.

More than 50 years of IPM have helped stabilize what could have been an otherwise dismal trajectory of crop devastation and environmental disasters resulting from the pesticide treadmill. IPM has allowed a major shift from the rapid facilitation of insect outbreaks, to delaying their occurrence for as long as decades by selecting for resistance more slowly and maintaining some level of biological control in crop fields. Though outbreaks are reduced in the short term and productivity increased, the eventual outcome can be an enduring insect outbreak that no longer responds to insecticides. Moving toward sustainable agriculture with an IPM model that avoids sustained insect outbreaks and makes food security independent of petrochemical inputs requires us to expand our conceptual and investigative scope, embrace biological and economic uncertainty, and take more progressive steps in science and politics.

Acknowledgments

I thank Robert van den Bosch and Andrew Gutierrez for my entry into the "late" UC Berkeley Division of Biological Control, and all my mentors there, primarily Ken Hagen and Miguel Altieri, for getting me started on an academic path that keeps me curious and inspires my students. I thank P. Barbosa, B. Ekbom, L. Ehler, T. Cornelisse, J. Jedlicka, T. Krupnik, and J. R. Sirrine for helpful suggestions on earlier versions of the chapter. Partial funding was provided by the USDA-NRI grant 2005-55302-16345.

References

Altieri, M. A., and D. K. Letourneau. 1984. Vegetation diversity and insect pest outbreaks. *Critical Reviews in Plant Sciences* 2:131–169.

Andow, D. A. 1991. Vegetational diversity and arthropod population response. *Annual Review of Entomology* 36:561–586.

Anjum-Zubair, M., M. H. Schmidt-Entling, P. Querner, and T. Frank. 2010. Influence of within-field position and adjoining habitat on carabid beetle assemblages in winter wheat. *Agricultural and Forest Entomology* 12:301–306.

Attwood S. J., M. Maron, A. P. N. House, and C. Zammit. 2008. Do arthropod assemblages display globally consistent responses to intensified agricultural land use and management? *Global Ecology and Biogeography* 17:585–599.

Avila, J., and L. A. Rodriguez-del-Bosque. 2005. Impact of a Brazilian nucleopolyhedrovirus release on *Anticarsia gemmatalis* (Lepidoptera: Noctuidae), secondary insect pests, and predators on soybean in Mexico. *Journal of Entomological Science* 40:222–230.

Badgley, D., J. Moghtader, E. Quintero, E. Zakem, M. J. Chappell, K. Aviles-Vasquez, A. Samulon, and I. Perfecto. 2006. Organic agriculture and the global food supply. *Renewable Agriculture and Food Systems* 22:86–108.

Bajwa, W. I., and M. Kogan. 2004. Cultural practices: springboard to IPM. Pages 21–38 *in* O. Koul, G. S. Dahliwal, and G.W. Cuperus (editors), *Integrated Pest Management: Potential, Constraints, and Challenges*. CABI Publishing, Cambridge, MA.

Barberi, P., G. Burgio, G. Dinelli, A. C. Moonen, S. Otto, C. Vazzana, and G. Zanin. 2010. Functional biodiversity in the agricultural landscape: relationships between weeds and arthropod fauna. *Weed Research* 50:388–401.

Benbrook, C. M., E. Groth, M. Hansen, J. Halloran, and S. Marquardt. 1996. *Pest Management at the Crossroads*. Consumers Union of US, Inc., Yonkers, NY.

Birch, A. N. E., G. S. Begg, and G. R. Squire. 2011. How agro-ecological research helps to address food security issues under new IPM and pesticide reduction policies for global crop production systems. *Journal of Experimental Botany* 62:3251–3261.

Broatch, J. S., L. M. Dosdall, J. T. O'Donovan, K. N. Harker, and G. W. Clayton. 2010. Responses of the specialist biological control agent, *Aleochara bilineata*, to vegetational diversity in canola agroecosystems. *Biological Control* 52:58–67.

Brust, G. E., and R. E. Foster. 1999. New economic threshold for striped cucumber beetle (Coleoptera:Chrysomelidae) in cantaloupe in the Midwest. *Journal of Economic Entomology* 92:936–940.

Butler, S. J., J. A. Vickery, and K. Norris. 2007. Farmland biodiversity and the footprint of agriculture. *Science* 315:381–384.

Cardinale, B. J., D. S. Srivastava, J. E. Duffey, J. P. Wright, A. L. Downing, M. Sankavan, and C. Jouseau. 2006. Effects of biodiversity on the functioning of trophic groups and ecosystems. *Nature* 443:989–992.

Castle, S., and S. E. Naranjo. 2009. Sampling plans, selective insecticides and sustainability: the case for IPM as "informed pest management." *Pest Management Science* 65:1321–1328.

Castle, S. J., P. B. Goodell, and J. C. Palumbo. 2009. Implementing principles of the integrated control concept 50 years later – current challenges in IPM for arthropod pests. *Pest Management Science* 65:1263–1264.

Chapman, R. N. 1939. Insect population problems in relation to insect outbreak. *Ecological Monographs* 9:261–269.

Clark, M. S., H. Ferris, K. Klonsky, W. T. Lanini, A. H. C. van Bruggen, and F. G. Zalom. 1998. Agronomic, economic, and environmental comparison of pest management in conventional and alternative tomato and corn systems in northern California. *Agriculture Ecosystems & Environment* 68:51–71.

Coble, H. D., and E. E. Ortman. 2009. The USA national IPM roadmap. Pages 471–478 *in* E. B. Radcliffe, W. D. Hutchison, and R. E. Cancelado (editors), *Integrated Pest Management*. Cambridge University Press, Cambridge.

Coupe, M. D., and J. F. Cahill. 2003. Effects of insects on primary production in temperate herbaceous communities: a meta-analysis. *Ecological Entomology* 28:511–521.

Crowder, D. W., T. D. Northfield, M. R. Strand, and W. E. Snyder. 2010. Organic agriculture promotes evenness and natural pest control. *Nature* 466:109–123.

Curtis, J. 1998. *Fields of Change: A New Crop of American Farmers Finds Alternatives to Pesticides*. Natural Resources Defense Council, New York.

Dalin, P., Kindvall, O., and Björkman, C. (2009) Reduced population control of an insect pest in managed willow monocultures. *PLoS ONE* 4(5):e5487. doi:10.1371/journal.pone.0005487

Deguine, J-P., P. Ferron, and D. Russell. 2008. Sustainable pest management for cotton production:a review. *Agronomy and Sustainable Development* 28:113–137.

De Schutter, O. 2010. Report submitted to the United Nations Human Rights Council. http://www.panap.net/sites/default/files/UNSR-Food_Report_Agroecology.pdf

Denno, R. F., C. Gratton, M. A. Peterson, G. A. Langellotto, D. L. Finke, and A. F. Huberty. 2002. Bottom-up forces mediate natural-enemy impact in a phytophagous insect community. *Ecology* 83: 1443–1458.

Dhaliwal, G. S., O. Koul, and R. Arora. 2004. Integrated pest management: retrospect and prospect. Pages 1–20 *in* O. Koul, G. S. Dahliwal, and G. W. Cuperus (editors), *Integrated Pest Management: Potential, Constraints, and Challenges*. CABI Publishing, Cambridge, MA.

Diaz-Montano, J., M. Fuchs, B. A. Nault, J. Fail, and A. M. Shelton. 2011. Onion thrips (Thysanoptera: Thripidae): a global pest of increasing concern in onion. *Journal of Economic Entomology* 104:1–13.

Dwyer, G., J. Dushoff, and S. H. Yee. 2004. The combined effects of pathogens and predators on insect outbreaks. *Nature* 430:341–345.

Eber, S., H. P. Smith, R. K. Didham, and H. V. Cornell. 2001. Holly leaf-miners on two continents:what makes an outbreak species? *Ecological Entomology* 26:124–132.

Ehler, L. E. 2006. Integrated pest management (IPM): definition, historical development and implementation, and the other IPM. *Pest Management Science* 62:787–789.

Ehler, L. E., and D. G. Bottrell. 2000. The illusion of integrated pest management. *Issues in Science and Technology* 16:61–64.

ENDURE. 2009. ENDURE Network: diversifying crop protection. European Union. http://www.endure-network.eu

Feeny, P. P. 1976. Plant apparency and chemical defense. *Recent Advances in Phytochemistry* 10:1–40.

Frantz, G., and H. C. Mellinger. 2009. Shifts in Western Flower thrips, *Frankliniella occidentalis* (Thripidae: Thysanoptera) population abundance and crop damage. *Florida Entomologist* 92:29–34.

Furlong, M. J., and M. P. Zalucki. 2010. Exploiting predators for pest management: the need for sound ecological assessment. *Experimentalis et Applicata* 135:225–236.

Geiger, F., J. Bengtsson, F. Berendse, W. W. Weisser, M. Emmerson, M. B. Morales, P. Ceryngier, J. Liira, T. Tscharntke, C. Winqvist, S. Eggers, R. Bommarco, T. Part, V. Bretagnolle, M. Plantegenest, L. W. Clement, C. Dennis, C. Palmer, J. J. Onate, I. Guerrero, V. Hawro, T. Aavik, C. Thies, A. Flohre, S. Hanke, C. Fischer, P. W. Goedhart, and P. Inchausti. 2010. Persistent negative effects of pesticides on biodiversity and biological control potential on European farmland. *Basic and Applied Ecology* 11:97–105.

Guedes, N. M. P., J. Tolledo, A. S. Correa, and R. N. C. Guedes. 2010. Insecticide-induced hormesis in an insecticide-resistant strain of the maize weevil, *Sitophilus zeamais. Journal of Applied Entomology* 134:142–148.

Gurr, G. M., and O. L. Kvedaras. 2010. Synergizing biological control: scope for sterile insect technique, induced plant defences and cultural techniques to enhance natural enemy impact. *Biological Control* 52:198–207.

Halaj, J., and D. H. Wise. 2001. Terrestrial trophic cascades: how much do they trickle? *American Naturalist* 157:262–281.

Hardin, M. R., B. Benrey, M. Colle, W. O. Lamp, G. K. Roderick, and P. Barbosa. 1995. Arthropod pest resurgence – an overview of potential mechanisms. *Crop Protection* 14:3–18.

Harrison, S., and J. L. Maron. 1995. Impacts of defoliation by tussock moths (*Orgyia vetusta*) on the growth and reproduction of bush lupine (*Lupinus arboreus*). *Ecological Entomology* 20:223–229.

Harrison, S., A. Hastings, and D. R. Strong. 2005. Spatial and temporal dynamics of insect outbreaks in a complex multitrophic system: tussock moths, ghost moths, and their natural enemies on bush lupines. *Ann. Zool. Fennici* 42:409–419.

Hawkins, B. A., Mills, N. J. Jervis, M. A., and Price, P. W. 1999. Is the biological control of insects a natural phenomenon? *Oikos* 86:493–506.

Haye, T., P. G. Mason, L. M. Dosdall, and U. Kuhlmann. 2010. Mortality factors affecting the cabbage seedpod weevil, *Ceutorhynchus obstrictus* (Marsham), in its area of origin: a life table analysis. *Biological Control* 54:331–341.

Hendrichs, J., P. Kenmore, A. S. Robinson, M. J. B. Vreysen. 2007. Area-wide integrated pest management (AW-IPM): principles, practice and prospects. Pages 3–33 *in* M. J. B. Vreysen, A. S. Robinson, and J. Hendrichs (editors), *Area-Wide Control of Insect Pests*. Springer, Dordrecht.

Higley, L. G., and R. K. D. Peterson. 2009. Economic decision rules for IPM. Pages 14–24 *in* E. B. Radcliffe, W. D. Hutchinson, and R. E. Cancelado (editors), *Integrated Pest Management: Concepts, Tactics, Strategies, and Case Studies*. Cambridge University Press, Cambridge.

Hill, S. B., and R. J. MacRae. 1995. Conceptual framework for the transition from conventional to sustainable agriculture. *Journal of Sustainable Agriculture* 7:81–87.

Hu, J. F., S. Angeli, S. Schuetz, Y. Luo, and A. E. Hajek. 2009. Ecology and management of exotic and endemic Asian longhorned beetle *Anoplophora glabripennis*. *Agricultural and Forest Entomology* 11:359–375.

Hull, R. B., D. P. Robertson, and D. Rich. 2002. Assumptions about ecological scale and nature knows best hiding in environmental decisions. *Conservation Ecology* 6:12. http://www.consecol.org/vol6/iss2/art12

Isaacs, R., J. Tuell, A. Fiedler, M. Gardiner, and D. Landis. 2009. Maximizing arthropod-mediated ecosystem services in agricultural landscapes: the role of native plants. *Frontiers in Ecology and the Environment* 7:196–203.

Jepsen, J. U., S. B. Hagen, S. R. Karlsen, and R. A. Ims. 2009. Phase-dependent outbreak dynamics of geometrid moth linked to host plant phenology. *Proceedings of the Royal Society B-Biological Sciences* 276:4119–4128.

Johnson, M. D., J. L. Kellermann, and A. M. Stercho. 2010. Pest reduction services by birds in shade and sun coffee in Jamaica. *Animal Conservation* 13:140–147.

Kenis, M., M. Auger-Rozenberg, A. Roques, L. Timms, C. Pere, M.J.W. Cock, J. Settele, S. Augustin, and C. Lopez-Vaamonde. 2009. Ecological effects of invasive alien insects. *Biological Invasions* 11:21–45.

Kogan, M. 1998. Integrated pest management: historical perspectives and contemporary developments. *Annual Review of Entomology* 43:243–270.

Lamine, C., and P. Ricci. 2010. Reducing dependency on pesticides: transitions toward sustainable solutions over the long term. *In* M. B. E. Labussière and P. Ricci (editors), *European Crop Protection in 2030: A Foresight Study*. ENDURE™ Diversifying Crop Protection Workshop. ENDURE, Brussels.

Landis, D. A., S. D. Wratten, and G. M. Gurr. 2000. Habitat management to conserve natural enemies of arthropod pests in agriculture. *Annual Review of Entomology* 45:175–201.

Letourneau, D. K. 1997. Plant–arthropod interactions in agriculture. Pages 239–290 *in* L. E. Jackson (editor), *Ecology in Agriculture*. Academic Press, New York.

Letourneau, D. K., and D. A. Andow. 1999. Natural-enemy food webs. *Ecological Applications* 9:363–364.

Letourneau, D. K., I. Armbrecht, B. Salguero Rivera, J. Montoya Lerma, E. Jiménez Carmona, M. Constanza Daza, S. Escobar, V. Galindo, C. Gutiérrez, S. Duque López, J. López Mejía, A. M. Acosta Rangel, J. Herrera Rangel, L. Rivera, C. A. Saavedra, A. M. Torres, and A. Reyes Trujillo. 2011. Does plant diversity benefit agroecosystems? A synthetic review. *Ecological Applications* 21:9–21.

Letourneau, D. K., and S. G. Bothwell. 2008. Comparison of organic and conventional farms: challenging ecologists to make biodiversity functional. *Frontiers in Ecology and the Environment* 6:430–438.

Letourneau, D. K., and B. Goldstein. 2001. Pest damage and arthropod community structure in organic vs. conventional tomato production in California. *Journal of Applied Ecology* 38:557–570.

Letourneau, D. K., J. A. Jedlicka, S. G. Bothwell, and C. R. Moreno. 2009. Effects of natural enemy biodiversity on the suppression of arthropod herbivores in terrestrial ecosystems. *Annual Review of Ecology Evolution and Systematics* 40:573–592.

Letourneau, D. K., and A. van Bruggen. 2006. Crop protection in organic agriculture. Pages 93–121 *in* P. Kristiansen, A. Taji, and J. Reganold (editors), *Organic Agriculture: A Global Perspective*. CSIRO Publishing, Australia.

Lichtfouse, E., M. Hamelin, M. Navarrete, P. Debaeke, and A. Henri. 2010. Emerging agroscience. *Agronomy for Sustainable Development* 30:1–10.

Losey, J. E., and M. Vaughan. 2006. The economic value of ecological services provided by insects. *Bioscience* 56:311–323.

Lotter, D. W. 2003. Organic agriculture. *Journal of Sustainable Agriculture* 21:59–128.

Macfadyen, S., and D. A. Bohan. 2010. Crop domestication and the disruption of species interactions. *Basic and Applied Ecology* 11:116–125.

Macfadyen, S., R. Gibson, A. Polaszek, R. J. Morris, P. G. Craze, R. Planque, W. O. C. Symondson, and J. Memmott. 2009. Do differences in food web structure between organic and conventional farms affect the ecosystem service of pest control? *Ecology Letters* 12:229–238.

Margosian, M. L., K. A. Garrett, J. M. S. Hutchinson, and K. A. With. 2009. Connectivity of the American agricultural landscape: assessing the national risk of crop pest and disease spread. *Bioscience* 59:141–151.

Matteson, P. C., Gallagher, K. D., and Kenmore, P. E. 1994. Extension of integrated pest management for planthoppers in Asian irrigated rice: empowering the user. Pages 656–685 *in* R. F. Denno and T. J. Perfect (editors), *Planthoppers: Their Ecology and Management*. Chapman and Hall, London.

May, R. M. 1986. The search for patterns in the balance of nature – advances and retreats. *Ecology* 67: 1115–1126.

Merrigan, K. A. 2000. Politics, policy and IPM. Pages 497–504 *in* G. G. Kennedy and T. B. Sutton (editors), *Emerging Technologies for Integrated Pest Management: Concepts, Research, and Implementation*. APS Press, St. Paul, MN.

Miles, H. W. 1921. Observations on the insects of grasses and their relation to cultivated crops. *Annals of Applied Biology* 8:170–181.

Moreno, C. R., S. A. Lewins, and P. Barbosa. 2010. Influence of relative abundance and taxonomic identity on the effectiveness of generalist predators as biological control agents. *Biological Control* 52:96–103.

Naranjo, S. E., and P. C. Ellsworth. 2009. The contribution of conservation biological control to integrated control of *Bemisia tabaci* in cotton. *Biological Control* 51:458–470.

National Research Council (NRC). 1996. *Ecologically Based Pest Management: New Solutions for a New Century*. National Research Council, National Academy of Sciences, Washington, DC.

Nicholls, C. I., and M. A. Altieri. 2007. *Agroecology: Contributions towards a Renewed Ecological Foundation for Pest Management*. Cambridge University Press, Cambridge.

Nothnagle, P. J., and J. C. Schultz. 1987. What is a forest pest? Pages 59–80 *in* P. Barbosa and J. C. Schultz (editors), *Insect Outbreaks*. Academic Press, San Diego.

Oerke, E. C. 2006. Crop losses to pests. *Journal of Agricultural Science* 144:31–43.

O'Rourke, M. E., K. Rienzo-Stack, and A. G. Power. 2011. A multi-scale, landscape approach to predicting insect populations in agroecosystems. *Ecological Applications* 21:1782–1791.

Osteen, C., and M. Livingston. 2006. Pest management practices. Pages 107–115 *in* K. Weibe and N. Gollehon (editors), *Agricultural Resources and Environmental Indicators*. US Department of Agriculture Economic Information Bulletin. US Department of Agriculture, Washington, DC.

Palumbo, J. C., and S. J. Castle. 2009. IPM for fresh-market lettuce production in the desert Southwest: the produce paradox. *Pest Management Science* 65:1311–1320.

Perkins, J. H. 1982. *Insects, Experts and the Insecticide Crisis: The Quest for New Pest Management Strategies*. Plenum Press, New York.

Pimentel, D. 1961. Species diversity and insect population outbreaks. *Annals of the Entomological Society of America* 54:76–86.

Pimentel, D., and N. Goodman. 1978. Ecological basis for management of insect populations. *Oikos* 30:422–437.

Pimentel, D., McLaughlin, L., Zepp, A., Latikan, B., Kraus, T., Kleinman, P., Vancini, F., Roach, W. J., Graap, E., Keeton, W. S., and Selig, G. 1991. Environmental and economic impacts of reducing US agricultural pesticide use. Pages 679–718 *in* D. Pimentel (editor), *Handbook on Pest Management in Agriculture*. CRC Press, Boca Raton, FL.

Poveda, K., M. I. Gomez, and E. Martinez. 2008. Diversification practices: their effect on pest regulation and production. *Revista Colombiana de Entomologia* 34:131–144.

Power, A. G. 2010. Ecosystem services and agriculture: tradeoffs and synergies. *Philosophical Transactions of the Royal Society B-Biological Sciences* 365:2959–2971.

Ripper, W. E. 1944. Biological control as a supplement to chemical control of insect pests. *Nature* 153:448–452.

Risch, S. J. 1987. Agricultural ecology and insect outbreaks. Pages 217–238 *in* P. Barbosa and J. C. Schultz (editors), *Insect Outbreaks*. Academic Press, San Diego.

Root, R. B. 1996. Herbivore pressure on goldenrods (*Solidago altissima*): its variation and cumulative effects. *Ecology* 77:1074–1087.

Roland, J. 1993. Large-scale forest fragmentation increases the duration of tent caterpillar outbreak. *Oecologia* 93:25–30.

Rosenthal, J. P., and R. Dirzo. 1997. Effects of life history, domestication and agronomic selection on plant defence against insects: evidence from maizes and wild relatives. *Evolutionary Ecology* 11:337–355.

Schellhorn, N. A., R. V. Glatz, and G. M. Wood. 2010. The risk of exotic and native plants as hosts for four pest thrips (Thysanoptera: Thripinae). *Bulletin of Entomological Research* 100:501–510.

Schott, T., S. B. Hagen, R. A. Ims, and N. G. Yocco. 2010. Are population outbreaks in sub-arctic geometrids terminated by larval parasitoids? *Journal of Animal Ecology* 79:701–708.

Scriber, J. M. 1984. Nitrogen nutrition of plants and insect invasion. Pages 441–460 *in* R. D. Hauck (editor), *Nitrogen in Crop Production*. American Society of Agronomy, Madison, WI.

Smith, R. F. 1962. Integration of biological and chemical control: introduction and principles. *Bulletin of the Entomological Society of America* 8:188–189.

Steingrover, E. G., W. Geertsema, and W. van Wingerden. 2010. Designing agricultural landscapes for natural pest control: a transdisciplinary approach in the Hoeksche Waard (the Netherlands). *Landscape Ecology* 25:825–838.

Stern, V. M., R. F. Smith, R. van den Bosch, and K. S. Hagen. 1959. The integrated control concept. *Hilgardia* 29:81–101.

Stireman, J., L. A. Dyer, and R. Matlock. 2005. Top-down forces in managed and unmanaged habitats. Pages 303–323 *in* P. Barbosa and I. Castellanos (editors), *Ecology of Predator–Prey Interactions*. Oxford University Press, Oxford.

Straub, C. S., D. L. Finke, and W. E. Snyder. 2008. Are the conservation of natural enemy biodiversity and biological control compatible goals? *Biological Control* 45:225–237.

Strong, D. R., J. L. Maron, P. G. Connors, A. Whipple, S. Harrison, and R. L. Jefferies. 1995. High mortality, fluctuation in numbers, and heavy subterranean insect herbivory in bush lupine, *Lupinus arboreus*. *Oecologia* 104:85–92.

Swezey, S. L., P. Goldman, J. Bryer, and D. Nieto. 2007. Six-year comparison between organic, IPM and conventional cotton production systems in the northern San Joaquin Valley, California. *Renewable Agriculture and Food Systems* 22:30–40.

Tamis, W. L. M., and W. J. van den Brink. 1999. Conventional, integrated and organic winter wheat production in the Netherlands in the period 1993–1997. *Agriculture Ecosystems & Environment* 76:47–59.

Trumble, J. T. 1998. IPM: overcoming conflicts in adoption. *Integrated Pest Management Reviews* 3:195–207.

Tscharntke, T., A. M. Klein, A. Kruess, I. Steffan-Dewenter, and C. Thies. 2005. Landscape perspectives on agricultural intensification and biodiversity – ecosystem service management. *Ecology Letters* 8:857–874.

Turchin P., A. D. Taylor, and J. Reeve. 1999. Dynamic role of predators in population cycles of a forest insect: an experimental test. *Science* 285:1068–1071

Tylianakis, J. M., E. Laliberte, A. Nielsen, and J. Bascompte. 2010. Conservation of species interaction networks. *Biological Conservation* 143:2270–2279.

UC-IPM. 2008. What is IPM? Statewide IPM Program, Agriculture and Natural Resources, University of California, Berkeley. http://www.ipm.ucdavis.edu/WATER/U/ipm.html

UNCED. 1992. Agenda 21: The United Nations Programme of Action from Rio. http://www.un.org/esa/dsd/agenda21/

Van den Bosch, R., and V. M. Stern. 1962. Integration of chemical and biological control of arthropod pests. *Annual Review of Entomology* 7:367–386.

Van Driesche, R. G., R. I. Carruthers, T. Center, M. S. Hoddle, J. Hough-Goldstein, L. Morin, L. Smith, D. L. Wagner, B. Blossey, V. Brancatini, R. Casagrande, C. E. Causton, J. A. Coetzee, J. Cuda, J. Ding, S. V. Fowler, J. H. Frank, R. Fuester, J. Goolsby, M. Grodowitz, T. A. Heard, M. P. Hill, J. H. Hoffmann, J. Huber, M. Julien, M. T. K. Kairo, M. Kenis, P. Mason, J. Medal, R. Messing, R. Miller, A. Moore, P. Neuenschwander, R. Newman, H. Norambuena, W. A. Palmer, R. Pemberton, A. P. Panduro, P. D. Pratt, M. Rayamajhi, S. Salom, D. Sands, S. Schooler, M. Schwarzlander, A. Sheppard, R. Shaw, P. W. Tipping, and R. D. van Klinken. 2010. Classical biological control for the protection of natural ecosystems. *Biological Control* 54:S2–S33.

Vanloqueren, G., and P. V. Baret. 2009. How agricultural research systems shape a technological regime that develops genetic engineering but locks out agroecological innovations. *Research Policy* 38:971–983.

Walker, G. P., T. J. B. Herman, A. J. Kale, and A. R. Wallace. 2010. Action thresholds with and without estimates of natural control: an adjustable action threshold using larval parasitism of *Helicoverpa armigera* (Lepidoptera:Noctuidae) in IPM for processing tomatoes. *Biological Control* 52:30–36.

Walter, G. H. 2008. Individuals, populations and the balance of nature: the question of persistence in ecology. *Biology & Philosophy* 23:417–438.

Werling, B. P., and C. Gratton. 2010. Local and broad-scale landscape structure differentially impact predation of two potato pests. *Ecological Applications* 20:1114–1125.

Weiss, A., J. E. Dripps, and J. Funderburx. 2009. Assessment of implementation and sustainability of integrated pest management programs. *Florida Entomologist* 92:24–28.

Wood, D. M., D. Parry, R. D. Yanai, and N. E. Pitel. 2010. Forest fragmentation and duration of forest tent caterpillar (*Malacosoma disstria* Hübner) outbreaks in northern hardwood forests. *Forest Ecology and Management* 260:1193–1197.

Zalom, F. G. 1993. Reorganizing to facilitate the development and use of integrated pest-management. *Agriculture Ecosystems & Environment* 46:245–256.

Zehnder, G., O. M. Gurr, S. Kiehn, M. R. Wade, S. D. Wratten, and E. Wyss. 2007. Arthropod pest management in organic crops. *Annual Review of Entomology* 52:57–80.

Zorner, P. S. 2000. Shifting agricultural and ecological context for IPM. Pages 32–41 *in* G. G. Kennedy and T. B. Sutton (editors), *Emerging Technologies for Integrated Pest Management: Concepts, Research, and Implementation*. American Phytopathological Society Press, St. Paul, MN.

19

Insect Invasions: Lessons from Biological Control of Weeds

Peter B. McEvoy, Fritzi S. Grevstad, and Shon S. Schooler

19.1   Introduction

Biological control introductions represent large-scale experiments in population ecology that potentially provide valuable information on the relevant independent and dependent variables in insect outbreaks and invasions. Contrasting outbreaks of insect pests and biological control organisms can be fruitful for understanding, predicting, and managing population growth, movement, spatial spread, and impacts of insect species with eruptive dynamics. Biological control systems offer unique opportunities to manipulate whole populations and communities through the intentional, sanctioned movement of organisms into new locations. In biological weed control, the accidental introduction of a plant

Insect Outbreaks Revisited, First Edition. Edited by Pedro Barbosa, Deborah K. Letourneau and Anurag A. Agrawal.
© 2012 Blackwell Publishing Ltd. Published 2012 by Blackwell Publishing Ltd.

species, followed by the deliberate introduction of an herbivore species, is an opportunity to study a system in flux. We can discover how populations in ecological networks are regulated, how they evolve, and how new equilibria are established.

We use the theory of invasions (Shigesada and Kawasaki 1997) as a framework for analyzing spatial and temporal fluctuations of arthropods with eruptive dynamics. Biological invasions develop by phases including (1) arrival (the process by which individuals are transported to new areas outside of their native range), (2) establishment (the process by which populations grow to levels where extinction is unlikely), and (3) spread (the expansion of an invading species' range into new areas). Studies of the three successive phases of invasion suggest ways to intervene to manage invasive species by preventing arrival through international quarantine and inspection, establishment by early detection and eradication, and spread by domestic quarantines and barrier zones (Liebhold and Tobin 2008). Outbreak species have the ability to increase when rare and spread rapidly across a landscape. Similarly, biological weed control develops by stages resembling an invasion process from release and establishment, to increase and spread, and to suppression of the target plant population and changes in the speed and direction of plant succession. A growing number of biological control releases have been monitored in sufficient detail to reveal patterns and processes at each step. In particular, observations in the early stages of a biological control program shed light on the transition in vital rates from *endemic* to *epidemic* phases of insect population dynamics.

While biological control introductions resemble an invasion process, we also review lessons from the more traditional approach of studying population dynamics of biological control introductions. Here the focus is on patterns in population fluctuations over time, the regulation of populations by density-dependent factors, and the abiotic and biotic factors that limit population abundance.

19.2   Population establishment

Population establishment is a stochastic process; an introduction will sometimes lead to permanent establishment and sometimes not. Estimates of rates of successful establishment for biological control projects (each typically involving multiple and repeated releases) have hovered around 30% for predators and parasitoids (Hall and Ehler 1979, Williamson and Fitter 1996) and around 63% for insect herbivores (Julien 1989, Lawton 1990, Williamson and Fitter 1996, Syrett *et al.* 2000), with particular country or regional success rates being higher or lower (McEvoy and Coombs 1999, Fowler *et al.* 2000). The success rate for individual releases ranges from 0% (never establishes) to 100% (always establishes), depending on the agent species and the region of introduction. Factors influencing population establishment include a wide variety of external environmental and species-specific factors, but also factors related to the colonization process itself such as the number and size of releases, demographic and environmental stochasticity, level of genetic variation, and Allee effects. Biological control introductions have offered the opportunity to study the factors that influence population establishment, both experimentally and comparatively.

19.2.1 Stochasticity, Allee effects, and the number of individuals released

Theoretical approaches reveal the potential vulnerabilities of small colonizing populations and the factors that influence the relationship between the size of a population and its persistence (Goodman 1987, Dennis 1989, Grevstad 1999b, Lockwood *et al.* 2005, Simberloff 2009). Small populations are susceptible to demographic stochasticity, or the chance variation in mean demographic rates resulting from individual differences in reproductive success due to such factors as sex, mating status, fecundity, and timing of mortality. Although most biological control releases are probably large enough to avoid the influence of demographic stochasticity, most accidental species introductions are not. In contrast to demographic stochasticity, environmental stochasticity arises from external factors such as weather and habitat. Environmental stochasticity is generally considered to affect all population sizes equally (density-independent factor), sometimes overwhelming density-dependent factors during colonization events. Environmental variability may also interact with density-dependent factors; for example, a storm may reduce a population to a level at which demographic stochasticity or Allee effects become important. An Allee effect is defined as a reduction in population growth rate at low densities. It may occur, for example, as result of mate limitation, the need to feed gregariously to overcome plant defenses, or through inbreeding depression. While much attention has been drawn to the role of Allee effects in biological invasions (Lewis and Kareiva 1993, Taylor and Hastings 2005, Drake and Lodge 2006), there are remarkably few examples of demonstrated Allee effects in insect populations. If present, an Allee effect will result in strong dependence of establishment probability on initial colony size, and there may be a threshold population size required for establishment to occur (Grevstad 1999b).

The outcomes of biocontrol introductions provide empirical evidence that the number of organisms introduced in the initial introduction affects the probability of establishment. Beirne (1985) reported that 60% of Canadian biological control introductions that used 800 or more individuals led to establishment but only 15% of smaller introductions did so. Hall and Ehler (1979) and Hopper and Roush (1993) report similar correlations. However, because these studies compare establishment rates *among species*, the effect of the number of individuals released might be confounded with traits that make some species easier to obtain in larger numbers, such as a high reproductive rate or abundance in their native range; both of these factors have been shown to correlate with a higher establishment rate (Crawley 1986, 1989).

Biocontrol introductions offer the opportunity to experimentally test the effect of initial colony size upon establishment using different sized releases of the same agent species. Grevstad (1999a) (Plate 19.1) tested the effect of release size on establishment for two species of leaf beetles (*Galerucella* spp.) introduced into 36 sites in central New York State for biological control of purple loosestrife

(*Lythrum salicaria*, Lythraceae). She found that establishment success (defined as persistence to the third generation) increased significantly with initial release size for both species. Moreover, she found that the population growth rate (defined as the slope of lines in Plate 19.1) increased with release size, suggesting the presence of an Allee effect. Memmott *et al.* (1998, 2005) had similar findings with experimental releases of a thrips (*Sericothrips staphylinus*) on gorse (*Ulex europaeus*) and a psyllid (*Arytainilla spartiophila*) on Scotch broom (*Cytisus scoparius*). But Fauverge *et al.* (2007) found a contrasting result with the parasitic wasp, *Neodryinus typhlocybae*, where establishment was unaffected by release size ranging from 1 to 100 females, and there was no Allee effect. Instead they found that the net reproductive rate was higher at low initial densities than at high densities and that low densities did not impede mate finding. Kindvall *et al.* (1999) found the bush cricket *Metrioptera roeseli* to be extremely adept at finding mates, even at densities as low as 5 individuals per hectare. As a result, crickets at low densities were as fecund as those at higher densities. More experiments of this type, for a variety of biocontrol or other insects, would be valuable in revealing generalities with regard to the prevalence of Allee effects and the role of initial colony size on establishment.

The presence or absence of Allee effects can be critical for managing populations – whether the goal is conserving, harvesting, or controlling the population (Caswell 2001). Allee effects may prevent initial establishment of invading organisms. For example, Liebhold and Tobin (2008) note the lack of invasion into North America by the European spruce bark beetle in spite of numerous known accidental introductions and hundreds of interceptions by port inspectors since 1985 (Haack 2001). This insect depends on high-density aggregations to overcome defenses of its host tree. Provided that only small numbers of insects are ever released, this species is unlikely to establish a reproductive population. For insect pests that are already established, Allee effects could make eradication possible without the need to kill 100% of the population. Reducing the population below a threshold density would result in negative population growth and a high probability of extinction. Liebhold and Bascompte (2003) examined data on eradication success rates for recurring colonies of gypsy moth (*Lymantria dispar*) in the state of Washington and found that smaller colonies or those reduced to low densities by a biological insecticide (Bt) spray program were more likely to go extinct on their own. The general applicability of this approach is unclear, but experiments using biocontrol introductions can help us understand the role of Allee effects in the dynamics of pest management and eradication programs.

Whereas biocontrol introductions have shown us that larger introduced populations are more likely to persist than small ones, they also have shown that populations can sometimes be founded by remarkably few individuals. Grevstad (1999a) tested the possibility for a population to be founded by a single female. Of 10 releases of single gravid females of each of leaf beetles *Galerucella pusilla* and *G. calmariensis*, one population of *G. calmariensis* was still present after three generations, a time frame typically considered as establishment in biocontrol programs. Memmott *et al.* (2005) recovered 2 of 10 populations of the broom psyllid *Arytainilla spartiophila* that were founded by one pair. The entire North American population of the encyrtid wasp *Pauridia peregrina*, a biocontrol agent of citrus mealybug, was propagated prior to release from just one gravid female (Clausen 1978). These examples suggest the real possibility that unintended pest invasions could also arise from very small founder populations, even individual insects. Of course, we do not know how often such founding events take place for accidental introductions because they tend to go undetected. However, Zayed *et al.* (2007) sampled the genetic diversity of the introduced solitary bee (*Lasioglossum leucozonium*) long after it invaded and used a stochastic genetic population model to demonstrate that the entire North American population was likely founded by just one singly mated female. All of these cases serve as a warning that effective quarantine policies must be able to detect and stop even single individuals of arthropod species from being imported with cargo or luggage.

19.2.2   Number of independent releases

A correlation between establishment success and number of attempted introductions can be assumed for any species for which the probability of an individual release establishing is greater than zero but less than one. Where each release is independent of the others, the probability of at least one population establishing is $1-(1-p)^N$, where p is the probability of establishment of a single introduction attempt and N is the number of introduction attempts. This is analogous to a lottery, where more tickets purchased can greatly increase the overall chance of winning. The probability that an individual release establishes (p) varies from species to species, as illustrated by the frequency distribution of outcomes for biocontrol agent species released into Oregon (Figure 19.1). Many species establish only some of the time, while some establish all of the time, and others never establish no matter how many attempts are made. The Oregon data include some noteworthy cases where insects were established only after many failed attempts. For example, 18 failed releases of *Diorhabda elongata* were made into Oregon between 2003 and 2006 before one release in 2006 led to establishment (E. Coombs, Oregon Department of Agriculture, personal communication). Biocontrol releases, like accidental pest introductions, may fail for a wide variety of reasons including not only the population processes described above, but also for reasons under human control such as release into the wrong habitat or at the wrong time of year, or using unhealthy stock or the wrong life stages (Coombs 2004). These factors could be lumped broadly into the category of environmental stochasticity. Biological control practitioners typically learn from mistakes to improve establishment rates through time. In contrast, accidental species introductions are more dependent on chance coincidence of conditions that allow establishment and thus have lower establishment rates overall than intentional introductions (Williamson 1996). In any case, the mixed outcomes of releases of many biological control insects offer an important lesson in invasive pest ecology. Just because one accidental introduction did not lead to an invasion does not mean that others will not. The overall probability of invasion is expected to increase with the number of attempts whenever there is a nonzero possibility for invasion.

   Another way that an increase in the number of introductions can increase the likelihood of invasion is through increased probability of populations entering

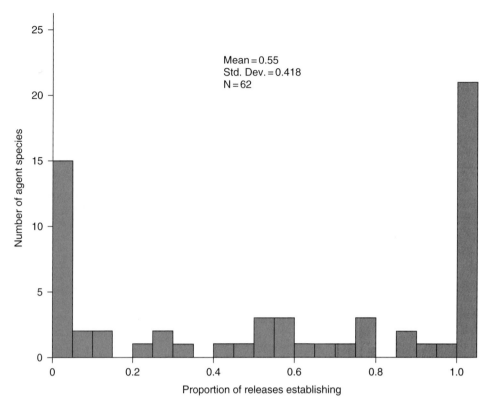

Figure 19.1 Histogram showing the number of weed biocontrol agents that established with different frequencies when released into the State of Oregon. Data are from the Oregon Department of Agriculture (E. Coombs, personal communication).

highly productive sites. Evidenced by replicated biocontrol introductions, intrinsic rates of growth following introduction can be highly variable among sites. For example, the initial population growth rate (R) varied by two orders of magnitude for *Galerucella calmariensis* and three orders of magnitude for *G. pusilla* (Grevstad 1999a). As a result of this variation, the number of independent introductions will affect the collective population growth of a biocontrol insect or invasive insect. The use of multiple introductions increases the likelihood that at least one rapidly growing population establishes to serve as a source for other locations, or a focal point for epidemic spread. Assuming continued exponential growth, the metapopulation will ultimately grow as fast as the fastest growing subpopulation.

Colony establishment is the first step in initiating invasions in new regions and also important during the spread phase after initial establishment. For many invading insects, spread occurs as the combination of local dispersal and rare long-distance dispersal resulting in outlying colonies which themselves become focal points of spread. Whether or not these distant colonization events result in population establishment can have a dramatic effect on the overall rate of spread

(Shigesada *et al.* 1995). Moreover, where host populations are spatially and temporally dynamic, the ability to colonize and establish populations on new host patches is critical to population level impacts on the host.

19.3 Population growth and spatial spread

19.3.1 Theoretical framework

A fundamental characteristic of any biological invasion is the speed at which the geographic range of the population expands, termed the "spread rate." Spread is usually measured as the increase in range radius over time, and spread rates vary considerably among invasive insects (Liebhold and Tobin 2008) and biological weed control organisms (Paynter and Bellgard 2011). The mathematical theory of invasions has been applied for analyzing spatial and temporal dynamics of insects in the context of invasions (Shigesada and Kawasaki 1997, Hastings *et al.* 2005) and biological control (Fagan *et al.* 2002). Invasion speed (1) depends on the environment and the life cycle of the species, (2) characterizes potential dynamics in simplified situations and provides a reference point for discussing more complicated scenarios, and (3) includes and extends population growth as a starting point for management, for conserving, harvesting, or controlling populations (Neubert and Caswell 2000). Weed biological control organisms make especially convenient experimental systems for studying each step in the invasion process because, in contrast to accidental introductions, many of the relevant details like the timing, location, population sizes, and species composition of introductions are well-documented.

The earliest studies of invasion speed used reaction–diffusion models of the form

Rate of change in density = dispersal rate + growth rate

$$\partial n / \partial t = D\nabla^2 n + f(n) \tag{19.1}$$

where dispersal enters through $D\nabla^2$, which denotes the diffusion operator; t is time; n(x, y, t) is the local population density at spatial coordinates x and y and time t; D is the coefficient of diffusion; and *f(n)* is a nonlinear function describing the net population change from birth and death. The basic reaction–diffusion model represented by Equation (19.1) assumes a homogeneous population (ignoring age or stage structure), continuous reproduction and movement (ignoring discrete generations and life cycle stages), a homogeneous environment (ignoring spatial or temporal heterogeneity in the environment), a deterministic process (ignoring stochastic variation in population growth and movement), and random movement (ignoring nonrandom movement in response to other organisms or environmental heterogeneity) and is essentially the result of a very large number of steps of arbitrarily small size (ignoring giant steps or jumps often associated with human transport). The model (Equation (19.1)) also assumes that the highest per capita growth rate occurs when the species is rare (ignoring Allee effects).

The null hypothesis that radial spread increases linearly with time as predicted by simple models has provided a useful starting point for analyzing the rate and pattern of invasion. While the linear spread predicted by the basic model in Equation (19.1) is a remarkably robust description of empirical spread, overall, empirical spread rates tend to be linear or faster (Williamson 1996, Shigesada and Kawasaki 1997). Spread of insects is often characterized by a combination of autonomous spread (usually, though not exclusively, a continuous process approximated by diffusion of an advancing front) and anthropogenic spread (a discrete process involving occasional long-distance dispersal or transport forming isolated colonies that grow, coalesce with each other and the main population, and greatly increase spread). Models have been developed that capture these components through stratified dispersal or fat-tailed dispersal kernels (Shigesada *et al.* 1995, Kot *et al.* 1996), but the chief problem in applying these models is estimating long-distance dispersal. In weed biological control, where long-distance dispersal is largely under the control of the investigator, both short-distance and long-distance dispersal, autonomous and anthropogenic (human-caused) components of spread, might be explicitly measured and modeled within the same analytical framework, as done for invasive ant species (Roura-Pascual *et al.* 2011).

19.3.2   Empirical analysis of spread of biocontrol agents

Several studies have investigated the statistical dependence of real invasions on potentially influential variables. Rates of spatial spread in insects and pathogens used for biological weed control can be predicted from statistical models (Paynter 2010, Paynter and Bellgard 2011). Paynter and Bellgard (2011) reviewed published dispersal data for 66 arthropod and 11 fungal pathogen weed biological control organisms. They screened seven factors that potentially affect spread rates and tested whether control organisms with rapid rates of spatial spread were more successful than those that spread slowly. Annual spread rates varied by four orders of magnitude from just tens of meters (e.g., the cochineal insect *Dactyplopius optunia*) to hundreds of kilometers per year (the moth *Arcola malloi*); the modal value in the distribution of spread rates was 1–10 km/yr. Successful agents were equally likely to be fast or slow dispersers, indicating that human-assisted redistribution of organisms can overcome the shortcomings of slow autonomous spread rates. Arthropod spatial spread rates were faster with increasing number of generations (voltinism) and fecundity, as expected if these contribute to increases in population growth rates. Spread rates were also correlated with mode of travel (for arthropods, flying is faster than crawling or passively dispersing by wind dispersal; wind-dispersed pathogens can travel long distances), taxon (among arthropods widely used in weed biological control, Diptera and Lepidoptera were the fastest and Coleoptera the slowest), lifestyle (miners, borers, and leaf rollers spread faster than external feeders, root and rosette feeders, and seed feeders, but not gall inducers), habitat (faster for aquatic and wetland than terrestrial), and the diversity of parasitoids attacking the agent in the native range (faster for high compared to low parasitoid diversity).

Paynter and Bellgard (2011) developed statistical models of spatial spread for biological control organisms that explained up to 73% of the variance in arthropod control organism spread rates and then applied these models to successfully predict the rate of invasion for six of seven species of invasive arthropod species. Model predictions for invasive species were generally close to observed rates (the same order of magnitude) except in the case of emerald ash borer *Agrilus planiennis* (Coleoptera), which disperses much faster than predicted. Human-mediated dispersal (e.g., in firewood) is known to transport significant numbers of individuals of this species considerably further than they can autonomously disperse. This highlights the need to take human-mediated spread into account when predicting the spread of an invasive species or planning the redistribution of a biological control organism; human assistance often provides the big jumps ahead of the advancing front of an invasion that lead to accelerating rates of spread. Such studies of dispersal by biological weed control organisms are helping to forecast how fast new organisms are likely to spread and thereby help (1) optimize release and redistribution strategies in biological control, and (2) optimize interventions to stop the spread of invading arthropods.

19.3.3   Local dynamics of biocontrol agents

Leaving the analysis of spatial spread, the more conventional framework for analyzing outbreaks is to analyze population fluctuations in time, more or less ignoring dispersal and spatial heterogeneity. Biological control systems analyzed in this way exhibit a range of dynamical behaviors under field conditions. Some control organisms have little discernible effect on their host's dynamics; in other interactions, the control organisms maintain their hosts at very low equilibrium levels (Huffaker and Kennett 1959); some control organisms and their hosts undergo cyclic oscillations (Moran and Zimmerman 1991, Bonsall *et al.* 2003); some interactions are characterized by frequent local extinction-and-recolonization events (McEvoy and Coombs 1999); and some interactions are characterized by episodic host outbreaks where the control organism ceases to limit host abundance (Schooler *et al.* 2011). Although the theory of multispecies interactions is relatively well developed, until recently there have been few tests of this theory using field data. Two cases are provided by plant–herbivore systems well studied in the context of biological control of weeds.

The first well-studied system involves interaction between the cinnabar moth (*Tyria jacobaeae* L., Lepidoptera: Arctiidae) and its food plant ragwort (*Jacobaea vulgaris* P.Gaertn. syn. *Senecio jacobaea* L., Asteraceae). At home in Europe and abroad (in North America, New Zealand, and Australia) where ragwort in invasive and a target for biological control, the essential features of this ecological system are a specialist, food-limited herbivore feeding on a plant that is often, though not always, limited by interspecific competition for recruitment sites. Bonsall *et al.* (2003) found the same insect–plant interaction exhibits very different dynamics in two locations with very similar climates by comparing time series of annual censuses of insect and plant populations in a mesic grassland in Silwood Park, United Kingdom (20 yr) and a coast sand dune in Meijendel, the Netherlands (26 yr). At Silwood Park, plant and insect populations do not cycle, and populations are relatively stable. At Meijendel, the interacting populations cycle, with amplitude fluctuations with a period of ~5 yr. Analyzing the time series revealed little if any effect of delayed density dependence on either the moth or the plant populations at Silwood Park. In contrast, significant lags at years 1 and 2 suggest that both direct and delayed density-dependent factors play a crucial role in the dynamics of the interaction at Meijendel. There was no evidence of entrainment of the fluctuations by external forcing at either sites; the two systems are at roughly the same latitude and experience similar weather. Instead, differences in dynamics are differences in the nonlinear feedbacks on the dynamics of plant and insect populations (i.e., differences in the form and strength of the density dependence). It is questionable whether either system of interactions, in which plant and insect populations remain at high levels, represents a desirable outcome for biological control. It may be possible to intervene in this system of interactions to produce more desirable outcomes – for example by coordinated reduction in disturbance, increase in plant competition, and addition of natural enemy individuals and species – as demonstrated in Life Table Response Experiments (combining factorial experiments and structured population models) conducted in Oregon, United States (McEvoy and Coombs 1999).

In a second well-studied example, Schooler *et al.* (2011) modeled population interactions in a plant–herbivore system consisting of a floating-aquatic weed *Salvinia molesta* and its biological control organism, the salvinia weevil *Cyrtobagus salviniae*, in four billabongs (oxbow lakes) sampled monthly over a period of 13 years. Global suppression of salvinia weed populations by the salvinia weevil is a classic example of successful biological control, but local outbreaks and local extinctions of host or insect can occur. Schooler *et al.* (2011) use a powerful blend of nonlinear dynamics, statistical inference, and stochastic time-series analysis applied to a large data set to show that incomplete control can be explained by alternative stable states. Their analysis shows that the system can alternate between a low, salvinia-free state, and a high state of dense salvinia weed biomass. Under some conditions, there are attractors associated with each of these two states and population trajectories may be attracted to either state; in other conditions, there may be only a single state that is attractive. Their modeling framework provides recommendations for management that would be difficult to identify using conventional perspectives and methods. Specifically, augmenting the system with weevils at key times might trap the system in a salvinia-free state. Such experiments could readily be undertaken using biological control organisms but might be undesirable or unethical with pest species. Their approach serves as a theoretical framework for analyzing outbreak dynamics of interacting populations in lakes, coral reefs, oceans, forests, and arid lands, where sudden changes occur between alternative, stable states (Scheffer *et al.* 2001).

19.3.4 Lessons learned

What general lessons can we draw for management of outbreak populations? First, biological control organisms and pest organisms in similar guilds have similar rates of population growth and dispersal. At first glance, there appears to be no ecological distinction between "pest" and "beneficial" organisms. However, insects used in biological weed control represent a biased sample from the pool of insect species that might become naturalized. Unlike weed biological control organisms, many accidentally introduced, invasive insects are extremely polyphagous. A study of invasive Lepidoptera established in New Zealand found that generalist insects did not establish more readily than specialists, but if they established, polyphagous species were more likely than specialists to become pests (Fowler *et al.* 2006). Some taxa that become invasive (e.g., ants and social wasps) are never used as weed biocontrol organisms, and they may behave differently or use different invasion pathways. Finally, a successful biological control organism must reduce weed populations, whereas a "successful" crop pest may cause economic damage that renders a crop unmarketable but does not cause the kind of ecological damage that would influence plant population dynamics. In sum, caution is advised when using the factors that predict biological control success to predict outbreaks and impacts of invasive insects.

With these caveats, additional lessons can be drawn. A second lesson is that rates of population growth, redistribution, and spatial spread may be incorporated in statistical models that predict rates of spread for biological control organisms; with allowance for the caveats given in this chapter, such models may be extrapolated to predict spread of invading and outbreaking insect species. Third, classical weed biological control programs aim to rapidly establish biological control organism populations throughout the range of the target organism. Optimal redistribution strategies require estimates of autonomous spread to know when and where human assistance is needed. Autonomous and anthropogenic effects on rates of population growth, movement, and spatial spread have rarely been modeled explicitly within the same framework in weed biological control. We can expect that harvesting of biological control organisms for redistribution or suppressing pest populations may change organism densities and thereby change local dynamics; in managing either biological control or arthropod pests, human-assisted redistribution may accelerate movement, allowing populations to spread at a much faster rate. In systems with alternative stable states, manipulation of density at key times can either promote or prevent outbreaks. And finally, analytic mathematical models of biological control systems promise to improve our ability to understand, predict, and manage insect outbreaks. To deliver on this promise, we need more effort to test predictions generated by models with subsequent observational and experimental studies of real ecological systems.

19.4 Abiotic influences on insect dynamics

Effects like outbreaks and invasions have multiple independent and interacting causes. Our analysis of population dynamics in earlier sections focuses on processes and parameters *internal* to the population (rates of development, growth, survival, reproduction, movement, population growth, and spatial spread) and how variation in these vital rates translates into impacts at the plant and herbivore trophic levels. Here we focus on *external* causes driving variation in the vital rates. How do abiotic (climate and weather) and biotic (resources, including host quantity and quality; and antagonists, including competitors and natural enemies – i.e., predators, parasites, or pathogens) factors act and interact to control the establishment, increase, and spread of biological control organisms; the suppression of plant abundance; and the successional changes in the community? We do not review mutualisms because, to our knowledge, no study has yet implicated mutualism as a form of interaction constraining the rates of increase or spatial spread of a biological weed control organism, although mutualists have been shown to influence feedbacks from weed biological control organisms to plant population levels (Paynter *et al.* 2010b). The role of facultative endosymbionts in increasing herbivore fitness on specific hosts is particularly ripe for further investigation (Himler *et al.* 2011).

19.4.1   Patterns and associations

Biological control scientists have tried to rank causes for success and failure in biocontrol, and study interactions among causes, by developing and analyzing large databases. In a pioneering effort, Crawley (1989) reviewed 627 biological weed control programs and found that few provided an objective measure of the depression in weed abundance. Subjective evaluation reveals little of how a population is affected by abiotic factors, by age or other aspects of its structure, by population density feedbacks, or by biotic interaction with other populations. However, subjective evaluation in biological control carries the weight of experienced observers. When experts were asked in an opinion survey in the Silwood International Project on Biological Control of Weeds (Moran 1985) what factors limit the effectiveness of control organisms in their weed biocontrol programs, they implicated a variety of possible causes. The percentage of all cases implicating each factor (some cases implicating more than one factor) was Climate 44%, Predators 22%, Parasitoids 11%, Disease 8%, Host incompatibility 33%, and Competition 12% (Crawley 1986). A common response to questions in this survey was "unknown" – the percentage of cases for which the influence of each factor was "unknown" was Climate 46%, Predators 56%, Parasitoids 49%, Host incompatibility 16%, and Competition 13%. The picture of what factors limit the effectiveness of control organisms when only the known cases are considered changes to Climate 81%, Predators 50%, Host incompatibility 39%, Parasitoids 22%, and Competition 14%. Reviewing these findings, McEvoy and Coombs (2000) earlier concluded that biocontrol scientists often have very limited knowledge of the factors that limit effectiveness of control organisms, and much of that knowledge is subjective.

Recent analyses of biological control databases (Syrett *et al.* 1996, Fowler *et al.* 2006, Martin and Paynter 2010, Paynter 2010, Paynter *et al.* 2010a, Paynter and Bellgard 2011) have partially redressed earlier shortcomings by emphasizing more objective and quantitative evaluation to create a more reliable basis for comparison, interpolation, and extrapolation and to develop and refine principles, concepts, and predictions regarding control organism specificity and effectiveness. These reviews have helped to synthesize information on the determinants of spatial spread (reviewed in this chapter) and the effects of the third trophic level on the density of herbivores and plants (reviewed in Section

19.5.1, "Predators, Parasitoids, and Pathogens"), and the insights gained have been extended to managing invading and outbreaking pest species.

19.4.2   Climate and seasonal adaptations

Like many pest insects, classical biocontrol organisms get moved from one part of the earth to another, or they experience climate change, and end up in a climate that differs to a greater or lesser degree from the one in which they evolved. In classical biological control, attempts are often made to use agents that come from a similar climate. However, for a variety of reasons, this does not always happen. Travel logistics, political access, or prior research can affect which source populations are used. Often, a single location is used as the source for introduction into a wide geographic area with many climates that cannot all be good matches. Thus, biocontrol systems can serve as examples of what happens when an insect population experiences a change in climate.

Approaches to predicting climatic effects on introduced organisms range from the simplest and most common approach of measuring the degree of similarity of climates in source and target regions (Wapshere 1974, 1983, Adair and Scott 1997, Goolsby *et al.* 2005), to measuring an organism's tolerance to critical extreme conditions (Good *et al.* 1997, De Clerck-Floate and Miller 2002), to degree-day phenology modeling (McClay and Hughes 1995, McClay 1996), to the comparison and modeling of species ranges (Julien *et al.* 1995, Robertson *et al.* 2008, Duncan *et al.* 2009, Sutherst *et al.* 2007). One shortcoming of climate matching as a predictive tool is that the native range of an organism may not accurately reflect its potential distribution in the new location because species ranges are also greatly influenced by biological interactions and source-sink dynamics (Davis *et al.* 1998). For example, the range of an insect may be smaller than the range it could physiologically occupy because natural enemies prevent its existence in certain areas. Alternatively, the current range may be greater than the physiological range if sink populations are sustained by immigration from source populations. In spite of the emphasis on predicting the suitability of climates for introduced insects, the empirical evidence of its importance for the establishment and performance of these insects remains sparse.

In the literature, one can find cases of biological control where climate matching appeared to be important in terms of population establishment and growth (Vogt *et al.* 1992, van Klinken *et al.* 2003, Bean *et al.* 2007) and other cases where it did not appear to be important (Hight *et al.* 1995, van Klinken *et al.* 2003). In fact, there are many examples of biocontrol and invasive pest species originating from one location (one climate) that have come to occupy wide geographic ranges with varying climates, suggesting the prevalence of plasticity or adaptability to new climates.

A few biocontrol studies have found evidence for specific climate or weather influences on particular insect populations. For example, Hill *et al.* (1993) constructed rain covers over gorse bushes to demonstrate that direct rainfall reduced gorse mite (*Tetranychus lintearius*) populations in New Zealand by 90%. By introducing a second strain of the mite from a part of Europe that had a better climate match to the rainiest regions of New Zealand, the mite reached field densities that were 30 times higher than the original strain. In another example, Cruttwell McFadyen (1992) found that long dry periods reduced densities of the moth *Epiblema strenuana*, limiting its effectiveness against parthenium weed. When periods of rain arrived, the insect population rebounded, but not as quickly as the plant population, so control was never effective.

Perhaps the best opportunities for investigating the general importance of climate match for introduced insect populations are from retrospective analysis of past biocontrol projects and their outcomes. Duncan *et al.* (2009) examined the ultimate distributions of five dung beetle species that were introduced to Australia to help reduce cattle dung and the nuisance flies associated with it. They compared the distribution of successful population establishment in Australia to the predicted species range based the range of climates occupied in the native range in South Africa (climate envelope model). They found that the predicted range from the climate envelope model was accurate for only two of the five species. In another retrospective study, Goolsby *et al.* (2005) examined the outcomes of five species of congeneric parasitic wasps (*Eretmocerus* spp.) introduced as biocontrol agents against the silverleaf whitefly (*Bemisia tabaci* biotype "B"), which is a pest across diverse climatic regions throughout the western United States. Each of the five agent species was from a different Eurasian source with a different climate, and all five species were released into each of four climatically different locations in the southwestern United States for a total of 20 climate match combinations. The outcomes were only partially consistent with climate match predictions. The five combinations with CLIMEX® software (Sutherst *et al.* 2007) "match index" of 64 or less did not establish (a score of 100 is a perfect match). But some populations with very good match scores also did not establish, suggesting other factors are also likely to be influencing establishment. Both of these studies involve limited data sets. Given the vast numbers of biocontrol projects on record, it is surprising that no one has carried out a comparative analysis of the relationship between climate match and project success using the larger collection of historic records of biocontrol introductions that have been made throughout the world (Julien and Griffiths 1998).

In understanding the effect of a new climate on an insect population, we need to consider not only the direct effects of the altered climate (e.g., reaction norms and tolerance limits), but also the way in which the new seasonal regime interacts with seasonal pre-adaptations. For example, many insects from temperate latitudes use critical photoperiod as a cue for timing either spring emergence or fall dormancy (Tauber and Tauber 1976). This is more reliable than cuing on temperature or rainfall directly, which are subject to irregularity. A locally adapted population will use a photoperiod that brings it into or out of diapause in synchrony with favorable conditions and resources. Since the timing of favorable conditions varies with geographic location, so too does the critical photoperiod used by populations in these different geographic locations (Masaki 1961). For introduced insects, seasonal pre-adaptations can have big consequences for establishment and/or population growth in the new location. For example, Bean *et al.* (2007) demonstrated how populations of the salt cedar leaf beetle *Diorhabda elongata* subsp. *deserticola* introduced into the Southwestern United States from northern China were able to have two generations north of 38° latitude, but just one generation in more southern latitudes where shorter day lengths triggered diapause in the first generation of adults. Early diapause meant that the insects did not have the energy reserves to survive the extended dormancy period until food resources again became available in the spring. As a result, releases that were made south of 38° did not establish. In other cases, a change in voltinism may not cause population failure altogether, but it is likely to have a large effect on population growth rates and outbreak potential. Shifts in phenology relative to the host could make introduced insects either more or less influential on the distribution and abundance of their hosts depending on the vulnerability of the plant stage attacked.

Climate, seasonal adaptations, host phenology, and other ecological inputs can all interact in complex ways to determine the performance and outbreak potential of an insect that has been moved to a new region. Intentional biological control introductions offer unique opportunities to study these interactions. Controlled experimental introductions of one population source into many climates or from many climates into one target location, experimental manipulations of local climate, and retrospective analyses of historic introductions are fruitful approaches that have been underutilized.

19.4.3   Resources

Insect herbivores are influenced by the nutritional quality of their host plants. As phytophagous insects are generally limited by nitrogen levels in their diets, many studies have found insect herbivores perform better when nitrogen levels in their host are increased through fertilizer addition (Mattson 1980). Other plant nutrients, salinity, and moisture can also influence plant quality for insect herbivores. Can biocontrol studies shed light on bottom-up causes for pest outbreaks?

In a literature review, we found 21 studies of weed biocontrol systems in which agent performance was compared over varying host quality (usually through manipulation of nitrogen or N–P–K fertilizer, but sometimes using natural variation in field plants) to examine the response of the weed biocontrol agent and/or the weed (Table 19.1). Of these studies, 19 involved insect biocontrol agents, one involved a pathogen, and one involved a combination of a pathogen and an insect. All but one study measured a response in the biocontrol agent population. In 15 cases (79%), there was a positive response of the agent to the increase in plant resources. In two cases (11%), there was a mixed or partial response. The response typically took the form of increased larval survival, faster and larger growth, and/or increased fecundity. In two cases (11%), there was little or no response of the insect. In the longest running study (nine populations studied for 4–6 yr), nitrogen appeared to cause increases in year-to-year variation in field densities (i.e., greater outbreak potential) (Myers and Post 1981).

A common conclusion from many of these studies is that using release sites high in nitrogen or adding fertilizer to release areas could improve establishment success. In the case of the weevil *Cyrtobagus salviniae*, introduced for the floating fern *Salvinia molesta*, the addition of nitrogen to release plots is reported to have turned a failing release population into an outbreaking one (Room and

Table 19.1 Outcomes of studies on the effects of plant resource addition and/or plant quality on weed biocontrol agents and the net effect of the added resource on the plant in the presence of biocontrol agents (if measured).

Reference	Biocontrol agent	Weed	Type of study	Resource that was varied	Effect on agent	Net effect on plant
Myers and Post 1981	Tyria jacobaeae (moth)	Seneceo jacobaea	Field experiment	N	Increased variance	Not measured
Room and Thomas 1985	Cyrtobagus sp. (weevil)	Salvinia molesta	Field experiment	N	Positive	Not measured
Room et al. 1989	Cyrtobagus salviniae (weevil), Samea multiplicalis (moth)	Salvinia molesta	Field experiment	N and latitude	Positive	Negative (reduction in undamaged meristems)
Kuniata 1994	Heteropsylla spinulosa (psyllid)	Mimosa invisa	Field experiment	N	Positive	Not measured
Wheeler and Center 1996	Hydrellia pakistanae (fly)	Hydrilla verticillata	Lab experiment with field-collected host material	Leaf toughness and leaf N	Positive	Not measured
Wheeler and Center 1997	Bagous hydrillae	Hydrilla verticillata	Lab experiment with field-collected host material	Stem toughness and stem N	Positive	Not measured
Hatcher et al. 1997	Uromyces rumicis (rust) and Gastrophysa viridula (leaf beetle)	Rumex obtusifolius	Lab experiment	N	Not measured	Positive
Wheeler et al. 1998	Lepidoptera Spodoptera pectinicornis	Pistia stratiotes	Lab experiment	N, P, and K	Partially positive (pupal weight but not fecundity)	Not measured
Grevstad 1999a	Galerucella pusilla, Galerucella calmariensis (leaf beetle)	Lythrum salicaria	Field observational	N	Neutral	Not measured

Reference	Biological control agent	Target plant	Study type	Nutrient/factor	Effect on agent	Effect on plant
Wheeler et al. 1998	Spodoptera pectinicornis (moth)	Pistia stratiotes	Lab experiment	N, P, and K	Positive	Not measured
Heard and Winterton 2000	Neochetina bruchi (weevil)	Eichhornia crassipes	Lab experiment	N, P, and K	Positive	Neutral
Heard and Winterton 2000	Neochetina eichhorniae (weevil)	Eichhornia crassipes	Lab experiment	N, P, and K	Positive	Positive
Hinz and Müller-Schärer 2000	Rhopalomyia n. sp. (gall midge)	Tripleurospermum perforatum	Lab experiment	N, P, K, and water	Positive	More damage, but plants more vigorous overall
Wheeler and Center 2001	Hydrellia pakistanae (fly)	Hydrilla verticillata	Lab experiment and field observation	N, P, and K	Positive	More damage, but net effect not measured
Wheeler 2001	Oxyops vitiosa (weevil)	Melaleuca quinquenervia	Lab experiment with field-collected host material	Leaf quality	Partially positive (larval survival, but not fecundity)	Not measured
Grevstad et al. 2003	Prokelisia marginata (planthopper)	Spartina alterniflora	Field observational	N	Positive	Not measured
LeJeune et al. 2005	Cuphocleonus achates, Sphenoptera jugoslavica, Larinus minutus, Urophora spp.	Centaurea diffusa	Field experiment	N, P	Mostly neutral	Positive (seed production)
Shearer et al. 2007	Mycoleptodiscus terrestris (pathogen)	Hydrilla verticillata	Lab experiment	sediment, and CO_2	Positive	Not measured
Van Hezewijk et al. 2008	Mogulones cruciger (weevil)	Cynoglossum officinale	Field and lab experiments	Nitrogen	Positive	Positive
Manrique et al. 2009	Episimus unguiculus (moth)	Schinus terebinthifolius	Lab experiment	N, P, K, and salinity	Positive	Not measured
Center and Dray 2010	Neochetina eichhorniae and Neochetina bruchi (weevils)	Eichhornia crassipes	Lab experiment	N, P, and K	Positive	Not measured

Thomas 1985). Once this population was established, feeding damage by the insect caused increases in plant leaf nitrogen, thereby supporting further growth and spread of the weevil and, in this case, successful biological control. The planthopper *Prokelisia marginata* introduced into estuaries of Washington for the control of *Spartina alterniflora* had higher population growth in sites with higher leaf nitrogen (Grevstad *et al*. 2004). However, the high nitrogen sites tended to also be those more exposed to wave and tide action resulting in lower winter survival of the planthopper and inferior establishment rates. In its native range, this species is a classic outbreaking species with population densities spanning three orders of magnitude (Denno *et al*. 1985). As a counterexample, establishment by *Galerucella* beetles was uncorrelated with levels of leaf nitrogen sampled at release sites (Grevstad 1999a).

In addition to affecting local population growth and establishment, host quality could affect spread rate. Reaction–diffusion models have illuminated the complex dynamics that may arise from interactions among herbivore density, host quality, dispersal frequency, and dispersal distance (Morris 1997, Dwyer and Morris 2006). The complex dynamics may include invasions with spread rates that vary through time.

Surprisingly few weed biocontrol studies have measured whether adding a plant resource increases herbivore effects on the plant population. Given that the plants themselves also respond positively to nutrient additions, can added nitrogen increase insect herbivore populations sufficiently to cause a decline in the plant population? In four cases where plants with and without added resources were compared in the presence of herbivores, those with resource additions tended to be more vigorous than control plants even if there was a higher herbivore load. One exception may be *Salvinia molesta*, where added nitrogen led to a reduction in the number of undamaged meristems (Room *et al*. 1989).

In summary, weed biocontrol systems support existing evidence that herbivorous insects generally respond positively to improved plant nutritional quality. Biocontrol systems also demonstrate that increases in plant resources can lead to insect outbreaks and may be critical in the initial establishment of insect populations in a region. However, there is little evidence that any outbreak that ensues would lead to reduced plant vigor or yield compared to a situation where resources were not manipulated. Weed biocontrol systems provide an untapped opportunity for field studies of the interactions of plant quality with spatial and temporal dynamics of invading populations.

19.5   Biotic interactions affecting insect dynamics

19.5.1   Predators, parasitoids, and pathogens

Weed biological control programs provide evidence that predators, parasitoids, and pathogens (natural enemies) can prevent outbreaks of introduced herbivores. It is important to understand the effect of natural enemies for several reasons: (1) to better understand the resistance of ecosystems to introduced herbivores, (2) to predict whether alien herbivores increase the populations of native and exotic natural enemies that "spill over" and affect populations of native prey species (Carvalheiro *et al.* 2008), and (3) to improve the effectiveness of biological control programs. Biological control agents are carefully screened in quarantine and reared through multiple generations before release, which means they are introduced without natural enemies. This is not necessarily the case for accidental introductions, because natural enemies, particularly parasites and pathogens, may arrive on or in the introduced phytophagous insect. Therefore, introduced biological control agents give us a better indication of the resistance of the recipient community to introduced herbivores than accidental introductions. It is perhaps no surprise that alien herbivores experience some mortality from natural enemies in their introduced range. The main questions examined regarding the effects of natural enemies on introduced herbivorous biological control agents are (1) do natural enemies attack introduced herbivores, (2) how and to what degree do natural enemies affect herbivore populations, and (3) does this mortality alter the effect of the herbivore on plant abundance? A review in 1976 (Goeden and Louda 1976) provides an examination of previous studies. We reviewed the subsequent literature on the subject to evaluate the evidence available to answer each of these questions.

Many studies simply report the taxonomic identity of the natural enemy. For predation, this is usually done in laboratory trials to determine whether a particular predator will feed on an herbivore. For parasitoids, parasites, and pathogens, this usually entails rearing field-collected eggs, larvae, and pupae and recording the parasite and pathogen species. Goeden and Louda (1976) do not include a review of these studies, but in our review we found that 9 of the 35 studies documented the occurrence of natural enemies, but did not quantify the proportion of herbivores affected (Table 19.2).

Accurately quantifying the effect of predators on herbivores requires separating the mortality associated with predation with that from other sources, typically using predator exclusion experiments in the field. Prior to Goeden and Louda's review, three studies used predator exclusion treatments to measure the extent to which the predators caused the reduction in biological control agent populations. Since their review, we found an additional nine studies that have used exclusion approaches to quantify the effect of predators on introduced herbivores (Table 19.2). Predation rates varied from 0% to 91% with an average of 40%. Only one study did not detect an effect of predation, suggesting that predation is expected to reduce herbivore abundance in 90% of introductions. However, this may be misleading, as there is a probable bias toward detecting predation due to selecting systems for experimentation where predation has previously been observed.

Quantifying the mortality associated with parasitoids is usually done through surveys; field collecting life stages of the herbivores, rearing them, and recording the proportion that have been parasitized. This method is cheaper than field exclusion experiments and perhaps as a consequence has been investigated more often. Goeden and Louda (1976) reported 20 studies that quantified the proportion of biological control organisms parasitized. Since then, 15 additional studies have examined the mortality associated with parasitism. Goeden and Louda

Table 19.2 Outcomes of studies on the effects of natural enemies on weed biocontrol agents and the net effect on the plant of the biological control agent in the presence of its natural enemies (if measured).

Reference	Biocontrol agent	Order	Weed	Method	% Parasitism	% Predation	% Pathogen	Effect on plant
Boughton and Pemberton 2008	Austromusotima camptozonale	Lepidoptera	Lygodium microphyllum	Field exclusion	0	20%	na	na
Briese 1986	Anaitis efformata	Lepidoptera	Hypericum perforatum	Field exclusion	14%	40%	na	na
Davalos and Blossey 2010	Galerucella calmariensis	Coleoptera	Lythrum salicaria	Field exclusion	0	0	na	No effect
Davalos and Blossey 2010	Galerucella pusilla	Coleoptera	Lythrum salicaria	Field exclusion	0	0	na	No effect
Hunt-Joshi et al. 2005	Galerucella calmariensis	Coleoptera	Lythrum salicaria	Field exclusion	na	13%	na	No effect
Pratt et al. 2003	Tetranychus lintearius	Acari	Ulex europaeus	Field exclusion	na	44%	na	na
Reimer 1988	Liothrips urichi	Thysanoptera	Clidemia hirta	Field exclusion	0	44%	na	na
Sebolt and Landis 2004	Galerucella calmariensis	Coleoptera	Lythrum salicaria	Field exclusion	0	39%	na	na
Wiebe and Obrycki 2004	Galerucella pusilla	Coleoptera	Lythrum salicaria	Field exclusion	0%	68%	na	na
Denoth and Myers 2005	Galerucella calmariensis	Coleoptera	Lythrum salicaria	Field exposure	na	91%	na	Reduced defoliation
Barratt and Johnstone 2001	Rhinocyllus conicus	Coleoptera	Thistles	Lab experiment	na	na	na	na
Costello et al. 2002	Oxyops vitiosa	Coleoptera	Melaleuca quinquenervia	Lab experiment	na	na	na	na
Davies et al. 2009	Tetranychus lintearius	Acari	Ulex europaeus	Lab experiment	na	na	na	na
Jacob et al. 2006	Zygina sp.	Hemiptera	Asparagus asparagoides	Lab experiment	na	na	na	na
Matos and Obrycki 2007	Galerucella calmariensis	Coleoptera	Lythrum salicaria	Lab experiment	na	17%	na	No effect
Nimmo and Tipping 2009	Boreioglycaspis melaleucae	Homoptera	Melaleuca quinquenervia	Lab experiment	na	na	na	na

Reference	Agent	Order	Weed	Method				
Broughton 2000	Various	Various	Lantana camara	Review	na	na	na	na
Hill and Hulley 1995	Various	Various	Various	Review and survey	varied	na	na	na
Paynter et al. 2010a	Various	Various	Various	Review and survey	varied	na	na	na
Baars 2003	Various	Lepidoptera	Lantana camara	Survey	na	na	na	na
Baars and Heystek 2003	Various	Various	Lantana camara	Survey	na	na	na	na
Carvalheiro et al. 2008	Mesoclanis polana	Diptera	Chrysanthemoides monilifera	Survey	12%	na	na	na
Dhileepan et al. 2005	Epiblema strenuana	Lepidoptera	Parthenium hysterophorus	Survey	13%	na	na	na
Edwards et al. 2009	Mesoclanis polana	Diptera	Chrysanthemoides monilifera	Survey	10%	na	na	na
Goeden et al. 1987	Coleophora klimeschiella	Lepidoptera	Salsola australia	Survey	66%	na	na	na
Heard et al. 2010	Macaria pallidata	Lepidoptera	Mimosa pigra	Survey	5%	na	na	na
Kula et al. 2010	Neomusotima conspurcatalis	Lepidoptera	Lygodium microphyllum	Survey	0–24%	25%	na	na
Lawton et al. 1988	Consercula cinisigna	Lepidoptera	Pteridium aquilinum	Survey	87%	na	na	na
Lawton et al. 1988	Panotima spp.	Lepidoptera	Pteridium aquilinum	Survey	50%	na	na	na
McFadyen 1997	Pareuchaetes pseudoinsulata	Lepidoptera	Chromolaena odorata	Survey	<10%	na	na	na
Muller and Goeden 1990	Coleophora parthenica	Lepidoptera	Salsola australia	Survey	2–45%	na	na	na
Murray et al. 2002	Rhinocyllus conicus	Coleoptera	Thistles	Survey	5%	na	na	na
Norman et al. 2009	Galerucella calmariensis	Coleoptera	Lythrum salicaria	Survey	na	na	na	na
Nuessly and Goeden 1984	Coleophora parthenica	Lepidoptera	Salsola australia	Survey	na	25%	na	na
Semple and Forno 1987	Samea multiplicalis	Lepidoptera	Salvinia molesta	Survey	22%	na	2%	na
van Klinken and Burwell 2005	Evippe sp. #1	Lepidoptera	Prosopis spp.	Survey	8%	na	na	na
Van Klinken and Flack 2008	Penthobruchus germaini	Coleoptera	Parkinsonia aculeata	Survey	47%	na	na	na

(1976) found that parasitism rates ranged from 1% to >90%, while subsequent studies have reported a range from 0% to 66%. Parasitism often varied widely within studies depending on site conditions.

Entomopathogens may affect populations of introduced herbivores. Two studies have examined the effect of entomopathogens on biological control agents (Table 19.2). Semple and Forno (1987) found three pathogens attacking a moth (*Samea multiplicalis*) which was introduced for the control of salvinia in Australia. However, these pathogens affected only 2% of the larvae collected, whereas 22% of the larvae were parasitized, suggesting a greater suppressive effect of the parasitoid than the pathogens. In contrast, McEvoy *et al.* (2008) showed that a host-specific microsporidian (*Nosema tyriae*) found on the cinnabar moth (*Tyria jacobaeae*), introduced for the control of ragwort (*Jacobaea vulgaris*), reduces the ability of the moth to effectively use a native plant (*Senecio triangularis*) as a host. Therefore, this pathogen may be preventing the introduced herbivore from switching hosts from an introduced weed to a native plant.

The most difficult question to answer is do natural enemies alter the effect of introduced herbivore outbreaks on plant abundance or populations? Goeden and Louda (1976) found that up to 1976, no study had followed the trophic cascade to its conclusion to measure the indirect effects of the predators or parasitoids on the plants. Since then four experimental studies have examined whether natural enemies affect the ability of a biological agent to control its target weed. A phytoseiid mite (*Phytoseiulus persimilis*) has been shown to suppress the populations of a tetranychid mite (*Tetranychus lintearius*) introduced for the control of gorse (*Ulex europeaus* L.) (Pratt *et al.* 2003). Exclusion experiments indicated that the predaceous mites were suppressing the populations of the biological control agents, but did not determine if the predation affected the herbivores' ability to suppress the target plant. A set of lab and field cage experiments found that a mirid predator (*Plagiognathus politis*) of the introduced chrysomelid control agent (*Galerucella calmariensis*) for purple loosestrife (*Lythrum salicaria*) greatly reduced agent populations and allowed the target plant to maintain greater leaf, stem, and reproductive biomass in the presence of predation (Hunt-Joshi *et al.* 2005). However, although this study detected an increase in plant biomass in cage experiments, it did not establish whether this led to an increase in field loosestrife populations. A study to detect an effect of predation and parasitism on plant condition in a field setting in British Columbia found that higher levels of predation were associated with reduced leaf defoliation by a beetle (*G. calmariensis*) introduced for the control of purple loosestrife (Denoth and Myers 2005). How changes in plant performance might translate into changes in plant population dynamics remains unresolved.

Although direct experimental evidence is lacking, conclusions based on the levels of predation and parasitism and reviews of biological control programs suggest predation and parasitism can reduce the effect of herbivore outbreaks on plant populations. A study on factors affecting survival of *Anaitis efformata*, a geometrid introduced into Australia to control St. John's wort (*Hypericum perforatum*), found that survival rates from egg to adult were <1.6% with mortality primarily due to egg parasitism and larval predation (Briese 1986). In addition, a study by van Klinken and Flack (2008) used mathematical models to determine the effect of parasitism on the effectiveness of the introduced bruchid seed feeder (*Penthobruchus germaini*) on an invasive tropical shrub (*Parkinsonia aculeata*). They found that egg parasitism by native Hymenoptera species varied between 10% and 70% and was density dependent, which greatly reduced the potential for the seed feeder to reduce populations of *Parkinsonia*. The conclusion of both of these studies was that this mortality would prevent successful control of the weed. Reviews of the success of weed biological control programs also indicate the ability of predation and parasitism to negate the effects of herbivore outbreaks on plant populations. A review of programs in New Zealand found that parasitism was significantly associated with the failure of the agents to suppress weed populations (Paynter *et al.* 2010a). This evidence indicates that predation and parasitism can prevent herbivore outbreaks from affecting plant populations.

From our review of the predation, parasitism, and pathogen studies, we conclude that there is ample evidence to indicate that natural enemies do reduce the potential for outbreaks of introduced herbivorous insects, but less evidence is available to document that natural enemies indirectly affect plant populations. This is due to a dearth of studies that have examined the full trophic cascade. The majority of the studies have documented the occurrence of natural enemies, fewer have measured the effect on survival, and four have examined whether there is a subsequent effect on the plant. The three studies that found no effect had low predation rates (>17%), while the study that found reduced defoliation recorded high predation rates (91%).

Recent research has also given us insights into more general ecological issues such as (1) can we predict susceptibility of an introduced herbivore to predation and parasitism, and (2) how quickly do introduced herbivores accumulate predators and parasites? The number of predators and parasites that attack an exotic organism may be related to the time since arrival (residence time and accumulation) and the relatedness of the exotic organism to native prey species (ecological analogues). In a survey of parasitism on 28 insects released for control of weeds in New Zealand, Paynter *et al.* (2010a) found that 19, mostly native, parasitoid species attacked 10 of the biological control agents. The attack rate on the introduced host species corresponded to the attack rate on the native hosts of the parasitoids; 15 of the parasitoids attacked five native herbivore species that were closely related to the introduced biological control agents. The authors also examined whether residence time, the time since the biological control agents were released, was positively correlated with the number of parasitoids the agents accumulated. They found no correlation, which indicates that the susceptible hosts rapidly accumulated the relevant parasitoids soon after release. This conclusion is supported by research on a stem-galling moth (*Epiblema strenuana*) that was introduced into Australia for the control of *Parthenium hysterophorus* (Dhileepan *et al.* 2005). The study found that 10 species of larval and pupal parasitoids attacked the moth in the initial stages of its field establishment, and the number had not increased when follow-up surveys were conducted 16 years later. In addition, a review found high rates of parasitism by parasitoids on weed biological control agents in South Africa soon after introduction, although they did not examine whether high levels of parasitism were maintained over the longer term (Hill and Hulley 1995). Therefore, alien herbivores that have closely related native species which experience parasitism in the introduced range are more likely to experience parasitism in the new range and the effect is likely to occur soon after their arrival.

These studies indicate that natural enemies have the ability to mitigate the adverse impacts of invading herbivores, either to the detriment of biological control programs or the benefit of reducing insect outbreaks and protecting native plant species and cropping systems. However, not all studies have detected an effect and in some cases effects may result from complex interactions. A study on a leaf-tying moth (*Evippe* sp.: Gelechiidae) introduced for the control of mesquite (*Prosopis* spp.) in Australia predicted a high parasitism rate due to a large number of closely related species of both the moth and the target plant (van Klinken and Burwell 2005). However, observed parasitism rates were actually very low, rarely above 2%. Predator–prey relationships may also be affected by the presence of other prey species. A study on the predation rates of the herbivorous chrysomelid (*Galerucella calmariensis*) introduced for the control of purple loosestrife found that predation from a coccinelid (*Harmonia axyridis*) had no overall effect on leaf tissue removed and survival of *G. calmariensis* was only reduced in the presence of an aphid (*Myzus lythri*) (Matos and Obrycki 2007). Possibly this is an example of apparent competition where *M. lythri* supplemented the low nutritional quality of *G. calmariensis*, which facilitated increased predation on *G. calmariensis* (Matos and Obrycki 2007).

In summary, research on weed biological control systems indicate that (1) natural enemies reduce the ability of introduced herbivores to outbreak and negatively affect plant populations; (2) accumulation of natural enemies may be rapid after introduction; and (3) alien herbivores with close, native relatives are more likely to be affected by natural enemies.

19.5.2 Competition

Introducing multiple control-organism species in biological control programs increases the chance that competitive interactions may reduce the rate of increase of a favored herbivore species and restrict the net effect of multiple herbivore species on plants. McEvoy and Coombs (2000) review competitive interactions that arise when one control organism (1) preys on another, (2) interferes with another's access to the shared resource, or (3) reduces the resources available to another. An assessment of the implications for weed control posed by antagonistic interactions must determine whether direct, negative effects on the plant by the antagonist outweigh the indirect, positive effect on the plant created by the antagonist acting via intermediate species. Figure 2 in McEvoy and Coombs (2000) portrays such direct and indirect effects. The relevance to insect outbreaks is that antagonistic interactions may prevent outbreaks and dampen the feedback from herbivores to plants.

One herbivore preys on another
Some species used to control plants function at two trophic levels, as carnivores and herbivores. Most scientists believe that facultative carnivores should not be

used for biological weed control, yet they have been introduced based on the argument that the control organism's direct, negative effect will outweigh its indirect, positive effect on the target plant. Take the case of spotted knapweed *C. maculosa* and several of its control organisms, a moth *Metzneria paucipunctella* Zeller (Lepidoptera: Gelechiidae) (the superior interference competitor that functions as a predator and a seed feeder) and two gall-flies *Urophora affinis* (Frauenfeld) and *U. quadrifasciata* (Meigen) (Diptera: Tephritidae) (the superior colonizers believed to be the more desirable biological control organisms). An observational study of the pattern of association among the three species within 10 field sites suggests that these species are distributed randomly and independently among seed heads (Story *et al.* 1991). Put the insect species singly or in combination on plants in cages, and the seed destruction by moth and fly larvae appears to be greater than with the moths or flies alone, despite the fact that *M. paucipunctella* larvae kill 19–67% of *U. affinis* larvae (Story *et al.* 1991). The evidence is equivocal because the artificial conditions of closed cages restrict movement, and the design confuses an increase in the number of consumer species with an increase in the number of consumer individuals.

One herbivore species interferes with another's access to the shared resource (Interference Competition)

If it is unclear what it means for biocontrol when control organisms eat each other, then it is even more difficult to discern the effects of more subtle forms of interference competition. Several categories of interaction along a continuum are best lumped here: competition that involves negative–negative interactions, asymmetric competition that implies unequal negative–negative interactions, and amensalism that involves negative–zero interaction. According to the opinion survey cited earlier in this chapter, competition limits the effectiveness of weed biocontrol organisms in only 13% of known cases (Crawley 1986), but his may be an underestimate. Successful biological control organisms, as well as invaders, are characterized by rapid population growth and high densities; high densities in turn increase the likelihood of interspecific competition. A meta-analysis showed that competition was stronger between invasive herbivores than between native herbivore species (Denno *et al.* 1995). Still, interspecific competition is notoriously difficult to document in the field, even when it occurs; this is one reason why there is a persistent debate in ecology concerning the importance of this force (Denno *et al.* 1995, Stewart 1996, Kaplan and Denno 2007). The opportunity to control the number of organisms and species released in biological control provides an underutilized test of the effects of competition in herbivorous insects.

One control organism species may interfere with another's access to shared resources such as hosts, mates, or microhabitats. Negative effects of one control organism on another's demography may include a decrease in reproduction or an increase in mortality or emigration. For example, at high density the weevil *Larinus minutus* Gyllenhal. (Coleoptera: Curculionidae) interferes with access to the flower heads of diffuse knapweed (*Centaurea diffusa*) by two gall fly species *Urophora affinis* and *U. quadrifasciata* (E. Coombs, unpublished observations). The implications of these interactions for the dynamics of interacting populations in this system are unknown.

One herbivore reduces the resources available to another
(Exploitation Competition)

It is generally assumed that if one control organism reduces host resources available to another, the likely outcome is a reduced host equilibrium. Most definitions of biocontrol assume that it is desirable to reduce pest equilibrium levels
while retaining sufficient stability in the pest–enemy interaction to prevent the
host from sporadically re-emerging as a pest. For the case of host–parasitoid
interactions, momentum is building once again to obtain a mathematical understanding of antagonistic interactions among control organisms, and their possible
adverse consequences for biocontrol success (May and Hassell 1981, Kakehashi
et al. 1984, Briggs 1993, Briggs *et al*. 1993). These investigations are relevant to
weed biocontrol because many insects that feed on plants (e.g., seed-slaying
insects) have lifestyles analogous to parasitoids (Price 1980), in which a female
lays an egg on a host and the larvae develop by feeding on the body of the host.
Briggs (1993) developed two versions of a stage-structured, delay-differential
equation model of two parasitoids attacking different developmental stages of a
single host species. She found realistic situations in which the best competitor is
not necessarily the parasitoid that produces the lowest adult host density and the
combination of two parasitoid species may yield higher host density than that
achieved by the single most effective parasitoid. This suggests that the lowest
host density may be achieved in biological control by the release of only the most
effective control organism rather than multiple control-organism species.

There is growing empirical evidence that competition among control organisms can reduce control success. Firstly, one control organism can undermine the
establishment of another organism species. Briese (1997) reports that the leaf
defoliator *Chrysolina quadrigemina* (Suffrian) (Coleoptera: Chrysomelidae) may
have inhibited the establishment of another, more effective control organism, a
root borer *Agrilus hyperici* (Liro) (Coleoptera: Buprestidae), by inducing boom-
and-bust fluctuations in populations of the target weed *Hypericum perforatum* L.
(Clusiaceae) in Australia. If the order in which control agents are established is
important for their eventual success, then we should try to predict the interactions that will occur between potential control agents, as well as their likely
individual impact on the weed, in the country of introduction, as currently practiced in New Zealand (Syrett *et al*. 1996). By analogy, the order of colonization
may affect the establishment and invasion of species.

Secondly, one established control organism may undermine the effectiveness
of another, more effective control organism. Woodburn (1996) reports that early,
aggregated attack on thistle seed heads by a univoltine, seed-feeding weevil,
Rhinocyllus conicus Frölich (Coleoptera: Curculionidae) hindered control of
nodding thistle *Carduus nutans* L. (Asteraceae) in Australia by decreasing the
numerical response of a more promising agent, a bivoltine tephritid seed fly
Urophora solstitialis (Linnaeus) (Diptera: Tephritidae). Also, plant changes
induced by one herbivore species may protect plants against a second, potentially
more damaging species. An example can be found in recent studies involving
plant-feeding mite species. Grape vines in the San Joaquin Valley of California
are protected from the highly virulent Pacific mite *Tetranychus pacificus*

McGregor by early inoculation with the more benign Willamette mite *Eotetranychus willametti* (McGregor) (Karban and English-Leob 1990, Karban *et al*. 1991, Karban *et al*. 1997). Plant pathologists have found similar cases in which "vaccination" with relatively benign pathogens reduces the negative impact of virulent viral, bacterial, and fungal pathogens (van Loon *et al*. 1998).

Another example, from the literature on pest outbreaks, shows how exploitive competition between two herbivorous insects benefits a native host plant as reported by Preisser and Elkinton (2008), who studied the invasive hemlock woolly adelgid (*Adelges tsugae*, HWA) and the more recently invading elongate Hemlock scale (*Fiorinia externa*, EHS); both feed on eastern hemlock (*Tsuga canadensis*) in the eastern United States. Their experiments indicated that competition from the benign herbivore species *F. externa* could mitigate the adverse effects of the more virulent *A. tsugae*.

A lesson for biological control is that releasing a species that is ineffective in weed control programs may reduce the impact of more virulent control species. For managing invasive species, introducing, augmenting, or conserving competitors of pathogens or weeds is a well-known practice; now the tactic of using competitors for control can be cautiously extended to biological control of arthropods.

19.6   Summary

We have applied two levels of explanation to arthropod species with outbreak dynamics. We have found strong evidence for effects of internal causes (changes in vital rates) on the fluctuations of arthropod species with eruptive dynamics. We have found less evidence to link variation in vital rates to variation and interaction in external causes (variation in abiotic factors, resources, competitors, mutualists, and natural enemies), and less still that characterizes the feedbacks from changes in insect populations to changes in plant populations and communities. Biological control practitioners often exaggerate the effects of abiotic factors on control organism establishment, increase, and spread; there are but a few studies of the contribution of abiotic factors to periodic eruptions in control organism populations. Statistical approaches to screening potentially influential variables have demonstrated the power to predict in some cases, but we need more work combining observations, experiments, and analytic mathematical models as a foundation for understanding, prediction, and management. The top-down effects of natural enemies on herbivore demography and population dynamics are well documented, but how these effects translate into changes in plant population dynamics is poorly documented. Despite renewed interest in competition among phytophagous arthropods, the issue of single versus multiple natural enemy introductions has not been resolved: when, where, and how does competition among multiple control-organism species lead to reduced population growth and/or reduced persistence in herbivore populations and reduced suppression of plant populations? Two unanswered questions deserve more attention for understanding, predicting, and managing biological control introductions and insect outbreaks. How do interactions of herbivores with resources, competitors, natural enemies, and mutualists feed back to the plants and change plant population dynamics, spatial spread, and succession? How do biological control organisms evolve as they increase and spread across the landscape? Biological control systems have provided very good examples of an invasion process, with measures of spread and dispersal that are difficult to obtain in established populations. They have provided evidence of the importance of release size and number of introduction attempts, the potential of invasion by very small founding populations, and the influence of bottom-up effects on establishment. Finally, they highlight the response and resilience of organisms to climate change, including the potentially complex roles of seasonal adaptations, in an era of global change.

References

Adair, R., and J. Scott. 1997. Distribution, life history and host specificity of *Chrysolina picturata* and *Chrysolina* sp. B (Coleoptera: Chrysomelidae), two biological control agents for *Chrysanthemoides monilifera* (Compositae). *Bulletin of Entomological Research* 87:331–341.

Baars, J. R. 2003. Geographic range, impact, and parasitism of lepidopteran species associated with the invasive weed *Lantana camara* in South Africa. *Biological Control* 28:293–301.

Baars, J. R., and F. Heystek. 2003. Geographical range and impact of five biocontrol agents established on *Lantana camara* in South Africa. *Biocontrol* 48:743–759.

Barratt, B. I. P., and P. D. Johnstone. 2001. Factors affecting parasitism by *Microctonus aethiopoides* (Hymenoptera : Braconidae) and parasitoid development in natural and novel host species. *Bulletin of Entomological Research* 91:245–253.

Bean, D. W., T. L. Dudley, and J. C. Keller. 2007. Seasonal timing of diapause induction limits the effective range of *Diorhabda elongata deserticola* (Coleoptera: Chrysomelidae) as a biological control agent for tamarisk (*Tamarix* spp.). *Environmental Entomology* 36:15–25.

Beirne, B. P. 1985. Avoidable obstacles to colonization in classical biological control of insects. *Canadian Journal of Zoology* 63:743–747.

Bonsall, M., E. Van der Meijden, and M. Crawley. 2003. Contrasting dynamics in the same plant–herbivore interaction. *Proceedings of the National Academy of Sciences of the USA* 100:14932–14936.

Boughton, A. J., and R. W. Pemberton. 2008. Efforts to establish a foliage-feeding moth, *Austromusotima camptozonale*, against *Lygodium microphyllum* in Florida, considered in the light of a retrospective review of establishment success of weed biocontrol agents belonging to different arthropod taxa. *Biological Control* 47:28–36.

Briese, D. T. 1986. Factors affecting the establishment and survival of *Anaitis efformata* (Lepidoptera: Geometridae) introduced into Australia for the biological control of St. John's wort, *Hypericum perforatum*. II. Field trials. *Journal of Applied Ecology* 23:821–839.

Briese, D. T. 1997. Biological control of St. John's wort: past, present, and future. *Plant Protection Quarterly* 12:73–80.

Briggs, C. J. 1993. Competition among parasitoid species on a stage-structured host and its effect on host suppression. *American Naturalist* 141:372–397.

Briggs, C. J., R. M. Nisbet, and W. W. Murdoch. 1993. Coexistence of competing parasitoid species on a host with a variable life cycle. *Theoretical Population Biology* 44:341–373.

Broughton, S. 2000. Review and evaluation of lantana biocontrol programs. *Biological Control* 17:272–286.

Carvalheiro, L. G., Y. M. Buckley, R. Ventim, S. V. Fowler, and J. Memmott. 2008. Apparent competition can compromise the safety of highly specific biocontrol agents. *Ecology Letters* 11:690–700.

Caswell, H. 2001. *Matrix Population Models: Construction, Analysis, and Interpretation*, 2nd ed. Sinauer, Sunderland, MA.

Center, T. D., and F. A. Dray Jr. 2010. Bottom up control of water hyacinth weevil populations: do the plants regulate the insects? *Journal of Applied Ecology* 47:329–337.

Clausen, C. P. (editor). 1978. *Introduced Parasites and Predators of Arthropod Pests and Weeds: A World Review*. USDA Agriculture Handbook No. 480. US Department of Agriculture, Washington, DC.

Coombs, E. M. 2004. Factors that affect successful establishment of biological control agents. Pages 85–94 *in* E. M. Coombs, J. K. Clark, G. L. Piper, and A. Cofrancesco (editors), *Biological Control of Invasive Plants in the United States*. Oregon State University Press, Corvallis.

Costello, S. L., P. D. Pratt, M. B. Rayachhetry, and T. D. Center. 2002. Morphology and life history characteristics of *Podisus mucronatus* (Heteroptera: Pentatomidae). *Florida Entomologist* 85:344–350.

Crawley, M. J. 1986. The population biology of invaders. *Philosophical Transactions of the Royal Society of London B Biological Sciences* 314:711–731.

Crawley, M. J. 1989. The successes and failures of weed biocontrol using insects. *Biocontrol News and Information* 10:213–223.

Cruttwell McFadyen, R. 1992. Biological control against parthenium weed in Australia. *Crop Protection* 11:400–407.

Davalos, A., and B. Blossey. 2010. The effects of flooding, plant traits, and predation on purple loosestrife leaf-beetles. *Entomologia Experimentalis et Applicata* 135:85–95.

Davies, J., J. Ireson, and G. Allen. 2009. Pre-adult development of *Phytoseiulus persimilis* on diets of *Tetranychus urticae* and *Tetranychus lintearius*: implications for the biological control of *Ulex europaeus*. *Experimental and Applied Acarology* 47:133–145.

Davis, A. J., L. S. Jenkinson, J. H. Lawton, B. Shorrocks, and S. Wood. 1998. Making mistakes when predicting shifts in species range in response to global warming. *Nature* 391:783–786.

De Clerck-Floate, R., and V. Miller. 2002. Overwintering mortality of and host attack by the stem-boring weevil, *Mecinus janthinus* Germar, on Dalmatian toadflax (*Linaria dalmatica* (L.) Mill.) in western Canada. *Biological Control* 24:65–74.

Dennis, B. 1989. Allee effects: population growth, critical density, and the chance of extinction. *Natural Resource Modeling* 3:481–538.

Denno, R. F., L. W. Douglass, and D. Jacobs. 1985. Crowding and host plant nutrition: environmental determinants of wing-form in *Prokelisia marginata*. *Ecology* 66:1588–1596.

Denno, R. F., M. S. McClure, and J. R. Ott. 1995. Interspecific interactions in phytophagous insects: competition reexamined and resurrected. *Annual Review of Entomology* 40:297–331.

Denoth, M., and J. H. Myers. 2005. Variable success of biological control of *Lythrum salicaria* in British Columbia. *Biological Control* 32:269–279.

Dhileepan, K., C. J. Lockett, and R. E. McFadyen. 2005. Larval parasitism by native insects on the introduced stem-galling moth *Epiblema strenuana* Walker (Lepidoptera: Tortricidae) and its implications for biological control of *Parthenium hysterophorus* (Asteraceae). *Australian Journal of Entomology* 44:83–88.

Drake, J. M., and D. M. Lodge. 2006. Allee effects, propagule pressure and the probability of establishment: risk analysis for biological invasions. *Biological Invasions* 8:365–375.

Duncan, R. P., P. Cassey, and T. M. Blackburn. 2009. Do climate envelope models transfer? A manipulative test using dung beetle introductions. *Proceedings of the Royal Society B-Biological Sciences* 276:1449–1457.

Dwyer, G., and W. F. Morris. 2006. Resource-dependent dispersal and the speed of biological invasions. The *American Naturalist* 167:165–176.

Edwards, P. B., R. J. Adair, R. H. Holtkamp, W. J. Wanjura, A. S. Bruzzese, and R. I. Forrester. 2009. Impact of the biological control agent *Mesoclanis polana* (Tephritidae) on bitou bush (*Chrysanthemoides monilifera* subsp. *rotundata*) in eastern Australia. *Bulletin of Entomological Research* 99:51–63.

Fagan, W. F., M. A. Lewis, M. G. Neubert, and P. van den Driessche. 2002. Invasion theory and biological control. *Ecology Letters* 5:148–157.

Fauvergue, X., J. C. Malausa, L. Giuge, and F. Courchamp. 2007. Invading parasitoids suffer no Allee effect: a manipulative field experiment. *Ecology* 88:2392–2403.

Fowler, S. V., H. M. Harman, R. Norris, and D. Ward. 2006. Biological control agents: can they tell us anything about the establishment of unwanted alien species? Pages 155–166 *in* R. Allen and W. G. Lee (editors), *Biological Invasions in New Zealand*. Springer Verlag, Berlin.

Fowler, S. V., P. Syrett, and R. L. Hill. 2000. Success and safety in the biological control of environmental weeds in New Zealand. *Austral Ecology* 25:553–562.

Goeden, R. D., and S. M. Louda. 1976. Biotic interference with insects imported for weed control. *Annual Review of Entomology* 21:325–342.

Goeden, R. D., D. W. Ricker, and H. Muller. 1987. Introduction, recovery, and limited establishment of *Coleophora klimeschiella* (Lepidoptera: Coleophoridae) on Russian thistles, *Salsola autralis*, in southern California. *Environmental Entomology* 16:1027–1029.

Good, W. R., J. M. Story, and N. W. Callan. 1997. Winter cold hardiness and supercooling of *Metzneria paucipunctella* (Lepidoptera: Gelechiidae), a moth introduced for biological control of spotted knapweed. *Environmental Entomology* 26:1131–1135.

Goodman, D. 1987. The demography of chance extinction. Pages 11–34 *in* M. E. Soulé (editor), *Viable Populations for Conservation*. Cambridge University Press, New York.

Goolsby, J. A., P. J. DeBarro, A. A. Kirk, R. W. Sutherst, L. Canas, M. A. Ciomperlik, P. C. Ellsworth, J. R. Gould, D. M. Hartley, K. A. Hoelmer, S. E. Naranjo, M. Rose, W. J. Roltsch, R. A. Ruiz, C. H. Pickett, and D. C. Vacek. 2005. Post-release evaluation of biological control of *Bemisia tabaci* biotype "B" in the USA and the development of predictive tools to guide introductions for other countries. *Biological Control* 32:70–77.

Grevstad, F. S. 1999a. Experimental invasions using biological control introductions: the influence of release size on the chance of population establishment. *Biological Invasions* 1:313–323.

Grevstad, F. S. 1999b. Factors influencing the chance of population establishment: implications for release strategies in biocontrol. *Ecological Applications* 9:1439–1447.

Grevstad, F. S., D. R. Strong, D. Garcia-Rossi, R. W. Switzer, and M. S. Wecker. 2003. Biological control of *Spartina alterniflora* in Willapa Bay, Washington using the planthopper *Prokelisia marginata*: agent specificity and early results. *Biological Control* 27:32–42.

Grevstad, F. S., R. W. Switzer, and M. S. Wecker. 2004. Habitat trade-offs in the summer and winter performance of the planthopper *Prokelisia marginata* introduced against the intertidal grass *Spartina alterniflora* in Willapa Bay, Washington. Pages 523–528 *in Proceedings of the XI International Symposium on Biological Control of Weeds*. April 27–May 3, 2003. CSIRO, Canberra, Australia.

Haack, R. A. 2001. Intercepted Scolytidae (Coleoptera) at U.S. ports of entry: 1985–2000. *Integrated Pest Management Reviews* 6:253–282.

Hall, R. W., and L. E. Ehler. 1979. Rate of establishment of natural enemies in classical biological control. *Bulletin of the Entomological Society of America* 25:280–282.

Hastings, A., K. Cuddington, K. F. Davies, C. J. Dugaw, S. Elmendorf, A. Freestone, S. Harrison, M. Holland, J. Lambrinos, U. Malvadkar, B. A. Melbourne, K. Moore, C. Taylor, and D. Thomson. 2005. The spatial spread of invasions: new developments in theory and evidence. *Ecology Letters* 8:91–101.

Hatcher, P. E., N. D. Paul, P. G. Ayres, and J. B. Whittaker. 1997. Added soil nitrogen does not allow *Rumex obtusifolius* to escape the effects of insect–fungus interactions. *Journal of Applied Ecology* 34:88–100.

Heard, T. A., L. P. Elliott, B. Anderson, L. White, N. Burrows, A. Mira, R. Zonneveld, G. Fichera, R. Chan, and R. Segura. 2010. Biology, host specificity, release and establishment of *Macaria pallidata* and *Leuciris fimbriaria* (Lepidoptera: Geometridae), biological control agents of the weed *Mimosa pigra*. *Biological Control* 55:248–255.

Hight, S. D., B. Blossey, J. Laing, and R. Declerck-Floate. 1995. Establishment of insect biological control agents from Europe against *Lythrum salicaria* in North America. *Environmental Entomology* 24:967–977.

Hill, M. P., and P. E. Hulley. 1995. Host-range extension by native parasitoids to weed biocontrol agents introduced to South Africa. *Biological Control* 5:287–302.

Hill, R. L., A. H. Gourlay, and C. J. Winks. 1993. Choosing gorse spider mite strains to improve establishment in different climates. Pages 377–383 *in Proceedings of the 6th Australian Conference on Grassland Invertebrate Ecology*. AgResearch, Ruakura Agricultural Research Center, Hamilton, New Zealand.

Himler, A. G., T. Adachi-Hagimori, J. E. Bergen, A. Kozuch, S. E. Kelly, B. E. Tabashnik, E. Chiel, V. E. Duckworth, T. J. Dennehy, E. Zchori-Fein, and M. S. Hunter. 2011. Rapid spread of a bacterial symbiont in an invasive whitefly is driven by fitness benefits and female bias. *Science* 332:254–256.

Hopper, K. R., and R. T. Roush. 1993. Mate finding, dispersal, number released, and the success of pbiological control introductions. *Ecological Entomology* 18:321–331.

Huffaker, C. B., and C. E. Kennett. 1959. A ten-year study of vegetational changes associated with biological control of Klamath Weed. *Journal of Range Management* 12:69–82.

Hunt-Joshi, T. R., R. B. Root, and B. Blossey. 2005. Disruption of weed biological control by an opportunistic mirid predator. *Ecological Applications* 15:861–870.

Jacob, H. S., A. Joder, and K. L. Batchelor. 2006. Biology of *Stethynium* sp (Hymenoptera : Mymaridae), a native parasitoid of an introduced weed biological control agent. *Environmental Entomology* 35:630–636.

Julien, M. H. 1989. Biological control of weeds worldwide: trends, rates of success and the future. *Biocontrol News and Information* 10:299–306.

Julien, M. H., and M. W. Griffiths (editors). 1998. *Biological Control of Weeds: A World Catalogue of Agents and Their Target Weeds*, 4th ed. CABI Publishing, Wallingford, UK.

Julien, M. H., B. Skarratt, and G. F. Maywald. 1995. Potential geographical distribution of alligator weed and its biological control by *Agasicles hygrophila*. *Journal of Aquatic Plant Management* 33:55–60.

Kakehashi, N., Y. Suzuki, and Y. Iwasa. 1984. Niche overlap of parasitoids in host–parasitoid systems: its consequence to single versus multiple introduction controversy. *Journal of Applied Ecology* 21:115–131.

Kaplan, I., and R. F. Denno. 2007. Interspecific interactions in phytophagous insects revisited: a quantitative assessment of competition theory. *Ecology Letters* 10:977–994.

Karban, R., and G. M. English-Leob. 1990. A "vaccination" of Willamette spider mites (Acari: Tetranychidae) to prevent large populations of Pacific spider mites on grapevines. *Journal of Economic Entomology* 83:2252–2257.

Karban, R., G. English-Loeb, and D. Hougen-Eitzmann. 1997. Mite vaccinations for sustainable management of spider mites in vineyards. *Ecological Applications* 7:183–193.

Karban, R., G. M. English-Loeb, and P. Verdigaal. 1991. Vaccinating grapevines against spider mites. *California Agriculture* 45:18–21.

Kindvall, O. 1999. Dispersal in a metapopulation of the bush cricket, *Metrioptera bicolor* (Orthoptera: Tettigoniidae). *Journal of Animal Ecology* 68:172–185.

Kot, M., M. A. Lewis, and P. vandenDriessche. 1996. Dispersal data and the spread of invading organisms. *Ecology* 77:2027–2042.

Kula, R. R., A. J. Boughton, and R. W. Pemberton. 2010. *Stantonia pallida* (Ashmead) (Hymenoptera: Braconidae) reared from *Neomusotima conspurcatalis* Warren (Lepidoptera: Crambidae), a classical biological control agent of *Lygodium microphyllum* (Cav.) R. Br. (Polypodiales: Lygodiaceae). *Proceedings of the Entomological Society of Washington* 112:61–68.

Kuniata, L. 1994. Biological control of *Mimosa invisa* in Papua-New-Guinea. *International Journal of Pest Management* 40:64–65.

Lawton, J. H. 1990. Biological control of plants: a review of generalisations, rules, and principles using insects as agents. Pages 3–17 *in Alternatives to the Chemical Control of Weeds*. Proceedings of an international conference, Rotorua, New Zealand, July 1989. Ministry of Forestry FRI Bulletin 155. Ministry of Forestry, Wellington.

Lawton, J. H., V. K. Rashbrook, and S. G. Compton. 1988. Biocontrol of British bracken: the potential of two moths from Southern Africa. *Annals of Applied Biology* 112:479–490.

LeJeune, K., K. Suding, S. Sturgis, A. Scott, and T. Seastedt. 2005. Biological control insect use of fertilized and unfertilized diffuse knapweed in a Colorado grassland. *Environmental Entomology* 34:225–234.

Lewis, M. A., and P. Kareiva. 1993. Allee dynamics and the spread of invading organisms. *Theoretical Population Biology* 43:141–158.

Liebhold, A., and J. Bascompte. 2003. The Allee effect, stochastic dynamics and the eradication of alien species. *Ecology Letters* 6:133–140.

Liebhold, A. M., and P. C. Tobin. 2008. Population ecology of insect invasions and their management. *Annual Review of Entomology* 53:387–408.

Lockwood, J. L., P. Cassey, and T. Blackburn. 2005. The role of propagule pressure in explaining species invasions. *Trends in Ecology and Evolution* 20:223.

Manrique, V., J. P. Cuda, and W. A. Overholt. 2009. Effect of herbivory on growth and biomass allocation of Brazilian peppertree (Sapindales: Anacardiaceae) seedlings in the laboratory. *Biocontrol Science and Technology* 19:657–667.

Martin, N., and Q. Paynter. 2010. Assessing the biosecurity risk from pathogens and herbivores to indigenous plants: lessons from weed biological control. *Biological Invasions* 12:1–12.

Masaki, S. 1961. Geographic variation of diapause in insects. *Bulletin of the Faculty of Agriculture, Hirosaki University* 7:66–98.

Matos, B., and J. J. Obrycki. 2007. Trophic interactions between two herbivorous insects, *Galerucella calmariensis* and *Myzus lythri*, feeding on purple loosestrife, *Lythrum salicaria*, and two insect predators, *Harmonia axyridis* and *Chrysoperla carnea*. *Journal of Insect Science* 7:1–8. http://www.insectscience.org/7.30

Mattson, W. J. 1980. Herbivory in relation to plant nitrogen content. *Annual Review of Ecology and Systematics* 11:119–162.

May, R. M., and M. P. Hassell. 1981. The dynamics of multiparasitoid–host interactions. *American Naturalist* 117:234–261.

McClay, A. S. 1996. Biological control in a cold climate: temperature responses and climatic adaptation of weed biocontrol agents. Pages 377–383 *in Proceedings of the IX International Symposium on Biological Control of Weeds*. 19–26 January, University of Cape Town, Stellenbosch, South Africa.

McClay, A. S., and R. B. Hughes. 1995. Effects of temperature on developmental rate, distribution, and establishment of *Calophasia lunula* (Lepidoptera: Noctuidae), a biocontrol agent for toadflax (*Linaria* spp). *Biological Control* 5:368–377.

McEvoy, P. B., and E. M. Coombs. 1999. Biological control of plant invaders: regional patterns, field experiments, and structured population models. *Ecological Applications* 9:387–401.

McEvoy, P. B., and E. M. Coombs. 2000. Why things bite back: unintended consequences of biological weed control. Pages 167–194 *in* P. A. Follett and J. J. Duan (editors), *Nontarget Effects of Biological Control*. Kluwer Academic Publishers, Boston.

McEvoy, P. B., E. Karacetin, and D. J. Bruck. 2008. Can a pathogen provide insurance against host shifts by a biological control organism? Pages 37–42 *in* Proceedings of the XII International Symposium on Biological Control of Weeds. CAB International, Wallingford, UK.

McFadyen, R. E. 1997. Parasitoids of the arctiid moth *Pareuchaetes pseudoinsulata* (Lep.: Arctiidae), an introduced biocontrol agent against the weed *Chromolaena odorata* (Asteraceae), in Asia and Africa. *Entomophaga* 42:467–470.

Memmott, J., P. G. Craze, H. M. Harman, P. Syrett, and S. V. Fowler. 2005. The effect of propagule size on the invasion of an alien insect. *Journal of Animal Ecology* 74:50–62.

Memmott, J., S. V. Fowler, and R. L. Hill. 1998. The effect of release size on the probability of establishment of biological control agents: gorse thrips (*Sericothrips staphylinus*) released against gorse (*Ulex europaeus*) in New Zealand. *Biocontrol, Science and Technology* 8:103–105.

Moran, V. C. 1985. The Silwood International Project on the biological control of weeds. Pages 65–68 *in Proceedings of the VI International Symposium on Biological Control of Weeds*. Agriculture Canada, Ottawa.

Moran, V. C., and H. G. Zimmerman. 1991. Biological control of jointed cactus, *Opuntia aurantiaca* (Cactaceae), in South Africa. *Agriculture, Ecosystems and Environment* 37:5–27.

Morris, W. F. 1997. Disentangling effects of induced plant defenses and food quantity on herbivores by fitting nonlinear models. *American Naturalist* 150:299–327.

Muller, H., and R. D. Goeden. 1990. Parasitoids acquired by *Coleophora parthenica* [Lepidoptera, Coleophoridae] 10 years after its introduction into southern California for the biological control of Russian thistle. *Entomophaga* 35:257–268.

Murray, T. J., B. I. P. Barratt, and C. M. Ferguson. 2002. Field parasitism of *Rhinocyllus conicus* froelich (Coleoptera : Curculionidae) by *Microctonus aethiopoides* loan (Hymenoptera : Braconidae) in Otago and South Canterbury. *New Zealand Plant Protection* 55:263–266.

Myers, J. H., and B. J. Post. 1981. Plant nitrogen and fluctuations of insect populations: a test with the cinnabar moth – tansy ragwort system. *Oecologia* 48:151–156.

Neubert, M. G., and H. Caswell. 2000. Demography and dispersal: calculation and sensitivity analysis of invasion speed for structured populations. *Ecology* 81:1613–1628.

Nimmo, K. R., and P. W. Tipping. 2009. An introduced insect biological control agent preys on an introduced weed biological control agent. *Florida Entomologist* 92:179–180.

Norman, K., N. Cappuccino, and M. R. Forbes. 2009. Parasitism of a successful weed biological control agent, *Neogalerucella calmariensis*. *Canadian Entomologist* 141:609–613.

Nuessly, G. S., and R. D. Goeden. 1984. Rodent predation on larvae of *Coleophora parthenica* (Lepidoptera, Coleophoridae), a moth imported for the biological control of Russian thistle. *Environmental Entomology* 13:502–508.

Paynter, Q. 2010. Tortoise or hare: which factors determine dispersal rate? *What's New in Biological Control of Weeds?* 53:2–3.

Paynter, Q., and S. Bellgard. 2011. Understanding dispersal rates of invading herbivorous arthropod and plant pathogen weed biocontrol agents. *Journal of Applied Ecology* 48:407–414.

Paynter, Q., S. V. Fowler, A. Hugh Gourlay, R. Groenteman, P. G. Peterson, L. Smith, and C. J. Winks. 2010a. Predicting parasitoid accumulation on biological control agents of weeds. *Journal of Applied Ecology* 47:575–582.

Paynter, Q., A. Main, A. Hugh Gourlay, P. G. Peterson, S. V. Fowler, and Y. M. Buckley. 2010b. Disruption of an exotic mutualism can improve management of an invasive plant: varroa mite, honeybees and biological control of Scotch broom *Cytisus scoparius* in New Zealand. *Journal of Applied Ecology* 47:309–317.

Pratt, P. D., E. M. Coombs, and B. A. Croft. 2003. Predation by phytoseiid mites on *Tetranychus lintearius* (Acari: Tetranychidae), an established weed biological control agent of gorse (*Ulex europaeus*). *Biological Control* 26:40–47.

Preisser, E. L., and J. S. Elkinton. 2008. Exploitative competition between invasive herbivores benefits a native host plant. *Ecology* 89:2671–2677.

Price, P. W. 1980. *Evolutionary Biology of Parasites*. Princeton University Press, Princeton.

Reimer, N. J. 1988. Predation on *Liothrips urichi* Karny (Thysanoptera: Phlaeothripidae): a case of biotic interference. *Environmental Entomology* 17:132–134.

Robertson, M. P., D. J. Kriticos, and C. Zachariades. 2008. Climate matching techniques to narrow the search for biological control agents. *Biological Control* 46:442–452.

Room, P. M., M. H. Julien, and I. W. Forno. 1989. Vigorous plants suffer most from herbivores: latitude, nitrogen and biological control of the weed *Salvinia molesta*. *Oikos* 54:92–100.

Room, P. M., and P. A. Thomas. 1985. Nitrogen and establishment of a beetle for biological control of the floating weed *Salvinia* in Papua New Guinea. *Journal of Applied Ecology* 22:139–156.

Roura-Pascual, N., C. Hui, T. Ikeda, G. Leday, D. M. Richardson, S. Carpintero, X. Espadaler, C. Gómez, B. Guénard, S. Hartley, P. Krushelnycky, P. J. Lester, M. A. McGeoch, S. B. Menke, J. S. Pedersen, J. P. W. Pitt, J. Reyes, N. J. Sanders, A. V. Suarez, Y. Touyama, D. Ward, P. S. Ward, and S. P. Worner. 2011. Relative roles of climatic suitability and anthropogenic influence in determining the pattern of spread in a global invader. *Proceedings of the National Academy of Sciences* 108:220–225.

Scheffer, M., S. Carpenter, J. A. Foley, C. Folke, and B. Walker. 2001. Catastrophic shifts in ecosystems. *Nature* 413:591–596.

Schooler, S. S., B. Salau, M. H. Julien, and A. R. Ives. 2011. Alternative stable states explain unpredictable biological control of *Salvinia molesta* in Kakadu. *Nature* 470:86–89.

Sebolt, D. C., and D. A. Landis. 2004. Arthropod predators of *Galerucella calmariensis* L. (Coleoptera : Chrysomelidae): an assessment of biotic interference. *Environmental Entomology* 33:356–361.

Semple, J. L., and I. W. Forno. 1987. Native parasitoids and pathogens attacking *Samea multiplicalis* Guenee (Lepidoptera, Pyralidae) in Queensland. *Journal of the Australian Entomological Society* 26:365–366.

Shearer, J. F., M. J. Grodowitz, and D. G. McFarland. 2007. Nutritional quality of *Hydrilla verticillata* (Lf) Royle and its effects on a fungal pathogen *Mycoleptodiscus terrestris* (Gerd.) Ostazeski. *Biological Control* 41:175–183.

Shigesada, N., and K. Kawasaki. 1997. *Biological Invasions: Theory and Practice*. Oxford University Press, New York.

Shigesada, N., K. Kawasaki, and Y. Takeda. 1995. Modeling stratified diffusion in biological invasions. *American Naturalist* 146:229–251.

Simberloff, D. 2009. The role of propagule pressure in biological invasions. *Annual Review of Ecology, Evolution, and Systematics* 40:81–102.

Stewart, A. J. A. 1996. Interspecific competition reinstated as an important force structuring insect herbivore communities. *Trends in Ecology and Evolution* 11:233–234.

Story, J. M., K. W. Boggs, W. R. Good, P. Harris, and R. M. Nowierski. 1991. *Metzneria paucipunctella* Zeller (Lepidoptera: Gelechiidae), a moth introduced against spotted knapweed: its feeding strategy and impact on two introduced *Urophora* spp. (Diptera: Tephritidae). *Canadian Entomologist* 123:1001–1007.

Sutherst, R. W., G. F. Maywald, and D. J. Driticos. 2007. *CLIMEX Version 3: User's Guide*. Hearne Scientific Software Pty Ltd. http://www.hearn.com.au

Syrett, P., D. T. Briese, and J. H. Hoffmann. 2000. Success in biological control of terrestrial weeds by arthropods. Page 448 *in* G. Gurr and S. Wratten (editors), *Biological Control: Measures of Success*. Kluwer, Dordrecht.

Syrett, P., S. V. Fowler, and R. M. Emberson. 1996. Are chrysomelid beetles effective agents for biological control of weeds? Pages 399–407 *in Proceedings of the IX International Symposium on Biological Control of Weeds*. University of Cape Town, Stellenbosch, South Africa.

Tauber, M. J., and C. A. Tauber. 1976. Insect seasonality: diapause maintenance, termination, and post diapause development. *Annual Review of Entomology* 21:81–107.

Taylor, C. M., and A. Hastings. 2005. Allee effects in biological invasions. *Ecology Letters* 8:895–908.

Van Hezewijk, B. H., R. A. De Clerck-Floate, and J. R. Moyer. 2008. Effect of nitrogen on the preference and performance of a biological control agent for an invasive plant. *Biological Control* 46:332–340.

van Klinken, R. D., and C. J. Burwell. 2005. Evidence from a gelechiid leaf-tier on mesquite (Mimosaceae: *Prosopis*) that semi-concealed Lepidopteran biological control agents may not be at risk from parasitism in Australian rangelands. *Biological Control* 32:121–129.

van Klinken, R. D., G. Fichera, and H. Cordo. 2003. Targeting biological control across diverse landscapes: the release, establishment, and early success of two insects on mesquite (*Prosopis* spp.) insects in Australian rangelands. *Biological Control* 26:8–20.

Van Klinken, R. D., and L. K. Flack. 2008. What limits predation rates by the specialist seed feeder *Penthobruchus germaini* on an invasive shrub? *Journal of Applied Ecology* 45:1600–1611.

van Loon, L. C., P. A. H. M. Bakker, and C. M. J. Pieterse. 1998. Systemic resistance induced by rhizosphere bacteria. *Annual Review of Phytopathology* 36:453–483.

Vogt, G. B., J. Quimby, P.C., and S. H. Kay. 1992. *Effects of Weather on the Biological Control of Alligatorweed in the Lower Mississippi Valley Region, 1973–1983, USA*. USDA-ARS Technical Bulletin No. 1766:143. USDA, Washington, DC.

Wapshere, A. 1974. The regions of infestation of wool by Noogoora burr (*Xanthium strumarium*), their climates and the biological control of the weed. *Australian Journal of Agricultural Research* 25:775–781.

Wapshere, A. J. 1983. Discovery and testing of a climatically adapted strain of *Longitarsus jacobaeae* [Col, Chrysomelidae] for Australia. *Biocontrol* 28:27–32.

Wheeler, G. S. 2001. Host plant quality factors that influence the growth and development of *Oxyops vitiosa*, a biological control agent of *Melaleuca quinquenervia*. *Biological Control* 22:256–264.

Wheeler, G. S., and T. D. Center. 1996. The influence of Hydrilla leaf quality on larval growth and development of the biological control agent *Hydrellia pakistanae* (Diptera: Ephydridae). *Biological Control* 7:1–9.

Wheeler, G. S., and T. D. Center. 1997. Growth and development of the biological control agent *Bagous hydrillae* as influenced by Hydrilla (*Hydrilla verticillata*) stem quality. *Biological Control* 8:52–57.

Wheeler, G. S., and T. D. Center. 2001. Impact of the biological control agent *Hydrellia pakistanae* (Diptera : Ephydridae) on the submersed aquatic weed *Hydrilla verticillata* (Hydrocharitaceae). *Biological Control* 21:168–181.

Wheeler, G. S., T. K. Van, and T. D. Center. 1998. Fecundity and egg distribution of the herbivore *Spodoptera pectinicornis* as influenced by quality of the floating aquatic plant *Pistia stratiotes*. *Entomologia Experimentalis et Applicata* 86:295–304.

Wiebe, A. P., and J. J. Obrycki. 2004. Quantitative assessment of predation of eggs and larvae of *Galerucella pusilla* in Iowa. *Biological Control* 31:16–28.

Williamson, M. 1996. *Biological Invasions*. Chapman and Hall, London.

Williamson, M., and A. Fitter. 1996. The varying success of invaders. *Ecology* 77:1661–1666.

Woodburn, T. L. 1996. Interspecific competition between *Rhinocyllus conicus* and *Urophora solstitialis*, two biocontrol agents released in Australia against *Carduus nutans*. Pages 409–415 *in Proceedings of the IX International Symposium on Biological Control of Weeds*. University of Cape Town, South Africa.

Zayed, A., S. A. Constantin, and L. Packer. 2007. Successful biological invasion despite a severe genetic load. *PLoS ONE* 2. http://www.plosone.org

20

Assessing the Impact of Climate Change on Outbreak Potential

Maartje J. Klapwijk, Matthew P. Ayres, Andrea Battisti, and Stig Larsson

20.1 Introduction

In this chapter we evaluate the potential effects of climate change on outbreak dynamics of forest insects. Accumulating evidence of climate change over the past decades has fuelled debate and research regarding its ecological consequences. Meteorological data confirm general increases in ambient temperatures and changes in precipitation patterns (Solomon *et al.* 2007). In some cases ecologists have connected increasing temperatures to changes in species ranges (Parmesan *et al.* 1999, Battisti *et al.* 2005, Hickling *et al.* 2006), advanced phenology (Sparks and Carey 1995, Sparks and Yates 1997, Walther *et al.* 2002, Cleland *et al.* 2007),

Insect Outbreaks Revisited, First Edition. Edited by Pedro Barbosa, Deborah K. Letourneau and Anurag A. Agrawal.
© 2012 Blackwell Publishing Ltd. Published 2012 by Blackwell Publishing Ltd.

and phenological asynchrony among interacting species (Visser *et al.* 1998, Visser and Holleman 2001, Dixon 2003, Visser and Both 2006, Memmott *et al.* 2007). Much less is known about consequences for population dynamics, but it seems likely from first principles that there would be effects for some species on reproductive success, life histories, mean abundance, and variation in abundance. Climate change involves almost enumerable features of global atmospheric systems. Here we concentrate on temperature, which is the aspect most likely to be a strong driver for arthropod populations, and is also that for which predictions are most robust. However, we also briefly consider the more numerous pathways by which climate change could affect herbivorous insects.

Insect populations provide many examples of dramatic fluctuations in abundance. For the purpose of this chapter, we accept the premise that most insect herbivore populations, most of the time, are maintained at low density by a combination of factors, some of which are dependent on density (feedback systems) and some of which are independent of density (exogenous effects) (Turchin 2003). We further assume that stochastic disturbances can sometimes result in deviations from relatively low equilibrium densities, with returns toward the equilibrium depending on density-dependent controls on survival, reproduction, and movements. Our focus will be on the build-up of an outbreak, when populations change from endemic to epidemic density. Even though there is no consensus on why certain populations outbreak, it has been frequently hypothesized that weather conditions can trigger the start of an outbreak, the so-called "population release" (e.g., Mattson and Haack 1987). Such effects could be relatively direct (e.g., winter temperatures influencing insect survival) or indirect via effects on interactions with host plants and natural enemies (Martinat 1987).

We take two approaches to investigate effects of climate warming on outbreak potential. First, we explore how a changing climate might act, directly and indirectly, on insect individuals, such that there can be qualitatively important changes in the dynamics of insect herbivore populations. Second, we present case studies of forest insects that have extended the range of their outbreaks, or expanded their range during outbreaks. Additionally, we attempt to scale up from individuals to populations to investigate the likelihood of qualitatively important changes in population dynamics due to climate warming.

20.2   Direct and indirect effects of climate warming on life history traits

In this section we emphasize the position of the herbivorous insect in the trophic web, and attempt to show the importance of integrating direct effects on the insect herbivore and indirect effects through host plant, natural enemies, and pathogens. Such an approach is necessary to fully understand the potential consequences that climate warming can have for outbreak risks.

20.2.1   Direct effects on the herbivore

Insect herbivores, being generally poikilothermic, are highly sensitive to changes in their thermal environment (Uvarov 1931). Furthermore, responses of insect herbivores are likely based on their life cycle in combination with the life cycle of their host plant (Bale *et al.* 2002). Increased temperatures have direct effects on insect development time, overwintering survival, voltinism, and diapause.

Development rate generally increases with increasing temperature to some maximum, above which there are frequently sharp increases in mortality associated with the decreasing development rate (Uvarov 1931, van Straalen 1983). The development rate of insects in mid to high latitude systems should increase with climate warming as ambient temperatures are well below that which permit maximum rates of feeding and growth (Hodkinson 1999).

Increased development rate could lead to increased voltinism in facultative multivoltine species (Tobin *et al.* 2008), as predicted for the bark beetle species *Ips typographus* (DeGeer) (Wermelinger and Seifert 1998, 1999, Faccoli 2009, Jonsson *et al.* 2009) and *Dendroctonus rufipennis* (Kirby) (Berg *et al.* 2006). Recent research has found increased voltinism in macrolepidopteran species thought to be strictly uni- or bivoltine, as a consequence of climate warming (Altermatt 2010). Species that employ photoperiodic cues to structure their life cycle should be least responsive to increases in the growing season from climate warming (Beaudoin *et al.* 1992, Fantinou *et al.* 2004).

Winter mortality is likely to decrease under increasing temperatures (Ayres and Lombardero 2000, Tran *et al.* 2007, Jepsen *et al.* 2008), although decreased snow cover (and therefore decreased insulation of overwintering sites) can reverse the pattern (Lombardero *et al.* 2000a), but details will depend on the overwintering life stage. Warmer winters may permit some nondiapausing species to continue feeding and development during months that were previously too cold (Bale *et al.* 2002). For example, larvae of the pine processionary moth (*Thaumetopoea pityocampa* (Schiffermüller)) have a higher probability of survival if winter temperatures do not often fall below specific feeding thresholds (Battisti *et al.* 2005). Overall, this will lead to higher year-to-year winter survival (Neuvonen *et al.* 1999, Friedenberg *et al.* 2008, Jepsen *et al.* 2008), but see Bale and Hayward (2010) for an extensive review on insect overwintering and the changing climate.

In some cases, warmer temperatures tend to produce larger and more fecund adults (Laws and Belovsky 2010, Yao *et al.* 2010) but the opposite has also been reported (Miller 2005, Costa *et al.* 2010). Increased fecundity might not be realized under warmer temperatures because adult life span tends to decrease with increasing temperatures (Uvarov 1931).

20.2.2   Indirect effects through the host plant

For the purpose of this discussion, we restrict ourselves to predicted changes in plant phenology and abiotic stress acting on the plant. These phenomena can potentially influence plant–insect interactions, both quantitatively and qualitatively.

Climate warming is already producing a noticeable lengthening of the growing season (especially in spring) for plants at mid- to high latitudes (Menzel *et al.* 2006, Morin *et al.* 2009). Many insect species evolved to match their feeding activity with certain developmental stages in the host plant. In particular, many insect species of deciduous trees feed during spring and early summer when they can exploit nutritious immature foliage (Feeny 1970, van Asch and Visser 2007,

Singer and Parmesan 2010). For example, larvae of the autumnal moth (*Epirrita autumnata* Borkhausen) feed during the short window when mountain birch leaves are expanding and of demonstrably high quality for larvae (Ayres and MacLean 1987, Riipi *et al.* 2002, Haukioja 2003). In at least some systems, the insect herbivores are quite efficient at matching the start of their feeding to the time of budburst (van Asch and Visser 2007). How well this synchrony is maintained under climate warming will depend on how well the physiological controls on insect development match those of their host plants (Vitasse *et al.* 2009, Valtonen *et al.* 2011). Changes in temperatures when leaves and larvae are developing could alter the outcome of the phenological race between them (Ayres 1993). A consequence of this could be correlated fluctuations (i.e., when the occurrence of one event increases the likelihood of the occurrence of a subsequent, similar event; in this case fluctuations) in spring-feeding lepidopteran species (Stange *et al.* 2011).

The primary and secondary metabolites of woody plants can be highly responsive to environmental factors (Herms and Mattson 1992, Stamp 2003). Scores of experimental studies show consequences of variation in phytochemistry for the performance of insect herbivores.

If the trees are reasonably well matched to historically favorable climatic conditions, then it is inevitable that changing climate will produce situations where trees are poorly matched to the new conditions. This predicts an increase in frequency of stressful conditions in plants. Stressed plants frequently undergo biochemical changes (Koricheva *et al.* 1998a). Stress-induced effects on survival and reproduction of some forest insects are well documented in experimental studies (Awmack and Leather 2002), and can have effects on the growth and survival of individuals, but the consequences for population dynamics are less clear (Koricheva *et al.* 1998b).

That plant stress can trigger insect outbreaks is a long-standing hypothesis in forest entomology (Craighead 1925, Rudinsky 1962). Insect outbreaks have been commonly correlated with stressful conditions of their host plant (e.g., Mattson and Haack 1987), and some have speculated that there is a causal link between stress-induced changes in plant quality, and thus insect performance, and the start of outbreaks (e.g., White 1974, Rhoades 1979). Experimental tests of the plant stress hypothesis have produced mixed results; insect species belonging to some feeding guilds respond to experimentally stressed trees with increased performance, some herbivores are unaffected, and some respond negatively (Larsson 1989, Koricheva *et al.* 1998b). The variable consequences for insects might be due partly to nonlinear responses of plant defenses to water and nutrient availability (Herms and Mattson 1992) and to differential effects of environmental conditions on constitutive versus inducible plant defenses (Lombardero *et al.* 2000b).

Bark beetles constitute a globally important group of insects for which plant stress seems relevant. Most bark beetle species only reproduce in dead, or almost dead, trees. However, a subset of aggressive bark beetle species regularly produce epidemics of tree mortality (Bentz *et al.* 2010) that collectively have enormous ecological and economic impacts in conifer forests around the world (Seppälä *et al.* 2009). A long-standing paradigm is that healthy trees are resistant to most bark beetle species (and other boring insects) (Larsson 1989), but that periods of stress can trigger outbreaks because lower attack densities are needed to overcome defense reactions in severely stressed trees, and subsequent beetle reproduction can be very high (Waring and Pitman 1985, Raffa *et al.* 2008). The resulting high numbers of bark beetles facilitate the successful attack of even healthy trees, resulting in self-propagating disturbances that continue to kill trees even after the initial episode of tree stress has ended (Berryman 1973); see the model of alternate attractors in the "Alternate Attractors" subsection of this chapter.

20.2.3 Indirect effects through natural enemies

Predators and parasitoids
Arthropod natural enemies can exert powerful forces on the population dynamics of herbivorous insects (Hawkins *et al.* 1997). Assessing their response to climate warming, and understanding the consequences for density-dependent feedback systems, is crucial to understanding possible shifts in outbreak frequencies in the future (Hoekman 2010). In this section, we discuss the indirect effects that climate warming can have on the insect herbivore through its natural enemies.

The distribution and abundance of natural enemies, like that of their prey, can be influenced by direct climatic effects (e.g., on overwinter survival) (Humble 2006, Hance *et al.* 2007). Temperature sensitivity might increase with trophic level (Berggren *et al.* 2009), in which case natural enemies would be more affected than the herbivores on which they feed (Davis *et al.* 1998). Also, the addition of new species to communities as various populations extend their poleward distribution limits could alter controls on herbivore density. If generalist natural enemies (e.g., ants) commonly control herbivore populations at low levels (Berryman 2002), then increases of generalist enemies should tend to reduce outbreak frequencies, and, conversely, poleward extensions of herbivores beyond the range of generalist enemies should tend to increase frequencies of outbreaks.

For arthropod parasitoids and predators, phenological synchrony can affect host or prey availability and host or prey size at the time of attack. Specialist enemies should be under strong selection to track phenological changes in their prey, which might make them less likely to become temporally uncoupled from their prey (Klapwijk *et al.* 2010). However, the evolution and adaptation of phenology are poorly understood, and thus the outcome of changes in interactions is hard to predict (Forrest and Miller-Rushing 2010). Increases or decreases in the phenological match between parasitoids and hosts could change the population dynamics of enemies (Thomson *et al.* 2010) through effects on prey attack rates and conversion of attacks into fecundity (Harvey *et al.* 1994).

Higher temperatures can influence parasitism and predation rates by increasing searching activity of individual parasitoids and predators (Dhillon and Sharma 2009). When the prey are relatively immobile (e.g., many immature insect herbivores), this should generally increase detection rates and attacks. In predators that hunt more mobile adult prey, the direction of change in predation success with increasing temperatures should depend more on hunting strategy because both predator and prey will increase their activity. With passive strategies, like ambushing or sit-and-wait, predation success depends on the activity of the prey. For active hunting strategies, predation success depends on both predator and prey agility (Kruse *et al.* 2008).

Pathogens

Insect pathogens (e.g., fungi, bacteria, and viruses), play an important role in the dynamics of many insect populations (Hajek 1997, Eldred *et al.* 2008). The infection rate of most pathogens is highly dependent on prey density and often barely detectable in low-density populations (Olofsson 1989a, 1989b, Roy *et al.* 2009). Temperature can be important in both infection rate and defense responses within the host (Carruthers *et al.* 1992). Different thermal optima for host and pathogen might lead to a situation where high temperatures favor the host by both optimizing defense responses and directly limiting pathogen growth (Blanford and Thomas 1999). The ability for pathogens to kill their host can depend on host body temperature (Thomas and Blanford 2003). Fungal pathogens tend to be favored by increases in humidity, especially when temperatures are high, but not too high (Ali *et al.* 1995, Filotas *et al.* 2006). Outside of their host, resting stages of nuclear polyhedral viruses are vulnerable to UV radiation, so increases in UV radiation could influence viral controls on some insect populations (Witt 1984, Olofsson 1989b, 1989a). As with specialist predators and parasitoids, the most pronounced consequences of climatic effects on herbivore–pathogen interactions may be in the decline phase of outbreaks (Anderson and May 1980).

20.2.4   Conclusions

Here we have summarized direct and indirect effects of climate warming on herbivorous insects, and focused on the performance of insect individuals (Figure 20.1):

- The direct effects on the herbivorous insect will be generally positive as a result of increased winter survival, faster development rates, and sometimes increased number of generations per year.
- Some insect herbivores will benefit from increased frequency of stressful events for plants. Insects that suffer from constitutive plant defenses will tend

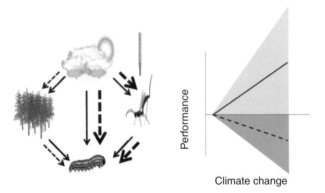

Figure 20.1   Schematic view of the direct and indirect effects of climate on herbivorous insects. The arrows refer to the interactions proposed by Martinat (1987) that has been revisited to document the expected effects of climate change (dashed arrows). The thickness of the dashed arrows indicates the magnitude of the effect on the different parts of the system. The graph to the right indicates the overall positive effect of climate change on individual performance of herbivorous insects, although a large uncertainty occurs for both positive and negative effects (light and dark fans).

to be favored when conditions promote increased plant growth and associated declines in secondary metabolism. Insects that exploit young, nutritionally favorable foliage may be sensitive to climatically driven changes at the start of the growing season and to temperatures that influence the relative rates of leaf maturation and insect development after budburst. The nature of physiological controls on plant versus insect phenology will influence the extent to which plant and insect phenology remains coupled under climate change.

- Increased activity in arthropod natural enemies could lead to increased strength of top-down control on the herbivorous insect. However, the magnitude of the effect will depend on attack strategy of predators and host usage in parasitoids. Pathogens are likely to increase their infection rate under increased humidity and temperature.

An important next step is to consider how climatic effects on insect individuals can be scaled up to effects on populations (Broekhuizen *et al.* 1993). Predictions regarding the dynamics of populations under climate change require integration of the direction and strength of direct and indirect effects. We first summarize empirical findings that indicate climate change effects on populations. Then we move on to a theoretical consideration of the general effects on population dynamics, and in particular the question of when and where changes in climate can be expected to increase or decrease the strength and frequency of outbreaks.

20.3   Climate and expansion of outbreak range

The geographic range of herbivorous insects can change over years to decades based on changes in climate, among other things (Gaston 2003). Although adequate data are only available for some species in some regions, there are quite clear signals of recent poleward range expansion in numerous insect species (Parmesan *et al.* 1999, Solomon *et al.* 2007), including outbreaking forest insects, such as the pine processionary moth (Battisti *et al.* 2005), the southern pine beetle (Tran *et al.* 2007), and the mountain pine beetle (Carroll *et al.* 2004, Robertson *et al.* 2009). Other species, such as birch geometrids (Jepsen *et al.* 2008) and the spruce budworm (Fleming and Candau 1998), are expanding the range of outbreaks within the current geographical range. Whether the new outbreaks occur within (Hodar *et al.* 2003, Kurz *et al.* 2008) or beyond (Tenow and Nilssen 1990, Jepsen *et al.* 2008) the current range, candidate explanations are the same: release from adverse climatic conditions, natural enemies, competition with other herbivores, and host plant resistance (Fleming 1996, Ayres and Lombardero 2000, Logan *et al.* 2003).

20.3.1   New areas of outbreaks within the historic
species distribution

Birch geometrids in northern Europe
Both the autumnal moth (*Epirrita autumnata*) and the winter moth (*Operophtera brumata* L.) have expanded their outbreak range in recent years (Jepsen *et al.* 2008), presumably as a result of improved overwinter survival of eggs, and

maintenance of synchrony (through adaptive phenological plasticity) with bud burst of their main host, the mountain birch (*Betula pubescens* Ehrhart) (Karlsson *et al.* 2003). Eggs of the autumnal moth are more cold tolerant than those of the winter moth affecting outbreak patterns in their recent range expansion. Winter moth populations show a pronounced north-eastern expansion of outbreaks toward the North-East into areas previously dominated by the autumnal moth, which in turn has expanded outbreaks into colder areas (Tenow 1996, Jepsen *et al.* 2008).

Spruce budworm in North America

Records of spruce budworm (*Choristoneura fumiferana* Clemens) defoliation and tree-ring analysis indicate that this insect has expanded its outbreak range towards the northwest limit of its distribution (central Alaska) since the early 1990s (Juday and Marler 1997). The simultaneous occurrence of a general increase in tempera-ture in northern latitudes indicates a relation with climate change (Candau and Fleming 2005). At least in the northern parts of its distribution, the frequency and duration of outbreaks appear to increase with climate warming due to the positive effects of temperature on insect metabolism and possibly a loss of synchronization between the spruce budworm and its natural enemies.

20.3.2   Outbreaks in recently invaded geographic areas

Pine processionary moth in southern Europe

In recent decades, the pine processionary moth has expanded its latitudinal and altitudinal limits. Improved survival during the feeding period in winter (Battisti *et al.* 2005) has contributed to outbreaks in pine forests previously unoccupied in France, Italy, Spain, and Turkey. Rapid range expansion is facilitated by warm summer nights that contribute to long-distance (i.e., more than 2 km) dispersal of female moths (Battisti *et al.* 2006).

Mountain pine beetle in Canada

In Canada, recent outbreaks of the native mountain pine beetle (*Dendroctonus ponderosae* Hopkins) have produced extensive tree mortality within at least 14 million hectares of lodgepole pine forest (*Pinus contorta* Douglas ex Loudon). The outbreak has extended into higher latitudes and elevations than previously recorded for this species (Aukema *et al.* 2006, Logan and Powell 2009). The start of the epidemic was facilitated by fire suppression during the last century that created vast tracts of over-mature pine stands (Raffa *et al.* 2008) in combination with recent climatic patterns, mild winters and warm dry summers (Logan and Powell 2001, Carroll *et al.* 2004, Régnière and Bentz 2007). The outbreak has been progressing as predicted based upon models of climatic effects on beetle physiology (Logan and Powell 2001). These same models project an eventual northern range expansion of 7° latitude (780 km) if mean annual temperature increases by 2.5°C. Furthermore, the range expansion of the mountain pine bee-tle into forests of jack pine (*Pinus banksiana* Lambert) in Alberta creates the potential for massive range expansions into north central and eastern North America (Logan *et al.* 2003, Logan and Powell 2009). Rapid population growth in the newly occupied areas may be promoted by low defenses and high nutritional suitability of populations of host trees that have not been exposed to bark beetle attack for many generations (Yanchuk *et al.* 2008, Cudmore *et al.* 2010).

20.3.3 Conclusions

Distinguishing between outbreak range expansion and geographical range expansion is partly artificial because distribution limits fluctuate and data on historical range limits are sparse. Nonetheless, the dichotomy may be of theoretical as well as practical value. When outbreaks occur within current distribution limits, we might expect shorter delays before negative feedbacks act on population density (Section 20.4). For example, specialist natural enemies were already present in areas with new outbreaks of the autumnal moth in Fennoscandia because they were within contemporary distribution limits (Neuvonen *et al.* 2005). Conversely, specialist natural enemies of the pine processionary moth take longer to colonize their host within newly invaded forests of northern Italy (Zovi *et al.* 2008). The definition of distribution limits becomes important to understand patterns and processes of outbreak range expansion. In most cases, range limits tend to be a belt of recurrent colonization and extinction events rather than a precise border (Gaston 2003). It is possible that many new outbreaks outside of the generally accepted range edge fall within this belt. The width of the belt would be defined by the dispersal potential of the species, including rare long-distance events. Climate change, however, may have an effect on dispersal *per se* (Battisti *et al.* 2006, Robertson *et al.* 2009), and thus, indirectly start outbreaks outside the range whenever conditions are suitable for insect herbivore performance. Climate-matching models frequently identify regions that seem suitable but are presently unoccupied – presumably because of dispersal limitation (Vanhanen *et al.* 2007). Probably, the extent of such suitable but unoccupied regions is increasing under climate warming, which increases the importance of understanding climatic effects on dispersal ability.

20.4 Principles of population dynamics as related to climate change

Outbreaks are a special case of population dynamics characterized by extreme population fluctuations. The tendency for population fluctuations is a function of how per capita population growth rate (R) behaves over time. Climate change can alter the behavior of R through either changes in the endogenous feedback system (the nature of density–dependence, $R = f(N)$), or exogenous effects (those that influence per capita growth rate but are not themselves influenced by abundance of the focal population; Turchin *et al.* 2003). A simple population system with immediate, linear density-dependence plus exogenous variation can be described as

$$R_t = r - \frac{r}{K} N_t + \varepsilon_t \qquad (20.1)$$

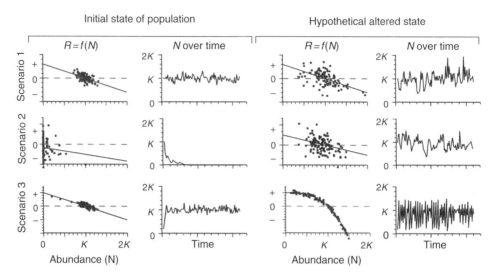

Figure 20.2   There are only a few general ways in which climate change can theoretically have an impact on the occurrence of population outbreaks. Three are illustrated here. Scenario 1: increase in exogenous (density-independent) variation. Scenario 2: an upward shift in the density-dependent function such that a stable equilibrium is created and the population can persist where it could not before. Scenario 3: decelerating nonlinearity in $R = f(N)$ moves the endogenous dynamics toward increasing complexity (tendency for oscillations), especially with high r. Model parameters: for scenario 1 (Equation (20.1)) $r = 0.3$ and $K = 100$ for both versions but $\sigma(\varepsilon) = 0.1$ versus 0.3; for scenario 2 $r = -0.1$ versus 0.3, $r/K = 0.003$ and $\sigma(\varepsilon) = 0.2$ for both versions; for scenario 3 (eq. 2) $r = 1$, $K = 100$, and $\sigma(\varepsilon) = 0.1$ for both models, but $b = 1$ versus 3. In the first and third columns, the solid line represents the density-dependent function, the dashed horizontal line indicates replacement reproductive rate, and their intersection represents the equilibrium abundance (K). In the second and fourth columns, the lines indicate a projected time series corresponding to the points in the figure to left.

where R_t = per capita growth rate at time t, r is the y intercept and represents the maximum per capita growth rate in the absence of resource limitation, K represents the equilibrium abundance, r/K represents density dependence, and ε_t, representing exogenous effects, is a random, normally distributed variable with $\mu = 0$ and σ as a function of the environment. Under this model (Equation (20.1), climate change could alter $\sigma(\varepsilon_t)$, or alter the feedback system by altering r and/or K. If climatic variables relevant to insect populations become more variable among years, this can be represented as an increase in $\sigma(\varepsilon_t)$, which produces increased variability in population size (Scenario 1 in Figure 20.2). In this sense, increases in $\sigma(\varepsilon_t)$ can move population dynamics towards outbreak behavior. If there is a decrease in average density-independent mortality (e.g., from decreased frequency of lethal winter temperatures), this would tend to shift the density-dependent function upward (increasing both r and K). Climatic ameliorations that permit poleward expansion of distributions are cases where populations invade regions where they were previously excluded (as in Scenario 2; Figure 20.2). This is most obviously consequential when the range expansions are of species with intrinsic tendencies for outbreaks, as in our examples of outbreak range expansions (Section 20.3). Less obvious changes in outbreak behavior can arise from effects on r, K, and ε of changes in generalist enemies or host plant suitability. For example, many generalist predators (e.g., ants, spiders, true bugs, and insectivorous birds), which respond in a density-independent manner to herbivore densities, are presumably also expanding their range poleward. The resulting increase in species richness of generalist predators will inevitably dampen interannual fluctuations in the abundance of predators (Doak *et al.* 1998), and therefore reduce the contributions of predators to exogenous interannual fluctuations in insect herbivores. Climatic effects on phytochemistry could affect dynamics as illustrated by either Scenarios 1 or 2 in Figure 20.2. At present, we know too little about the responses of insect populations to phenotypic variation in phytochemistry (but see Stiling and Cornelissen 2007). It could be that environmental effects on phytochemistry contribute some exogenous variation to insect population dynamics but do not substantially alter the system of population regulation and do not usually produce outbreaks or other qualitative shifts in dynamics.

20.4.1 Changes in endogenous feedback can change tendencies for outbreaks

Population outbreaks frequently arise from complex endogenous dynamics (Turchin *et al.* 2003), most commonly a tendency for population cycles. Three general features of negative feedback systems can move a population system toward complex endogenous dynamics: high intrinsic growth potential (high r), nonlinearity (deceleration of the density-dependent function), and delays in the negative feedback.

A rearrangement of Equation (20.1) can allow for nonlinearity in density dependence (May 1976):

$$R_t = r\left(1 - \frac{N_t}{K}\right)^b + \varepsilon_t \tag{20.2}$$

Equation (20.2) reduces to Equation (20.1) when $b = 1$, but becomes a decelerating function with $b > 1$ (scramble competition). Scenario 3 in Figure 20.2 illustrates the tendency for increased fluctuations from endogenous dynamics when the negative feedback is a decelerating function rather than linear. Scramble competition tends to result from rapid colonization of finite resource patches (e.g., blowflies colonizing carrion, or bark beetles colonizing a susceptible tree) (Hassell 1975, Reeve *et al.* 1998). If increasing temperatures increase the efficacy of resource location by insects (see Section 20.2), then climate warming could increase the potential for scramble competition and therefore complex endogenous dynamics (i.e., tendency for cyclical population dynamics). However, some bark beetles become developmentally synchronized by cool winters and can become less synchronous in their attack dynamics with climate warming, which makes it less likely that they can attain critical numbers to overwhelm tree defenses (Powell and Logan 2005, Friedenberg *et al.* 2007).

Delayed density dependence

Delayed density dependence is a powerful force in generating outbreak behaviour via population cycles (Turchin *et al.* 2003). Equation (20.3) is a simple but reasonably satisfying general expression of delayed negative feedback in population dynamics (Royama 1992).

$$R_t = r + \beta_1 \cdot N_t + \beta_2 \cdot N_{t-1} + \varepsilon_t \tag{20.3}$$

Equation (20.3) matches Equation (20.1) but with the addition of a term for negative feedback from previous population abundance ($\beta_2 \cdot N_{t-1}$). $\beta_2 = 0$ in the absence of delayed feedback, β_2 / β_1 is a measure of the relative strength of the delayed feedback. Increased strength of delayed density dependence can move a population from monotonic convergence on a point equilibrium toward damped oscillations, stable limit cycles, or perhaps even chaotic dynamics ((Turchin and Taylor 1992), Figure 20.3). In insect populations, delayed density-dependent feedback is most commonly attributed to numerical responses of specialist natural enemies (e.g., parasitoids or pathogens) (Berryman 2002), but it also can arise from long-term inducible resistance in host plants, degradation in resources that recover slowly relative to potential rebounds in the insect population, and probably some other mechanisms (Haukioja 1980, Økland and Bjørnstad 2006). Regardless of the mechanism, delayed negative feedbacks move populations in the direction of complex endogenous dynamics, and longer delays yield more extreme fluctuations and longer periods between peaks.

Increases in r (maximum per capita growth rate) have the general effect of moving populations in the direction of more complex endogenous dynamics, especially in the presence of decelerating or delayed density dependence (May 1976, Turchin *et al.* 2003). Insects have high potential fecundity compared with many other taxa, and therefore have generally high r. Dynamic feedback systems naturally produce bifurcations in system behavior such that complex dynamics can appear or disappear relatively suddenly with modest changes in the form of density dependence (Hassell 1975, Royama 1992). If climate warming commonly decreases density-independent mortality in high latitude populations of insect herbivores (e.g., from increased winter survival), then it will commonly increase r which will shift some poleward populations toward more complex endogenous dynamics, and therefore greater tendencies for outbreaks. Further effects can be anticipated from any environmental changes that alter population growth potential in specialist enemies (Section 2). When such effects on enemy populations decrease the delay between insect outbreaks and the development of negative feedbacks (e.g., increases in specialist enemies or decreases in host plant suitability), there will be a decreased tendency for cyclical population dynamics, and vice versa. There is a need for studies of how climatic variation influences the timing of demographic feedbacks in forest insects prone to cyclical outbreaks.

Alternate attractors

Another long-standing hypothesis for outbreak behavior is the existence of alternate attractors in population dynamics (i.e., a tendency for populations to exist at either low or high abundance; Berryman 1973, Holling 1973, Southwood and

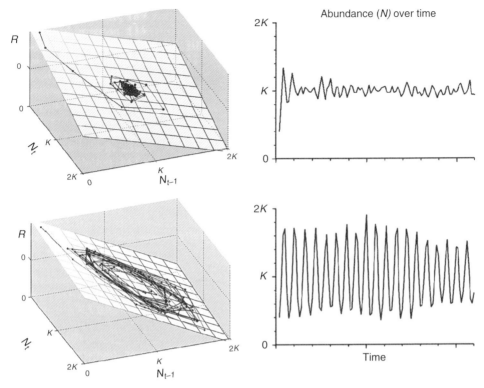

Figure 20.3   The most frequently demonstrated general mechanism for population outbreaks is cyclical behaviour due to delayed density dependence (where per capita population growth is a function of both immediate and previous abundance as illustrated in left-hand figures). Modest delayed density dependence has little effect on population fluctuations (upper two panels), but as the strength of delayed feedback increases (lower), the endogenous dynamics cross a bifurcation into stable limit cycles (lower panels). Model parameters (Equation (20.3)): r = 1.2, K = 100, and σ(ε) = 0.05 for both models but β_2 / β_1, the relative strength of delayed feedback, = 1 versus 6 in upper versus lower.

Comins 1976, May 1977). In general, alternate attractors require some form of positive density dependence (social facilitation; Berryman 2003). Examples include (1) larger populations of bark beetles have higher per capita success in overwhelming tree defenses (Raffa and Berryman 1983); (2) larger populations experience less per capita mortality from generalist enemies due to predator swamping (Holling 1965); (3) herbivore defense mechanisms are more effective with larger group size (e.g., pine sawflies) (Björkman *et al.* 1997, Larsson *et al.* 2000), and (4) more efficient modification of thermal environment by tent-making herbivores such as pine processionary moths (Pimentel *et al.* 2010). In terms of Equation (20.1), social facilitation can be thought of as systems where K increases as a function of N. For example, in Equations (20.4) and (20.5), K ranges from low to high, as a logistic function of N. Facilitation (F) ranges from 0 to 1 with F_{mid} equal to the value of N where F = 0.5, and β represents the slope of the facilitation function.

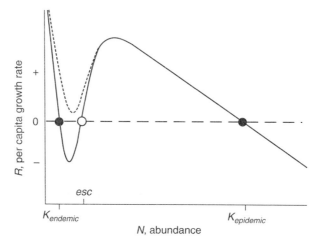

Figure 20.4   A general theoretical mechanism for outbreak dynamics that is a good match with some notoriously outbreaking populations of forest insects. There are two locally stable attractors (one at endemic abundance and the other at epidemic abundance) separated by an escape threshold. The model is illustrated above in terms of the replacement function (R as a function of N). The solid line shows a population model with alternate attractors (closed circles) at $K_{endemic}$ and $K_{epidemic}$ separated by a region of positive feedback and an unstable equilibrium (open circle, escape threshold). The dotted line represents a change in the form of the density-dependent function that would eliminate the lower stable equilibrium and permit deterministic increases in abundance to outbreak levels. Direct or indirect effects of climate on the shape of the function can theoretically produce switches in the tendency for population outbreaks in forest insects.

$$K_N = (1 - F) \cdot K_{low} + K_{high} \cdot F \tag{20.4}$$

$$F = \frac{1}{1 + e^{\beta(F_{mid} - N)}} \tag{20.5}$$

Such a system can have a natural tendency to be regulated at low or high abundance with occasional switches between attractors (Figure 20.4). Other things being equal, increases in $\sigma(\varepsilon_t)$ will increase transitions between endemic and epidemic abundances. Furthermore, environmental variation could alter the form of the density-dependent function. For example, large blow-downs of conifers can ease resource limitations for endemic populations of bark beetles (increase K_{low}) such that populations increase deterministically to abundances near F_{mid} where they are then able to overcome the defense of living trees instead of being restricted to weakened or dead trees (Wermelinger 2004).

As the unstable equilibrium becomes more or less separated from the stable equilibria, transitions between alternate states will tend to become less or more frequent, respectively. If the density-dependent function slides above the line where $R = 0$ (as in the dotted line in Figure 20.3), then the lower equilibrium disappears and populations will tend to grow deterministically to the higher equilibrium. This is a general mechanism by which a population with stable, moderate abundance could switch into one with a strong tendency for outbreaks.

In terms of our model, such alterations of the density-dependent function could arise in a few different ways from changes in F_{mid}, B, K_{low}, or K_{high}. For example, climatic effects that reduce tree defenses against bark beetles would tend to lower F_{mid}, and decreased abundance of generalist predators could increase K_{low} enough to erase the "predator pit" in the density-dependent function (Elkinton and Liebhold 1990).

20.4.2 Conclusions

The potential effects of climate change on population dynamics are considerably richer than implied by the oft-repeated simplification that weather is density independent. It can be this simple, as in the case where the only effects are on $\sigma(\varepsilon t)$ (Scenario 1, Figure 20.2), but climate can also alter the form of density-dependent feedbacks. Poleward expansion of distribution limits, of which there is a rapidly growing list of examples, can be understood as the creation of a stable equilibrium where there was not one before (Scenario 2, Figure 20.2). In some cases, such range expansions are resulting in outbreaks of familiar pests in unfamiliar locations. Probably there will also be cases of insects that have not previously been noted for outbreaks becoming pests in a manner similar to some non-native invasive species (Seppälä *et al.* 2009).

In general, persistent tendencies for outbreaks can arise via (1) high stochastic fluctuations around a noisy point equilibrium, (2) cyclical population dynamics, or (3) switches between alternate attractors.

- Mechanisms resulting in high stochastic fluctuations include increased climate variation that yields increased fluctuations via increases in $\sigma(\varepsilon_t)$. Such stochastic fluctuations in abundance could change from direct (autecological) effects of climate change on the insect herbivore or less directly via climatic effects on host plant or natural enemies (Figure 20.1). However, population dynamics are not strikingly sensitive to changes in $\sigma(\varepsilon_t)$ and we doubt that this by itself will be a frequent cause of changes in outbreak tendencies.
- Mechanisms for changing cyclical tendencies include increases in intrinsic growth rates (r), or increases in delayed negative feedback. Changes in the latter typically involve community interactions, especially natural enemies. Changes in the geographic distribution of specialist versus generalist enemies (e.g., from climate change) are likely to change the feedback system for the insect herbivore; in particular, it is reasonable to expect increased tendencies for population cycles toward the poleward limits of contemporary distributions and decreased tendencies for cycles in the warmer edge of current outbreak distributions. Climatic effects on delayed inducible resistance in host plants is another mechanism by which feedback systems could change in ways that alter outbreak dynamics.
- Mechanisms by which climate change could affect outbreaks that arise from alternate attractors include (1) changes in stochastic variation ($\sigma(\varepsilon_t)$), which influences transitions between attractors; (2) increased incidence of acute

stress in host trees from mismatch of existing forests with new climatic regimes; and (3) changes in the abundance or functional response of generalist predators that influence separation between the lower attractor and the escape threshold (Figure 20.4).

20.5 Synthesis

Population ecology is among the most theoretically mature fields of biology, and has been evolving for many decades via studies of forest insects in particular. It is clear that we should expect important changes in outbreak dynamics in some forest systems as a result of climate change. Furthermore, there is considerable empirical knowledge of factors, including climate, that influence the survival and reproduction of forest insects. However, for most systems, we still need more explicit understanding of how climatic effects on physiology, life histories, and species interactions translate into birth rates, death rates, and demographic feedback systems (i.e., how to connect our understanding of effects on individuals to anticipate the consequences for populations). At present, theory permits a quite satisfying understanding of the general factors that influence qualitative population dynamics (e.g., delayed feedbacks, high potential growth rates, and the creation or elimination of equilibria). This can be matched with growing knowledge of particular ecological systems to permit the identification and testing of the most realistic scenarios by which climate change can be transduced into changing patterns of insect outbreaks. We anticipate that climate change will lead to both increases and decreases in outbreak frequencies depending on the herbivore species, the forest types, and the historical climatic conditions. The frontier for further research lies in developing the knowledge to make sufficiently detailed predictions that are satisfying not only to ecologists but also to forest managers and other stakeholders.

Acknowledgments

Research was funded through Future Forests (a multidisciplinary research program supported by Mistra, the Swedish Forestry Industry, and Swedish University of Agricultural Sciences, Umeå University, Skogforsk) (MJK and SL), the BACCARA project which received funding from the European Community's Seventh Framework Programme (FP7/ 2007–2013) under the grant agreement no. 226299a (MJK, AB, and SL), and USDA NRI/AFRI 2009–65104–05731 and a cooperative agreement with the Southern Research Station, US Forest Service (MPA). The manuscript benefited from the input of Helena Bylund, Jean-Noel Candau, Jeff Garnas, Sharon Martinson, and Erik Stange.

References

Ali, M. A., El-Khouly, A. S., Moawad, G. M., and Ragab, G. M. 1995. Seasonal abundance of cabbage looper, *Trichoplusia ni* as related to temperature, relative humidity and the natural incidence of a nuclear polyhedrosis virus. *Acta Agronomica Hungarica* 43:331–339.

Altermatt, F. 2010. Climatic warming increases voltinism in European butterflies and moths. *Proceedings of the Royal Society B – Biological Sciences* 277:1281–1287.

Anderson, R. M., and May, R. M. 1980. Infectious diseases and population cycles of forest insects. *Science* 210:658–661.

Aukema, B. H., Carroll, A. L., Zhu, J., Raffa, K. F., Sickley, T. A., and Taylor, S. W. 2006. Landscape level analysis of mountain pine beetle in British Columbia, Canada: spatiotemporal development and spatial synchrony within the present outbreak. *Ecography* 29:427–441.

Awmack, C. S., and Leather, S. R. 2002. Host plant quality and fecundity in herbivorous insects. *Annual Review of Entomology* 47:817–844.

Ayres, M. 1993. Plant defense, herbivory and climate change. In: Kareiva, P., Kingsolver, J. G. and Huey, R. B. (eds.), *Biotic Interactions and Global Change*, pp. 75–94. Sinauer Associates Inc.

Ayres, M. P., and Lombardero, M. J. 2000. Assessing the consequences of global change for forest disturbance from herbivores and pathogens. *Science of the Total Environment* 262:263–286.

Ayres, M. P., and MacLean, S. F. 1987. Development of birch leaves and the growth energetics of *Epirrita autumnata* (Geometridae). *Ecology* 68:558–568.

Bale, J. S., and Hayward, S. A. L. 2010. Insect overwintering in a changing climate. *Journal of Experimental Biology* 213:980–994.

Bale, J. S., Masters, G. J., Hodkinson, I. D., Awmack, C. T., Bezemer, M., Brown, V. K., Butterfield, J., Buse, A., Coulson, J. C., Farrar, J., Good, J. E. G., Harrington, R., Hartley, S., Jones, T. H., Lindroth, R. L., Press, M. C., Symrnioudis, I., Watt, A. D., and Whittaker, J. B. 2002. Herbivory in global climate change research: direct effects of rising temperature on insect herbivores. *Global Change Biology* 8:1–16.

Battisti, A., Stastny, M., Buffo, E., and Larsson, S. 2006. A rapid altitudinal range expansion in the pine processionary moth produced by the 2003 climatic anomaly. *Global Change Biology* 12:662–671.

Battisti, A., Stastny, M., Netherer, S., Robinet, C., Schopf, A., Roques, A., and Larsson, S. 2005. Expansion of geographic range in the pine processionary moth caused by increased winter temperatures. *Ecological Applications* 15:2084–2096.

Beaudoin, L., Geri, C., and Allais, J. P. 1992. The effects of photoperiodicity, temperature, food and endogenous processes on the determination of diapause of *Diprion pini*. *Bulletin De La Societe Zoologique De France – Evolution Et Zoologie* 117:351–363.

Bentz, B. J., Régnière, J., Fettig, C. J., Hansen, E. M., Hayes, J. L., Hicke, J. A., Kelsey, R. G., Negron, J. F., and Seybold, S. J. 2010. Climate change and bark beetles of the western United States and Canada: direct and indirect effects. *BioScience* 60:602–613.

Berg, E. E., Henry, J. D., Fastie, C. L., De Volder, A. D., and Matsuoka, S. M. 2006. Spruce beetle outbreaks on the Kenai Peninsula, Alaska, and Kluane National Park and Reserve, Yukon Territory: relationship to summer temperatures and regional differences in disturbance regimes. *Forest Ecology and Management* 227:219–232.

Berggren, Å., Björkman, C., Bylund, H., and Ayres, M. P. 2009. The distribution and abundance of animal populations in a climate of uncertainty. *Oikos* 118:1121–1126.

Berryman, A. A. 1973. Population dynamics of the fir engraver, *Scolytus ventralis* (Coleoptera: Scolytidae). I. Analysis of population behavior and survival from 1964 to 1971. *Canadian Entomologist* 105:1465–1488.

Berryman, A. A. (ed.). 2002. *Population Cycles: The Case for Trophic Interactions*. Oxford University Press, Oxford.

Berryman, A. A. 2003. On principles, laws and theory in population ecology. *Oikos* 103:695–701.

Björkman, C., Larsson, S., and Bommarco, R. 1997. Oviposition preferences in pine sawflies: a trade-off between larval growth and defence against natural enemies. *Oikos* 79:45–52.

Blanford, S., and Thomas, M. B. 1999. Host thermal biology: the key to understanding host–pathogen interactions and microbial pest control? *Agricultural and Forest Entomology* 1:195–202.

Broekhuizen, N., Evans, H. F., and Hassell, M. P. 1993. Site characteristics and the population-dynamics of the pine looper moth. *Journal of Animal Ecology* 62:511–518.

Candau, J. N., and Fleming, R. A. 2005. Landscape-scale spatial distribution of spruce budworm defoliation in relation to bioclimatic conditions. *Canadian Journal of Forest Research* 35:2218–2232.

Carroll, A. L., Taylor, S. W., Régnière, J., and Safranyik, L. 2004. Effects of climate change on range expansion by the mountain pine beetle. In: Shore, T. L., Brooks, J. E., and Stone, J. E. (eds.), *Proceedings of the Mountain Pine Beetle Symposium: Challenges and Solutions 2003*, pp. 223–232. Canadian Forestry Service, Ottawa.

Carruthers, R. I., Larkin, T. S., Firstencel, H., and Feng, Z. D. 1992. Influence of thermal ecology on the mycosis of a rangeland grasshopper. *Ecology* 73:190–204.

Cleland, E. E., Chuine, I., Menzel, A., Mooney, H. A., and Schwartz, M. D. 2007. Shifting plant phenology in response to global change. *Trends in Ecology and Evolution* 22:357–365.

Costa, E. A. P. d. A., Santos, E. M. d. M., Correia, J. C., and Albuquerque, C. M. R. d. 2010. Impact of small variations in temperature and humidity on the reproductive activity and survival of *Aedes aegypti* (Diptera, Culicidae). *Revista Brasileira De Entomologia* 54:488–493.

Craighead, F. C. 1925. Bark beetle epidemics and rainfall deficiency. *Journal of Economic Entomology* 18:577–586.

Cudmore, T. J., Björklund, N., Carroll, A. L., and Lindgren, B. S. 2010. Climate change and range expansion of an aggressive bark beetle: evidence of higher beetle reproduction in naive host tree populations. *Journal of Applied Ecology* 47:1036–1043.

Davis, A. J., Jenkinson, L. S., Lawton, J. S., Shorrocks, B., and Wood, S. 1998. Making mistakes when predicting shifts in species range in response to global warming. *Nature* 391:783–786.

Dhillon, M. K., and Sharma, H. C. 2009. Temperature influences the performance and effectiveness of field and laboratory strains of the ichneumonid parasitoid, *Campoletis chlorideae*. *Biocontrol* 54:743–750.

Dixon, A. F. G. 2003. Climate change and phenological asynchrony. *Ecological Entomology* 28:380–381.

Doak, D. F., Bigger, D., Harding, E. K., Marvier, M. A., O'Malley, R. E., and Thomson, D. 1998. The statistical inevitability of stability–diversity relationships in community ecology. *American Naturalist* 151:264–276.

Eldred, B. D., Dushoff, J., and Dwyer, G. 2008. Host–pathogen interactions, insect outbreaks, and natural selection for disease resistence. *American Naturalist* 172:829–842.

Elkinton, J. S., and Liebhold, A. M. 1990. Population dynamics of the gypsy moth in North America. *Annual Review of Entomology* 35:571–596.

Faccoli, M. 2009. Effect of weather on *Ips typographus* (Coleoptera Curculionidae) phenology, voltinism, and associated spruce mortality in the Southeastern Alps. *Environmental Entomology* 38:307–316.

Fantinou, A. A., Perdikis, D. C., and Zota, K. F. 2004. Reproductive responses to photoperiod and temperature by diapausing and nondiapausing populations of *Sesamia nonagrioides* Lef. (Lepidoptera-Noctuidae). *Physiological Entomology* 29:169–175.

Feeny, P. 1970. Seasonal changes in oak leaf tannins and nutrients as a cause of spring feeding by winter moth caterpillars. *Ecology* 51:565.

Filotas, M. J., Vandenberg, J. D., and Hajek, A. E. 2006. Concentration–response and temperature-related susceptibility of the forest tent caterpillar (Lepidoptera : Lasiocampidae) to the entomopathogenic fungus *Furia gastropachae* (Zygomycetes : Entomophthorales). *Biological Control* 39:218–224.

Fleming, R. A. 1996. A mechanistic perspective of possible influences of climate change on defoliating insects in North America's boreal forests. *Silva Fennica* 30:281–294.

Fleming, R. A., and Candau, J-N. 1998. Influences of climatic change on some ecological processes of an insect outbreak system in Canada's boreal forests and the implications for biodiversity. *Environmental Monitoring and Assessment* 49:235–249.

Forrest, J., and Miller-Rushing, A. J. 2010. Toward a synthetic understanding of the role of phenology in ecology and evolution. *Philosophical Transactions of the Royal Society B: Biological Sciences* 365:3101–3112.

Friedenberg, N. A., Powell, J. A., and Ayres, M. P. 2007. Synchrony's double edge: transient dynamics and the Allee effect in stage-structured populations. *Ecology Letters* 10:564–573.

Friedenberg, N. A., Sarkar, S., Kouchoukos, N., Billings, R. F., and Ayres, M. P. 2008. Temperature extremes, density dependence, and southern pine beetle (Coleoptera: Curculionidae) population dynamics in east Texas. *Environmental Entomology* 37:650–659.

Gaston, K. J. 2003. *The Structure and Dynamics of Geographic Ranges*. Oxford University Press, New York.

Hajek, A. E. 1997. Ecology of terrestrial fungal entomopathogens. *Advances in Microbial Ecology* 15:193–249.

Hance, T., van Baaren, J., Vernon, P., and Boivin, G. 2007. Impact of extreme temperatures on parasitoids in a climate change perspective. *Annual Review of Entomology* 52:107–126.

Harvey, J. A., Harvey, I. F., and Thompson, D. J. 1994. Flexible larval growth allows use of a range of host sizes by a parasitoid wasp. *Ecology* 75:1420–1428.

Hassell, M. P. 1975. Density-dependence in single-species populations. *Journal of Animal Ecology* 44:283–295.

Haukioja, E. 1980. On the role of plant defences in single-species populations. *Oikos* 35:202–213.

Haukioja, E. 2003. Putting the insect into the birch–insect interaction. *Oecologia* 136:161–168.

Hawkins, B. A., Cornell, H. V., and Hochberg, M. E. 1997. Predators, parasitoids, and pathogens as mortality agents in phytophagous insect populations. *Ecology* 78:2145–2152.

Herms, D. A., and Mattson, W. J. 1992. The dilemma of plants: to grow or defend. *Quarterly Review of Biology* 67:283–335.

Hickling, R., Roy, D. B., Hill, J. K., Fox, R., and Thomas, C. D. 2006. The distributions of a wide range of taxonomic groups are expanding poleward. *Global Change Biology* 12:450–455.

Hodar, J. A., Castro, J., and Zamora, R. 2003. Pine processionary caterpillar, *Thaumetopoea pityocampa*, as a new threat for relict Mediterranean Scots pine forests under climatic warming. *Biological Conservation* 110:123–129.

Hodkinson, I. D. 1999. Species response to global environmental change or why ecophysiological models are important: a reply to Davis *et al*. *Journal of Animal Ecology* 68:1259–1262.

Hoekman, D. 2010. Turning up the heat: temperature influences the relative importance of top-down and bottom-up effects. *Ecology* 91:2819–2825.

Holling, C. S. 1965. The functional response of predators to prey density and its role in mimicry and population regulation. *Memoirs of the Entomological Society of Canada* 45:3–60.

Holling, C. S. 1973. Resilience and stability of ecological systems. *Annual Review of Ecology and Systematics* 4:1–23.

Humble, L. M. 2006. Overwintering adaptations in Arctic sawflies (Hymenoptera: Tenthredinidae) and their parasitoids: cold tolerance. *Canadian Entomologist* 138:59–71.

Jepsen, J. U., Hagen, S. B., Ims, R. A., and Yoccoz, N. G. 2008. Climate change and outbreaks of the geometrids *Operophtera brumata* and *Epirrita autumnata* in subarctic birch forest: evidence of a recent outbreak range expansion. *Journal of Animal Ecology* 77:257–264.

Jonsson, A. M., Appelberg, G., Harding, S., and Bärring, L. 2009. Spatio-temporal impact of climate change on the activity and voltinism of the spruce bark beetle, *Ips typographus*. *Global Change Biology* 15:486–499.

Juday, G. P., and Marler, S. A. 1997. Tree-ring evidence of climatic warming stress in Alaska: variation and stand history context. *Bulletin of the Ecological Society of America* 78:119.

Karlsson, P. S., Bylund, H., Neuvonen, S., Heino, S., and Tjus, M. 2003. Climatic response of budburst in the mountain birch at two areas in northern Fennoscandia and possible responses to global change. *Ecography* 26:617–625.

Klapwijk, M. J., Grobler, B. C., Ward, K., Wheeler, D., and Lewis, O. T. 2010. Influence of experimental warming and shading on host–parasitoid synchrony. *Global Change Biology* 16:102–112.

Koricheva, J., Larsson, S., Haukioja, E., and Keinänen, M. 1998a. Regulation of woody plant secondary metabolism by resource availability: hypothesis testing by means of meta-analysis. *Oikos* 83:212–226.

Koricheva, J., Larsson, S., and Haukioja, E. 1998b. Insect performance on experimentally stressed woody plants: a meta-analysis. *Annual Review of Entomology* 43:195–216.

Kruse, P. D., Toft, S., and Sunderland, K. D. 2008. Temperature and prey capture: opposite relationships in two predator taxa. *Ecological Entomology* 33:305–312.

Kurz, W. A., Dymond, C. C., Stinson, G., Rampley, G. J., Neilson, E. T., Carroll, A. L., Ebata, T., and Safranyik, L. 2008. Mountain pine beetle and forest carbon feedback to climate change. *Nature* 452:987–990.

Larsson, S. 1989. Stressful times for the plant stress – insect performance hypothesis. *Oikos* 56:277–283.

Larsson, S., Ekbom, B., and Björkman, C. 2000. Influence of plant quality on pine sawfly population dynamics. *Oikos* 89:440–450.

Laws, A. N., and Belovsky, G. E. 2010. How will species respond to climate change? Examining the effects of temperature and population density on an herbivorous insect. *Environmental Entomology* 39:312–319.

Logan, J. A., and Powell, J. A. 2001. Ghost forests, global warming, and the mountain pine beetle (Coleoptera: Scolytidae). *American Entomologist* 47:160–173.

Logan, J. A., and Powell, J. A. 2009. Ecological consequences of climate change: altered insect disturbance regime. In: Wagner, F. H. (ed.), *Climate Change in Western North America: Evidence and Environmental Effects*. University of Utah Press, Salt Lake City.

Logan, J. A., Régnière, J., and Powell, J. A. 2003. Assessing the impacts of global warming on forest pest dynamics. *Frontiers in Ecology and the Environment* 1:130–137.

Lombardero, M. J., Ayres, M. P., Ayres, B. D., and Reeve, J. D. 2000a. Cold tolerance of four species of bark beetle (Coleoptera:Scolytidae) in North America. *Environmental Entomology* 29:421–432.

Lombardero, M. J., Ayres, M. P., Lorio, P. L. Jr., and Ruel, J. J. 2000b. Environmental effects on constitutive and inducible resin defences of *Pinus taeda*. *Ecology Letters* 3:329–339.

Martinat, P. J. 1987. The role of climatic variation and weather on forest insect outbreaks. In: Barbosa, P., and Schultz, J. C. (eds.), *Insect Outbreaks*, pp. 241–268. Academic Press, San Diego.

Mattson, W. J., and Haack, R. A. 1987. The role of drought stress in provoking outbreaks of phytophagous insects. In: Barbosa, P., and Schultz, J. C. (eds.), *Insect Outbreaks*, pp. 365–394. Academic Press, San Diego.

May, R. M. 1976. Simple mathematical models with very complicated dynamics. *Nature* 261:459–467.

May, R. M. 1977. Thresholds and breakpoints with multiplicity of stable states. *Nature* 269:471–477.

Memmott, J., Craze, P. G., Waser, M. W., and Price, M. V. 2007. Global warming and the disruption of plant–pollinator interactions. *Ecology Letters* 10:710–717.

Menzel, A., Sparks, T. H., Estrella, N., Koch, E., Aasa, A., Ahas, R., Alm-Kubler, K., Bissolli, P., Braslavska, O., Briede, A., Chmielewski, F. M., Crepinsek, Z., Curnel, Y., Dahl, A., Defila, C., Donnelly, A., Filella, Y., Jatcza, K., Mage, F., Mestre, A., Nordli, O., Penuelas, J., Pirinen, P., Remisova, V., Scheifinger, H., Striz, M., Susnik, A., Van Vliet, A. J. H., Wielgolaski, F. E., Zach, S., and Zust, A. 2006. European phenological response to climate change matches the warming pattern. *Global Change Biology* 12:1969–1976.

Miller, W. E. 2005. Extrinsic effects on fecundity–maternal weight relation in capital-breeding Lepidoptera. *Journal of the Lepidopterists' Society* 59:143–160.

Morin, X., Lechowicz, M. J., Augspurger, C., O' Keefe, J., Viner, D., and Chuine, I. 2009. Leaf phenology in 22 North American tree species during the 21st century. *Global Change Biology* 15:961–975.

Neuvonen, S., Bylund, H., and Tømmervik, H. 2005. Forest defoliation risks in birch forest by insects under different climate and land use scenarios in Northern Europe. *Ecological Studies: Plant Ecology, Herbivory, and Human Impact in Nordic Mountain Birch Forests*: 125–138.

Neuvonen, S., Niemela, P., Virtanen, T., Hofgaard, A., Ball, J. P., Danell, K., and Callaghan, T. V. 1999. Climatic change and insect outbreaks in boreal forests: the role of winter temperatures. *Ecological Bulletins: Animal Responses to Global Change in the North*: 63–67.

Økland, B., and Bjørnstad, O. N. 2006. A resource-depletion model of forest insect outbreaks. *Ecology* 87:283–290.

Olofsson, E. 1989a. Transmission agents of the nuclear polyhedrosis virus of *Neodiprion sertifer* (Hym, Dirpionidae). *Entomophaga* 34:373–380.

Olofsson, E. 1989b. Transmission of the nuclear polyhedrosis virus of the European pine sawfly from adult to offspring. *Journal of Invertebrate Pathology* 54:322–330.

Parmesan, C., Ryrholm, N., Stefanescu, C., Hill, J. K., Thomas, C. D., Descimon, H., Huntley, B., Kaila, L., Kullberg, J., Tammaru, T., Tennent, W. J., Thomas, J. A., and Warren, M. 1999. Poleward shifts in geographical ranges of butterfly species associated with regional warming. *Nature* 399:579–583.

Pimentel, C., Ferreira, C., and Nilsson, J. A. 2010. Latitudinal gradients and the shaping of life-history traits in a gregarious caterpillar. *Biological Journal of the Linnean Society* 100:224–236.

Powell, J. A., and Logan, J. A. 2005. Insect seasonality: circle map analysis of temperature-driven life cycles. *Theoretical Population Biology* 67:161–179.

Raffa, K. F., Aukema, B. H., Bentz, B. J., Carroll, A. L., Hicke, J. A., Turner, M. G., and Romme, W. H. 2008. Cross-scale drivers of natural disturbances prone to anthropogenic amplification: the dynamics of bark beetle eruptions. *BioScience* 58:501–517.

Raffa, K. F., and Berryman, A. A. 1983. The role of host plant resistence in the colonization behavior and ecology of bark beetles (Coleoptera: Scolytidae). *Ecological Monographs* 53:27–50.

Reeve, J. D., Rhodes, D. J., and Turchin, P. 1998. Scramble competition in the southern pine beetle, *Dendroctonus frontalis*. *Ecological Entomology* 23:433–443.

Régnière, J., and Bentz, B. 2007. Modeling cold tolerance in the mountain pine beetle, *Dendroctonus ponderosae*. *Journal of Insect Physiology* 53:559–572.

Rhoades, D. F. 1979. Evolution of plant chemical defence against herbivores. In: Rosenthal, G. A., and Janzen, D. H. (eds.), *Herbivores: Their Interaction with Secondary Plant Metabolites*, pp. 3–54. Academic Press, New York.

Riipi, M., Ossipov, V., Lempa, K., Haukioja, E., Koricheva, J., Ossipova, S., and Pihlaja, K. 2002. Seasonal changes in birch leaf chemistry: are there trade-offs between leaf growth, and accumulation of phenolics? *Oecologia* 130:380–390.

Robertson, C., Nelson, T. A., Jelinski, D. E., Wulder, M. A., and Boots, B. 2009. Spatial-temporal analysis of species range expansion: the case of the mountain pine beetle, *Dendroctonus ponderosae*. *Journal of Biogeography* 36:1446–1458.

Roy, H. E., Hails, R. S., Hesketh, H., Roy, D. B., and Pell, J. K. 2009. Beyond biological control: non-pest insects and their pathogens in a changing world. *Insect Conservation and Diversity* 2:65–72.

Royama, T. 1992. *Analytical Population Dynamics*. Chapman and Hall, London.

Rudinsky, J. A. 1962. Ecology of Scolytidae. *Annual Review of Entomology* 7:327–348.

Seppälä, R., Buck, A., and Katila, P. 2009. *Adaptation of Forests and People to Climate Change: A Global Assessment Report*. International Union of Forest Research Organisations, Vienna.

Singer, M. C., and Parmesan, C. 2010. Phenological asynchrony between herbivorous insects and their hosts: signal of climate change or pre-existing adaptive strategy? *Philosophical Transactions of the Royal Society B – Biological Sciences* 365:3161–3176.

Solomon, S., Qin, D., Manning, M., Chen, Z., Marquis, M., Averyt, K. B., Tignor, M., and Miller, H. L. 2007. *Contribution of Working Group I to the Fourth Assessment Report of the Intergovernmental Panel on Climate Change*. Intergovernemental Panel on Climate Change, Geneva.

Southwood, T. R. E., and Comins, H. N. 1976. A synoptic population model. *Journal of Animal Ecology* 65:949–965.

Sparks, T. H., and Carey, P. D. 1995. The responses of species to climate over 2 centuries: an analysis of the Marsham phenological record, 1736–1947. *Journal of Ecology* 83:321–329.

Sparks, T. H., and Yates, T. J. 1997. The effect of spring temperature on the appearance dates of British butterflies 1883–1993. *Ecography* 20:368–374.

Stamp, N. 2003. Out of the quagmire of plant defense hypotheses. *Quarterly Review of Biology* 78:23–55.

Stange, E. E., Ayres, M. P., and Bess, J. A. 2011. Concordant population dynamics of Lepidoptera herbivores in a forest ecosystem. *Ecography* 34(5):772–779.

Stiling, P., and Cornelissen, T. 2007. How does elevated carbon dioxide (CO2) affect plant–herbivore interactions? A field experiment and meta-analysis of CO2-mediated changes on plant chemistry and herbivore performance. *Global Change Biology* 13:1823–1842.

Tenow, O. 1996. Hazards to a mountain birch forest – Abisko in perspective. *Ecological Bulletins* 45:104–114.

Tenow, O., and Nilssen, A. 1990. Egg cold hardiness and topoclimatic limitations to outbreaks of *Epirrita autumnata* in Northen Fennoscandia. *Journal of Applied Ecology* 27:723–734.

Thomas, M. B., and Blanford, S. 2003. Thermal biology in insect–parasite interactions. *Trends in Ecology and Evolution* 18:344–350.

Thomson, L. J., Macfadyen, S., and Hoffmann, A. A. 2010. Predicting the effects of climate change on natural enemies of agricultural pests. *Biological Control* 52:296–306.

Tobin, P. C., Nagarkatti, S., Loeb, G., and Saunders, M. C. 2008. Historical and projected interactions between climate change and insect voltinism in a multivoltine species. *Global Change Biology* 14:951–957.

Tran, J. K., Ylioja, T., Billings, R. F., Régnière, J., and Ayres, M. P. 2007. Impact of minimum winter temperatures on the population dynamics of *Dendroctonus frontalis*. *Ecological Applications* 17:882–899.

Turchin, P. 2003. Ecology: evolution in population dynamics. *Nature* 424:257–258.

Turchin, P., and Taylor, A. D. 1992. Complex dynamics in ecological time series. *Ecology* 73:289–305.

Turchin, P., Wood, S. N., Ellner, S. P., Kendall, B. E., Murdoch, W. W., Fischlin, A., Casas, J., McCauley, E., and Briggs, C. J. 2003. Dynamical effects of plant quality and parasitism on population cycles of larch budmoth. *Ecology* 84:1207–1214.

Uvarov, B. P. 1931. Insects and climate. *Transactions of the Royal Entomological Society of London* 79:1–232.

Valtonen, A., Ayres, M. P., Roininen, H., Pöyry, J., and Leinonen, R. 2011. Environmental controls on the phenology of moths – predicting plasticity and constraint under climate change. *Oecologia* 165:237–248.

van Asch, M., and Visser, M. E. 2007. Phenology of forest caterpillars and their host trees: the importance of synchrony. *Annual Review of Entomology* 52:37–55.

van Straalen, N. M. 1983. Physiological time and time-invariance. *Journal of Theoretical Biology* 104:349–357.

Vanhanen, H., Veleli, T. O., Paivinen, S., Kellomaki, S., and Niemela, P. 2007. Climate change and range shifts in two insect defoliators: gypsy moth and nun moth – a model study. *Silva Fennica* 41:621–638.

Visser, M. E., and Both, C. 2006. Shifts in phenology due to global climate change: the need for a yardstick. *Proceedings of the Royal Society of London Series B – Biological Sciences* 272:2561–2569.

Visser, M. E., and Holleman, L. J. M. 2001. Warmer springs disrupt the synchrony of oak and winter moth phenology. *Proceedings of the Royal Society of London Series B – Biological Sciences* 268:289–294.

Visser, M. E., van Noordwijk, A. J., Tinbergen, J. M., and Lessells, C. M. 1998. Warmer springs lead to mistimed reproduction in great tits (*Parus major*). *Proceedings of the Royal Society of London Series B – Biological Sciences* 265:1867–1870.

Vitasse, Y., Delzon, S., Bresson, C. C., Michalet, R., and Kremer, A. 2009. Altitudinal differentiation in growth and phenology among populations of temperate-zone tree species growing in a common garden. *Canadian Journal of Forest Research–Revue Canadienne De Recherche Forestiere* 39:1259–1269.

Walther, G. R., Post, E., Convey, P., Menzel, A., Parmesan, C., Beebee, T. J. C., Fromentin, J. M., Hoegh-Guldberg, O., and Bairlein, F. 2002. Ecological responses to recent climate change. *Nature* 416:389–395.

Waring, R. H., and Pitman, G. B. 1985. Modifying lodgepole pine stands to change susceptibility to mountain pine beetle attack. *Ecology* 66:889–897.

Wermelinger, B. 2004. Ecology and management of the spruce bark beetle *Ips typographus* – a review of recent research. *Forest Ecology and Management* 202:67–82.

Wermelinger, B., and Seifert, M. 1998. Analysis of the temperature dependent development of the spruce bark beetle *Ips typographus* (L) (Col, Scolytidae). *Journal of Applied Entomology* 122:185–191.

Wermelinger, B., and Seifert, M. 1999. Temperature-dependent reproduction of the spruce bark beetle *Ips typographus*, and analysis of the potential population growth. *Ecological Entomology* 24:103–110.

White, T. C. R. 1974. Hypothesis to explain outbreaks of looper caterpillars, with special reference to population of *Selidosema suavis* in a plantation of *Pinus radiata* in New-Zealand. *Oecologia* 16:279–301.

Witt, D. J. 1984. Photoreactivation and ultravoilet-enhanced reactivation of ultraviolet-irradiated nuclear polyhedroses virus by insect cells. *Archives of Virology* 79:95–107.

Yanchuk, A. D., Murphy, J. C., and Wallin, K. F. 2008. Evaluation of genetic variation of attack and resistance in lodgepole pine in the early stages of a mountain pine beetle outbreak. *Tree Genetics and Genomes* 4:171–180.

Yao, S. L., Huang, Z., Ren, S. X., Mandour, N., and Ali, S. 2010. Effects of temperature on development, survival, longevity, and fecundity of *Serangium japonicum* (Coleoptera: Coccinellidae), a predator of *Bemisia tabaci* Gennadius (Homoptera: Aleyrodidae). *Biocontrol Science and Technology* 21:23–34.

Zovi, D., Stastny, M., Battisti, A., and Larsson, S. 2008. Ecological costs on local adaptation of an insect herbivore imposed by host plants and enemies. *Ecology* 89:1388–1398.

Subject Index

Insect Outbreaks Revisited, First Edition. Edited by Pedro Barbosa, Deborah K. Letourneau
and Anurag A. Agrawal.
© 2012 Blackwell Publishing Ltd. Published 2012 by Blackwell Publishing Ltd.

Taxonomic Index

Note: Page references in **bold** refer to Tables; those in *italics* refer to Figures

Insect Outbreaks Revisited, First Edition. Edited by Pedro Barbosa, Deborah K. Letourneau
and Anurag A. Agrawal.
© 2012 Blackwell Publishing Ltd. Published 2012 by Blackwell Publishing Ltd.